INDUSTRIAL BIORENEWABLES

INDUSTRIAL BIORENEWABLES

A Practical Viewpoint

Edited by

PABLO DOMÍNGUEZ DE MARÍA

WILEY

Published by John Wiley & Sons, Inc., Hoboken, New Jersey
Published simultaneously in Canada

Library of Congress Cataloging-in-Publication Data applied for:

Hardback: 9781118843727

Typeset in 10/12pt TimesLTStd by SPi Global, Chennai, India

Printed in the United States of America

10 9 8 7 6 5 4 3 2 1

CONTENTS

LIST OF CONTRIBUTORS

Carlos Abreu, Grove Advanced Chemicals SA. Rua Barão S. Januário 33, 5°, sala 5.3-5.4, 4470-473 Maia, Portugal. www.grovechemicals.com

Niuris Acosta, Sustainable Physical-Chemical innovation S.L. (InFiQuS). Paseo Juan XXIII, n°1. Madrid, Spain. www.infiqus.es

João Monnerat Araujo Ribeiro de Almeida, School of Chemistry, Department of Organic Processes, Centro de Tecnologia, Bloco E, Ilha do Fundão, Federal University of Rio de Janeiro (UFRJ), Rio de Janeiro, Brazil. www.petrobras.com

Gary Anderson, Department of Agricultural and Biosystems Engineering, South Dakota State University, Box 2120, Brookings, SD 57007, USA. www.phytonix.com

Inmaculada Aranaz, Sustainable Physical-Chemical innovation S.L. (InFiQuS). Paseo Juan XXIII, n°1. Madrid, Spain. www.infiqus.es

Alex Berlin, Novozymes, Inc. 1445 Drew Ave. Davis CA 95618, USA. www.novozymes.com

Stefan Buchholz, Evonik AG. Paul-Baumann-Strasse. 45772 Marl, Germany. www.evonik.com

Godfried J. H. Buisman, Arizona Chemical B.V. 1225 W Lathrop Avenue Savannah, GA 31415, USA. www.arizonachemical.com

Armando M. Byrne, E.I. DuPont de Nemours & Co. 200 Powder Mill Road, Wilmington, DE 19803, USA. www.dupont.com

Laura Calderón, Sustainable Physical-Chemical innovation S.L. (InFiQuS). Paseo Juan XXIII, n° 1. Madrid, Spain. www.infiqus.es

Pedro Carvalho, Grove Advanced Chemicals SA. Rua Barão S. Januário 33, 5°, sala 5.3-5.4, 4470-473 Maia, Portugal. www.grovechemicals.com

Yuri Carvalho, School of Chemistry, Department of Organic Processes, Federal University of Rio de Janeiro (UFRJ), Centro de Tecnologia, Bloco E, Ilha do Fundão, Rio de Janeiro, Brazil. www.petrobras.com

Armando Córdova, Department of Natural Sciences, Engineering and Mathematics, Mid Sweden University, SE-851 70 Sundsvall, Sweden; Berzelii Center EXSELENT on Porous Materials, Arrhenius Laboratory, Stockholm University, SE-106 91 Stockholm, Sweden. www.organoclick.com

Stefan Cornelissen, Evonik AG. Rellinghauser strasse 1-11. 45128, Essen, Germany. www.evonik.com

Georg Dahmen, Greasoline GmbH. Osterfelder Str. 3. D-46047. Oberhausen. Germany. www.greasoline.com

Bruce Dannenberg, Phytonix Corporation. One Town Square Blvd., Suite 315. Asheville, NC 28803, USA. www.phytonix.com

Jean-Luc Dubois, ARKEMA, 420 Rue d'Estienne d'Orves, 92705 Colombes, France. www.arkema.com

Maria Fé Elía, Cepsa Research Centre, Avenida Punto Com, nº 1. 28805. Alcalá de Henares. Madrid. Spain. www.cepsa.com

Brandon Emme, Novozymes North America, Inc., Franklinton, NC, USA. www.novozymes.com, Present address: ICM Inc. 2811 S 11th Street Rd, St Joseph, MO 64503, United States

Nadia Evremova, Phytowelt Green Technologies GmbH. Kölsumer Weg 33. D-41334 Nettetal, Germany. www.phytowelt.com

Stefano Facco, Novamont SpA. Via Fauser 8. 28100 Novara, Italy. www.novamont.com

Gunter Festel, Festel Capital, Mettlenstrasse 14. CH-6363, Fueringen, Switzerland. www.festel-capital.com; Technical University Berlin. Chair of Entrepreneurship and Innovation Management. Straβe des 17. Juni 135. D-10623 Berlin. Germany. www.entrepreneurship.tu-berlin.de; University of Basel. Chair of Financial Management. Peter Merian-Weg 6. CH-4002 Basel. Switzerland. wwz.unibas.ch

Juana Frontela, Cepsa Research Centre, Avenida Punto Com, nº 1. 28805. Alcalá de Henares. Madrid. Spain. www.cepsa.com

Eneko Ganuza, Heliae Development, LLC. 578 E. Germann Road. Gilbert, AZ 85297. USA. www.heliae.com

Henrike Gebhardt, Evonik AG. Rellinghauser strasse 1-11. 45128, Essen, Germany. www.evonik.com

Philippe Geyr, Grove Advanced Chemicals SA. Rua Barão S. Januário 33, 5°, sala 5.3-5.4, 4470-473 Maia, Portugal. www.grovechemicals.com

Johan Willem Gosselink, Shell Technology Centre Amsterdam. Grasweg 31, 1031 HW. Amsterdam, the Netherlands. www.shell.com

Jonas Hafrén, Department of Forest Products, Swedish University of Agricultural Sciences, 750 07 Uppsala, Sweden. www.organoclick.com

Ruth Harris, Sustainable Physical-Chemical innovation S.L. (InFiQuS). Paseo Juan XXIII, n°1. Madrid, Spain. www.infiqus.es

Peter Haug, Greasoline GmbH. Osterfelder Str. 3. D-46047. Oberhausen. Germany. www.greasoline.com

Evert van der Heide, Shell Technology Centre Amsterdam. Grasweg 31, 1031 HW. Amsterdam, the Netherlands. www.shell.com

Volker Heil, Fraunhofer UMSICHT. Osterfelder Str. 3. D-46047. Oberhausen. Germany. http://www.umsicht.fraunhofer.de/

Ángeles Heras, Sustainable Physical-Chemical innovation S.L. (InFiQuS). Paseo Juan XXIII, n°1. Madrid, Spain. www.infiqus.es

Guido Jach, Phytowelt Green Technologies GmbH. Kölsumer Weg 33. D-41334 Nettetal, Germany. www.phytowelt.com

Ethel N. Jackson, E.I. DuPont de Nemours & Co. 200 Powder Mill Road, Wilmington, DE 19803, USA. www.dupont.com

Ed de Jong, Avantium Chemicals. Zekeringstraat 29. 1014 BV Amsterdam, the Netherlands. www.avantium.com

Kaisa Karhumaa, C5 Ligno Technologies AB, P.O. box 124, SE-22100 Lund, Sweden. www.c5lt.se

Axel Kraft, Fraunhofer UMSICHT. Osterfelder Str. 3. D-46047. Oberhausen. Germany. http://www.umsicht.fraunhofer.de/

Jos H. M. Lange, Arizona Chemical B.V., Transistorstraat 16, 1322 CE Almere, The Netherlands. www.arizonachemical.com

Jean-Paul Lange, Shell Technology Centre Amsterdam. Grasweg 31, 1031 HW. Amsterdam, the Netherlands. www.shell.com

Rafael Larraz, Cepsa Research Centre, Avenida Punto Com, nº 1. 28805. Alcalá de Henares. Madrid. Spain. www.cepsa.com

Rob Lee, Shell Technology Center, 3333 Highway 6 South , Houston, TX 77082, USA. www.Shell.com

Peter Lindblad, Department of Chemistry – Angstrom Laboratory, Uppsala University, Box 523. SE-751 20 Uppsala, Sweden. www.phytonix.com

Martijn van Loon, AkzoNobel Center, Christian Neefestraat 2, 1077 WW Amsterdam, the Netherlands

Renate Luehrs, Phytowelt Green Technologies GmbH. Kölsumer Weg 33. D-41334 Nettetal, Germany. www.phytowelt.com

Rafael Luque, Green Applied Solutions S.L.U., Av. Virgen de Fatima 50, E14014, Cordoba, Spain. www.greenappliedsolutions.com.es

Paulo Martins, Grove Advanced Chemicals SA. Rua Barão S. Januário 33, 5°, sala 5.3-5.4, 4470-473 Maia, Portugal. www.grovechemicals.com

Achim Marx, Evonik AG. Rellinghauser strasse 1-11. 45128 Essen, Germany. www.evonik.com

Andreas Menne, Fraunhofer UMSICHT. Osterfelder Str. 3. D-46047. Oberhausen. Germany. http://www.umsicht.fraunhofer.de/

Edward S. Miller Jr., E.I. DuPont de Nemours & Co. 200 Powder Mill Road, Wilmington, DE 19803, USA. www.dupont.com

Molly Morse, Mango Materials. 490 Lake Park Avenue #16098, Oakland, CA 94610, USA. www.mangomaterials.com

Peter Nagler, Evonik AG. Paul-Baumann-Strasse. 45772 Marl, Germany. www.evonik.com

Mark J. Nelson, E.I. DuPont de Nemours & Co. 200 Powder Mill Road, Wilmington, DE 19803, USA. www.dupont.com

Peter Nieuwenhuizen, AkzoNobel Center, Christian Neefestraat 2, 1077 WW Amsterdam, the Netherlands

Ann Oakenfull, Mango Materials. 490 Lake Park Avenue #16098, Oakland, CA 94610, USA. www.mangomaterials.com

Allison Pieja, Mango Materials. 490 Lake Park Avenue #16098, Oakland, CA 94610, USA. www.mangomaterials.com

Joseph B. Powell, Shell Technology Center, 3333 Highway 6 South , Houston, TX 77082, USA. www.shell.com

Alistair Reid, AkzoNobel Center, Christian Neefestraat 2, 1077 WW Amsterdam, the Netherlands

Pedro Nothaft Romano, School of Chemistry, Department of Organic Processes, Centro de Tecnologia, Bloco E, Ilha do Fundão, Federal University of Rio de Janeiro (UFRJ), Rio de Janeiro, Brazil. www.petrobras.com

Michael A. Saltzberg, E.I. DuPont de Nemours & Co. 200 Powder Mill Road, Wilmington, DE 19803, USA. www.dupont.com

Violeta Sànchez i Nogué, C5 Ligno Technologies AB, P.O. box 124, SE-22100 Lund, Sweden. www.c5lt.se

Filipe Santos, Grove Advanced Chemicals SA. Rua Barão S. Januário 33, 5°, sala 5.3-5.4, 4470-473 Maia, Portugal. www.grovechemicals.com

Rui Santos, Grove Advanced Chemicals SA. Rua Barão S. Januário 33, 5°, sala 5.3-5.4, 4470-473 Maia, Portugal. www.grovechemicals.com

Anne Schauer-Gimenez, Mango Materials. 490 Lake Park Avenue #16098, Oakland, CA 94610, USA. www.mangomaterials.com

Colin John Schaverien, Shell Technology Centre Amsterdam. Grasweg 31, 1031 HW. Amsterdam, the Netherlands. www.shell.com

Katrin Schullehner, Phytowelt Green Technologies GmbH. Kölsumer Weg 33. D-41334 Nettetal, Germany. www.phytowelt.com

Edda Schulze, Evonik AG. Rellinghauser strasse 1-11. 45128 Essen, Germany. www.evonik.com

Elisabete Sepúlveda, Grove Advanced Chemicals SA. Rua Barão S. Januário 33, 5°, sala 5.3-5.4, 4470-473 Maia, Portugal. www.grovechemicals.com

Richard Solal, Grove Advanced Chemicals SA. Rua Barão S. Januário 33, 5°, sala 5.3-5.4, 4470-473 Maia, Portugal. www.grovechemicals.com

Eduardo Falabella Sousa-Aguiar, Petrobras Research Centre (CENPES), Ilha do Fundão, Q7, Cidade Universitária, CEP 21949-900, Rio de Janeiro, Brazil; Federal University of Rio de Janeiro (UFRJ), School of Chemistry, Department of Organic Processes, Centro de Tecnologia, Bloco E, Ilha do Fundão, Rio de Janeiro, Brazil. www.petrobras.com

Henrik Thunman, Chalmers University of Technology. Hörsalsvägen 7B. Göteborg, Sweden. www.goteborgenergi.se

Sara Tollin, AkzoNobel Center, Christian Neefestraat 2, 1077 WW Amsterdam, the Netherlands

Anna Lee Tonkovich, Heliae Development, LLC. 578 E. Germann Road. Gilbert, AZ 85297. USA. www.heliae.com

Olalla de la Torre, Cepsa Research Centre, Avenida Punto Com, nº 1. 28805. Alcalá de Henares. Madrid. Spain. www.cepsa.com

Bjorn D. Tyreus, E.I. DuPont de Nemours & Co. 200 Powder Mill Road, Wilmington, DE 19803, USA. www.dupont.com

Christoph Unger, Fraunhofer UMSICHT. Osterfelder Str. 3. D-46047. Oberhausen. Germany. http://www.umsicht.fraunhofer.de/

Jorge Vasconcelos, Grove Advanced Chemicals SA. Rua Barão S. Januário 33, 5°, sala 5.3-5.4, 4470-473 Maia, Portugal. www.grovechemicals.com

Henrique Villas-Boas, Grove Advanced Chemicals SA. Rua Barão S. Januário 33, 5°, sala 5.3-5.4, 4470-473 Maia, Portugal. www.grovechemicals.com

Jan C. van der Waal, Avantium Chemicals. Zekeringstraat 29. 1014 BV Amsterdam, the Netherlands. www.avantium.com

Peter Welters, Phytowelt Green Technologies GmbH. Kölsumer Weg 33. D-41334 Nettetal, Germany. www.phytowelt.com

Chunping Xu, School of Food and Biological Engineering, Zhengzhou University of Light Industry, Dongfeng Road 5, Zhengzhou, Henan, 450002, P.R. China. www.greenappliedsolutions.com.es

Quinn Zhu, E.I. DuPont de Nemours & Co. 200 Powder Mill Road, Wilmington, DE 19803, USA. www.dupont.com

Eric Zinn, Göteborg Energi AB. P.O. Box 53, 401 20 Göteborg, Sweden. www.goteborgenergi.se

Bárbara van Asch, Grove Advanced Chemicals SA. Rua Barão S. Januário 33, 5°, sala 5.3-5.4, 4470-473 Maia, Portugal. www.grovechemicals.com

Marian Mengibar, Sustainable Physical-Chemical innovation S.L. (InFiQuS). Paseo Juan XXIII, n°1. Madrid, Spain. www.infiqus.es

PREFACE

The use of biogenic resources for the production of biofuels and chemicals is presently a tendency worldwide. Different largely available raw materials – lignocellulosic derivatives, agricultural residues, starch, CO_2, or algae, to cite some of them – are being thoroughly investigated as potential renewable resources to reach sustainability targets while keeping (and extending) high life quality standards. Reasons for this trend can be found in the need of replacing fossil-based resources – due to environmental issues, availability, and geopolitical strategies – while tackling challenges of an increasing (over) population in the planet at the same time.

Being quite a broad and heterogeneous field, research on renewables is currently performed at academia from different perspectives, with emphasis on (bio) catalysis, process development, and feedstock valorization, among other important topics. Interestingly, the industry has also taken the lead in research, development, innovation, and process implementation using renewable resources. Actually, the industrial penetration for renewable raw materials is larger than what one could primarily expect, with a variety of outstanding examples of well-integrated biorefineries at industrial level already in use since decades. Novel products and processes are being implemented each year.

There are many excellent and comprehensive books and review articles covering the key aspects of the use of renewables to deliver biofuels and chemicals. To complement these, the motivation of this book is to give a more practical perspective to the renewables field. Thus, this book has been envisioned to be directly written by industrialists to firsthand describe their own current and underdevelopment (technical) processes, as well as their techno-economic goals. Likewise, their critical reflections and challenge identification lead to a real vision on how renewables must/should be set up on fluctuating markets in unstable global frameworks.

As a result, 23 diverse companies comprising large and medium established ones and a number of emerging and highly innovative start-ups have contributed with outstanding chapters. Some firms describe their current processes, products, and the establishment of novel ones. Other firms have focused on reflecting some challenges frequently encountered for renewable penetration – mostly dealing with tough economics – or on the need of more in-depth cooperations with academia to secure success in process implementation. All in all, the book depicts a rather heterogeneous picture – as it is in real world in renewables – in which interdisciplinary research and economic constraints are always carefully interconnected for successful applications.

As the ultimate goal, this book will provide readers with very valuable information on how industries think and develop options in renewables. I do hope that this book will serve as an inspiring basis for students, academic researchers, and other industrialists to get hints on how practical processes are being envisioned, implemented, and developed, contributing to bridge the gap between academic targets and industrial ones. As it may be anticipated, this is not an academic book on renewables, and thus the scientific approach of each topic is not comprehensive, albeit the chapters actually report on quite a broad state of the art of their different fields, emphasizing the need of a strong academic expertise to pave the way for innovation at a practical (economic) level.

I would like to thank warmly all contributors of the book, as they have invested quite an amount of time and high commitment in delivering wonderful and very useful chapters. Moreover, I would like to thank my colleagues Dr. Fabrizio Sibilla and Dr. Andreas Buthe for the fruitful and stimulating discussions on the topics of the book, as well as Mrs. Anita Lekhwani, Mrs. Cecilia Tsai, and the rest of Wiley team for the patience and support throughout the preparation of this book.

Dr. Habil Pablo Domínguez de María
Founder, CEO, CSO
Sustainable Momentum, SL
www.sustainable-momentum.net
dominguez@sustainable-momentum.net
Las Palmas de Gran Canaria, Spain

1

AKZONOBEL: BIOBASED RAW MATERIALS

ALISTAIR REID, MARTIJN VAN LOON, SARA TOLLIN, AND PETER NIEUWENHUIZEN

AKZONOBEL, Amsterdam, The Netherlands

1.1 AKZONOBEL'S BIOBASED RAW MATERIALS STRATEGY IN CONTEXT

This chapter sets out AKZONOBEL as a company, its position in the value chain, and sustainability approach and describes how the company has responded to recent

Industrial Biorenewables: A Practical Viewpoint, First Edition. Edited by Pablo Domínguez de María.

developments in biobased materials. AKZONOBEL has developed and implemented a white biotechnology strategy to help respond to the sustainability challenges of its customers and to contribute toward its corporate sustainability goals. Guided by this strategy the company is now partnering with innovators and leading companies in the rapidly developing field of industrial biotechnology and biobased raw materials to bring economic competitive developments delivering meaningful sustainability improvements to our markets.

AKZONOBEL is a committed and recognized leader in chemical industry sustainability and our proactive effort has been recognized through our presence in the top three of the Materials Industry group of Dow Jones sustainability index (DJSI) for each of the last 5 years. We have even been Materials Industry group leader for each of the last three years. The DJSI benchmarks the sustainability performance of leading companies based on environmental, social, and economic performance, including forward-looking indicators, and is regarded as the world's foremost sustainability index. It assesses various criteria, including supply chain management, operational eco-efficiency, product stewardship, human capital development, and occupational health and safety. AKZONOBEL is building on the basis of active health, safety, and environment (HSE) programs internally and through supplier support visits (SSV) and works with our key suppliers to improve HSE performance though our value chain. AKZONOBEL also takes a leading position in cross-sectoral initiatives such as the World Business Council for Sustainable Development and Together for Sustainability.

We are further building on this base with eco-premium solutions (EPS) and our Carbon Strategy. As part of our sustainability effort, AKZONOBEL is helping its customers to address their sustainability challenges by offering products that enable them to reduce their environmental footprint at competitive price points. To add impetus to this effort, we developed the concept of EPS in 2007 to translate the eco-innovation challenge into an operational target for our company. This eco-innovation challenge was defined by the World Business Council for Sustainable Development as the introduction of any new or significantly improved product – which can be either goods or services – process, organizational change, or marketing solution that reduces the use of natural resources (including materials, energy, water, and land) and decreases the release of harmful substances across the whole life cycle.

EPS are measured using a quantitative analysis or a qualitative assessment of performance in seven categories: toxicity, energy efficiency, use of natural resources/raw materials, emissions and waste, land use, and risks (e.g., accidents) and health and well-being (added in 2013) against the most commonly available equivalent commercial products or industrial processes (mainstream solutions) from a life cycle (value chain) perspective. The EPS must be significantly better with respect to at least one criterion and have no significant adverse effects with respect to any of the other criteria. The assessment is carried out by an experienced cross-functional group, including experts in R&D, marketing and sales, purchasing, manufacturing, and eco-efficiency. Since 2008, the EPS assessment has been audited as part of a broader sustainability audit.

Our 2020 target is to achieve 20% of revenue from products and services that provide customers and consumers in our downstream value chain with a significant sustainability advantage. This is in addition to our target of increasing revenue share from EPS (with benefits at any stage of the value chain) to 30% by 2015. Both are challenging goals because the assessments are made against equivalent mainstream or standard commercial products and as such are an upward moving target, as both we and our competitors introduce new and more sustainable products into the market.

Addressing concerns about contribution to and the impact of climate change prompted AKZONOBEL to establish a Carbon Policy in 2009. In this we recognize the need to move beyond controlling emissions from our own operations toward – throughout our product chain – identifying and addressing both the opportunities for more sustainable sources of materials and the strategic risks arising from dependence on fossil fuels and fossil-based raw materials. In line with our commitment to develop eco-efficient solutions for customers, the company acknowledges that managing our carbon footprint through innovative products, technology, and energy management constitutes both a business opportunity and a social imperative. Our target is to reduce our cradle-to-grave carbon footprint by 25–30% per ton of sales between 2012 and 2020. The cradle-to-grave footprint adds the impact from our customer applications and end of life of our products to the cradle-to-gate measure we have used since 2009. Since 2007, we publicly disclose our Policy, Management and Performance on Energy Efficiency and Carbon Footprint by our annual reporting to the Carbon Disclosure Project (CDP).

To support these sustainability efforts, AKZONOBEL set out a biobased strategy that identified opportunities in biobased materials where these might be delivered economically and maximize the environmental impact reduction. This strategy aims to forge supply chain relationships that connect these innovations and opportunities to our markets and application areas.

We have meanwhile struck a number of partnerships as a result of this strategy and are working through our value chains to bring these to market. Developing and implementing this strategy have required considerable thought on selection of the target molecules, engagement with current and emerging suppliers in new ways, and significant attention on how to address sustainable sourcing and integrating these materials into existing value chains.

1.2 AKZONOBEL IN THE VALUE CHAIN

AKZONOBEL is a leading global paints and coatings company and a major producer of specialty chemicals that holds leadership positions in many markets. In 2013 the company employed approximately 50,000 people and reported revenue of €14.6B with high-growth markets in Latin America and Asia Pacific representing 44% of this revenue.

The company is divided into three business areas, each of which represents approximately one third of our revenue:

- *Performance coatings*. Performance coatings are used predominantly for protection, mainly to industrial consumers having hundreds of uses across a wide range of industries and sectors including automotive, consumer electronics, aviation, shipping and leisure craft, sport equipment, construction, furniture, and food and beverage.

- *Decorative paints.* Our decorative paints business is the world's leading decorative architectural paints company supplying a full range of interior and exterior decoration and protection products for both the professional and the do-it-yourself markets. The product range includes paints, lacquers, and varnishes as well as products for surface preparation (predeco products). We also supply building adhesives and floor leveling compounds for tile, floor, and parquet layers.
- *Specialty chemicals.* AKZONOBEL is a major supplier of specialty chemicals with leading positions in selected market segments. Our products are used in a wide variety of everyday products such as ice cream, soups, disinfectants, plastics, soaps, detergents, cosmetics, paper, and asphalt.

AKZONOBEL by and large is not vertically integrated back to the primary extraction of resources but occupies a position in the value chain as a formulator and convertor of chemicals. While AKZONOBEL as the corporation is not widely known, many of our brands are household names, trusted by customers to brighten, protect, and preserve their homes, buildings, offices, and factories. These brands include Sikkens, Dulux, Jozo, International Paints, and Interpon, among others.

1.3 DRIVERS BEHIND DEVELOPMENT OF THE BIOBASED RAW MATERIAL STRATEGY

1.3.1 Background

AKZONOBEL has, for some years, been actively monitoring progress in the field of industrial biotechnology, white biotechnology, and biobased raw materials and recognized the advances being made in both industry and academia. Considerable progress has been made in developing fermentation, catalytic, and thermochemical technologies that could convert biomass from a variety of sources into both energy/fuels and materials useful to the chemical industry. These new developments have opened up new ways to produce drop-in replacements for existing and interesting novel materials from biomass.

Notable examples of progress in existing or drop-in materials we have noted include exploration of several routes to production of acrylic acid (e.g., through thermochemical conversion of glycerol or fermentation of sugars to 3-hydroxypropionic acid and subsequent dehydration) and a number of other routes and *n*-butanol (revisiting the acetone/butanol/ethanol (ABE) fermentation applied some 50–100 years ago applying new knowledge to improve yield, process, and product mix). Examples in novel materials include such 2,5-furandicarboxylic acid (FDCA) produced from sugars and proposed as an alternative to terephthalic acid in resins and PET packaging and polylactic acid proposed as an alternative to packaging and wrapping material (e.g., polyethylene/polypropylene).

1.3.2 Existing Use of Biobased Raw Materials

In assessing how to address and respond to these and other developments, our first step was to better understand what use AKZONOBEL already makes of biobased materials. Doing so it was found that biobased materials represented a considerably higher proportion of our raw material base than we expected, wherein 9% by value of spend raw material has a renewable element (Figure 1.1). We believe that this compares favorably against our estimate of a chemical industry average of ~3%, of which most are traditional biobased materials with only 0.3% coming from recent developments.

This use of biobased materials is in part ascribable to the long history in coatings and surfactants, using materials such as tallow, natural oils, rosins, natural waxes, gums, polysaccharides, etc. These still represent the main use within AKZONOBEL of biobased materials. In addition, while looking at our use of biobased materials, we found that we had more knowledge and experience represented inside the company than we expected and through this traditional base and the efforts of our R&D teams we held a significant but dispersed body of knowledge. Bringing these experts and knowledge base together was a significant step in allowing us to recognize and build on the potential.

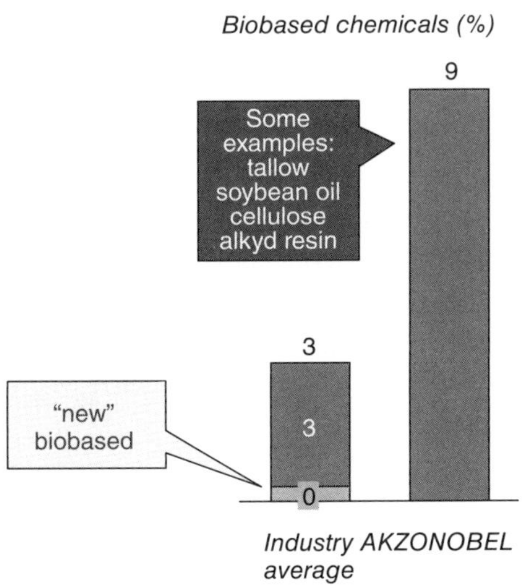

Figure 1.1 Wider chemical industry and AKZONOBEL use of renewable raw materials.

1.3.3 Emerging Products from Biobased Raw Materials

Building on this strong base, AKZONOBEL has also been active in developing and delivering to our markets a number of new biobased products and materials. These include examples such as Dissolvine GL chelating agent and our hybrid polymer technologies for formulation of cleansers and detergents. Dissolvine GL is part of our range of chelating agents wherein the majority of the molecule originates from a natural, renewable source. It is a safe and readily biodegradable alternative for phosphates, NTA, and EDTA and can be used in a number of applications, such as detergents, personal care and cosmetics, hard surface cleaning, automatic and mechanical dishwashing, oilfield chemicals, etc. Hybrid polymer technology is used in cleansers and detergents replacing the typically used synthetic water-soluble polymers. AKZONOBEL has developed a way to modify a natural polysaccharide "backbone" to make hybridized polymers that incorporate over 60% renewable ingredients and are nontoxic and readily biodegradable yet match or surpass the performance of the wholly synthetic alternatives.

1.3.4 Triggers to Developing an AKZONOBEL Biobased Strategy

As we monitored the sector, we noted three significant developments that triggered a more detailed assessment on whether to engage and respond and if so how:

- *Supply.* The supply of biobased materials is growing rapidly and many useful materials are becoming cost competitive.
- *Demand.* There is a growing customer interest and business drive for greener products.
- *Carbon footprint.* There is an increasing focus on reduction of greenhouse gas emissions.

Furthermore, three possible drivers were identified for our company to more actively address this area:

- *Security of supply.* Can renewable raw materials create alternative sources of supply addressing current or future bottlenecks in our value chain?
- *Market position.* Can renewable raw materials strengthen our position as world's leading coatings and specialty chemicals company?
- *Sustainability.* Can renewable raw materials contribute toward realizing our sustainability targets?

It is worth noting that we did not base our analysis on the assumption of a green premium or niche market segments. It was, and still is, our clear belief that to be meaningful and make a genuine difference in sustainability, biobased materials will have to achieve economics and volume that allow them to compete on level terms with the incumbent petrochemicals and in mainstream markets.

Biobased raw materials do not sit isolated from the wider chemical market. Petrochemicals are primarily utilized for the production of energy and changes in supply or demand through resource or technology developments in energy will have a significant knock-on effect on the chemical sector including biobased materials. This can already be seen with the emergence of shale gas in the United States both as an energy source and petrochemical feedstock, since this has had the effect of reducing the prices of C1 and C2 molecules, for example, methanol, acetic acid, and ethylene, while reduced use of naphtha as a refinery input has created pressures on propylene and aromatic value chains. This also creates significant intraregion market differences and is reshaping the political discourse. The full impact of disruptive developments of shale gas is still to be seen, but one could imagine that further fossil fuel developments, for example, shale development in other regions, fuel cell, solar, or battery technologies, or even significant shift in regulatory environments, would have considerable impact on the chemical markets. This would, of course, also impact development of the market for biobased materials possibly to the extent of quashing the sector outright.

1.3.5 Developing the Biobased Chemicals Strategy

In developing our white biotechnology strategy, we first looked at our existing value chains, formulations, and ingredients to identify those areas where biobased materials could make most meaningful impact and then address these allowing us to direct our efforts appropriately. Building on the work carried out to examine AKZONOBEL's carbon footprint, we see that of the 16 million tons per year cradle-to-gate carbon footprint total less than one third of the cradle to gate is determined by our processes and operations, while approximately 11 million tons is upstream from our operations and embedded in the material and chemicals we buy. This emphasizes the importance of working with our value chain if we are to achieve our sustainability ambitions.

We focused on organic raw materials in spend areas that cover ~35% of our purchase volume, excluding materials not amenable to renewable replacement such as pigments, additives, salt, and sulfur. We mapped out these organic raw materials in our value chain and then assessed each material both its direct use and indirect through its derivatives in terms of volume, value, and environmental impact. For example, we looked at *n*-butanol not just for its use as a solvent but also as butyl acetate solvent, butyl acrylate monomer, and acrylic resins and latexes that contain butyl acrylate. We then worked backward through the chemical value chain step-by-step till we had an overall picture back to the primary fossil and biobased feedstocks. Mapping our value chain and calculating the indirect uses were also valuable in highlighting those materials, value chains, and businesses most exposed to disruptive developments in petrochemical sector (e.g., aromatic and C3/4 naphtha fractions). Having identified the ~50 key building blocks and chemical conversions that make up our value chain, we assessed the possibilities for biobased processes and materials in three streams (Figure 1.2):

- *Drop-in materials*. Wherein a like-for-like replacement is possible (e.g., *n*-butanol produced in fermentation rather than derived from propylene).

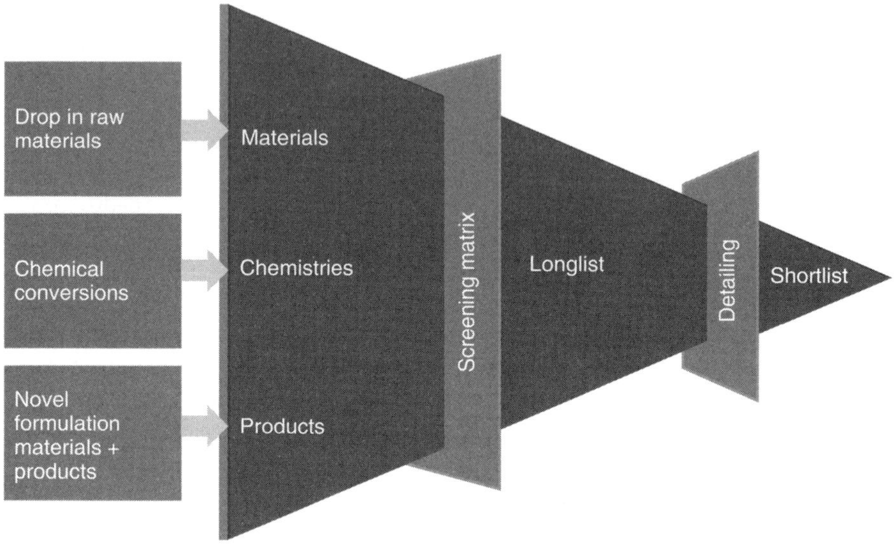

Figure 1.2 Three-workstream approach.

- *Novel formulation materials and products.* Where we find a material that offers similar or better functionality than an existing material (e.g., replacing synthetic adipic acid with a similar bioderived material such as succinic acid).
- *Chemical conversions.* In which we looked at the chemical reactions, conversions, and processes we use and tried to find examples of these conversions being achieved more efficiently or effectively using bioprocesses (e.g., esterification reactions utilizing lipase enzymes).

Looking at this map and with the support of RD&I chemists and chemical engineers in our business units, we were able to identify some 50 materials with potential drop-in opportunities from renewable raw materials, 50 novel renewable materials with potential relevance in our applications and markets, and 20 chemical conversions with some analogue in biochemistry.

Plotting our 50 drop-in, 50 novel, and process opportunities on matrices (Figure 1.3) and giving these a qualitative rating on technical and economic viability and potential impact – economic, volume, and sustainability – allowed us to prioritize and draw up a long list of materials and processes of interest. This rating was based on discussions with prospective suppliers, industry analysis, and patent/academic literature.

Based on this long list, the most viable and impactful of these opportunities were then examined in more detail. Each opportunity was assessed on common criteria basis, asking five questions:

- Could this material be cost competitive within five years (or provide a cost advantage)?

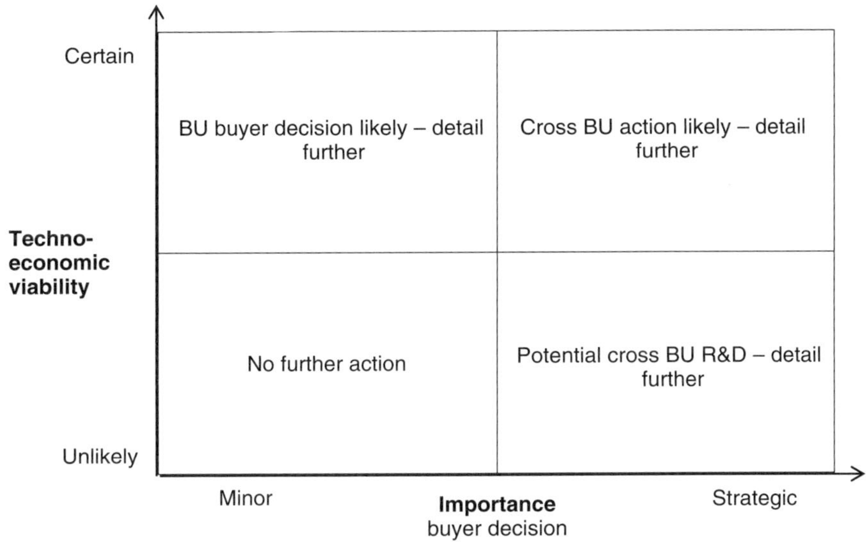

Figure 1.3 Example matrix.

- Could the material help address security of supply concerns for fossil material subject to current or anticipated disruption or uncertainty?
- Can the material help market AKZONOBEL's products and brand?
- How much could the material help reduce AKZONOBEL's carbon footprint and support its sustainability goals?
- Are there other drivers for adopting this material (e.g., HSE impacts with the fossil material)?

This long list was further refined and detail added working with potential suppliers and academic experts. Information was gathered on price both now and long-term expectation, life cycle analysis compared to the incumbent technology and current and anticipated use of that technology, both direct and through derivatives. This detailing then allowed opportunities to be scored on a standard template against consistent criteria (Figure 1.4).

- *Potential for CO_2 savings*. Potential annual CO_2 savings obtainable by replacing total volume (direct and indirect spend) of fossil-based material with biobased product.
- *Strategic supply issues*. Will biotech material help alleviate a strategic supply issue associated with the fossil material?
- *Marketing potential*. Is there any evidence that customers will value the use of this material from a renewable source?
- *Other*. Does the biotech material have any other drivers for action?

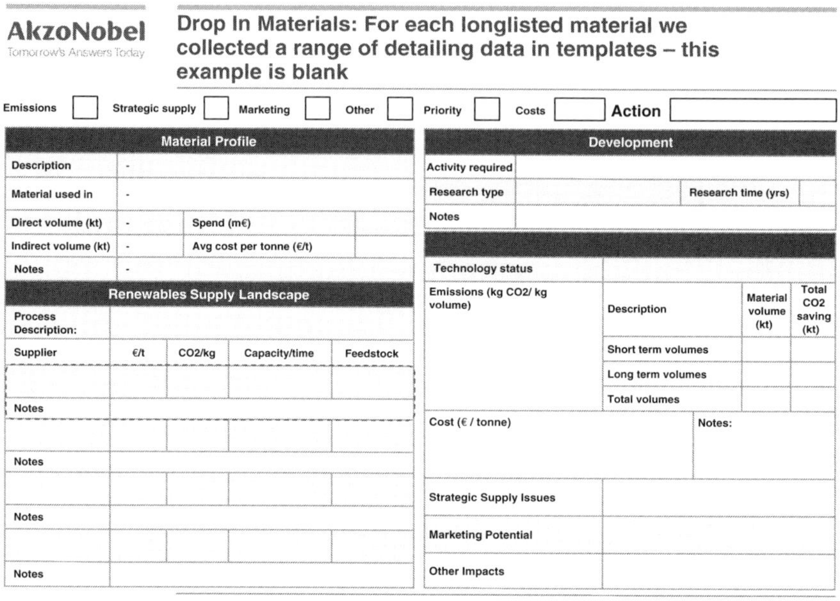

Figure 1.4 Example of a detail scorecard.

In parallel with these assessments of materials, the social, moral, and economic considerations around biobased materials were debated and their impact on any actions in this area was weighed.

1.4 CONCLUSIONS OF THE BIOBASED CHEMICALS STRATEGY

After consideration a strategy was approved that AKZONOBEL would make "A proactive effort to lead the deployment of selected, relevant biobased chemicals in our core markets and applications"; that is, AKZONOBEL will connect and work with innovators in the value chain rather than try to compete and would use its size, experience, and expertise in formulations to help provide a route to market for biobased chemicals.

Supporting this conclusion four guiding principles were defined:

1. *Leverage market innovation and focus on development of applications for renewables.* Rather than attempting in-house development of biotech processes and products ourselves, we will work with innovative suppliers (both new and established) and work with them bringing our expertise and insight in conversion, formulation, and application.
2. *Focus on cost competitive, high impact.* We believe that the greatest chance of success and opportunity to maximize our impact is to direct our efforts to a relatively few areas guided by supply issues, sustainability, and customer

insight. Where correctly chosen these will have more effect than many piecemeal efforts and make a meaningful contribution toward our sustainability goals.

3. *Work across the supply chain.* We will work together with multiple value chain partners to remove barriers and integrate renewable raw materials not just to utilize renewable raw materials as simple drop-ins but to look more broadly at possible derivatives and polymers wherein we can find larger volumes and markets, thereby maximizing impact and quickly building toward efficiencies of scale.
4. AKZONOBEL will utilize biobased materials utilizing first-generation feedstocks as a stepping stone but *insist suppliers work toward second generation.* We acknowledge the issues and concerns around first-generation renewable raw materials that could adversely impact food availability and prices and have recognize that there are significant environmental impacts associated with agriculture through land, irrigation water, and fertilizer use. We believe that for renewable raw materials to be accepted by the public without accusations of greenwashing, we will need to work toward the use of second-generation cellulosic feedstocks derived from municipal wastes, agricultural by-products, and other underutilized resources. We, however, also believe that to build capacity and develop the industry a direct leap to second-generation feedstocks might prove a significant technical or economic barrier.

Scoring against our five criteria, we prioritized materials and assigned actions for each. These actions covered a range of options ranging from more proactive efforts to a watch-list monitoring future technical or economic developments.

Those drop-in materials we prioritized included (in alphabetical order) and a nonexhaustive selection of the biobased routes we considered:

- *Acetic acid.* This is used by AKZONOBEL both for use in production of monochloroacetic acid and through acetate solvents, vinyl acetate monomer, and latexes. Acetic acid could be produced from biomass by feeding anaerobic digestion or syngas-derived methanol into an existing carbonylation process or by fermentation of sugars with an acetogenic organism.
- *Acetone.* We use this as a solvent but also see it in our value chain as a precursor to methyl methacrylate, bisphenol A, and a number of other chemicals. This is expected to be a by-product of ABE fermentation.
- *Adipic acid.* While not a high-volume constituent of our coating polyester resins we understand it has a relatively high carbon footprint per kg. Some routes by fermentation or production by catalytic conversion of natural fatty acids have been proposed.
- *Acrylic acid.* This is used to some extent in surface chemistry products and also a key precursor to acrylate monomers and acrylic resins and latexes. A number of potential routes from biomass have been proposed including catalytic conversion of glycerol and routes fermenting sugars to either 3- or 2-hydroxypropionic acid and then dehydrating.

- *Ammonia*. Our use of ammonia is mainly through the production of ethylene amines and fatty amines. Although not a carbon-containing molecule (and therefore not meeting any current or proposed standard for biobased certification), it has been proposed that ammonia might be produced from biobased syngas.
- *n-Butanol*. AKZONOBEL uses this as a solvent both directly and through butyl acetate. Indirect volumes also include substantial quantities through butyl acrylate monomer and containing resins. The principal biobased route is through the fermentation of sugars in ABE fermentation, but other routes have also been proposed.
- *Ethylene*. This is used by AKZONOBEL directly for ethylene oxide and ethylene amines but is also a precursor to many chemicals including vinyl acetate monomer and latexes. Biobased routes include dehydration of fermentation-derived bioethanol.
- *Epichlorohydrin*. AKZONOBEL used this material directly in some pulp and performance chemicals applications, but by far our largest use is as a precursor to epoxy resins. Production based on glycerol rather than propylene is now commercial and cost competitive.
- *Methyl methacrylate*. MMA is an important monomer and constituent of acrylic resins and latexes. While we believe that this is probably difficult to produce by a biobased route, some initial work has been reported both in incorporating renewable precursors into existing processes and in novel routes.
- *Terephthalic acid (and other aromatic diacids)*. These diacids are key monomers in coating polyester resins. Driven primarily by actors in PET packaging, some routes have been developed including hydrothermal reforming of sugars or conversion of fermentation-derived isobutanol.
- *Xylenes*. These aromatic molecules are both important as solvents but are also a potential precursor to the aromatic acids described above. Main routes from biomass suggested have been the hydrothermal reforming of sugars or catalytic conversion of syngas.

In these areas we had seen some external research and activity which we judged as having some potential to change the palette of chemical building blocks. Of this list for action acetone, *n*-butanol and epichlorohydrin were expected to be well developed, close to commercialization and cost competitive (or close to). On these materials we are working toward supply agreements and integration into our supply chain. Ethylene, ammonia, and acetic acid as high-volume commodities available at relatively low-cost form natural gas were expected to be technically possible but facing considerable economic hurdles. We would explore these materials with suppliers and look closely at economic viability and if feasible work toward supply agreements and integration into our supply chain over the longer term. Adipic acid, acrylic acid, methyl methacrylate, aromatic diacids, and xylenes were all expected to be in development but to still be some time away from reaching market. In these we would have open discussions with potential suppliers and explore the options on an ongoing basis.

Excluded were materials that were too fragmented or low volume for substitution to make a real impact or those where the petrochemical material is so cheap as to make the biobased route uncompetitive regardless of potential impact.

Moreover, we also identified some areas for potential substitution by novel monomers or resins:

- Novel monomers for polyester resins
- Alternatives to epoxy resins
- Monomers for acrylic resins and latexes

Finally some areas for further RD&I attention were defined. One area identified of particular potential interest is around the concept of the integrated biorefinery. There are a number of competing and complementary definitions of biorefinery, but by this we mean a facility designed and integrated for maximum valorization and utilization of biomass perhaps in combination or tandem with food, forestry, or waste processing. We believe that a biorefinery might offer both new raw material sources where integration offers the most competitive economics for all products but also find new applications for our BU knowledge in pulping chemistry, chlorine products, peroxides, process engineering, and optimization. Both thermochemical and fermentation-based process infrastructure might also find benefit from our knowledge in protecting chemical, oil, and gas installations through our coating products.

1.5 IMPLEMENTING THE STRATEGY: STRIKING PARTNERSHIPS

To implement the strategy we are bringing together cross-functional expertise in RD&I, sourcing, sustainability, and our value chain partners in derivatives to assemble the necessary technoeconomic assessment, validation, and life cycle analysis for biobased materials and to connect new companies into our supply chain. To this end we assembled a small team with responsibility for implementing the strategy reporting into our corporate directors of future proof supply chains and open innovation. This team is charged with connecting the right business units and responsible persons within AKZONOBEL with the selected opportunities in biobased materials either with new companies or established suppliers. These opportunities are then assessed (as far as is possible) for technical and economic feasibility and working with business and spend area managers we work on such sampling and testing as is necessary to validate the materials and work toward a definitive commercial agreement.

This process is generally supported by a nondisclosure agreement (NDA) and project memorandum of understanding (PMU), which sets out a project structure and timetable to allow both parties to be clear on steps necessary to achieve a commercial supply relationship. While nonbinding in nature this structure allows both parties clarity and mechanisms for building confidence. This process, while supported from a central team, is not a discrete activity and has to work in concert with the business units and spend area managers who drive the sourcing decisions of the company.

Spend areas coordinate across AKZONOBEL on material categories such as solvents, solid resins, colored pigments, etc.

Since establishing the strategy we have been able to announce several partnerships in succession:

- *Biobased solvents in Latin America.* This is a partnership with Solvay–Rhodia, which will target volumes of up to 10 ktpa of biobased solvents by 2017, through:
  - Joint reformulation of coatings to incorporate a novel glycerol-derived solvent into our coatings matching existing formulations both in performance and economically
  - Incorporation of drop-in *n*-biobutanol and acetone
- *Algae-derived fatty acids.* We have signed a joint development agreement with Solazyme to develop novel algae-derived fatty acids, which offer novel tailored compositions.
- *Biobased epichlorohydrin.* We are working in partnership with Solvay to develop a "chain of custody" methodology to integrate biobased epichlorohydrin used in AKZONOBEL's epoxies working through supply chain epoxy producers with Ernst & Young (EY) acting as an independent party for data handling, addressing the validation of evolving volumes. This effort has the target to increase the proportion of biobased ECH in AKZONOBEL's indirect use through epoxy resins to 20% by 2016.
- *Cellulose-derived acetic acid.* We are working with ZeaChem to explore the potential for acetogenic organisms to ferment cellulosic sugars to acetic acid and connect with other complementary users to growth volume and possibly obtain derivatives.

In theory all current petrochemicals could be derived from biomass, but in practice this would be inefficient and uneconomic. In practice a few drop-ins are being studied or applied where these are competitive. In fact with our activities and partnerships, we can start to see a path over time to significantly improve the sustainability of our supply chain (Figure 1.5).

In addition, we expect shortly to be able to announce several further partnerships addressing more of our target materials. This has not, however, been a straightforward process.

1.6 EXPERIENCE TO DATE

Particular experiences that we have identified over the 18 months since we launched this strategy include:

- *This incorporation of biobased materials needs considerable commitment to be seen through.* This is a long-term effort. Many technologies are only at demonstration scale. While first facilities are being planned, it may take years before

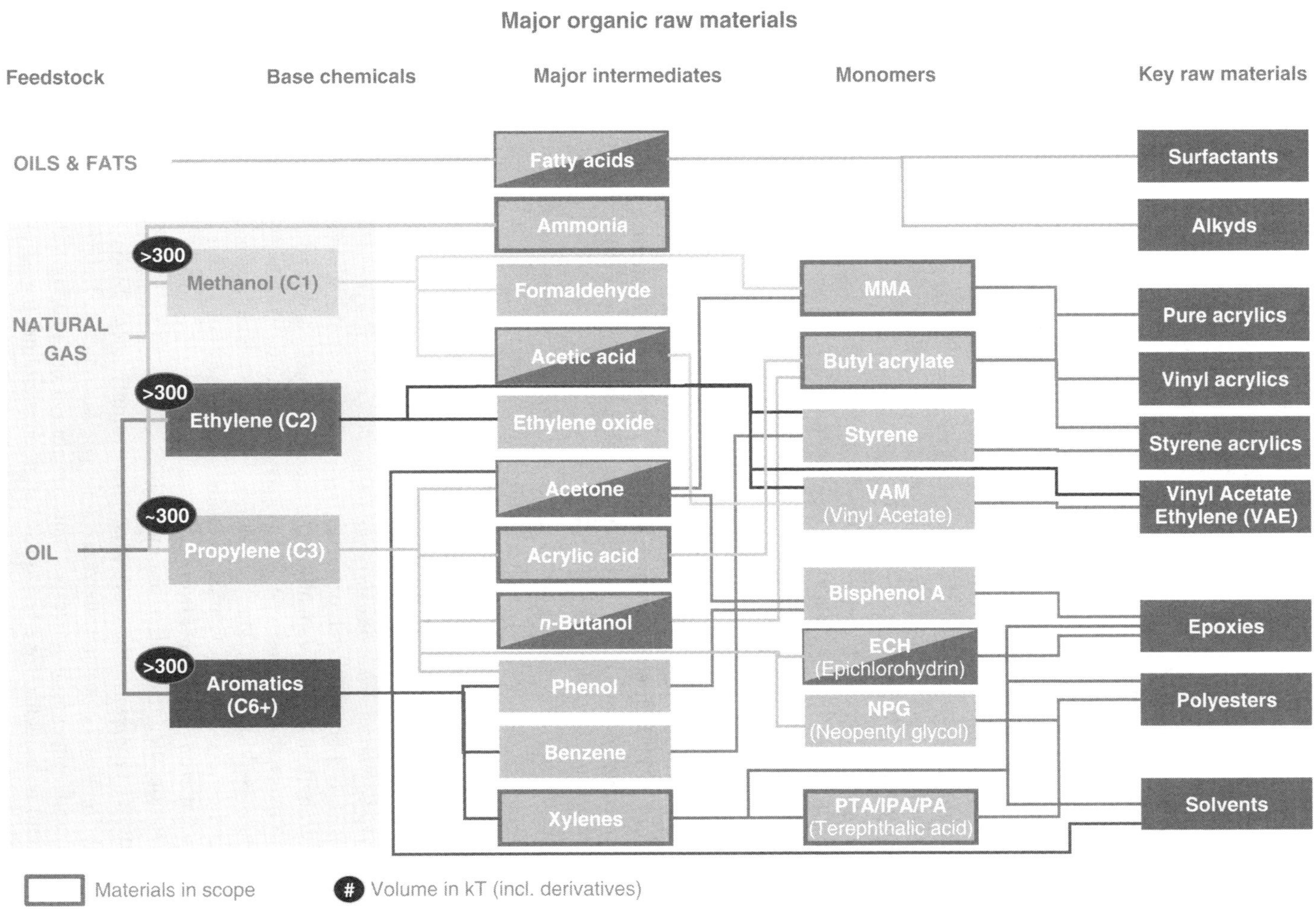

Figure 1.5 Potential green supply chain.

more than test samples will be available. Patience and long-term planning from both supplier and user need to come together.

- *A clear internal strategy is vital to move from analysis to action.* With all of the hype and rapid advances in the field, it would be easy but ineffectual to fall into the trap of trying to cover everything and be involved somehow in every consortium. Identifying a few addressable but impactful actions and seeing these through are critical.
- *Even with the best efforts of all parties surprises can come up*, which cause issues both technical and operational, and these need to be addressed by both parties in an open and honest fashion. Internally these also need to be communicated in such a fashion that these are put in context and build trust in a robust and responsive process rather than tarnishing all efforts.
- *We do not see a willingness from the broad market end users to pay a "green premium,"* but partners can frequently be found who can compete. A small segment of the market especially in brands or consumer products/marketing may find a willingness to pay a premium for a "green" image, but this is a limited segment of the market and likely temporary. For biobased materials development/producers, it is vital to be cost competitive at a mature stage. Although definitely not all will be competitive, numerous can or could be.
- *Feedstock availability, pricing, government support with investments, and integration will be critical.* Paramount aspects to success are to decide where these technologies are applied. There is little point in public investments and drive toward biotech and biobased materials if regional economics (either scarcity of raw materials or very low-cost petrochemical feedstock) prevent this from becoming feasible.
- *Attention needs to be paid to the logistical and business models*, which will eventually bring these materials and technologies to market. While petrochemicals are well defined, are energy dense, and have an established infrastructure, biomass is locally and seasonally variable, is geographically dispersed, and contains a lot of water, so aggregation, drying, and transport will play a large part in determining economic success.
- *Participants need to be clear about what they bring to a partnership.* AKZONOBEL can be a channel to market, guarantee of offtake, and bring considerable strength in downstream and formulation chemistry but is generally not aiming to make equity investments or to license technology to integrate upstream into biobased materials. Many small firms develop technology and are looking for either investors or eventual licensors – if there is not a clear understanding of what each party wants from a discussion and the eventual path to commercialization, then frustrations are inevitable.
- *Win-win opportunities do exist.* Increasingly these biobased materials are becoming economic alternatives to petrochemicals. Improved fermentation yields, separation processes, and integration are making these real. There is still no evidence for a broad market green premium, but partners can be found where this is not needed.

- *Genuine interest to make it work for both parties*. There is no one-size-fits-all model – every opportunity and partner requires a different approach. As in real life, partner choice is a high-stakes matter. It is not always easy – no partner is perfect right away – the strategy sees you through. Each situation and partnership is unique – this can be due to the challenges of working with new processes and materials but also the complexities of a value chain. Indirect use might be a bigger absolute volume but extremely challenging to achieve substitution. For both participants in a partnership, making the right choice is an important decision. You need a robust process to see the work through to conclusion.

1.7 MEASURING, REPORTING, AND ENSURING SUSTAINABLE SOURCING OF BIOMASS

In developing a strategy we also had to consider how best to measure, report, and communicate this effort. Setting a strategy is one thing, and quantitatively monitoring progress another one. We had to define internally the language and terminology, that is, what we mean by biobased, renewable, etc., as these might be interpreted differently as questions arise over materials such as water, inorganic materials that might contain carbon (e.g., carbonate minerals), and materials that are derived from fossil resources but do not contain carbon (e.g., ammonia or hydrogen). The terms green and sustainable were also examined and discarded for internal reference to biobased or renewable raw materials. With agricultural land, materials, and energy use, it was recognized that it was possible for material to be biobased but to have a higher environmental impact than the fossil alternative.

For clarity and reporting we have defined biobased materials as raw materials that are wholly or partly derived from a biomass source that is continually replenished (typically well within 100 years). The biomass can have undergone physical, chemical, or biological treatment(s). Peat and natural forest are not defined as renewable raw materials due to the fact that these materials cannot be continually replenished well within 100 years. Definition includes materials derived from trees, crops, fats, etc. and excludes peat and natural forest.

Having measured our baseline use of biobased raw materials, we then considered how we should measure this for the future. A line-by-line interrogation of sourcing volumes is not an efficient approach. AKZONOBEL has implemented the following working procedure to ensure that the reporting is more easily and reproducibly done in the future. A report of spend by weight and value is generated from our purchasing database – this is then matched with material listings of biobased and organic raw materials. The spend report can be generated on several levels: AKZONOBEL, business unit, and region. The business units receive their report with the draft biobased volumes for review and sign-off. This setup ensures that each business unit owns and is responsible for this measure with support from the sourcing community. In a nutshell, business units are data owners and accountable for the material listings and sign of procedure in place with checks and balances. Central support is given in setting up and identifying renewable and organic materials and in regularly providing the results. This procedure should ensure cross-AKZONOBEL consistency as well as accuracy.

A number of ways to measure biobased materials were reviewed each approach with its pros and cons:

- *Biobased materials reported as a fraction of organics*. This is used by some peers, provides a bigger number, gives the opportunity for 100% biobased to be theoretically possible, and is not impacted by variation in inorganics. It does however raise some concerns over potential to be accused of "greenwashing" and could be cumbersome to determine/maintain.
- *Biobased materials as a proportion of all raw materials*. This is used by most other organizations and easiest to determine and maintain. This is absolutely the easiest to communicate externally requiring no caveats or explanation. On the downside it is impossible to reach 100% substitution and will be impacted by changes in inorganic product value chains, which are not affected by biobased programs (e.g., salt chain, clays).
- *Biobased plus abundant materials as a proportion of all raw materials*. This provides a bigger number and stimulates the use of abundant materials. On the downside it is cumbersome to determine and explain the definition of abundant materials (e.g., salt, calcium carbonate, etc). It also loses the focus on biobased materials.
- *Spend fluctuation in raw material pricing*. Advantages more expensive materials.

While a case can be made for excluding all inorganic materials and reporting biobased as a fraction of carbon-containing (organic) materials, we decided that it was more transparent and easily communicated externally to report as a fraction of all of our raw materials:

- *External reporting*. Report renewable raw materials as % total purchased materials alongside the AKZONOBEL KPI for comparability in the industry:

$$\frac{\text{mass (renewable RMs)}}{\text{mass (total RMs)}} = \%\ \text{Renewables of total}$$

- *AKZONOBEL KPI*. Renewable raw materials as % organic materials is the internal measure to drive improvement:

$$\frac{\text{mass (renewable RMs)}}{\text{mass (organic RMs)}} = \%\ \text{Renewables of organics}$$

1.8 BOOK AND CLAIM

While it would be desirable for transparency biobased materials to be separated and incorporated into products as wholly biobased, this is in many cases unrealistic.

A fatty acid or carbohydrate will be 100% biobased from the start of the value chain to final formulated product and be traceable by carbon isotope measurement in keeping with standards such as the US BioPreferred program, but dogmatic adherence to this approach will present a significant barrier for future growth of the sector.

Dedicated supply chains would be very inefficient. If a supplier needs to specially segregate batches or set up separate storage and production lines, for example, keeping biobased *n*-butanol from petrochemical in the production of butyl acrylate or derived resins, then we believe that this will create an impassable hurdle for acceptance.

In order to see this sector grow, we should take some cue from the energy sector wherein renewable electricity or gas is fed into the general grid system and the customer can choose supplier but not require separate supply (e.g., book and claim). A comparable book and claim approach would be desirable and we are in the process of discussing mechanisms with value chain partners and an independent auditor who could verify volume and ensure that competitive information is not shared inappropriately.

1.9 SUSTAINABILITY IN THE VALUE CHAIN: LCA

AKZONOBEL ensures that suppliers endorse our environmental and social standards as formulated in our code of conduct by asking suppliers to sign a vendor policy declaration. This procedure applies also to suppliers of biobased RMs. Biobased RMs often have a lower carbon footprint than the fossil alternative but that is not always the case. Renewable raw materials are often connected to use of fossil fuels due to harvesting, handling, transport, processing, etc., which also need to be included when compared to the fossil alternative.

There are also a number of potential concerns that should be addressed in assessing whether renewable RMs offer a genuine sustainability advantage:

- *Land transformation and biodiversity.* Cultivation replaces natural forests (land transformation, biodiversity, ecosystem values).
- *Food competition and indirect land use change (ILUC).* Competitive uses of the land (food security in region, rights of indigenous peoples) and the environmental impact connected to that other land use interest are moved to other regions (ILUC).
- *Poor land management practices.* Cultivation without good land management practices (pesticide use, soil degradation, direct land use change leading to emission of greenhouse gases, conservation of rare, threatened, or endangered species, and other high conservation value habitats, etc.).
- *Water stress.* Cultivation takes place in area with serious water issues.
- *Social issues.* For example, not allowing unions, child labor, etc.

These concerns have already been raised due to the growth in biofuels using first-generation technologies like corn ethanol and palm oil biodiesel, which have both seen significant negative attention. It is, however, important to note that still by far the biggest demand for materials of concern such as palm oil comes from consumer use as cheap oils and fats for soaps and foodstuffs. This is a nuanced area and in some cases the answer will not be clear – it might seem unsustainable to grow more sugarcane land in Brazil to expand in biobased materials, but if the main sugarcane-growing areas are a long way from rainforest and can make fresh use of land, which was degraded by pasture and abandoned, this can provide new income and fresh opportunities for local farmers. Consequently in addition to our existing sourcing standards, we are reviewing what additional standards/assessments will be appropriate for the future.

2

ARIZONA CHEMICAL: REFINING AND UPGRADING OF BIO-BASED AND RENEWABLE FEEDSTOCKS

GODFRIED J. H. BUISMAN[1] AND JOS H. M. LANGE[2]

[1]*Arizona Chemical B.V., Savannah, GA, USA*
[2]*Arizona Chemical B.V., Almere, The Netherlands*

Industrial Biorenewables: A Practical Viewpoint, First Edition. Edited by Pablo Domínguez de María.

2.1 COMPANY INTRODUCTION

Pine trees from sustainably managed forests are the basis of much of today's pulp and paper production in Europe and North America. During the Kraft pulp production process, two major pine chemical product streams are generated: crude tall oil (CTO) and crude sulfate turpentine (CST). CTO and CST are refined and subsequently upgraded into a variety of specialty chemicals. These chemicals are utilized in an extensive range of essential products and applications across diverse industrial and consumer markets (Figure 2.1). Examples include adhesives, tires, rubber, road marking, coatings, fuel additives, mining processing fluids, and lubricants.

Established in 1930, Arizona Chemical is the world's leading biorefining company of pine chemicals. By virtue of these feedstocks, our efforts are dedicated to providing innovative and sustainable solutions that satisfy the performance demands of these markets and offer green alternatives to fossil oil-derived chemicals. Typically, Arizona Chemical's products have a high biorenewable content (BRC). For instance, pine chemical-based tackifiers that are used in hot-melt and pressure-sensitive adhesive (PSA) formulations have 50–99% BRC. Petrochemical-based alternatives have a negligible BRC. What seems like an obvious choice today – to increase the use of renewable resources – has been the mainstay of our business since its founding. We produce a wide product range from biorenewable raw materials originating from sustainable forestry that do not compete with global food supplies. A carbon footprint analysis of our operations has shown that emissions resulting from our products amount to less than half of that of the existing alternatives. This enables our customers to drive sustainability objectives while enhancing the properties of thousands of products for use in daily life.

2.2 HISTORY OF PINE CHEMICALS

2.2.1 Overview

Harvesting pine tree resins and the use of the obtained products date back to ancient times. Pine tar and pitch were used for caulking the seams of wooden ships and vessels.[1] In reference to these first and vitally important maritime applications, the term "naval stores" was often used to products originating from pine trees.

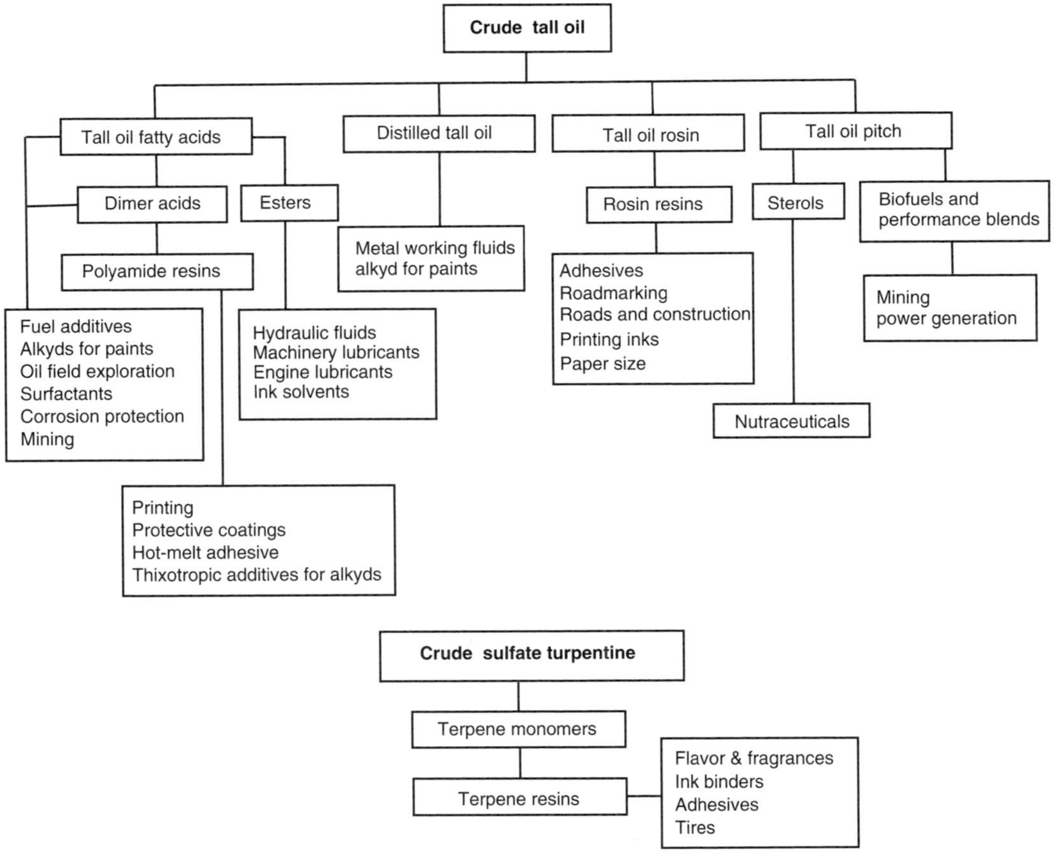

Figure 2.1 Crude tall oil products and its applications.

For centuries, pine-based products have also found some utility in pharmaceutical applications; one example is the use of coal tar to combat chronic dermatitis.[2] Pine tree derivatives have been widely used in adhesives, soaps, and water-repellant surface coatings for ropes and construction.[3]

The first tapping of pine trees was achieved by making cuts in the bark of the tree. The collected crude gum was processed and distilled into products such as rosin (colophony) and turpentines. However, when the labor costs of recovering rosin and turpentine by tapping living trees increased in developed countries, gum naval stores became less profitable. This shift took place around the beginning of the 20th century and consequently boosted the development of alternative techniques to efficiently produce pine chemicals. These processes were important sources of numerous useful products, including not only wood and cellulose but also nonwood products utilized by the chemical, food, flavor and fragrance, and pharmaceutical industries.

Destructive methods to distill pine stumps, left after cutting the forests for timber, became a popular alternative to pine oleoresin tapping. Products such as rosin oils, turpentine, monoterpenes, and creosote oil became widely available and found many applications as binders and solvents for ink and paints, insecticides, and pharmaceuticals.[4] Since 1930, the research on wood chemicals and the understanding of molecular structure and performance properties have resulted in the design and synthesis of many tailor-made products with optimal properties for specific applications.[5a] A comprehensive overview of the origins and history of the pine chemical industry can be found in the publication *Naval Stores*.[5b]

Pine chemicals have been used for many decades. The continuous application of new scientific insights and technical developments in processing optimization and purification techniques has resulted in an integrated biorefinery infrastructure over the past 80 years. The availability of industrial quantities of bio-based chemical building blocks, concerns over the depletion of fossil-based raw materials, and growing awareness of the negative impact of global warming have all led to an increased interest in forestry-based biomass as an alternative source for renewable chemical building blocks. In this book chapter, a description is provided of the main principles of the modern biorefinery for pine chemicals. Several examples illustrate the importance of pine chemicals in a range of applications that we encounter daily.

2.2.2 Terpenes

Pine oleoresins contain a complex mixture of volatile and nonvolatile terpenes. Terpenes can be chemically viewed as oligomers of five-carbon isoprene building blocks.[6] Terpenes biochemically originate from the common C_5 building blocks isopentenyl pyrophosphate (IPP) and its regioisomer dimethylallylpyrophosphate (DMAPP), which are synthesized from the biogenetic precursor acetyl-CoA. IPP/DMAPP is converted into geranyl pyrophosphate (GPP), which subsequently is converted by monoterpene synthase enzymes into a variety of monoterpenes, such as α-pinene, β-pinene δ-3-carene, limonene, and many others (Figure 2.2).[7] α-Pinene and β-pinene are the main products of many monoterpene synthases, especially in pine trees. It should be noted that the formed α-terpinyl cation can undergo either

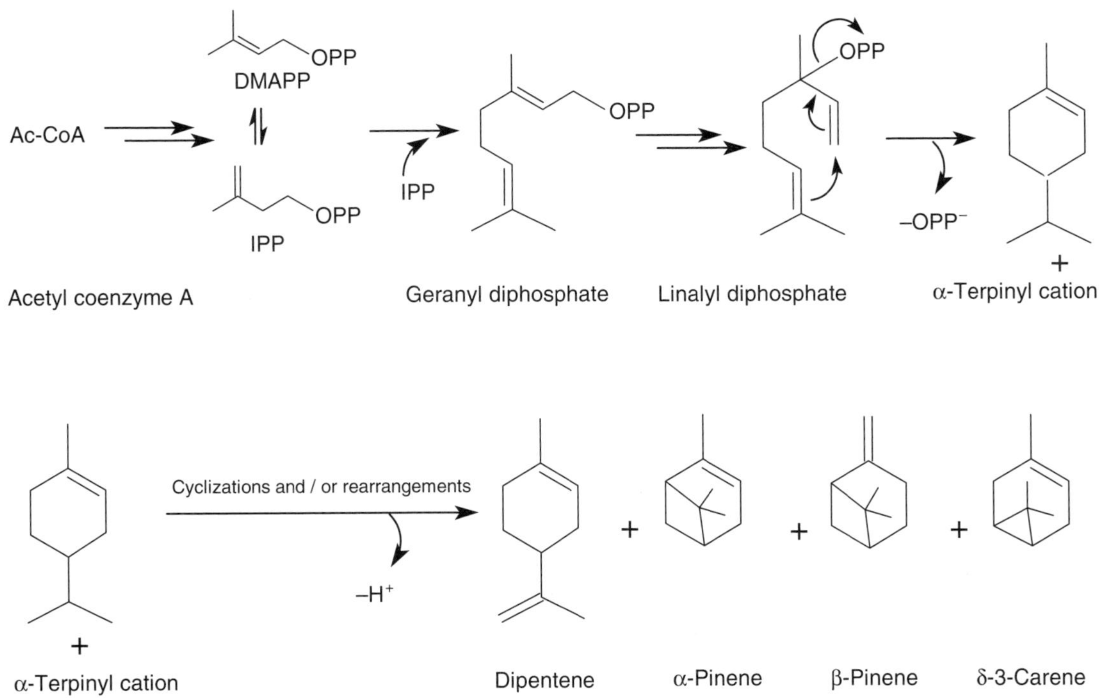

Figure 2.2 Monoterpene biosynthesis route (stereochemistry omitted).

cyclizations by electrophilic attack of the carbocationic center on one of the carbon atoms of the remaining double bonds or alternatively can undergo hydride shifts as well as other molecular rearrangements. These reactions all contribute to the formed monoterpene variety. The monoterpene formation is typically terminated by proton loss. The formed terpenes are present in several plants, as part of a defense mechanism against their major predators (such as bark beetles and pathogenic fungi).

The volatile monoterpene (C_{10}) and sesquiterpene (C_{15}), along with the non-volatile diterpenes (C_{20}, rosin) built up from several isoprene units, make up a valuable natural source of materials for the pine chemical industry.[8] It should be noted that isoprene itself is produced at a scale of 800,000 tons per year as a by-product of naphtha cracking or during the production of ethylene. This volume is almost completely utilized for the production of *cis*-1,4-polyisoprene or synthetic rubber.[9]

In commercial resin-tapping operations (e.g., China, Indonesia, Vietnam, Brazil), pine oleoresins are tapped from living trees (e.g., *Pinus pinaster*, *Pinus halepensis*, *Pinus nigra*) by means of a repeated wounding process.[10] The collected crude oleoresins are purified by steam distillation into gum turpentines and gum rosin. The physical requirements for turpentines are typically related to their relative density; refractive indices; acid, iodine, and hydroxyl values; and flash points.[11] It has been estimated that the total world production of turpentine is around 350,000 tons per annum, of which around 100,000 t is believed to be gum turpentine.[12] This distinguishes it from turpentine recovered as a by-product from chemical pulping of pines and that is referred to as CST with an estimated annual production volume of about 235,000 t.[13] The chemical composition of turpentine varies depending on the species of the tree, the geographical location, and the applied isolation procedures. For example, the monoterpene composition of the trees harvested in the United States versus Sweden is significantly different (Table 2.1).

The main components of turpentine are unsaturated hydrocarbon monoterpenes such as α-pinene, β-pinene, and δ-3-carene. It should be noted that many monoterpenes are chiral compounds. It has been found that the enantiomeric compositions of monoterpenes in *Pinus sylvestris* varied considerably, not only between different trees but also between different tissues of an individual plant.[14] The chirality aspect is not necessarily relevant for industrial tackifier applications, but it is essential for the use of monoterpene feedstock in the fragrance industry (Figure 2.3).

Wood turpentine is obtained from aged pine stumps. Pine stump processing was a major source of pine chemicals in the early years of the industry but has declined and

TABLE 2.1 Major Crude Sulfate Turpentine (CST) Monomer Composition Differences by Comparison of the United States with Swedish CST.[5b]

	α-Pinene (%)	β-Pinene (%)	δ-3-Carene (%)	Dipentene (%)
Sweden	45	5	41	3
The United Sates	64	21	2	4

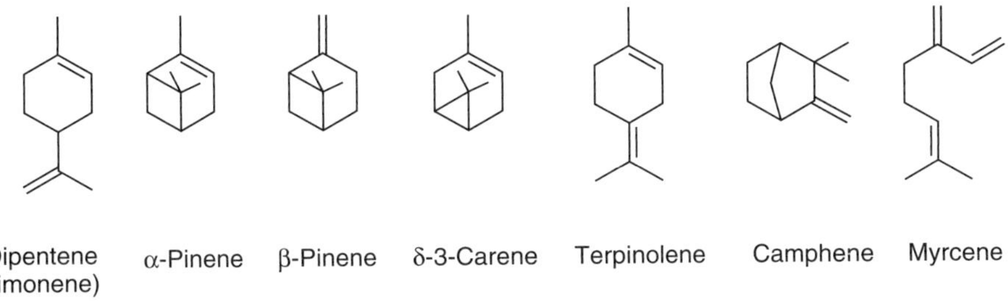

Figure 2.3 Chemical structures of common unsaturated monoterpenes (stereochemistry omitted).

is now of minor commercial significance. Most collected turpentine streams yield α-pinene and/or β-pinene as the major components, and it is estimated that around 50,000–100,000 t of pinenes is used as starting material for the flavor and fragrance industry per year.[15] Pinenes can also be chemically converted to other valuable products and are used as building blocks for pharmaceuticals, plasticizers, repellents, insecticides, and antiviral or antimicrobial moieties. Alternatively, terpenes undergo Friedel–Crafts reactions and (Lewis acid-catalyzed) cationic oligomerization and polymerization reactions to form branched resins. Resin production can be practiced with one or more terpene monomers, and copolymerization is often practiced with monomers such as phenols or styrenes. Monomer choices and the applied specific reaction conditions dictate the structure, molecular weight, polymer compatibility, and physical properties, which allow these resins to be used as rheology modifiers for rubbery materials such as adhesives and tires.[16–19]

2.2.3 Rosins

Rosin is a collective term for plant resins derived primarily from coniferous plants; it is a solid and brittle material at room temperature. Gum rosin is obtained from the tapping of pine trees, whereas tall oil rosin (TOR) is obtained from the Kraft pulping process. These two variants are the principal sources of rosin. Wood rosin is extracted from pine stumps and today constitutes a negligible contribution to the overall pine rosin supply. Rosin typically has a color that can vary from pale yellow to dark red, depending on its origin and isolation/purification.[20] Acid number, saponification number, color, and softening point are the most important properties to the industrial chemist or engineer. TOR is primarily an isomeric mixture of fused tricyclic diterpenoid (C_{20}) resin acids such as abietanes (abietic, neoabietic, palustric, levopimaric, and dehydroabietic acid), pimaranes, and isopimaranes (pimaric, isopimaric, and sandaracopimaric acid). The chemical structures of the most common rosin acid regioisomers are depicted in Figure 2.4.

Rosins or their derivatives are important components of industrial products, such as adhesives, coatings, printing inks, emulsifiers and surfactants,[21] polymer precursors,[22,23] performance additives for tires, and various road and construction materials.

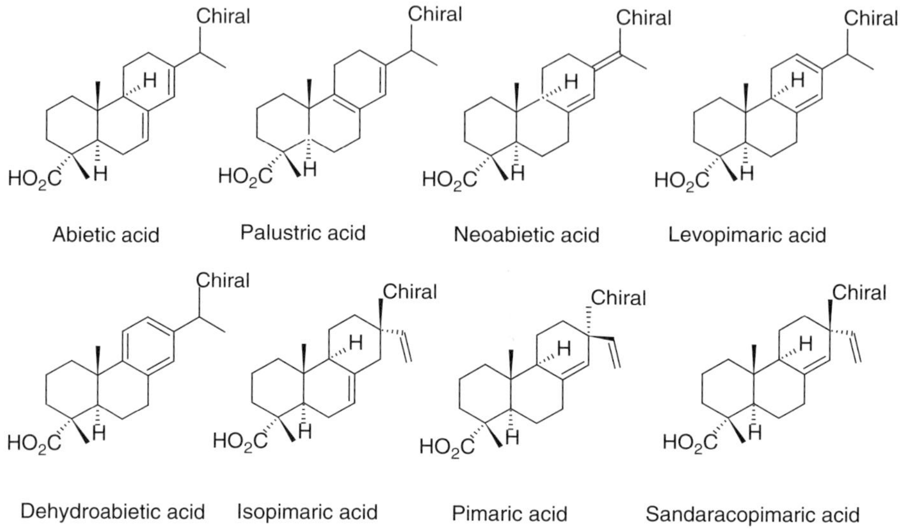

Figure 2.4 Chemical structures of most common rosin acids.

2.3 MODERN BIOREFINING

During the oil crises in the 1970s, we were reminded just how dependent our industrialized civilization is upon fossil oil. The majority of earth-extracted oil is utilized in energy fuel production, and the remaining is used for the production of a large variety of bulk and specialty chemicals. The global chemical industry and the energy supply and transportation fuel supply chain are dependent on the same petroleum, motivating the search for renewable. The increasing needs for energy and chemicals mirrors the global growth in economic development and populations. In particular, the demand for polymeric materials is expected to increase considerably during the coming decades. Raw material demands are partially offset by positive contributions of polymeric materials to energy efficiency. Polymers assist in conservation thanks to their lower weight relative to metal-based materials. Consequently, plastics and polymers continue to displace heavier metals and play an increasing role in transport vehicles like cars, planes, and trains. Furthermore, their lower thermal conductivity and their insulation properties often boost energy saving.

Due to the eventual and inevitable limits of fossil fuel availability,[24] the associated increasing costs, the desire to be less dependent on oil-exporting countries, and a growing sense of responsibility to our planet, research programs – often heavily subsidized by governments – have been put in place to search for alternatives. The efficient conversion of biomass into chemicals and energy will be essential to sustain our present way of life in light of a growing world population and increasing demands for raw materials. Solar, nuclear, and wind energy research projects have received significant attention and have resulted in several successful launches for fossil-based energy alternatives. Solar panels and electric/hybrid cars have become increasingly available, and wind mill parks can be found all over the globe as green alternatives to fossil fuel-derived electricity. Over the last decade, major public and political debates have argued the merits of alternatives for fossil transportation fuels to reduce greenhouse gas (GHG) emissions. Several national and international policies have been put in place aiming at reducing GHG.[25] Policies requiring the mandatory blending of biofuels are having a large impact on production, utilization, and long-term availability of biomass.[26] The introduction of biofuels is not alone, but it is a major influence in the search for petroleum alternatives.[27] However, the use of biomass as a long-term resource for energy and fuel production is expected to be limited by production rates and the supply of biomass rather than the demand for energy and fuel.[28] The relatively low energy content of biomass – due to the higher oxygen content the energy needed to generate dry content – has been recognized as a barrier for large-scale utilization in energy and fuels.

Specific energy (SE) (sometimes called energy density) is an important metric for comparison of fuel or feedstock purposes in this respect. SE is the amount of energy stored in a given system or region of space per mass unit. For example, hydrogen has a very high SE value of 142 MJ/kg. The gasoline/diesel SE value amounts to 46 MJ/kg, whereas the SE values of materials like wood, sugar, and proteins range from 16 to 17 MJ/kg. The relatively low SE values for wood, sugar, and proteins reflect their limited value for fuel purposes.

Recent studies have shown that several economic drivers exist for the development of bio-sourced chemical products. For example, in the United States in 2007, the chemical industry was responsible for just over 3% of the total US petroleum consumption, worth a value of $ 375 billion USD (including capital and labor costs).[29] By comparison, the transportation segment accounted for over 70% of the US petroleum consumption having a value worth of $385 billion USD. The remaining 27% of the US petroleum usage was allocated to nontransportation energy usage, for a value-added worth of $135 billion USD. This analysis shows that the value of the chemical industry is comparable to the fuel industry but requires only a fraction of the available energy to satisfy the economic demand. Oxygen-containing functionalities make biomass suitable for industrial chemical building blocks if they can be made available in significant commercial quantities at acceptable processing costs. Similar to petroleum, the composition of biomass can be complex, and as such, its primary fractionation into its main components allows for treatment and processing into a wide range of products.[30] Most of the petrochemicals are produced from crude oil refineries fed by fossil fuel, whereas in the future, it is anticipated that many bio-based products or chemical building blocks will be increasingly manufactured from biorefineries fed with biomass. However, unlike fossil-based feedstocks, biomass feedstocks typically contain oxygen content between 10% and 45%, introducing a high degree of unique functionalities in the molecular backbones, which in turn makes them interesting building blocks for the chemical industry. In general, the oxygen content of biomass should make it less favorable for burning and potentially more interesting for derivatization. There has been considerable debate about the definition of the term biorefinery.[31] The overall objective of a biorefinery production approach is to manufacture a wide range of goods from biomass feedstocks through a combination of separation and purification technologies.[32] Ideally, renewable biomass is converted into one or more (co)products rendering no waste, as each production stream has the potential to be converted into products to the maximum extent. The biorefinery approach therefore generally involves multi-step processes to process the preferred feedstock. Different integrated processing steps can be applied to optimize product yield, quality, and product consistency. However, unlike petroleum feedstocks, biomass feedstocks typically are more complex due to their higher degree of functionality. This makes the purification steps in most cases more sophisticated. Lower thermal stabilities of natural products will often complicate the separation into their individual chemical components. Therefore, in a true integrated biorefinery concept, the biomass components are in general subject to a combination of sequential purification and/or chemical conversion steps in order to maximize economic efficiency and improve end product robustness.

Purposeful biological production of chemicals through the harnessing of microbes is not new but is getting considerable attention today because of its potential to manufacture high value chemicals. The material demands for this industry are relatively low. Since 1950, several commodity chemicals like acids and alcohols were produced by fermentation technology including lactic acid, levulinic acid, acetic acid, citric acid, itaconic acid, and ethanol. These initiatives can be considered a form of biorefineries.[33] Projects to utilize biomass on a large scale as a source for biofuels have been demonstrated at pilot and industrial scales. Importantly, forests work as sinks of atmospheric carbon, thereby contributing to greenhouse effect mitigation. Biomass fixes carbon dioxide by photosynthesis and therefore contributes to a reduction in GHGs if the biorefining process steps are fully optimized with minimized emissions.

2.3.1 Challenges of Biorefining

The related social and economic discussions of evolving from a fossil-based to a bio-based economy continue.[34] It has become clear that a large-scale utilization of biofuels will compete with food production. Changes in crop production and harvesting, in some cases leading to deforestation, are of concern. Several studies reported on the feasibility of large-scale production of biofuels to meet the increasing energy demand and also to replace the use of fossil fuels.[35–37] In general, it is impractical to replace the current petroleum consumption with biomass in a significant manner; these studies have shown that most countries simply do not have the land, water, and labor capacity available for the production of large quantities of biofuels.[38] A heavy reliance on biofuel production without consideration of environmental and longer-term economic factors could have serious negative consequences compared to the use of fossil reserves. Studies have also shown that the estimated total work force required for the biofuel industry is less than 1% of the available work force of the nation and does not contribute significantly to more employment.[39] By contrast, the expansion of today's pulp and paper industry, having significant opportunities for biomass optimization, could create thirteen times more jobs than the biofuel industry.[40] These factors indicate that the current approach toward biofuel production needs to be strategically improved to make it more sustainable and to have more potential in creating job opportunities.

Legislative remedies for biomass utilization are most progressive in the European Union. EU policies provide substantial subsidies with mandatory fuel blending programs to facilitate the increased use of biofuels at the expense of fossil fuels. The European Union has targeted that by 2020, 20% of the energy consumption will be met by renewable energy sources including gaseous, liquid (e.g., biofuels), and solid biomass. Biofuel production is generally not economically competitive with the production of fossil fuels at their current market prices. Therefore, subsidies and tax benefits are legislated to promote the utilization of biofuels as energy source. These are often introduced via complex tax counting rules, but these well-intentioned initiatives create a nonlevel playing field for companies already having established a solid business based on biomass refining and valorization.[41] Subsidies have a positive impact in mitigating early enterprise investment risk but have negative consequences when the enterprise depends upon the subsidies for sustaining their business model. These should be invoked cautiously and sparingly with clear and finite time limits.

The real challenge for developing a balanced bio-based economy through the biorefinery concept depends on the long-term availability, sustainability, and efficient use of biomass. This means that high value applications need to be identified and that products should be brought to the market, which are based on solid business cases without the continuing need of subsidies or other state aid supports. As for any successful project, biorefinery operations should align with standard business concepts. A key driver is continuous innovation to meet dynamic market needs with due consideration to the current social, political, and economic conditions. Fundamental changes in the way sustainability is regarded among all stakeholders will be required to create a common vision how biomass should be optimally used in a sustainable manner. Guidelines are currently being developed by politicians and decision makers. However, it is crucial that decisions eventually will be made based on solid market analyses and customer demands, economic viability models and scenarios, and proven technology platforms. A sustainable business should be viable without relying on subsidies and/or tax benefits.

2.3.2 Types of Biorefining

The biological production of chemicals and polymer building blocks is not a new field of technology. The capture, refining, and modification of naturally occurring feedstocks predate the evolution of the modern chemical industry, reliant upon cheap and abundant oil and gas. Increasingly, nonpetroleum routes to useful chemicals are being investigated and developed. In the first half of the twentieth century, several commodity products were produced by fermentation including bulk carboxylic acids like acetic acid, citric acid, lactic acid, and itaconic acid as well as other basic chemicals like *n*-butanol and acetone.[42] Classical chemical methods to derive polymers from biomass involve the conversion of biologically derived fatty acids or other functional chemicals into polymer building blocks. An example is the application of metathesis chemistry to interchange olefinic bonds between unsaturated chemical entities.[43] Readily available monounsaturated fatty acids have been used in a self-metathesis reaction to produce long-chain unsaturated α,ω-dicarboxylic acids.[44] Alkyds form an important industrial product group wherein biologically derived unsaturated fatty acids such as tall oil fatty acid (TOFA) are cross-linked by air oxidation with polyesters to enable their application in a variety of paints and coatings.[45]

The biorefinery concept describes the collection of processes used to convert biomass into materials, chemicals, and/or energy. The conceptual major steps in a modern biorefinery resemble those applied in conventional refinery settings, such as oil refineries.[46] Feedstock is collected as raw material input and processed by one or a combination of technologies to afford one or more products. The obtained product(s) can be either chemical materials or energy. A major difference is the level of maturity of oil refining versus biorefining processes. The production of bio-based chemicals in biorefineries is in various respects still a nascent field. The lack of a clear description of the term biorefinery and a clear classification of biorefinery types and efforts is in line with a high level of immaturity and uncertainty. Several descriptions of the biorefinery definitions and classifications have been provided and are expected to be subject to rephrasing during the further development of this dynamic area.[47]

Biorefineries have been classified into:

1. Green biorefineries.[48]

   These use primarily starch and sugar materials like cereals and sugar cane, including wet biomass such as green plants, grasses, and bioenergy crops to produce biofuels and biomaterials.
2. Forest and lignocellulose biorefineries.[28]

   These recover and refine plant saps (e.g., *f*atty acids or esters, resins, and terpenoids), cellulose, hemicellulose, and lignin. Wet chemical processes and thermochemical processes like fractionation are applied to the biomass to obtain the main components based on efficient, economical, robust, and reliable processing. Trees and forests are available year-round, and therefore a consistent renewable feedstock supply can be guaranteed.
3. Aquatic or algae-based biorefineries.[49–51]

   Examples are the use of algae as feedstock for the product of biodiesel, biopolymers, proteins, amino acids, and polysaccharides. Algae can be grown in brackish and even saline water and also in photobioreactors set up at deserts and arid lands.

The industrial production of chemicals from biomass via a biorefinery can meet ecological and sustainability demands since the quantity of the feedstock is relatively small compared to the volume of the raw material required for fuel production. One of the main enablers for running a profitable biorefinery is the cascading use principle of renewable biomass. The cascading use principle aims to integrate the manufacture of several value-added chemicals including coproducts as well as waste streams. Key objectives of this principle include maximizing profitability and minimizing wastes and emissions.[52] Numerous breakthrough technologies have recently been reported for the transformation of biomass raw materials into functional chemicals.[53–56] However, only a very limited number of biorefineries have made it to a commercially viable business at a relevant scale. In general, bio-based commodity chemicals are more expensive to produce as compared to fossil-based analogues. Two main reasons are the availability and the logistics of aggregating the biomass, which have not been optimized yet to produce at the desired scale. In principle, a modern biorefinery should run comparable to today's crude oil refineries producing several products simultaneously at high atom efficiency (atom economy).[57]

The "*E* factor" constitutes a related metric, which refers to the summed amounts of by-product and waste, divided by the amount of product.[58] More refined metrics such as *environmental quotient* (EQ) have also been proposed, which take the energy consumption and CO_2 waste as well as the nature of the waste into account. This shows the complexity of the matter but will not be detailed further in this manuscript. Current oil refineries exhibit very low *E* factor values (<0.1), which reflects the high level of established process optimization. It should be noted that most of the current biorefineries have a relatively short history and have not yet benefitted from extensive process optimization. Therefore, significant improvements in many biorefinery *E* factor values in the course of time can be anticipated.

Most biorefineries still have a long way to go since fossil fuels and shale gas are still available at relative low cost. Countries like Brazil have several drivers like country location (i.e., climate), economic development, and natural resources, which have contributed to the development of alternative large-scale investment programs. These countries have implemented renewable energy in the form of bioethanol made from sugar cane. With bioethanol abundantly available locally, other chemical companies (i.e., Braskem) have developed processes to transform the commodity chemical ethanol into ethylene monomer for green polyethylene production. Investors are often hesitant to invest in biorefinery scale-up for a number of reasons. The long-term availability of raw materials is uncertain, and process robustness on commercial scale requires further efficiencies since alternatives based on fossil fuels are currently more cost competitive. The long-term success of developing and producing sustainable bio-based chemicals is not trivial and will require continuous optimization and development.

2.3.3 Biorefining at Arizona Chemical

Arizona Chemical has embraced the concepts of a true biorefinery since its beginnings over eighty years ago. The company has taken a proactive approach in developing sustainable products and refining processes leading either directly or indirectly to lower GHG emissions by offering value-added bio-based chemicals. Our company has also embraced the concepts outlined in the twelve principles of green chemistry,[59] focusing on the use of renewable feedstock, maximization of atom efficiency, and waste prevention.

A profitable business based on CTO and CST fractionation is operational for more than 80 years. This little known mature subsector of the chemical industry represents one of the first and successful examples of the biorefinery concept. Utilizing captive assets, a crude renewable feedstock is fractionated into several chemical intermediates, each having valuable applications and allowing substituting nonrenewable (i.e., fossil oil-based)-derived materials.[60] CTO and CST constitute two of the many outputs originating from the Kraft wood pulping process.[61]

2.4 THE KRAFT PULPING PROCESS

As outlined above, our biorefinery includes one or more fractionation steps for the separation of primary refinery products from crude biomass. Wood biomass consists of several components. Its three main structural macromolecular substances are cellulose, hemicelluloses, and lignin. In addition, wood also contains low-molecular-weight substances, which can be classified as either organic or inorganic. The pulp and paper manufacturing industry is a well-established industry using sustainable forestry biomass (i.e., wood) to produce pulp and biochemicals.[62] The throughput of a single pulp mill is in the order of thousands of tons per day. Prior to the 1940s, the primary target for recovery was cellulose. Other valuable macromolecular components such as lignin, hemicellulose, and low-molecular-weight organic substances (e.g., TOR, fatty acids, and turpentines) were ignored. In the modern chemical pulp mill, most of the wood and cooking chemicals are recovered and have become part of the integrated biorefinery concept. Additionally, the energy required to run pulp mills is no longer fully dependent on fossil fuels but can run on renewable energy generated by the refinery itself.[63]

The current global production of Kraft pulp is about 130 million tons per year, accounting for two-thirds of the global virgin pulp production and for over 90% of chemical pulp. The efficiency of the Kraft process has been steadily optimized for many years. The flexibility and efficiency of the current mature Kraft process – about 97% chemical recovery efficiency – and the ability to handle almost all species of softwood and hardwood render advantages to the Kraft process over alternative pulping processes.[64] The Kraft process recovers and recycles virtually all of the pulping chemicals. The separated fibers show excellent paper-making properties. Besides the production of paper and viscose staples, the Kraft process delivers valuable chemicals and the usage has been converted into a sustainable and profitable business.

In a Kraft pulp mill, wood chips are digested at high temperature and pressure in an aqueous mixture of sodium hydroxide (NaOH) and sodium sulfide (Na_2S) – the "white liquor" – to convert wood into pulp and steam.[65,66] The white liquor overall chemical reaction can be summarized as

$$NaOH + Na_2S + H_2O \rightarrow 3Na^+ + 2OH^- + HS^-$$

A simplified scheme of the Kraft process is shown in Figure 2.5. The wood chips are cooked in a digester where their polymeric lignin and hemicellulose constituents are

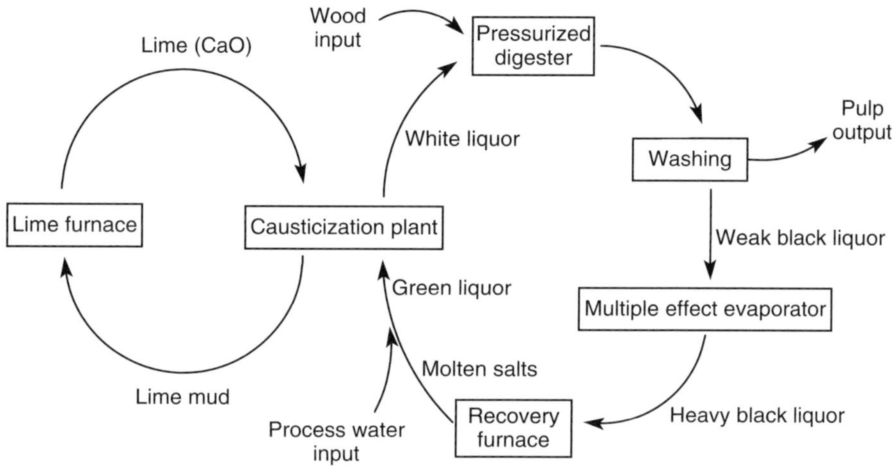

Figure 2.5 Kraft chemical recovery process.

chemically degraded into much smaller fragments and dissolved in the strongly basic delignification solution.

2.4.1 Crude Sulfate Turpentine (cst)

During the digestion steps, the volatile CST fraction is collected for further fractionation. Turpentines can also be extracted from gum tapping from living trees, followed by distillation. Wood stump solvent extraction is also applied to obtain turpentines, albeit in much lower quantities. Today, turpentine is produced primarily as a coproduct from the pulp and paper industry and isolated as a volatile fraction during the Kraft pulping of pine and spruce (Figure 2.6).

The collected turpentine can then be used as a source of energy for the mill but preferably is used as a versatile renewable feedstock. Arizona Chemical and others refine the CST into its main components – unsaturated hydrocarbon monoterpenes such as α-pinene, β-pinene, and δ-3-carene, dipentene, and other *p*-menthadienes – by high-efficiency vacuum fractionation. This fractionation process delivers besides the main monoterpene outputs as described above also a smaller fraction (1-2%) of less volatile components comprising aromatics (anetholes and methyl chavicol), sesquiterpenes (e.g. the bicyclic caryophyllene) and monoterpene alcohols such as borneol. Fractionation of CST provides also a highly volatile lights fraction (1-2%) which contains dimethyl sulfide, dimethyl disulfide and other low-boilers as well as a higher-boiling residue (±2%).[5b] It should be noted that the chemical composition of the terpene isomers can be different from CST and gum turpentine. The chemical composition of gum turpentine is known to vary depending on the geographical location. For example, the monoterpene composition of the trees harvested in the United States *versus* Russia has been reported as considerably different in β-pinene and δ-3-carene content.[5b].

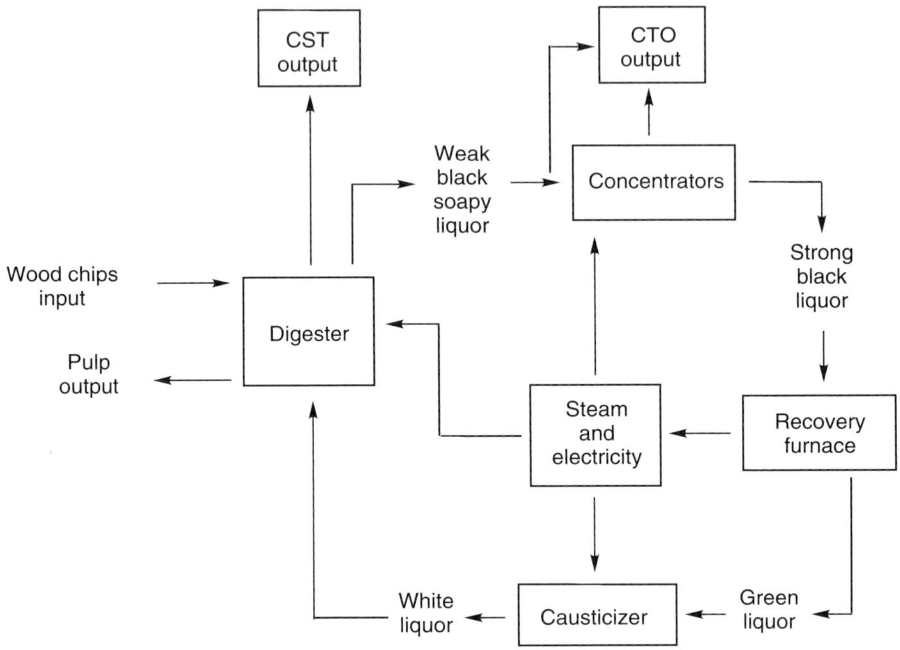

Figure 2.6 Kraft pulping process.

For several reasons, the α-pinene and β-pinene streams obtained from the sulfate process are desulfurized prior to further processing. Desulfurization techniques involve fractionation, absorption, hypochlorite or metal treatments, or a combination of techniques. Several patents describe these techniques.[67,68]

The reactor output from the delignification stage consists of a suspension of cellulose fiber. The fibers are washed to remove lignin residues and process chemicals. Subsequently, they are bleached and dried with steam and hot air to deliver the pulp for the papermaking process.

The active cooking chemicals are recovered in order to optimize the efficiency of the Kraft process. This proceeds via several evaporation steps to concentrate the residual weak black liquor (BL), which contains several valuable dissolved wood-based chemicals. The inorganic sodium and sulfur are (partly) recovered as a melt of sodium sulfide (Na_2S) and sodium carbonate (Na_2CO_3) and subsequently dissolved in water to form the green liquor. The green liquor is eventually sent to the causticizing plant, where it is reacted with lime (CaO), to convert the sodium carbonate back into sodium hydroxide. The precipitated calcium carbonate from the caustic reaction is washed and sent to a lime kiln to be converted into CaO for its reuse.[69]

2.4.1.1 Green Liquor

$$Na_2S + Na_2CO_3 + CaO + H_2O \rightarrow Na_2S + 2NaOH + CaCO_3$$

$$CaCO_3 \rightarrow CaO + CO_2$$

The sodium sulfide passes through the causticizing plant unchanged and is recycled with the sodium hydroxide into the pulping process.

2.4.2 Black Liquor (BL) and Black Liquor Soap (BLS)

Most of the BL was traditionally burned in recovery boilers to generate energy for the pulping mills. However, the integrated approach of a forest product refinery has resulted into the implementation of other conversion units into these pulp mills to diversify the chemical product mix to serve new applications and markets.

Instead of burning the BL for fuel value, the installation of other integrated (chemical) processes (e.g., soap skimming and acidulation) has contributed to the development of even more efficient wood-based biorefineries (Figure 2.7). The strong alkaline pulping of pine softwoods in the Kraft process converts wood constituents like resin acids and fatty acids into their sodium salts or soaps. Neutral or unsaponifiable compounds that are present in the spent cooking liquor (BL) are collected with the salts. Both the salts and unsaponifiables naturally phase separate from the BL as black liquor soap (BLS) at various points in the pulping process. The wood extractives in BL are partially solubilized, and as a result, the weak BL is a complex colloidal system. The extractives may exist in several different physical states including true solution but also in the form of micelles, liquid crystalline, or as neat sodium soaps. The aggregation state of the soap depends on the exact processing conditions (i.e., salt content, solids content) of the weak BL. As the weak BL is concentrated, the colloid equilibrium is stressed, which results in an increasing solid content due to water evaporation from the BL. Typically, 40–80% of the soap is already recovered from the weak BL fraction (before fractionation) by the physical process of soap skimming. The salt concentration increases further during the sequential water evaporation steps, thereby causing a natural phase separation of the organic species at the surface of the tanks. In addition, the rosin acid and fatty acid salts form micelles that solubilize the neutral fraction of the wood extractives. During these steps, more soap quantities are mechanically skimmed. Typical recoveries of pine wood extracts vary between 40 and 50 kg/1000 kg pine wood. Approximately 70–75% of the extractable wood chemicals are recovered; losses occur in boilers, pulp, and waste streams.[70] The exact degree of extraction depends on the Kraft process specifications and the efficiency of the integrated processing. In a modern integrated biorefinery, an effective tall oil soap recovery is important to maximize the cost-efficiency of the Kraft process, and therefore, it is crucial to maximize the collection of the tall oil soap, which is a scarce valuable coproduct of the refinery process.[71]

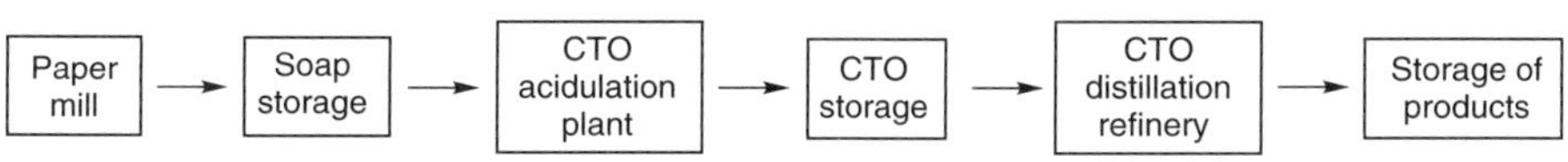

Figure 2.7 Flow of products: acidulation and distillation plant.

2.4.3 Acidulation Process

Kraft BLS collected from the weak BL fraction as well as during the sequential concentration steps can be converted to CTO in a variety of ways. Reaction with sulfuric acid (H_2SO_4) in batch, semicontinuous, and continuous processes to convert the soaps into distillable oils is the most common method.[72] For each 1000 kg of tall oil soap, approximately 100 kg of concentrated sulfuric acid is added with about 600 kg of water and steam, and the reaction mixture is heated to about 95–100 °C. Due to these corrosive reaction conditions, the reactor and metal pipes need to be fabricated from corrosion-resistant stainless steel.

Upon completion of the acidulation, the reaction mixture typically has a pH value of approximately 2–4 and separates into a precipitated lignin layer, a CTO layer, and a water fraction containing the inorganic process chemicals.[73] The calcium ions that are present in the soap react with the sulfate anions. The resulting poorly soluble $CaSO_4$ precipitates during the settling period. The sulfide ions, present in the alkaline BL and tall oil soap, are protonated during the acidulation steps, and the formed hydrogen sulfide can either be captured by scrubbers or burned. However not all sulfur ions (SO_4^{2-} and S^{2-}) are removed. Studies have revealed a direct proportionality between the formation of organic sulfur compounds and the concentration of sulfur derivatives used in the Kraft and acidulation processes (e.g., sulfides and sulfuric acid).[74] As a result, methyl mercaptan, dimethyl sulfide, and dimethyl disulfide species have been detected during the Kraft process, contributing to the typical odor of Kraft mills and products produced via this process.[75,76] In order to reduce the emission of sulfur species and to reduce the typical odor in the products, improved deodorization processes have been put in place.[77]

Historically, it was common to separate these layers by a gravity separation extractive process. More recently, the CTO layer is being isolated by centrifugation. The remaining desoaped "heavy BL" is burned, thereby generating heat to produce steam and electricity to fuel the Kraft pulping plant. Also the combined sodium sulfate and lignin fractions are usually treated with caustic soda and recycled. As a result, the Kraft process represents a closed loop system that enables a substantial recycling of most of the inorganic process chemicals and energy to produce the two valuable raw material streams: CST and CTO.[78] Both CTO and CST are rich sources of valuable chemicals and contain useful building blocks for further derivatization. On average, the treatment of 1000 kg of dry pulp yields the following coproducts:

- 3–10 kg of CST
    - CST consists of a mixture of volatile monoterpenes such as α-pinene and β-pinene and in some cases also considerable amounts of δ-3-carene. These products are either used as such or applied as chemical building blocks in polymerization reactions. Another economically viable option is their chemical conversion into a wide range of aroma chemicals for the flavor and fragrance industry.
- 10–60 kg of CTO

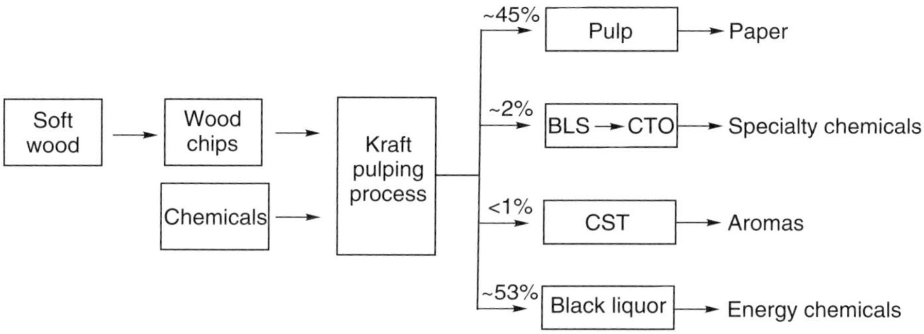

Figure 2.8 Kraft pulping process and the resulting products.

- CTO contains various chemical components like C_{16-18} fatty acids (or TOFAs), diterpenoid C_{20} fatty acids (rosin acids), sterols, and a low percentage of unspecified other components.

- 1200 kg of desoaped heavy BL
  - Highly viscous material used as energy source (fuel) in the recovery plant.

A simple schematic of the primary derivatives from the Kraft process is shown in Figure 2.8.

The tree species, climate conditions, and age all have an impact on the quality and molecular composition of tall oil. These main variables define the ratios of rosin acids, fatty acids, and unsaponifiables, which in turn define the compositions and isolated amounts of tall oil.[79] Moreover, the process conditions (pH, time, and temperature) during and after the acidulation reaction effect on the product distribution. For example, an acid-induced esterification of rosin acids with alcoholic by-products (e.g., sterols, higher alcohols) may lead to ester formation during processing and storage.

2.4.4 Crude Tall Oil (CTO) Distillation

CTO derived from the acidulation process is a mixture of fatty acids and tall oil resin acids along with various other materials such as unsaponifiables. The current practice is to separate and refine the different chemical components on a large scale by continuous fractionated distillation. In general, more than one distillation column is utilized in order to achieve an effective separation of the main CTO components. The following five stages occur in CTO fractionation:

- Dehydration
  - Removal of water and volatile chemical entities from the CTO feedstock, yielding a heads stream.
- Depitching
  - Removal of nonvolatile materials from the CTO to provide a product stream and to minimize operating problems in subsequent stages.

- Rosin separation
  - Recovery of rosin as the bottoms product in the first distillation column, which typically is approximately 40 m high by 1.5 m diameter width.
- Heads separation
  - Removal of impurities (heads) from the rosin column prior to recovering the fatty acids. These heads are the top product from the second distillation column, typically 30 m high and 0.6 m width.
- Fatty acid separation
  - Recovery of high-grade fatty acids as top product and distilled tall oil (DTO) as bottoms in the final distillation column.

Two basic types of CTO distillation processes are currently practiced: steam distillation and dry (inert gas) distillation processes. The latter is the most widely applied technique. In the steam injection process, the partial pressures of the organic components are considerably reduced, and effective separations can be achieved at much lower temperatures than the normal boiling point of the occasionally sensitive chemical product fractions. However, a large volume of water is quite often contaminated with organic chemicals and has to be treated before release from the plant site involving extra capital investments and operational costs for the tall oil distillation plant. The more frequently used dry distillation process uses no direct steam in the process and therefore minimizes the generation of waste water and uses less energy.

Each column has a feed, top, and bottom offtake, thereby providing two product streams at each stage of the distillation process. One fraction from each stage provides the feed to the next fractionated distillation stage, while the other fraction yields either a product or a recycle stream to the CTO feed. In general, these processes have the following main characteristics:

- Thin film evaporators for dehydration, depitching, and reboiler duties
- No steam stripping
- Typically three distinct towers to collect rosin acids, DTO, and TOFAs, respectively
- Several light fractions (heads) are collected at each column

Heat is provided to the distillation towers from circulating hot oil, heated by fossil fuels or recovered pitch. The columns are operated under partial reflux conditions to deliver the driving force for an effective separation of the different fractions.

A major processing challenge for the distillation of CTO and its products is the sensitivity of some of to the species to heat and oxygen (air). Additionally, the relatively high temperatures during distillation may result in partial decarboxylation of the rosin carboxylic acid-containing products, like rosin and fatty acids. The reduction in acid functionality is an undesirable side reaction because it will negatively impact the further upgrading into higher-value chemical products, such as rosin esters.

Excessively high temperatures or relative long residence times in the distillation columns may lead to degradation, discoloration, and subsequent loss of product yield. Measures have been taken in modern tall oil refineries to minimize operating temperatures and heat exposure times, for example:

- One-pass processing with minimum recycling.
- Short residence time in columns, using wiped film evaporators (WFEs).
- High temperature in combination with high vacuum conditions.
- Optimum reflux conditions with either liquid or vapor takeoff.
- Oxygen exposure is minimized through the following steps:
  - Distillation under vacuum
  - Inert gas blanketing at ambient pressures
  - Closed process loops, including transfers to tanks and further reactors

2.4.5 CTO Fractionation Products

CTO comprises approximately 10% tall oil heads (TOH), 30% tall oil fatty acids (TOFA), 10% distilled tall oil (DTO), 30% tall oil rosin (TOR) and 20% tall oil pitch (TOP). Each of the fractions obtained from CTO has found useful applications – some unique and others as an alternative to petrochemical or oleochemical products. The amount and composition of each product stream depend on the tree species and other factors such as the geographic origin and climate conditions. Practically, the refinery fractions may consist of overlapping blends of these molecules as a result of the distillation process. The refinery typically yields these fractions (Figure 2.9).

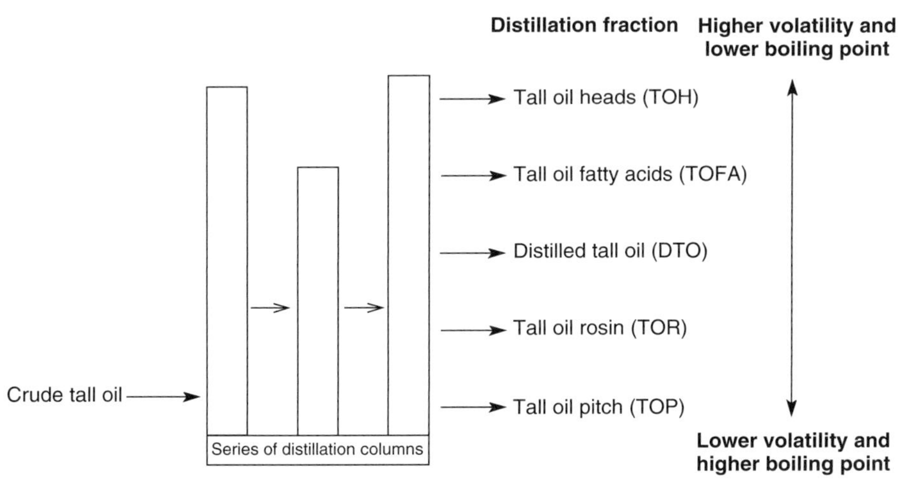

Figure 2.9 Typical refinery main fractions of CTO distillation. The dashed lines schematically represent inter-distillation column mass transfers.

- *Tall oil heads*

  This fraction contains fatty acids (40–75%) and also neutral components (15–60%). Palmitic acid C_{16} is the major fatty acid component in tall oil heads but also nonacidic components like terpenic compounds derived from turpentine components or decarboxylated and decomposed rosin (e.g., retene and pimanthrene precursors).
- *TOFAs*

  TOFA mainly consists of C_{16-18} fatty acids containing high levels of single monounsaturated oleic acid (C18:1) and double unsaturated linoleic acid (C18:2) (Figure 2.10). Both constituents occur naturally in CTO as free fatty acids, which is in contrast to other vegetable sources where they are present as esters (triglycerides). TOFAs are used in alkyd paint applications and are chemically reacted into the alkyd binder. This provides the desired (drying) performance properties of the paint. They are also used for the formulation of biolubricants and water-based metal working or drilling fluids. TOFAs are also used as fuel additives for low sulfur diesel. In these outlets, they compete with other vegetable oil-based fatty acids or triglycerides like soybean or cotton oil as well as fossil oil-based derivatives such as petroleum sulfonates or C_4/C_5-based synthetic lubricants.
- *Tall Oil Rosin (TOR)*

  Rosin acids are actually present as a mixture of $C_{19}H_{29}CO_2H$ isomers, and the most typical isomers are abietic acid (27–47%), levopimaric, palustric, neoabietic acid (4–5%), dehydroabietic acid (29%), and dehydroabietic and tetrahydroabietic acid.[80] It has been shown that the resin acid composition between CTO and TOR changes during the high-temperature fractionation process. Furthermore, some of the rosin acids originally not present in the CTO are found after distillation. This is a consequence of the decomposition, dehydrogenation, isomerization, dimerization, and disproportionation reactions taking place during high-temperature fractionation.[81] Rosin acids are used as sizing agents in papermaking where they compete with synthetic fossil-based derivatives made from maleic anhydride or diketene. Rosin acids are also esterified with polyols such as pentaerythritol to yield rosin esters used as tackifiers for the adhesives industry, binders for ink, and road-marking resins and tread enhancement additives for tires. The use of renewable rosin-based products enables in all these applications the substitution of petrochemicals such as hydrocarbon-based C_5 or C_9 resins.[82]
- *Distilled Tall Oil (DTO)*

  DTO is actually a mixture of TOR and TOFAs. The composition varies depending on the ratios between the fractions as collected during the fractionation step.
- *Sterols*

  These components are present in the tall oil pitch (TOP) fraction (*vide infra)* and are used as fermentation feedstock to obtain steroid intermediates like androstenedione or are esterified to yield phytosterol esters whose main

Name	Abundance	Structural formula
Palmitic acid	~1%	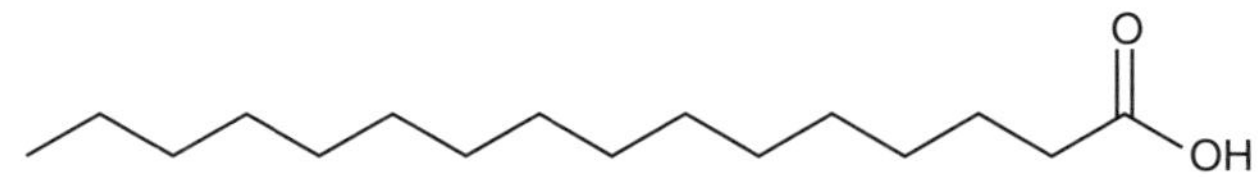
Stearic acid	~2%	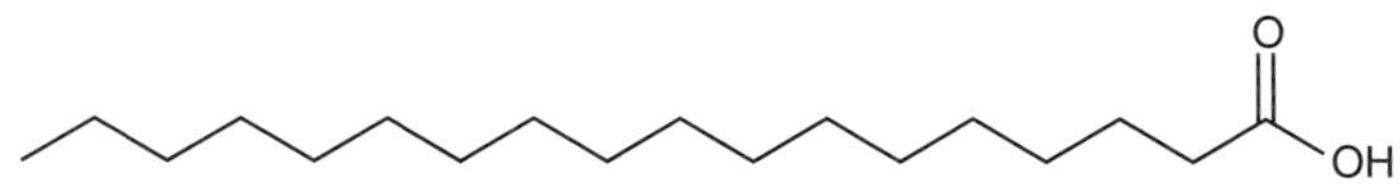
Oleic acid	45–55%	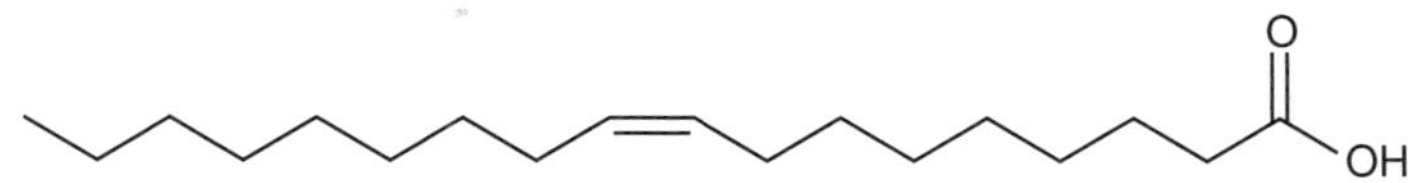
Linoleic acid	35–40%	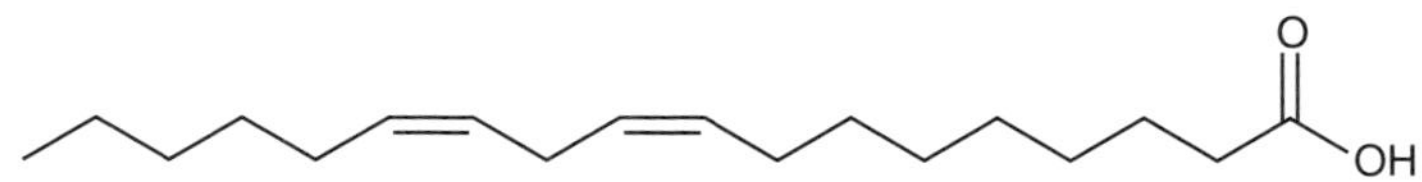
Conjugated octa-decadienoic acids	6–8%	$CH_3(CH_2)_4CH{=}CHCH{=}CH–(CH_2)_8–CO_2H$ or $CH_3(CH_2)_5CH{=}CHCH{=}CH–(CH_2)_7–CO_2H$

Figure 2.10 Major tall oil fatty acid (TOFA) components. Abundance percentages were taken from Zinkel and Russell.[5b]

applications are functional food ingredients or nutraceuticals. Nutraceuticals such as β-sitosterol can decrease elevated blood cholesterol levels and thereby contribute to cardiovascular risk mitigation. In both outlets, CTO-derived sterols compete with material extracted from soybean oil distillate residues mostly obtained from genetically modified (GM) soybeans.

- *Tall Oil Pitch (TOP)*

  TOP has a very dark color and variations in chemical composition. The TOP residue is typically burned for steam-generating purposes in boilers to substitute natural gas or heavy fuel oil. TOP is mainly an aliphatic material. Sterols such as β-sitosterol and campesterol and their dehydrated analogs are TOP components.[5b] In addition, TOP contains resin acids such as abietic acid as well as fatty acids such as oleic acid. Unsaponifiables such as diterpene alcohols and long-chain fatty alcohols are present in TOP. Esters that are based on the sterols and the abovementioned aliphatic alcohols on the one hand and the acidic components on the other hand are also chemical constituents in TOP. TOP also contains higher molecular weight polymeric components.

2.5 CRADLE-TO-GATE

As a result of the technology advances during the 1930s, the Kraft pulping process was widely adopted. In Europe, approximately 65% of the wood pulping production is based on the Kraft process to deliver about 39 million tons of dry pulp annually. On average the Kraft process requires 2 t of dry wood per ton of dry pulp output. The other outputs from the process include CST (3–10 kg), weak BLS, and heavy BL (abt. 1200 kg), based on 1000 kg dry pulp produced. The BLS can then be acidulated and converted into CTO (abt. 10–60 kg). CTO represents on average no more than 3–4% weight fraction of the BL but nevertheless accounts for a yearly 600,000 t quantity in the EU region. Although accounting only for a small fraction of its total organic content, CTO is recovered from the weak BLS for its substantial economic value from multiple technical applications in various outlets. The distilled fractions obtained from the CTO fractionation process are used for their functional properties in a number of different applications, and in many cases, they compete with products that are manufactured from nonrenewable fossil feedstocks (Figure 2.11).

A cradle-to-gate evaluation of the total carbon footprint and GHG emissions has been conducted to compare the environmental impact of producing CTO and its derivatives, gum rosin and its derivatives, and their hydrocarbon analogs.[83,84] The carbon footprint expressed per unit of functionality associated with CTO derivatives, like TOR and TOFAs, is substantially lower in comparison with their alternatives including vegetable oils and petroleum-derived products. For example, TOFAs have a carbon footprint that is six times lower in comparison with the quantity of soybean oil to obtain a similar functionality in alkyd paints (Table 2.2).

For CTO a detailed analysis has demonstrated that its fractionation and processing significantly contribute to sustainability and lower GHG emissions. As calculated for the EU region, outputs of the CTO fractionation provide attractive alternatives

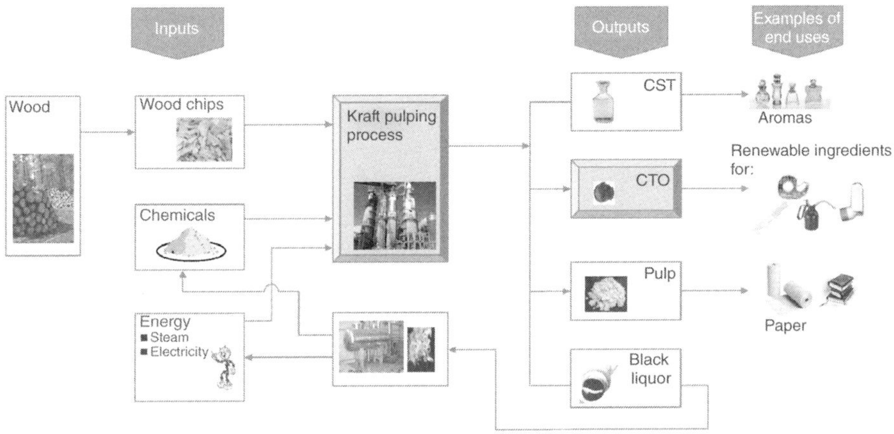

Figure 2.11 CTO distillation into its main components and their derivative products.

TABLE 2.2 Carbon Footprint as a Multiple of CTO-Based Equivalents.[60]

Product Description	Carbon Footprint as a Multiple of CTO-Based Equivalents*
CTO	1[a]
Soybean oil	6
Rapeseed oil	9.4
Petroleum sulfonates	13.9
Gum rosin	2.5–7.2
Heavy fuel oil	3.1

[a] Cradle-to-gate emissions expressed per unit of functionality.

for about 170 kt of edible vegetable oils and about 500 kt of crude oil equivalents. Since most CTO refining operations are located in areas where limited other industrial activities exist, they are often important for the local economies and communities and thereby contribute to both employment and social cohesion. The estimated GHG emissions for CTO refining operations in the northern part of Europe are in the order of 40–50 kg CO_2/t of output per manufacturing plant.

The emissions of both Arizona plants (abt. 40 kg CO_2/t averaged) are modest in comparison with equivalent operations producing fossil-based products and using heavy fuel for heating purposes (abt. 160 kg CO_2/t averaged). This example illustrates that the use of CTO derivatives substantially reduces GHG emissions (Table 2.3).

The availability of CTO is finite and is linked to the output of the Kraft pulping industry and as a result depends upon the overall paper demand. The total global CTO availability, assuming 100% capacity utilization and optimized recovery rates, is about 2.1–2.3 million tons per annum. In Europe, the availability of CTO is already inadequate to meet the demand of the local refining industry. The installed

TABLE 2.3 GHG Emissions for CTO versus Fossil-Based Chemical Companies.[82]

Company Name	Emissions/kg of Output	Emissions/MJ Consumed
Arizona Chemical, Sweden	38	9
Arizona Chemical, Finland	43	11
Company A[a]	123	N.A.[b]
Company B[a]	200	206

[a] Companies operating comparable chemical activities to Arizona Chemical.
[b] N.A = not available.

fractionation capacity significantly exceeds the local CTO availability, which needs to be supplemented by imports from other regions such as Brazil, Russia, and the United States. Within this framework, it is worth mentioning that a given product based on renewable feedstock or manufactured through biorefining operations is not necessarily associated with a superior sustainability or a lower carbon footprint. This can be illustrated by the unsettled controversy around the impact and carbon footprint of bioethanol, biodiesel, and other renewable-derived chemicals where conclusions differ widely.[85,86] Allocating CTO to the renewable biomass energy pool together with the BL contributes little to renewable energies and sustainability. The energy content of CTO and heavy BL (80% solids) is 36 and 17 GJ/t, which equals to 6.1 and 2.9, respectively, barrels of oil per ton of material. The overall quantities of CTO and BL obtained per ton of dry pulp wood are 30–50 and 1200 kg/t, respectively. In terms of crude oil equivalents, this corresponds to 0.23 and 3.5 barrels. The CTO energy content therefore is only 6–7% of the total value equivalents obtained from the Kraft pulping process. Considering the different CTO distillation fractions, pitch (i.e., distillation bottoms) has an approximate energy content of 38 GJ/t, corresponding to 6.5 barrels of oil equivalents or 0.9 t of fuel oil per ton material. Given its energy content, pitch is used as bioenergy, but less than 4% of the renewable energy content of the original BLS is removed from the BL energy pool in the form of TOFAs, rosin acids, DTO, and sterols. Therefore, it can be concluded that the amount of renewable energy contained in CTO is of limited significance. This calculation has been based on an annual European availability of 650,000 t and assuming an average recycling rate for bioliquids of 40% as an average.[82] Based on the total estimated European energy consumption of 1825 million ton oil, only 0.015% can be provided by CTO. Obviously, this tiny fraction will be negligible at a macrolevel. Despite this low percentage, it is surprising that incentives to use biomass include CTO as a renewable energy source for the production of biofuels.

2.6 OUTLOOK

Several important lessons and sources of inspiration can be learned from CTO refining. This is a good example of successfully applying the value cascading principles by the use of biomass and its coproducts, sometimes being considered as residues or

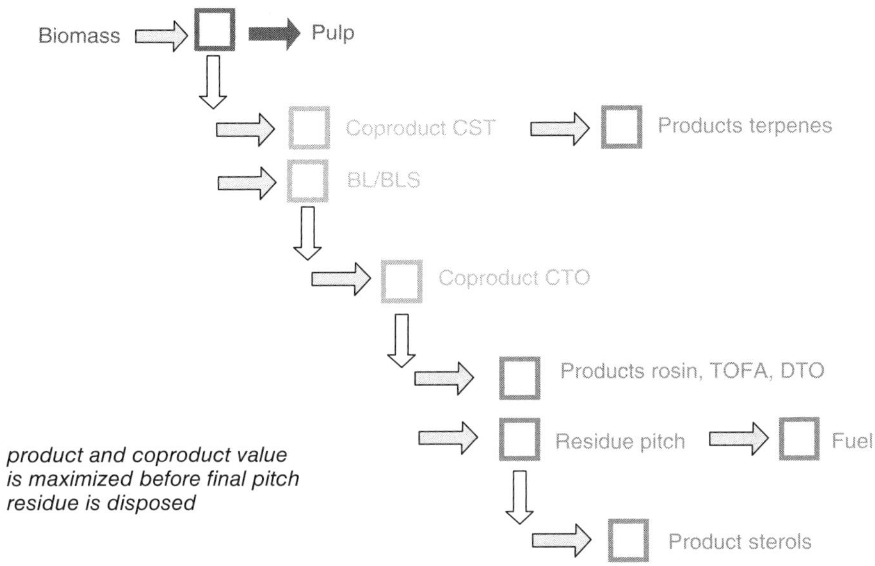

Figure 2.12 Cascading use of scarce raw materials to maximize value.

waste, as valuable sources, and inputs for further processing (Figure 2.12). Valuable coproducts like CST and CTO can be fractionated for a range of applications. For CTO, the industrial pretreatment processing steps have resulted in maximum recovery with minimum degradation of the chemical components and commercial products for many decades. Value has been created for turpentines where terpene building blocks are widely used in the fragrance and flavor industry. Terpenes are also chemically modified into tackifiers to serve the adhesives industry. Tackifiers constitute an essential ingredient of adhesives wherein they among others act as viscoelastic property modifiers. A case study at the end of this chapter describes the use of renewable terpene-based tackifiers in adhesives in more detail.

Recent diesel engine tests confirmed that turpentine oils have properties similar to those of diesel oil. However turpentine oils yield considerably lower exhaust emissions such as carbon monoxide, unburned hydrocarbons, and nitrogen oxides.[87] CSTs from Kraft mills are used as additives to liquid fuels with positive effects on engine performance like reduced fuel consumption and decreased carbon dioxide and sulfur exhaust.[88] It is impractical to replace current petroleum consumption with biomass in significant quantities. Therefore, it seems more suitable to use pine tree chemicals as additives to improve performance properties and reduce pollution related to diesel engines.

In the modern Kraft pulping process, only 42–44% of the woody biomass is converted into pulp. After recovery of CST and CTO, most of the lignocellulosic material and hemicelluloses are currently burned in the recovery boilers. To find new high-value marketable products, a significant amount of research is ongoing to increase the revenue margin of the pulp mills. Alternative process optimization and purification techniques like chemical pretreatment and organosolvent, ionic liquid, and supercritical fluid extractions are currently under development but have not matured to a practical industrial and economically viable technical solution yet.[89,90] The pretreatment processes should result in maximum recovery yields, with minimum concomitant degradation of the end products. The current commercial sales of lignin are mainly as lignosulfonate. Isolation and utilization of lignin in its natural state are difficult because it shows a high level of variation in its chemical composition and structure. A considerable research and development effort will be required for the practical and economically viable conversion of natural lignin into value-added chemicals like syngas, aromatic acids, xylenes, phenol derivatives and resins, vanillin, carbon fibers, and composites.[54,55,91]

Several countries are currently revisiting their renewable resources and seek alternatives to develop sustainable and profitable businesses. Economically viable business activities have been developed for creating value by offering customers functional products with low carbon footprint. Such products can often compete favorably with fossil-based products or vegetable edible oils while exhibiting a comparable or even superior performance profile. Arizona Chemical is a successful model of a resource efficient company in the bio-based economy, operating a biorefinery for more than 80 years with a proven track record advocating four fundamental principles:

- A business aims at maximizing its value creation.
- Resource efficiency is a necessity in creating value.
- A precondition for resource efficiency is a level playing field between users of the same resource, also when they compete in different market areas.
- Cascading use is an important – but not the only – instrument to be resource efficient.
- Sound business cases are based on market economics and not on subsidies.

Continuous progress in resource efficiency will create healthy and competitive businesses and will lead to an increased use of renewable raw materials. Resource efficiency should be regarded as a key paradigm, and policy makers should embrace important principles such as cascading use, a level playing field between applications, and a focus on value creation, including creating meaningful employment for their citizens. A true bio-based economy consists of much more than solely bioenergy. It has to be born in mind that incentives, applied without a holistic perspective, will counteract resource efficiency. If energy policies and incentives are put in place, with the goal to encourage the burning of biomass prematurely in the cascading use, that is, before the value creation has been maximized, a sound bio-based economy will not develop. Moreover, such subsidies and tax benefits eventually can aggravate a precarious governmental financial situation and can aid in increasing national debts, which constitute a serious issue for many countries nowadays. It is important to realize that biomass is abundantly available, but it should be produced, harvested, traded, shipped, and processed in a sustainable way to have a positive effect on GHG emissions. Biomass is seen as the only renewable organic resource and more evenly distributed compared to fossil fuel with a lot of potential, if properly used.

2.7 CASE STUDY: TACKIFIERS FROM RENEWABLE PINE-BASED CRUDE TALL OIL AND CRUDE SULFATE TURPENTINE FOR ADHESIVE APPLICATIONS

2.7.1 Introduction

CTO and CST are refined and upgraded by Arizona Chemical on an industrial scale with concomitant focus on optimizing energy efficiency and minimizing emissions and waste. The industrial demand by the adhesives industry for tackifying resins in various applications such as tapes, labels, and packaging market segments has increased in time, in terms of volumes, performance, and quality.

2.7.2 Rosin Ester Tackifiers

Rosin is made available by refining CTO as described in more detail herein above. Rosin generally consists of about twenty tricyclic carboxylic acid isomers having the molecular formula $C_{19}H_{29}CO_2H$ (regardless of whether it is TOR or gum rosin). The isomer ratios depend heavily on the rosin source. In general, it can be stated that approximately 50% of the tall oil-based rosin consists of palustric, abietic, and neoabietic (PAN) acids (Figure 2.4). The PAN content of gum rosin is typically about 10% higher.

The PAN acid level is of importance since a higher level will cause the rosin to be more susceptible to radical-mediated oxidation reactions. Such a lack of oxidative stability will lead to an undesired rapid yellowing of the rosin and molecular degradation.[92] Such a discoloration is associated with a loss of product quality in certain industrial adhesive applications where a light color is required.

Four major synthetic methods have been developed on an industrial scale in order to produce more oxidatively stable rosins:

1. By reacting the cisoid diene moiety of levopimaric acid (Figure 2.4) in a Diels–Alder reaction with a dienophile, for example, fumaric acid or acrylic acid, a thermally more stable rosin acid is formed. In addition, more than one carboxylic acid functionality will be present in the resulting fortified rosin. Following esterification with a polyol, this fortified rosin will result in a stabilized rosin ester product with a significantly higher molecular weight and higher softening point than a conventional nonfortified rosin ester.[18,93] High softening point rosin esters are applied in certain adhesive applications, which require an improved high-temperature performance.
2. The disproportionation reaction of the PAN acids is another important method to stabilize rosins against oxidation.[94–97] Dehydroabietic acid and dihydroabietic acid isomers are formed from abietic acid in a catalyzed, thermal conversion (Figure 2.13). During such a rosin disproportionation reaction, the three

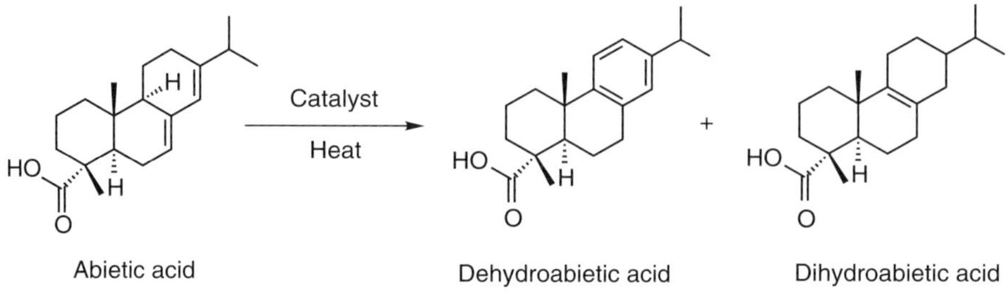

Figure 2.13 Disproportionation of abietic acid.

abietadienoic PAN acids and levopimaric acid exist in a dynamic equilibrium, and as a result, the rosin PAN number will be considerably reduced. The rosin disproportionation reaction involves several chemical conversions, including dehydrogenation, hydrogenation, aromatization, and olefinic bond isomerization reactions. Pimaric rosin types are also partly hydrogenated herein at their less sterically hindered exocyclic vinylic moieties. Disproportionation will not lead to an increased softening point, but it reduces the level of conjugated unsaturation.

3. Heterogeneous catalytic hydrogenation is another industrial method to stabilize rosin. The most common hydrogen source in such reactions is hydrogen gas (H_2). Common hydrogenation catalysts are palladium catalysts, such as Pd supported on carbon (Pd/C), platinum catalysts, (PtO_2), and nickel catalyst, such as Raney nickel (Ra-Ni).[98] Hydrogenation proceeds in two stages. First, the rosin diene moiety is reduced to an alkene, and concomitant double bond isomerizations will occur. This mainly leads to the formation of dihydroabietic acid and some of its regioisomers. Under more harsh conditions (elevated temperature and pressure), the remaining olefinic bond of dihydroabietic acid will also be reduced to deliver tetrahydroabietic acid.[99,100] Here too, the impact on the derived rosin ester softening point will be limited.
4. Rosin can dimerize and polymerize to a considerable extent by reaction with a Brønsted acid catalyst such as sulfuric acid or in the presence of a Lewis acid like $AlCl_3$. The reaction typically delivers a mixture of rosin dimer isomers and some starting rosin. Three types of rosin dimers can be formed. A dimerized rosin molecule bears two acid functionalities and is highly rigid. Analogously to the rosin fortification reaction, these features will eventually result in a stabilized rosin ester product with a significantly higher molecular weight and higher softening point than a conventional rosin ester (*vide supra*) (Figures 2.14 and 2.15).[18,101]

The rosin Diels–Alder fortification reaction, disproportionation, hydrogenation, and dimerization reactions will all result in modified rosin streams with better thermal stability properties. In order to effectively act as a tackifier in an adhesive, the rosin acids need to be esterified with a polyol.[102] The reaction of a carboxylic acid such as rosin acid and a (poly)alcohol results in a (poly)ester.

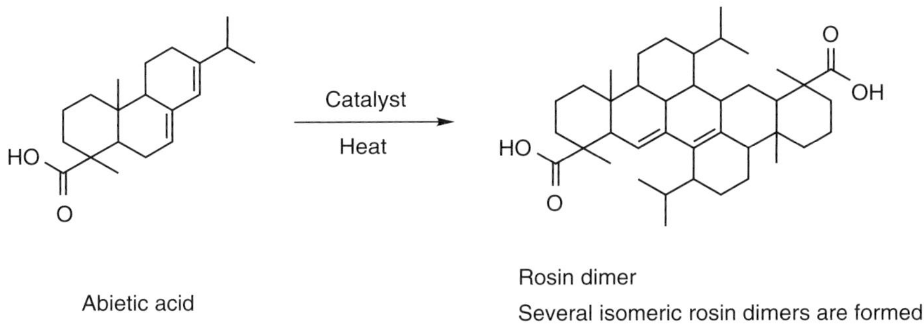

Figure 2.14 Rosin dimerization (stereochemistry omitted).

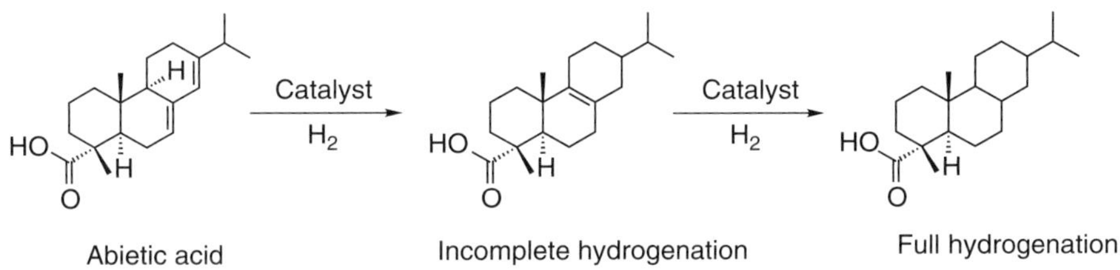

Figure 2.15 Rosin acid hydrogenation (stereochemistry in the fully hydrogenated product is partly omitted).

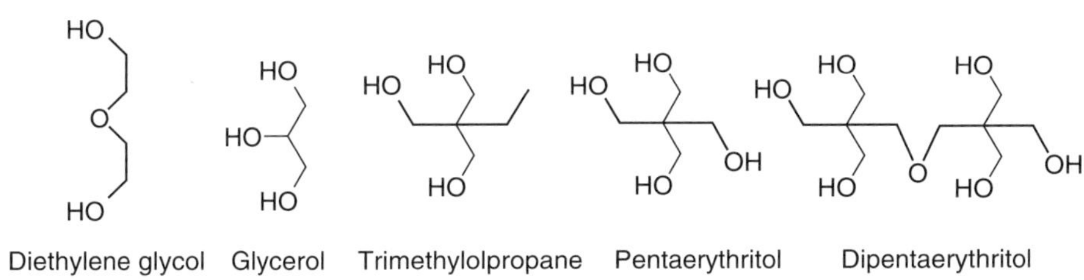

Figure 2.16 Examples of commonly used polyols in rosin ester syntheses.

Rosin ester formation is a condensation oligomerization reaction, which proceeds via an esterification reaction mechanism. Esterification is an equilibrium reaction. Active removal of water will be required in order to drive the equilibrium to the right and the reaction to completion, thereby producing the required rosin ester as the main product. Softening points can also be tuned by the choice of polyol and range from liquid (diethylene glycol ester) to solids having a softening point of 70–90 °C (glycerol or trimethylolpropane-based esters) to higher than 90 °C (pentaerythritol ester) or even higher in the case of the application of dipentaerythritol (see Figure 2.16).

In the rosin esterification reactions with a polyol, the reaction is sequential, that is, the polyol reacts with the rosin molecules one by one. Multiple equilibrium reactions

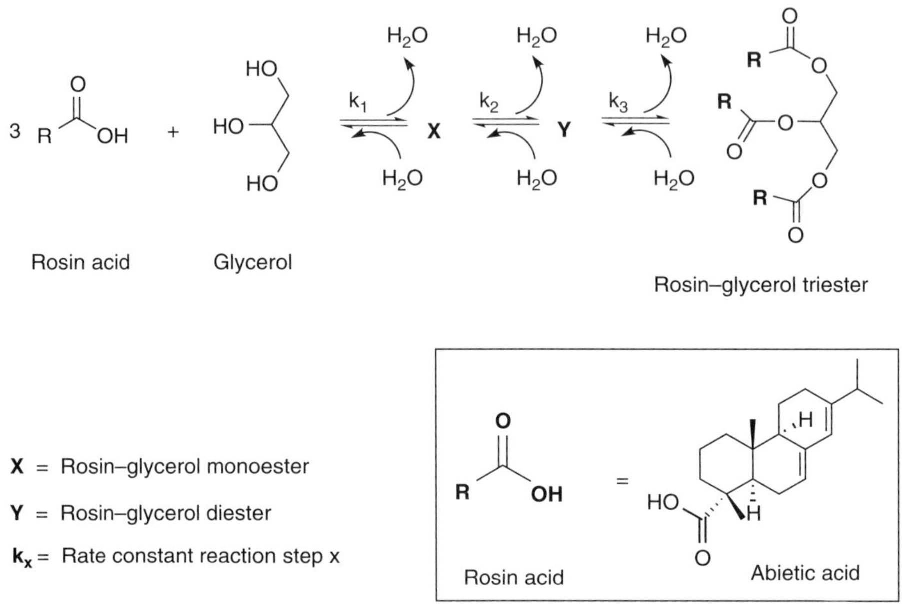

Figure 2.17 Three distinct equilibria in the esterification reaction of rosin acid with a three-functional-group polyol (glycerol).

will exist during such a rosin ester synthesis. For example, the esterification of the trifunctional glycerol with a rosin acid like abietic acid will result in three equilibrium stages, each having a specific reaction rate (Figure 2.17).[102] In general, such an esterification reaction will not go to completion, and typically, a mixture of products will be formed. For example, glycerol, its monoester, diester, and the fully reacted triester will be formed in the abovementioned case. Gel permeation chromatography (GPC) is an adequate analytical method to establish the formed product ratios in such rosin–polyol esterification reactions.

The acid number is a measure for the amount of free carboxylic acid. The rosin esters are typically esterified to an acid number of approximately 20 mg KOH per gram sample, starting at approximately 170 mg KOH per gram sample for the applied rosin. By converting the rosin acids with different polyols into esters with varying acid number, a range of tackifiers can be obtained, which are able to tackify a broad range of polymers. In case of a high softening point requirement, the abovementioned Diels–Alder fortification reaction or dimerization strategy can be applied.

The rosin ester product range of Arizona Chemical (SYLVALITE™, SYLVATAC™, and SYLVAMELT™) allows excellent specific adhesion to a wide range of substrates due to their polarity and polymer compatibility. Their low molecular weight and narrow molecular weight distribution, combined with their cycloaliphatic–aromatic structure, make rosin esters broadly compatible. They are used in a wide range of adhesive polymers, including ethylene vinyl acetate

(EVA), acrylics, ethylene butyl acrylate (EBA), and block copolymers such as styrene–isoprene–styrene (SIS) and styrene–isoprene–styrene (SBS).

Rosin ester-based high solid dispersions (AQUATAC™) can be easily formulated to achieve optimum performance properties for both pressure-sensitive and non-PSAs. They are compatible with a broad range of waterborne polymers, including natural rubber, styrene–butadiene rubber (SBR), and acrylics. Their solids content is approximately 60%.

2.7.3 Monoterpene-Based Tackifiers

CST is a rich source of valuable monoterpenes, having $C_{10}H_{16}$ as common molecular formula. The main constituents are α-pinene, β-pinene, δ-3-carene, and to a lesser extent dipentene (Figure 2.3). CST composition varies considerably by source (Table 2.1).

The monoterpenes from CST can be used to synthesize a variety of tackifiers, such as polyterpenes, terpene phenols, and styrenated terpene resins. Depending on the adhesive application and the applied adhesive polymer, one can select any of these resins, based on their compatibility with the desired polymer (Figure 2.18). It should be noted that Arizona Chemical also applies refined limonene as a monoterpene building block in some terpene resins. D-Limonene is also a renewable feedstock, which is commercially obtained from citrus fruits. The dipentene feedstock from CST has the same chemical formula as racemic limonene.

2.7.4 Monoterpene Tackifier Chemistry

The synthesis of terpene-containing tackifiers proceeds via an addition polymerization mechanism (cationic polymerization), using a strong Lewis acid such as BF_3 or $AlCl_3$ and a cocatalyst.[103] A number of cationic rearrangements of α-pinene and

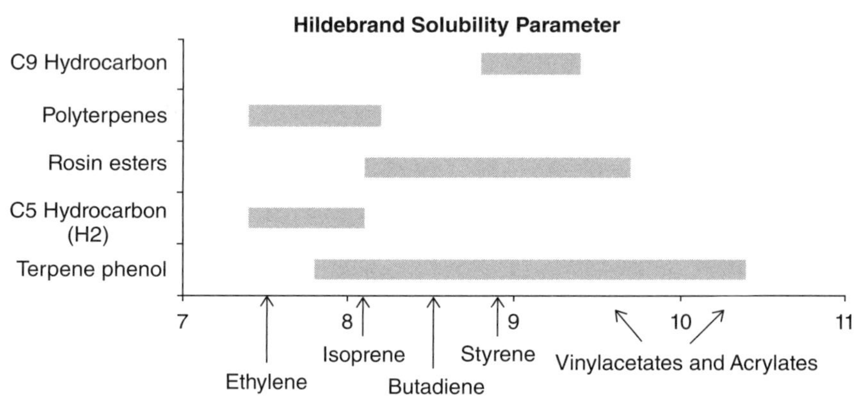

Figure 2.18 Hildebrand solubility parameters of different tackifiers in relation to most common adhesive polymers.

β-pinene are involved in such polymerization and oligomerization reactions.[104,105] In general, this will lead to a complex mixture of products. Typically, number average molecular weights (M_n) of the obtained polyterpene resins range from approximately 350 to 1150, thereby exhibiting softening points in the 25–115 °C range.

Terpene phenol resins are prepared by the oligomerization of a monoterpene unit (such as α-pinene) and phenol streams.[106] The polarity of the resulting resins can be tuned by the relative amount of added polar phenol versus the nonpolar monoterpenes. As such they can be tailored to be highly compatible with a range of polymers.

Terpene phenols are low molecular weight ($M_n \sim$ 800–1300 range) oligomers, which are prepared by a Friedel–Crafts alkylation reaction.[107] Protonation of α-pinene results in a cationic species, which reacts with phenol. The phenol acts herein as an *ortho*-directing and in particular *para*-directing group. Once alkylated, the phenol moiety becomes more susceptible to react with monoterpene cations at the phenolic hydroxyl group as well to give also some O-terpinylated product. Pendant *ortho*- and *para*-terpene adducted phenols can react further to produce higher molecular weight oligomeric terpene phenols.

Polyterpene resins (SYLVARES™ TR and SYLVAGUM™ TR) are highly compatible with polyolefins and the midblock of SIS block copolymers and as a result provide excellent tack and peel properties to pressure sensitive systems. The high softening point grades combine high heat resistance and superior adhesive performance. These resins meet the broadest range of the Food and Drug Administration (FDA) requirements, including direct food contact.

Styrenated terpenes (SYLVARES™ ZT and ZONATAC™ NG 98) are light-colored, low-odor resins that are the right choice for premium hot-melt adhesive applications. The incorporation of an aromatic component renders the resin somewhat more polar than the polyterpene resins. The aliphatic–aromatic nature of these resins gives them excellent adhesion to a range of substrates. Styrenated terpenes are highly compatible in EVA and styrene block copolymer (SBC) systems, such as SIS and SBS.

Terpene phenols (SYLVARES™ TP) are widely regarded as having the highest polarity of all tackifiers. As explained above, their polarity can be tuned by the amount of polar phenolic hydroxyl groups. Terpene phenols provide outstanding specific adhesion in EVA and SBC systems. Their polarity and high softening points combine to provide excellent hot tack in bookbinding and packaging applications. These versatile resins can even be used in bookbinding adhesives that require resistance to ink oil. They are known to improve the adhesion of hot-melt packaging and PSAs to difficult-to-bond substrates like coated and recycled paper, glass, and metal foils. Terpene phenols exhibit a very broad polymer compatibility window.

2.7.5 Adhesives and Adhesive Composition Background

Adhesives are substances capable of holding at least two surfaces together in a strong and permanent manner.[108–111] Adhesive compositions can be very complex and heavily depend on the area of application. Adhesive materials in general consist of a polymer (e.g., polyolefin, SIS, acrylics), a tackifier (e.g., rosin ester, terpene oligomer, hydrocarbon resin), and an oil or wax as well as small amounts of antioxidants and other additives. The choice of polymer is leading and depends on the final application of the adhesive, for example, the applied polymer is likely to be a water-based acrylic copolymer or can be a hot-melt PSA based on SIS in case of a general-purpose paper labeling of bottles. Based on a given polymer choice, the appropriate tackifier and wax or oil are selected. Each component in the adhesive has its specific function. The polymer primarily acts as a backbone. It provides cohesive strength and a limited degree of flexibility and thereby in most cases will render a limited degree of immediate adhesion, the so-called tack. The tackifying resin provides or improves the required tack property by increasing the flexibility of the polymeric backbone within a defined temperature range, that is, raising the glass transition temperature (T_g) of the adhesive mixture and lowering of its Young's modulus. Otherwise stated, the tackifier balances cohesion and adhesion by affecting viscoelastic and rheological properties.[112] The wax is a processing aid that quickly crystallizes upon hot-melt adhesive cooling and therefore is used to adjust the open and setting times by providing initial adhesive strength. The wax also serves to considerably reduce the viscosity in the molten state of the HMA to the desired operational level. Waxes are generally not applied in PSA formulations since this would go at the expense of their pressure sensitivity. Oils are used as a diluent instead to modify the viscosity of an adhesive formulation.

Arizona Chemical has the capability to characterize and modify adhesives based on their rheology profile and supply the adhesive producers with high-performance tackifying resins from renewable CTO and CST feedstocks.

The compatibility or miscibility of the components in an adhesive is an important feature that governs its adhesion properties.[113] The lack of compatibility will result in phase separation of certain components in the adhesive formulation and thereby lower product performance. Compatibility among two or more materials is governed by their molecular weights, molecular weight distributions (polydispersities), and solubility parameters. Compatibility in adhesive formulations correlates with the Hildebrand solubility parameter (δ). This parameter provides a numerical estimate of the degree of interaction between materials and is a good indicator for compatibility, particularly for nonpolar materials such as many polymers.[114] The Hildebrand solubility parameter can be described as $\delta = [(\Delta H_v - RT)/V_m]^{1/2}$, that is, the square root of the cohesive energy density: ΔH_v is the heat of vaporization and V_m the molar volume. R represents the gas constant and T the absolute temperature.

The cohesive energy density is the amount of energy needed to completely remove the unit volume of molecules from their neighbors to infinite separation (an ideal gas), which is equal to the heat of vaporization divided by molar volume. For a material to dissolve, the same interactions need to be overcome as the molecules are separated from each other and surrounded by the solvent. Materials with similar values of δ are likely to be miscible or compatible – "like dissolves like" (Figure 2.18). It is interesting to note that Hansen proposed an extension of the Hildebrand parameter to estimate the relative miscibility contributions of apolar, polar, and hydrogen bonding systems. This led to a more accurate miscibility prediction tool.[115,116]

The Gibbs free energy of mixing (ΔG_{mix}) can be described as $\Delta G_{mix} = \Delta H_{mix} - T\Delta S_{mix}$, wherein T represents the absolute temperature. The positive entropy of mixing (ΔS_{mix}) contribution to ΔG_{mix} diminishes with increasing molecular weights of the mixed ingredients (cf. Flory–Huggins lattice model theory). Therefore, a favorable enthalpy of mixing (ΔH_{mix}) factor (which is reflected in similar values of δ) will be a key parameter in the production of compatible polymer–resin-based adhesive formulations.

The solubility parameter of the most common polymers used to formulate adhesives ranges from 7 (less polar) to about 10.5 (more polar). The tackifiers that can be used to effectively tackify a given polymer have to match its solubility parameter value. It should be noted that the polarity of random copolymers like EVA or acrylates such as EBA can vary. For example, the weight percent vinyl acetate in EVA usually varies from 10 to 40%. A higher vinyl acetate content will lead to a more polar copolymer and will also affect other physical properties.

In the case of polar adhesive polymers like solvent acrylics (copolymers of several acrylates such as methyl methacrylate, 2-ethylhexyl acrylate, etc.), tackifiers that can be applied are a terpene phenol resin and a rosin ester. The ability of the resulting adhesive formulation to form more and stronger bonds (cf. the polar hydrogen bonding impact of the phenolic hydroxyl moiety) is enhanced, which is visible in an increased peel performance while tack is simultaneously enhanced (Figure 2.19).

In packaging applications – wherein hot-melt adhesives are used to seal boxes – EVA polymers are often used, and these can be tackified using a rosin ester

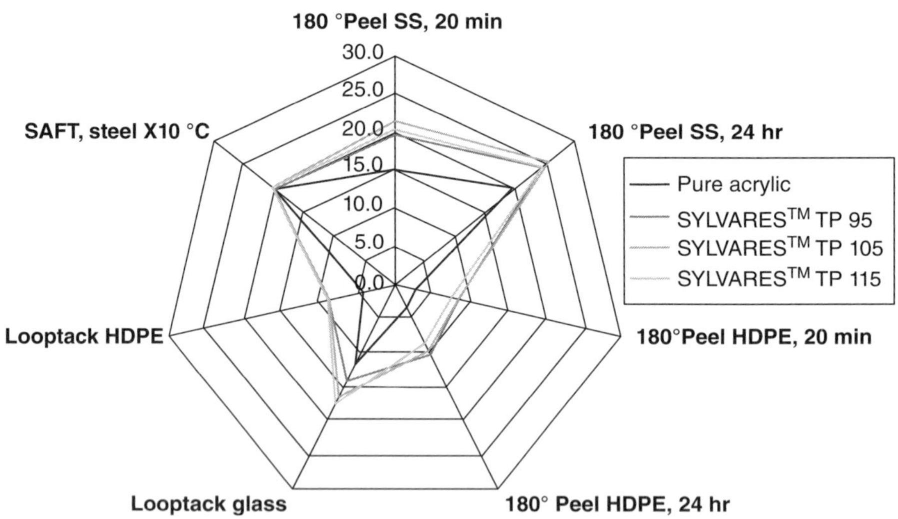

Figure 2.19 Application performance of different SYLVARES™ grades in Newton in a foam tape application, coat weight 60 g/m^2, formulation based on industry standard acrylic polymer tackified with 25 w/w% terpene phenol resin. HDPE = high density polyethylene, SS = stainless steel, TP = terpene phenol resin, SAFT = shear adhesion failure temperature, and Peel = peel force.

from the SYLVALITE™ RE series. The choice for a specific adhesive depends on the technical demands in terms of application performance, color, and color stability.[117,118]

2.7.6 Packaging Adhesives Market

Two main adhesives systems are available that are used for packaging adhesives, namely, EVA based and metallocene/polyolefin based. The global hot-melt adhesive market for packaging applications has been estimated to be 715 kt in 2011. This market comprises EVA-based hot-melts (~75%, 589 kt) and metallocene-based hot-melts (~25%, 126 kt). The overall market is growing approximately 4% every year.

2.7.7 Conclusions

Tackifier technologies are manifold and enable a broad range of adhesive technologies. Since an increasing number of polymers and copolymers need to be tackified, the provision of the right tackifier solution to produce optimal adhesive formulations remains challenging and will continue to demand R&D attention. Arizona Chemical offers renewable resins suitable for a range of polymer systems and with properties commensurate to petrochemical alternatives. Since the final application of the formulated adhesive often dictates the polymer type of use, the selection of the proper tackifier is crucial.

Our mission is to offer sustainable tackifying solutions for a wide range of polymer-based adhesives, by further expanding and improving our rosin- and monoterpene-based chemistries as well as our technical and analytical capabilities.

ACKNOWLEDGMENTS

Carl Bilgrien, Peter Flippo, Ronnie Zeeman, and Anna Holmberg are gratefully acknowledged for their helpful suggestions.

REFERENCES

[1] J. Drew, In "*Naval Stores*" (Eds. D.F. Zinkel, J.R. Russell). Pulp Chemicals Association, New York, **1989**, 3–9, and references cited therein.

[2] Anonymous In "*The United States Dispensatory and Physician's Pharmacology*", 26th Ed., (Eds. R. Pratt, M.D. Altschule, A. Osol). J. B. Lippincott Company, Philadelphia, **1967**.

[3] J.S. Mills, R. White, *Stud. Conserv.* **1977**, *22*, 12–31.

[4] J. Drew, In "*Naval Stores*", (Eds. D.F. Zinkel, J.R. Russell), Pulp Chemicals Association, New York, **1989**, 3–29, and references cited therein.

[5] (a) H. Panda, *Handbook on Tall Oil Rosin, Production, Processing and Utilization*, Asia Pacific Business Press Inc., New Delhi, **2013**, 339–367. (b) "Naval stores", In, "*Production, Chemistry, Utilization*" (Eds. D.F. Zinkel, J.R. Russell), Pulp Chemicals Association, New York, **1989**.

[6] E. Breitmaier, *Terpenes, Flavors, Fragrances, Pharmaca, Pheromones*, Wiley, Weinheim, **2006**, 1–23.

[7] J. Degenhardt, T.G. Kollner, J. Gershenzon, *Phytochemistry* **2009**, *70*, 1621–1637.

[8] J. Bohlman, C.I. Keeling, *Plant J.* **2008**, *54*, 656–669.

[9] D. McPhee, In "*Catalytic Process Development for Renewable Materials*", (Eds. P. Imhof, J.C. van der Waal). Wiley, Weinheim, **2013**, 51–80.

[10] J.H. Langenheim, *Plant Resins: Chemistry, Evolution, Ecology, and Ethnobotany*, Timber Press, Portland, **2003**.

[11] J.J.W. Coppen, G.A. Hone, *Non-Wood Forest Products, Vol. 2*, Natural Resources Institute, FAO, Rome, **1995**.

[12] A. Gandini, *Green Chem.* **2011**, *13*, 1061.

[13] K.A.D. Swift, *Top. Catal.* **2004**, *27*, 143–155.

[14] K. Sjödin, M. Persson, A.K. Borg-Karlson, T. Norin. *Phytochemistry* **1996**, *41*, 439–445.

[15] K.C. da Silva Rodrigues-Correa, J.C. de Lima, A.G. Fett-Neto, *Food and Energy Security*, John Wiley & Sons and the Association of Applied Biologists, **2013**, 81–93.

[16] F. Macchioni, P.L. Cioni, G. Flamini, I. Morelli, S. Maccioni, M. Ansaldi, *J. Flavour Frag.* **2003**, *18*, 139–143.

[17] A.C.R. Silva, P.M. Lopes, M.M.B. Azevedo, D.C.M. Costa, C.S. Alviano, D.S. Alviano, *Molecules* **2012**, *17*, 6305–6316.

[18] A.J.D. Silvestre, A. Gandini In "*Monomers, Polymers and Composites from Renewable Resources*", (Eds. M.N. Belgacem, A. Gandini). Elsevier, Amsterdam, **2008**, 17–38.

[19] (a) P.A. Wilbon, F. Chu, C. Tang, *Macromol. Rapid Commun.* **2013**, *34*, 8–37. (b) K.D. Kumar, S. Gupta, A.H. Tsou, A.K. Bhowmick, *J. Appl. Polym. Sci.* **2008**, *110*(3), 1485–1497.

[20] G.C. Harris, *Encyclopedia of Chemical Technology*, Interscience Encyclopedia, Inc., New York, **1953**, *11*, 779–810.

[21] X. Rao, In *Rosin-based Chemicals and Polymers* (Ed. J. Zhang). Smithers Rapra Publishing, Shawbury, **2012**, 129–172.

[22] A.J.D. Silvestre, A. Gandini, In *Monomers, Polymers and Composites from Renewable Resources* (Eds. M.N. Belgacem, A. Gandini). Elsevier, Amsterdam, **2008**, 67–88.

[23] S. Maiti, S.S. Ray, A.K. Kundu, *Prog. Polym. Sci.* **1989**, *14*, 297–338.

[24] D.L. Klass, *Biomass for renewable energy, fuels, and chemicals*, Academic Press, New York, **1998**.

[25] *Directive **2009**/28/EC* of the European Parliament and of the Council of 23 April 2009 on the promotion of the use of energy from renewable sources and amending and subsequently repealing *Directives **2001**/77/EC and **2003**/30/EC.*

[26] S. Octave, D. Thomas, *Biochimie* **2009**, *91*, 659–664.

[27] A. Demirbas, *Energy Convers. Manage.* **2009**, *50*, 2782–2801.

[28] M.F. Patrick, P. Champagne, M.F. Cunningham, R.A. Whitney, *Bioresource Technol.* **2010**, *101*, 8915–8922.

[29] P.B. Smith, ADM: *Building a future in renewable industrial chemicals*. Lecture at BIOP: The Canadian Conference on Industrial Bioproduct Innovation, Montreal, 2008, November 5–6.

[30] B. Kamm, M. Kamm, In "*Advances in Biochemical Engineering/Biotechnology*", *Vol. 105* (Eds. R. Ulber, D. Sell). Springer, Heidelberg, **2007**, 175–204.

[31] (a) A.J. Ragauskas, C.K. Williams, B.H. Davison, G. Britovsek, J. Cairney, C.A. Eckert, W.J. Frederick Jr., J.P. Hallett, D.J. Leak, C.L. Liotta, J.R. Mielenz, R. Murphy, R. Templer, T. Tschaplinski, *Science* **2006**, *311*, 484–489; (b) B. Kamm, P.R. Gruber, M. Kamm, *Biorefineries – Industrial Processes and Products: Status Quo and Future Directions*, Wiley-VCH, Weinheim, **2010**.

[32] J.K. Kurian, G.R. Nair, A. Hussain, G.S.V. Raghavan, *Renew. Sustainable Energy Rev.* **2013**, *25*, 205–219.

[33] B. Kamm, P.R. Gruber, M. Kamm, *Encyclopedia of Industrial Chemistry, Biorefineries – Industrial Processes and Products*, Wiley-VCH, Weinheim, **2012**, 659–688.

[34] J.W.A. Langeveld, J.P.M. Sanders, In *The Biobased Economy: Biofuels, Materials And Chemicals In The Post-Oil Era* (Eds. J.W.A. Langeveld, J.P.M. Sanders, M.J.G. Meeusen). Taylor & Francis, Oxford, **2010**, 3–17.

[35] A. Demirbas, *Energy Educ. Sci. Technol.* **2007**, *18*, 117–132.

[36] M.K. Deshmukh, S.S. Deshmuk, *Energy. Educ. Sci. Technol.* **2007**, *18*, 1–15.

[37] A. Demirbas, *Energy Educ. Sci. Technol.*, **2008**, *21*, 1–59.

[38] M. Giampietro, S. Ulgliati, D. Pimentel, *BioScience*, **1997**, *47*, 587–600.

[39] G. Berndes, C. Azar, T. Kåberger, D. Abrahamson, *Biomass Bioenergy* **2009**, *20*, 371–383.

[40] W.E. Mabee, P.N. McFarlane, J.N. Saddler, *Biomass Bioenergy* **2011**, *35*, 4519–4529.

[41] J. Ruane, A. Sonnino, A. Agostini, *Biomass Bioenergy* **2010**, *34*, 1427–1439.

[42] D.R. Dodds, R.A. Cross, *Science* **2007**, *318*, 1250–1251.

[43] S. Kotha, M.K. Dipak, *Tetrahedron* **2012**, *68*, 397–421.

[44] H.L. Ngo, K. Jones, T.A. Foglia, *J. Am. Oil Chem. Soc.* **2006**, *83*, 629–634.

[45] K. Holmberg, In "*Coatings Technology Handbook*", 3rd Ed., (Ed. A.A. Tracton). Taylor & Francis, Oxford, **2006**, 51–1–51–12.

[46] F. Cherubini, *Energy Convers. Manage.* **2010**, *51*, 1412–1421.

[47] B. Kamm, M. Kamm, *Appl. Microbiol. Biotechnol.* **2004**, *64*, 137–145.

[48] B. Kamm, M. Kamm, P.R. Gruber, S. Kromus, In "*Biorefineries, Industrial Processes and Products. Status Quo and Future Directions*", *Vol. 1.* (Eds. B. Kamm, P.R. Gruber, M. Kamm). Wiley, Weinheim, **2006**, 3–11.

[49] K.A. Jung, S.R. Lim, Y. Kim, J. M. Park. *Bioresource Technol.* **2013**, *135*, 182–190.

[50] R.H. Wijffels, M.J. Barbosa, M.H.M. Eppink, *Biofuels Bioprod. Bioref.* **2010**, *4*, 287–295.

[51] J. Singh, S. Gu, *Renew. Sust. Energy Rev.* **2010**, *14*, 2596–2610.

[52] A.R.C. Morais, R. Bogel-Lukasik, *Sustain. Chem. Process.*, **2013**, *1*:18.

[53] T. Dickerson, J. Soria, *Energies* **2013**, *6*, 514–538.

[54] Q. Bu, H. Lei, A.H. Zacher, L. Wanga, S. Ren, J. Liang, Y. Wei, Y. Liu, J. Tang, Q. Zhang, R. Ruan, *Bioresource Technol.* **2012**, *124*, 470–477.

[55] J. Zakzeski, P.C.A. Bruijnincx, A.L. Jongerius, B.M. Weckhuysen, *Chem. Rev.* **2010**, *110*, 3552–3599.

[56] S.N. Naik, V.V. Goud, P.K. Rout, A.K. Dalai, *Renew. Sust. Energy Rev.* **2010**, *14*, 578–597.

[57] B.M. Trost, *Science* **1991**, *254*, 1471–1477.

[58] (a) R.A. Sheldon, *Pure Appl. Chem.* **2000**, *72*, 1233–1246; (b) D.J.C. Constable, A.D. Curzons, V.L. Cunningham, *Green Chem.* **2002**, *4*, 521–527.

[59] P.T. Anastas, J.C. Warner, *Green Chemistry: Theory and Practice*, Oxford University Press, New York, **1998**.

[60] E. Polastro, J. Tuovinen, *Chem. Today* **2010**, *28*(2), 39–41.

[61] M. Moshkelani, M. Marinova, M. Perrier, J. Paris, *Appl. Therm. Eng.* **2013**, *50*, 1427–1436.

[62] B. Thorp, *Pulp Paper* **2005**, *79*(11), 35–39.

[63] A. Johansson, *Biomass* **1982**, *2*, 103–113.

[64] (a) D.W. Reeve, *The Kraft Recovery Cycle*, Tappi Kraft Recovery Operations Short Course, Tappi Press, Atlanta, **2002**; (b) D.W. Reeve, *The Kraft Recovery Cycle,* from Course Notes - Kraft Recovery Short Course, Orlando, Jan. 17–20, **2000**, 1–16.

[65] H. Sixta, G. Schild, *Lenziger Ber.* **2009**, *87*, 26–37.

[66] G. Smook, *Handbook for Pulp and Paper Technologists*, Angus Wilde, Vancouver, **2003**.

[67] F.S. Starr, J.M. Wong, N.L. Kennedy (Westvaco Corporation), US 5,543,110 A1, **1996**.

[68] M. Inomata, K. Hanabusa (Air Liquide), US 3,857,902, **1974**.

[69] M. Marinova, E. Mateos-Espejel, N. Jemaa, J. Paris, *Chem. Eng. Res. Des.* **2009**, *87*, 1269–1275.

[70] M. Propst, J. Drew, *Tall Oil*, Tappi Press, Atlanta, **1987**, pp. 109–112.

[71] (a) C.D. Foran, *Tall Oil Soap Recovery in Tall Oil*, Tappi Press, Atlanta, **1996**, Session 4, Paper No. 3, 1–31; (b) C.D. Foran, *Tall Oil Soap Recovery* from Course Notes – Kraft Recovery Short Course, Orlando, Jan. 17–20, **2000**, 279–297; (c) C.D. Foran, *Pulp Paper* **1984**, *58*, 104–108; (d) R.W. Ellerbe, *J. Am. Oil Chem. Soc.* **1973**, *50*, 501–513.

[72] S.Z.Q. Song, E. Zavarin, D.F. Zinkel, *J. Wood Chem. Tech.* **1985**, *5*(4), 535–542.

[73] L. da Costa Sousa, S.P. Chundawat, V. Balan, B.E. Dale, *Curr. Opin. Biotech.* **2009**, *20*, 339–347.

[74] K. Andersson, J.G.T. Bergström, *Svensk Papperstidning*, **1969**, *11*, 375–379.

[75] D.F. Wilson, B. Hrutfiord, *Tappi J.* **1971**, *54*, 1094–1098.

[76] F. Andrawes, T. Chang, *J. Chromatogr. A*, **1989**, *468*, 145–155.

[77] K. Andersson, *Svensk Papperstidning*, **1970**, *1*, 1–10.

[78] D.C. Foran, *Tall Oil Soap Recovery*, Kraft Recovery Operation Short Course 3.7-5, **2006**.

[79] B. Holmbom, A. Pekkala, A. Pekkala, *J. Am. Oil. Soc.*, **1974**, *51*, 397.

[80] R.V. Lawrence, *Tappi J.* **1962**, *45*, 654.

[81] G.-F. Chen, *Progr. Org. Coat.* **1992**, *20*, 139–167.

[82] Internal Study Arizona Chemical – Arthur D. Little, E. Polastro, July, 2009. *Annual availability of European CTO estimated abt. 650.000 metric tons*. Source: FAO-UNECE

[83] *Greenhouse gasses – Specification with guidance at the organization level of quantification and reporting greenhouse gas emissions and renewal* – ISO 14060 – *March,* **2006**.

[84] R. Fornell, T. Berntsson, A. Asblad, *Energy* **2013**, *50*, 83–92.

[85] M. Hamaguchi, M. Cardoso, E. Vakkilainen, *Energies* **2012**, *5*, 2288–2309.

[86] B.P. Anand, C.G. Saravanan, C.A. Srinivasan, *Renew. Energy* **2010**, *35*, 1179–1184.

[87] R. Yumrutas, M.H. Alma, H. Ozcan, O. Kaska, *Fuel* **2008**, *87*, 252–259.

[88] D.J. Hayes, *Catal. Today* **2009**, *145*, 138–151.

[89] G. Brodeur, E. Yau, K. Badal, J. Collier, K.B. Ramachandran, S. Ramakrishnan, *Enzyme Res.* **2011**, *2011*, 787532.

[90] J.J. Bozell, G. R. Petersen, *Green Chem.* **2010**, *12*, 539–554.

[91] (a) F. Mustata, I. Bicu, *Polimery* **2005**, *50*, 176–181; (b) W.O.S. Doherty, P. Mousavioun, C.M. Fellows, *Ind. Crops Prod.* **2011**, *33*, 259–276; (c) A.E. Effendi, H. Gerhauser, A.V. Bridgewater, *Renew. Sustainable Energy Rev.* **2008**, *12*, 2092–2116; (d) T. Yoshikawa, T. Yagi, S. Shinohara, T. Fukunaga, Y. Nagasaka, T. Tago, T. Masuka, *Fuel Process Technol.* **2013**, *108*, 69–75; (e) P. Azadi, O.R. Inderwildi, R. Farnood, D.A. King, *Renew. Sust. Energy Rev.* **2013**, *21*, 506–523.

[92] J. Minn. *Thermochim. Acta* **1985**, *91*, 87–94.

[93] (a) Z.-Q. Song, E. Zavarin, D.F. Zinkel, *J. Wood Chem. Technol.* **1985**, *5*, 535–542; (b) A.M. Atta, S.M. El-Saeed, R.K. Farag, *React. Funct. Polym.* **2006**, *66*, 1596–1608.

[94] L. Wang, X. Chen, J. Liang, Y. Chen, X. Pu, Z. Tong, *Chem. Engin. J.* **2009**, *152*, 242–250.

[95] J.C. Souto, P. Yustos, M. Ladero, F. Garcia-Ochoa, *Biores. Technol.* **2011**, *102*, 3504–3511.

[96] M. Ladero, M. de Gracia, J.J. Tamayo, I. Lopez de Ahumada, *Chem. Engin. J.* **2011**, *169*, 319–328.

[97] L. Wang, X. Chen, W. Sun, J. Liang, X. Xu, Z. Tong, *Ind. Crops Prod.* **2013**, *49*, 1– 9.

[98] L. Lloyd, *Handbook of Industrial Catalysts*, Springer, Heidelberg, **2011**, 73–117.

[99] (a) J.B. Montgomery, A.N. Hoffmann, A.L. Glasebrook, J.I. Thigpen, *Ind. Eng. Chem.* **1958**, *50*, 313–316; (b) H. Jiang, S. Liu, Z. Song, S. Shang, *J. Chem. Soc. Pak.* **2012**, *34*, 429–432; (c) Y.J. Lu, R.S. Xu, Z.D. Zhao, P.H. Zhang, M.X. Wang, *Adv. Mat. Res.* **2013**, *785–786*, 1111–1116.

[100] D.P. Duncan (Westvaco Corporation) US 4,507,228, **1985**.

[101] (a) R. Fujii, K. Arimoto, D.F. Zinkel, *J. Am. Oil Chem. Soc.* **1987**, *67*, 1144–1149; (b) R.G. Sinclair, D.A. Berry, W.H. Schuller, R.V. Lawrence, *Ing. Eng. Chem. Prod. Res. Dev.* **1970**, *9*, 60–65; (c) S. Liu, C. Xie, S. Yu, F. Liu, *Catal. Commun.* **2008**, *9*, 2030–2043; (d) B.A. Parkin, W.H. Schuller, *Ind. Eng. Chem. Prod. Res. Dev.* **1972**, *11*, 156–158; (e) D.F. Zinkel, *Naval Stores Rev.* **1987**, *97*, 5–8.

[102] L. Li, S. Yu, F. Liu, C. Xie, S. Zhang, J. Yang, *Catal. Commun.* **2007**, *8*, 719–724.

[103] G. Odian, *Principles of polymerization*, Wiley, New Jersey, **2004**, 372–463.

[104] F. Ebmeyer, *J. Mol. Struct. (Theochem)* **2002**, *582*, 251–255.

[105] N. Flores-Holguin, A. Aguilar-Elguezabal, L.M. Rodrıguez-Valdez, D. Glossman-Mitnik, *J. Mol. Struct. (Theochem)* **2008**, *854*, 81–88.

[106] K.L. Thompson, A.K. Deshpande (Arizona Chemical Company), US 6,160,083, **2000**.

[107] R.M. Roberts, A.A. Khalaf, *Friedel-Crafts Alkylation Chemistry: A Century of Discovery*, Marcel Dekker, New York, **1984**.

[108] E.M. Petrie. *Handbook of Adhesives and Sealants*, McGraw-Hill Professional, New York, **2007**.

[109] S. Sun, M. Li, A. Liu, *Int. J. Adhes. Adhes.* **2013**, *41*, 98–106.

[110] T.M. Goulding, In "*Handbook of Adhesive Technology*" (Ed. A. Pizzi, K.M. Mittal). Taylor & Francis, Oxford, **2003**, 839–854.

[111] J. Komornicki, G. Marin, I. Leclere, *Tappi J.* **1991**, 233–237.

[112] E.P. O'Brien, L.T. Germinario, G.R. Robe, T. Williams, D.G. Atkins, D.A. Moroney, M.A. Peters, *J. Adhes. Sci. Technol.* **2007**, *21*, 637–661.

[113] L.H. Sperling, *Introduction to physical polymer science*, Wiley, New Jersey, **2006**, 145–195.

[114] J.H. Hildebrand, *The solubility of non-electrolytes, 2nd ed. Am. Chem. Soc. Monograph No. 17,* Reinhold, New York, **1936**.

[115] C.M. Hansen, *J. Paint Technol.* **1967**, *39*, 104–117.

[116] C.M. Hansen, *J. Paint Technol.* **1967**, *39*, 505–510.

[117] M.L. Barrueso-Martinez, T.D.P. Ferrandiz-Gomez, J.M. Martin-Martinez, F. Aran-Ais, A.M. Torro-Palau, A.C. Orgiles-Barcelo, *Adhesives Age* **2001** (April), 32–37.

[118] Y.-J. Park, H.-S. Joo, H.-S. Do, H.-J. Kim, *J. Adhes. Sci. Technol.* **2006**, *20*, 1561–1571.

3

ARKEMA: CASTOR REACTIVE SEED CRUSHING PROCESS TO PROMOTE CASTOR CULTIVATION

JEAN-LUC DUBOIS
ARKEMA, France

Industrial Biorenewables: A Practical Viewpoint, First Edition. Edited by Pablo Domínguez de María.

3.1 ARKEMA: CONTEXT FOR BIORENEWABLES

Arkema is a major chemical company which has already a long history of products made from renewable resources. These products and their processes were developed and commercialized, since they were solving technical problems faced by the fossil-carbon based solutions. For instance, Arkema introduced Rilsan® PA11 more than 60 years ago, a technical polyamide made from castor oil, as well as other castor oil derivatives having various applications such as aviation lubricants and fragrances. In 2007, Arkema also introduced Pebax® Rnew, the first thermoplastic elastomer partially based on castor oil. Currently Arkema uses bioresources not only for the production of its polyamides (including PA11, and PA10.10), but also for the production of epoxidized soy/lin oils, fatty amines, activated carbon, alkyd resins emulsions, and some other minor products.

Arkema operates currently 5 "biorefineries" which have always been considered as all its other chemical plants. Globally products derived from renewable resources contribute to 11% of Arkema's turnover, while it is expected to grow to reach 15% in 2016. The first biorefinery is located in Marseille/Saint-Menet in south of France. This plant processes Castor Oil, extracted from Castor seeds, a non-edible tropical oil crop. The first processing step is transesterification much like in biodiesel processes, co-producing methyl ester of castor oil and crude glycerine. After a thermal cleavage step, heptanaldehyde is obtained along with methyl-10-undecenoate, and a blend of non-converted esters valorized as a renewable solvent. Later, methyl-10-undecenoate is converted in several steps to a high value monomer for long chain polyamide: amino-undecanoic acid. Heptanaldehyde, used in Flavor & Fragrance industry is also converted into heptanol and heptanoic acid. The latter is being used in the synthesis of aviation lubricants. A second biorefinery, located in China, also uses castor oil, and through a different process produces sebacic acid, a 10 carbon atoms long linear dicarboxylic acid, but also glycerine, 2-octanol and a blend of fatty acids. Sebacic acid is a valuable monomer for polyamides and polyesters but also a raw material for dimethyl and dibutylsebacate produced on-site.

A third biorefinery located in Blooming Prairie, USA, is processing linseed oil and soybean oil to produce epoxidized oils used as polymer additives. The plant also processes terpenes, limonenes, pinenes and other unsaturated materials. A forth biorefinery is located in Parentis, France, and process locally sourced pine wood to produce activated carbons through different processes. The fifth biorefinery is located in Feuchy, France, and processes fatty acids to produce fatty nitriles and fatty amines and derivatives. These products are used in many applications such as surfactants.

More recently, Arkema introduced new materials derived from supramolecular chemistry, such as a self-healing rubber, and developed new technologies for the conversion of glycerine to acrolein (a useful chemical intermediate in the fine chemical area). Small dedicated units based on the now ubiquitous glycerine would avoid the transportation of a dangerous chemical, offering a unique technical solution based on renewable resources.

In the present Chapter, novel developed strategies for Castor oil business – a core competence of Arkema – will be discussed in detail as "case study" on biorenewables. Figures 3.1 and 3.2 provide an overview of Arkema and its bio-based products.

3.2 INTRODUCTION TO CASTOR OIL

Castor (*Ricinus communis*) is a well-established oilseed crop with many industrial applications (Figure 3.3). It is mainly cultivated in tropical climates; India, Brazil and China being the main producers with respectively 1630, 170 and 62 kt of seeds in 2012 according to FAOStat database. Castor oil is very rich in ricinoleic acid (12-hydroxy 9-octadecenoic acid) with more than 85%, and the seeds have an oil content of about 48%.[1] Castor oil is mainly used for lubricants but also for polymers such as polyurethanes. It also has a history of more than 50 years to be used to produce the polyamide-11 known as Rilsan® PA11, a highly technical polyamide. Castor oil is also used to produce sebacic acid (a 10 carbon atoms linear diacid) which can also be used as a monomer, for example in polyamide and polyesters, or to produce solvents for example. The industrial applications of castor are numerous, and this crop is now very attractive since it is known to be non-edible, non-GMO, and to survive where many other crops would die.

Castor is an important oil crop for Europe which imported 234 kt, 143 kt, 212 kt and 205 kt in 2008, 2009, 2010, and 2011, respectively, for corresponding values of 347, 192, 332 and 474 M$. The Castor oil market is dominated by India, which has been exporting 331 kt, 368 kt, 424 kt, and 406 kt over the same period, while Brazil and China where net importers. For consumers, there are economic risks associated to the fact that a single country can be dominating a market; and one of them is a climatic risk, especially in a place that can be affected by monsoon. Therefore, a current challenge is to find technical solutions that could promote Castor cultivation in other parts of the world. There are several criteria that should be met in order to generate appropriate conditions to create a new "Value Chain" in which all the players would be satisfied:

- As mentioned above it is advantageous to be in a different climatic zone, to level-off production and prices on a year-to-year basis.

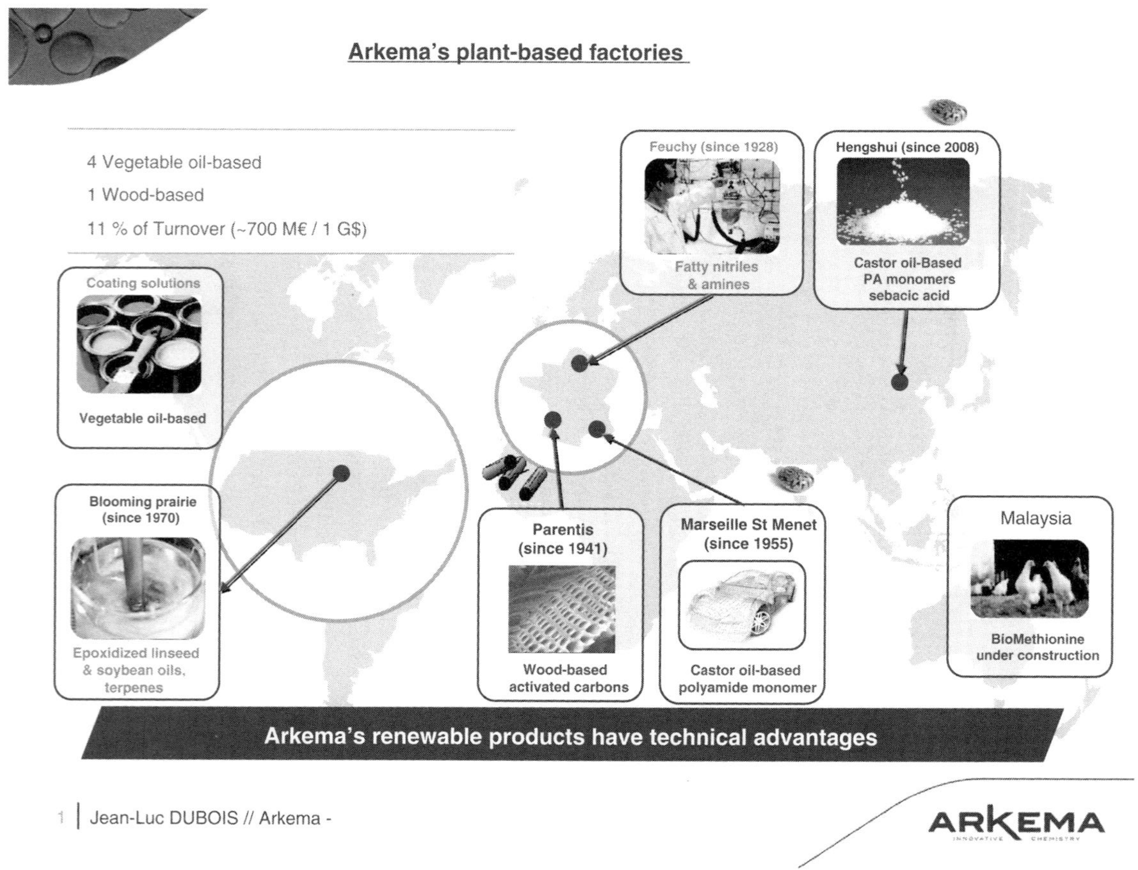

Figure 3.1 Current operating bio-based products and plants. (*See color plate section for the color representation of this figure.*)

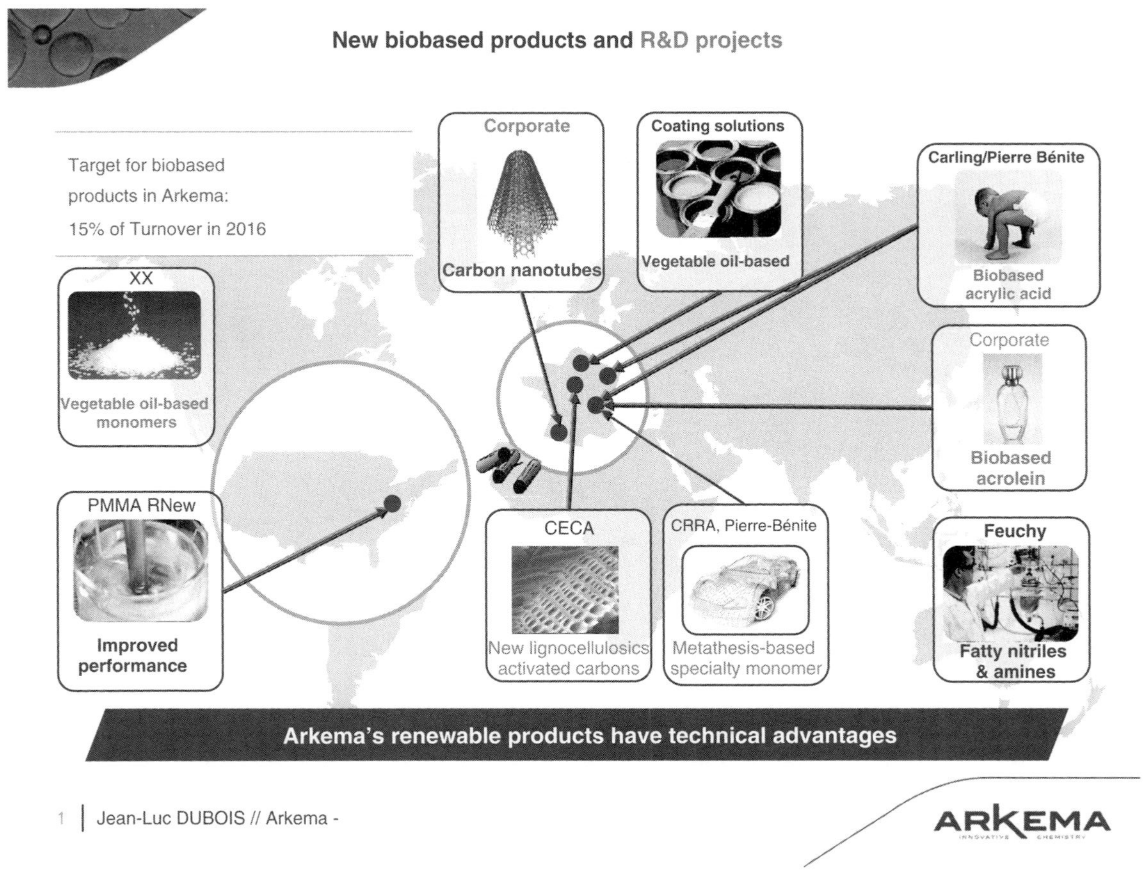

Figure 3.2 On-going developments. (*See color plate section for the color representation of this figure.*)

Figure 3.3 *Ricinus communis* plant. *Source*: Copyright by Dr. Jean-Luc Dubois. (*See color plate section for the color representation of this figure.*)

- Castor should be native from the country. For example it would be impossible to start castor production in Australia.
- Castor being a tropical crop, it is currently not suited for all climates, and for example, Northern Europe is not appropriate, since Castor vegetative cycle is too long.
- Castor production is mostly done manually (harvesting) since all the fruits are not mature at the same time. So that requires countries with low labour cost. With some farming techniques, it is however possible to mechanize castor production.

Appropriate locations would include China, Brazil, Argentina, but also a significant part of Africa. Places in the southern hemisphere would benefit of a 6 month delay on plantation/harvest that would be useful to stabilize world production.

How to promote the creation of a new "Value Chain" around castor? Being a non-edible crop, castor beans cannot be processed in conventional seed crushing unit, since there would be a high risk of contamination with food products. So a dedicated crushing unit is highly desirable. A conventional oil production unit for most vegetable oil is including a crushing unit to extract oil at low or high temperature while pressing the seeds, and a solvent extraction unit which most of the time uses hexane as a solvent to extract the remaining few percent of oil which is left in the meal after the first step; and then there is an oil refining unit. Since Castor oil has a rather high value current processors have installed solvent extraction units, in order to recover as much oil as possible. The high capital cost of the solvent extraction units is justified by large plant size and a high oil value. Many oil crops have seen their development enhanced recently by the demand for biodiesel, which also means that an additional capital cost (investment) is needed to prepare the methyl ester of vegetable oil also known as Fatty Acid Methyl Ester (FAME).

So, in order to reach the final market, for Castor or Jatropha – another tropical crop that can grow in the same areas and which is also non-edible – a large amount of capital would be needed to implement a new value chain in a new area. Therefore, there is a need to develop alternative technologies that would be able to facilitate the introduction of these crops in new areas, while reducing the capital cost, but also the operating cost, and that additionally would be able to reduce the toxicity of the meal after oil extraction.

Castor is a non-edible crop, mostly because the seed contains several toxic compounds. The oil itself has only a laxative effect, but the meal contains a highly toxic protein, ricin; less toxic *R. communis* agglutinin (RCA) a low molecular weight alkaloid, ricinine (MW = 164.2 g/mol), and an allergen compound named CB-1A. In its crude form, ricin is often found in association with RCA. Both ricin and RCA are highly concentrated in the seed meal and they comprise up to 5% of the weight of the seed meal,[2] which makes castor a possible threat and weapon for terrorist attacks. Ricin is also known as the poison used in the "Bulgarian umbrella" which killed a dissident many years ago. Antibody-based immunoassays are standard technologies for detection of ricin in castor meals, but one of the drawbacks is that they cannot differentiate different ricin isoforms and RCA due to the high homology of their sequence. We assume that the content and toxicity of ricin determined in this study were from a mixture of different ricin isoforms and RCA. The allergen might be somewhat more critical, since the 8–20% of the population might be allergic to castor depending on the reports, with in some cases anaphylactic shocks have been described.[3,4] After oil extraction, these toxic compounds remain in the meal.

Various methods including physical, chemical, and biological treatments have been employed to detoxify the castor meal,[5,6] but the detoxification process is not always effective.[7,8] For these reasons, castor meal cannot be used for animal feed and is most often used as a natural fertilizer since it is rich in Nitrogen. So far, there is no international regulation limiting the amount of *R. communis* in fertilizers. However, there are national regulations, e.g., Germany allows *R. communis*–residual material in fertilizer if no acute oral toxicity in rats is observed after uptake of 2000 mg material per kg body weight.[9]

In order to reduce the cost of production of castor oil, and facilitate the introduction of a new Value Chain in an area where no appropriate infrastructures exist, we need to reduce the capital costs needed especially at small production scale. Currently there are many projects discussing the implementation of new BioRefineries of world scale capacities, all over the world. Some of these projects are targeting "Integrated Biorefineries," meaning plants which mimic petroleum refineries, and which use all the raw materials, make multiple types of products, address different markets, have no residues and are tuned to make the best value from their starting materials. But this high level of integration can only be achieved by integrating one product at a time, since it would be impossible to generate so many new markets at the same time, and all growing at the same rate. It is basically the same here, if we want to generate a new value chain on Castor. Initial production will be small since it is difficult to consider that straight away 60,000 ha will be planted the first year. During a first phase, the seeds produced could be exported to other countries having the appropriate crushing units. In this case the meal, which is non-edible but which can be used as a fertilizer since it is rich in Nitrogen (proteins), is produced in a different location than the castor seeds. So, chemical fertilizers would have to be imported. An alternative is to invest in seed crushing units only that would produce castor oil and would leave a meal still containing some valuable oil. Then a compromise has to be found between the lower capital cost and operating cost of these crushing units but the advantage to locally source a natural fertilizer, and the loss of oil (value) in the meal. So obviously a new technology that would allow to recover more oil, while reducing capital cost for new equipment would be attractive. Such a new technology would be even more attractive for a region without pre-existing facilities than for an area already equipped with castor seed crushing and solvent recovery units, where the heavy capital investments have already been done.

With all what has been stated before, we get a first sketch of how such a technology should look like:

- It should extract as much oil as a crushing unit combined with a solvent extraction unit.
- It should be appropriate for small plant capacities, of about 10,000–15,000 t oil equivalent.
- The defatted meal should be less toxic than that obtained with conventional technologies, facilitating the use of the meal as fertilizer in organic farming and/or for local uses.
- It should avoid the use of solvents like hexane, which is contributing to the emissions of Volatile Organic Compounds (VOC).

- It should combine several steps in a single plant and/or avoid energy/water consumption incurred by conventional technologies.

To satisfy all these criteria we designed what we later called a Reactive Seed Crushing process, on purpose for Castor. Variations of that process were later designed for Jatropha, and other crops.

The basic idea, in developing a new process, was to identify a technology – through Process Intensification principles – that would reduce the number of major equipments, thereby reducing the Capital cost (CAPEX) and operating cost (OPEX). The conventional process requires the following main equipments to get to the final methyl ester:

1. A drying unit – to reduce moisture content in the seeds.
2. A screw press (cold and/or hot) – to get a first pressure oil.
3. A solvent extraction unit (including a toaster to dry the cake) – to recover the remaining oil in the cake.
4. A solvent recovery section.
5. An oil refining unit – for refining, degumming, and deodorization.
6. A transesterification unit, including the methanol recovery section and glycerine recovery.

Typical cost figure for a Seed crushing unit including a solvent extraction unit would be in the range of 250–500 €/t oil capacity (based on rapeseed or soybean) and for units of 100–400 kt capacity (2011 figures). Typical Capex figures for a biodiesel (transesterification unit) would be in the range of 150–200 €/t ester capacity for units in the range of 100–200 kt capacity. Therefore, the objective in the Reactive Seed Crushing Process was to avoid the units 1, 2, 5, and 6 in the list above, and while adding only minimum equipment to significantly lower the production cost. In our reactive seed crushing method, methanol (or another alcohol) is used as a solvent for the extraction of the triglycerides, and at the same time as a reactant for a transesterification reaction. In this purpose, methanol is used in a large excess, to facilitate the extraction. Indirectly it facilitates the transesterification reaction. A basic catalyst, such as sodium hydroxide is used in the process, but since the seeds are used as received, their enzymes (lipases) cannot be excluded to play a role in the reaction, especially since the reaction takes place at low temperature.

The method described reacts “in planta” (meaning that the chemistry is carried on the seeds rather than on the oil), the light alcohol with the oil contained in the seed. In this method, the alcohol plays the role of both a solvent and a reactant.

Methanol is soluble with hydroxyfatty acids (HFA)-rich oils, although it is not soluble in other oils such as soybean oil, rapeseed oil and jatropha oil. For this reason, in a biodiesel-like process, the reaction starts more easily with hydroxyfatty acids containing oils, but as the reaction proceeds, glycerol is produced and it is also more soluble in hydroxyfatty acids (HFA)-rich esters and do not separate readily. The higher solubility of methanol in HFA rich oils might explain why the extraction of the castor oil proceeds smoothly (Figures 3.4 and 3.5).

Figure 3.4 Obtained castor seeds via the conventional process (a) or the novel approach (b). *Source*: Copyright by Dr. Jean-Luc Dubois.

3.3 EXPERIMENTAL DETAILS

3.3.1 Castor Seeds Origin

Three batches of castor seeds were evaluated for their levels of volatile matter (method NF ISO 665) and fat content (method NF ISO 659), their degrees of acidity of the fat (expressed in mg KOH/g) (method EN 14104), and their ricin contents (expressed in mg ricin/kg seeds) ELISA test. Results are summarized in Table 3.1. Batch 1 and Batch 3 were received from India. With this later batch a sample of an industrial castor seed meal, after oil extraction and solvent extraction, was also received and used as such.

These results show that batches 1 and 3 have low acidities (Acid Number < 1 mg KOH/g) whereas batch 1B, which corresponds to batch 1 after 12 months of storage, has become highly acidic (Acid Number > 2 mg KOH/g). Basically with such an acid number the oil would be considered as inappropriate for a conventional biodiesel process. But as it will be seen in the following results, these seeds still lead to appropriate results through the reactive seed crushing method.

Samples of castor meals obtained after oil production by low temperature seed crushing, without any solvent extraction, were obtained from SOABE in Madagascar. Right after oil expression, the seed meals were dried and sealed in plastic bags and then shipped. The seed meals were found in both cases to contain high level of free fatty acids, probably produced by hydrolysis during transportation but anyway used

Figure 3.5 Obtained methyl ricinoleate via the conventional process (a) or the novel approach (b). *Source*: Copyright by Dr. Jean-Luc Dubois.

TABLE 3.1 Features of Three Batches of Castor Seeds Processed Under Different Methods.

Characteristics	Method	Batch 1	Batch 1B[(2)]	Batch 3
Volatile matter, wt%	NF ISO 665	5.2	5.6	5.2
Fat, % DM	NF ISO 659	52.2	54.0	52.2
Acidity of the fat, mg KOH/g	EN 14104	0.6	2.6	0.6
Ricin content, mg/kg	ELISA[(1)]	Not investigated	18 000[(3)]	Not investigated

[(1)]CEA ELISA method (see hereinafter, Section 3.3.5.1).
[(2)]Batch 1B corresponds to batch 1 after 12 months of storage.
[(3)]Measurement carried out on defatted seed flakes obtained according to the procedure described hereinafter in Section 3.3.

to evaluate the efficiency of the reactive seed crushing method on seed meals after a simple cold press step aimed to produce oil.

3.3.2 Seed Pretreatment: Preparation of Castor Seed Flakes

The castor seed (a kernel itself covered with a thin skin) is flattened as it is (whole seed) using a serrated-roller flattener, according to a double-flattening method

TABLE 3.2 Flattening of Castor Seeds.

Equipment	First Flattening		Second Flattening	
	Rotational Speed (rpm)	Interroller Space (mm)	Rotational Speed (rpm)	Interroller Space (mm)
Flattener with rollers Ø 240 mm, flute pitch: 1 striation/cm	430	0.1	430	0.1
Average flake thickness	0.4–0.6 mm		0.3–0.5 mm	
Length	3–4 mm		2–3 mm	

described in Table 3.2. This operation is carried out by passing through the flattener twice to obtain a castor seed flake. After flattening, the flakes are dried under a flow of hot air at 100 °C for 16 h to achieve a residual moisture content of less than 2 wt%.

The residual moisture of the seed is determined by thermogravimetric analysis. The seed is ground beforehand, and then the ground material obtained is dried at 105 °C in a thermobalance until stabilization of the weight. The water content is expressed as percentage of the crude material.

3.3.3 Defatted Flakes

The castor seed flakes were defatted by hexane extraction in a Soxhlet according to the standardized method NF ISO 659. The defatted flakes are subsequently de-solved under a stream of hot air in a ventilated oven at 60 °C for 10 h.

3.3.4 Reactive Seed Crushing Methods

The previously dried flakes are fed into a thermoregulated percolation column fitted with a support grid (fixed bed). In a typical experiment, the column temperature is regulated at 50 °C. A pump feeds the column with the methanol–sodium hydroxide reaction mixture or with anhydrous methanol. The feed is addressed to the top of the bed. The reaction liquid then percolates through the bed and is recovered in a flask located downstream, under the bed. The liquid is sent back to the top of the bed, by pumping, so as to again diffuse through the flake bed. The duration of the mixture recirculation cycle is 30 min and is considered to be equal to the duration of the reaction. This procedure has been designed to mimic a counter current extraction like a De Smet or a Crown continuous extractor. At the end of the cycle, the liquid feed is stopped. The liquid still present in the soaked flake is then recovered by draining (duration 15 min).

In a second step, the flake is extracted and washed. For this, the column is fed with anhydrous methanol which again diffuses by percolation without subsequent recirculation of the methanol. The required amount of solvent is injected for a period of 5 min, and the liquid is then drained for 15 min. 5 washing cycles are thus carried out. The defatted cake soaked with alcohol is then dried in a ventilated oven at 120 °C for 4 h.

The liquid phases obtained either after reaction or following the washing of the oil cake with anhydrous methanol, are combined in order to be purified. The methanol is first of all removed by vacuum distillation using a rotary evaporator (90 °C in a final vacuum at 40 mbar). The fatty phase recovered is then centrifuged (4500 rpm) for 5 min in order to separate the ester from the crude glycerine. The ester is finally washed with demineralized water at a temperature of 90 °C until the washing waters are neutral. Finally, the ester is dried at a temperature of 90 °C under a vacuum of 40 mbar.

In some cases, the mixture of liquid phases collected is subjected to an additional step of neutralizing esterification, which consists in:

- Adding an amount of strong acid, such as sulfuric acid, to the pool of liquid phases.
- Heating the resulting acidified liquid phase, and
- Purifying the ester phase as described above.

The other product directly resulting from the process is the castor oil cake. The defatted oil cake soaked with alcohol is dried in a ventilated oven for 4 h at a temperature of less than 200 °C. The aim of this drying step is to remove from the seed cake, the remaining solvent (methanol) used during the extraction. In parallel, this drying completes the decomposition of the ricin and of the CB-1A allergen remaining in the seed cake.

3.3.5 Ricin Content and Activity Measurements

Three different methods were used to determine the activity of the ricin in the seed cakes and also in the liquid phases recovered after the extraction/transesterification step.

3.3.5.1 Ricin Determination by ELISA (Immunoaffinity Assay) 250 mg of previously ground (average particle size: Ø = 1 mm) castor seed flake or seed cake are accurately weighed out. 5 mL of an extraction buffer composed of 100 mM of potassium phosphate (pH 7.4), 0.15 M of sodium chloride, 0.1% of BSA (bovine serum albumin) and 0.01% of sodium azide are added to these 250 mg. The solution obtained is then incubated for 1 h with agitation every 10 min. The supernatant is recovered by centrifugation for 15 min at 15,000 rpm. The supernatant is finally recovered for assaying of the lectins using an assay kit developed by the Commissariat à l'Energie Atomique (C.E.A.) as described by Becher et al. and allowing the quantitative determination of ricin.[10] This method has a detection limit of less than 0.1 ng/mL:

- Serial 10-fold divisions of the supernatants in a diluting buffer having a composition identical to the extraction buffer.

- First dilution 1/100, 6 dilutions per sample.
- Once this concentration is determined, a further assay is carried out with 3-fold dilutions and 8 dilutions per sample.

This method is named in the text continuation as "CEA method." The two other methods described below are also referred afterwards as USDA methods.

3.3.5.2 ***Cell-Based Assay*** HeLa cells were cultured in Dulbecco's modified Eagle medium supplemented with 10% fetal calf serum and were maintained in a humidified incubator (37 °C, 5% CO_2). For cytotoxicity assays, the cells were trypsinized, adjusted to approximately 0.5×10^5 to 1.0×10^5 cells per mL, seeded (100 μL/well) onto white 96-well plates (Corning), and allowed to adhere overnight. HeLa cells were then treated with pure ricin or castor meal extracts at the concentration indicated for 1 h at 4 °C. The cells were washed to remove non-internalized toxin and then incubated for 24 h at 37 °C. Cell viability was assessed using CellTiter-Glo® reagent (Promega) according to the manufacturer's instructions, except that the reagent was diluted 1:5 in PBS prior to use. Luminescence was measured with Victor 2 plate reader (PerkinElmer). All treatments were performed in triplicate, and 100% viability was defined as the average value obtained from wells in which cells were treated with medium only.

3.3.5.3 ***Cell-Free Translational Assay*** The translation lysate mixture consisting of nuclease-treated rabbit reticulocyte lysate, amino acids, RNasin, nuclease-free water and luciferase mRNA was prepared in a ratio (v/v) of 35:1:1:33:2. After mixing, the lysate was distributed (15 μL/tube) to 1.5 mL micro tubes containing 3 μL PBS buffer (as a control) or castor meal extracts containing 50 ng/mL protein. The CFT reaction was incubated at 30 °C for 90 min with gentle shaking (60 rpm). Aliquots of the reaction were then transferred to a black microtiter plate (5 μL/well). The Bright-Glo Luciferase Assay Buffer (100 μL) containing the Bright-Glo Luciferase Assay Substrate was added to each well and luminescence was measured as counts per second (cps) in a Victor 2 plate reader (Perkin Elmer, Shelton, CT). Translation lysate with PBS in lieu of sample was used as a control. Toxicity was presented as the percentage of translational inhibition [(cps from PBS control–cps from castor meal sample)/cps from PBS control × 100]. All data represent the mean of three replicates from a representative experiment ± standard deviation (SD). Three individual experiments were performed.

3.3.6 Allergenicity Test/Sensitizing Capacity

Moreover, the CB-1A allergen can be assayed by means of the precipitin test, described in the literature.[11] The skin sensitizing capacity of the oil cake can be advantageously measured by the qualitative local lymph node assay (LLNA) method, which method is validated and recommended by EC Directive No. 440/2008 (REACH regulation) in the context of the evaluation of chemical substance toxicity.

The sensitizing capacity of the defatted flakes and oil cakes of castor seeds was evaluated according to the *LLNA (Local Lymph Node Assay)* test adapted from the OECD Guidelines No. 429, of April 24, 2002 and Council Regulation (EC) No. 440/2008, B. 42, of May 3, 2008. A good correlation of the allergenic capacity was found between the thresholds determined experimentally in mice by the *LLNA* method and those determined in humans by the *Human Repeated Insult Patch Test* method (HRIPT).[12]

The LLNA method is based on the application of the test substance to the ear of a group of mice on days 0, 1 and 2 and the injection of titrated thymidine on day 5. Measurement of the incorporation of radioactivity in the draining lymph nodes makes it possible to measure the cell proliferation following exposure to the substance tested.

The female mice selected for the *LLNA* test are of the standardized species CBA/j. Their age on the first day of treatment is between 8 and 12 weeks and their weight between 18 and 25 g $\pm$ 20%. For each dose, the cohort consists of 3 mice developing in the same cage.

The reactant used with a view to causing proliferation is [^{3}H]methyl-thymidine (^{3}H-TdR; GE Healthcare, Les Ulis, France). At least 3 days before the injections, the required amount of ^{3}H-TdR is diluted with 0.9% NaCl (20 μCi of ^{3}H-TdR in 250 μL and 0.9% NaCl per animal). The solution obtained is stored at +4 °C in the dark.

A preliminary test is carried out on each animal by measuring, with a micrometer, the thickness of the ear before each application and 72 h after the final application. The vehicle most suitable for evaluating the sensitizing capacity of the defatted flakes and oil cakes of castor seeds is a homogeneous suspension of these materials at 10% and 25% in propylene glycol.

3.3.7 Evaluation of Acute Toxicity

The acute toxicity of the defatted flakes and oil cakes of castor seeds was evaluated according to the OECD Guidelines No. 423, of December 17, 2001, and Council Regulation (EC) No. 440/2008, B.1 tris of May 3, 2008. The female rats selected for the acute toxicity test are of the Sprague-Dawley standardized species. The vehicle used during the evaluation of the defatted flakes and oil cakes of castor seeds is a methylcellulose powder (0.5%). The product is administered from D0 to D8. The administered daily doses of defatted flakes and oil cakes of castor seeds are 50, 300 and 2000 mg/kg of body weight. For each dose, a group of 3 female rats was formed.

3.4 RESULTS

3.4.1 Seed Processing Step

The first step of the Reactive Seed Crushing Process consists in processing the whole castor seeds. It is not necessary in this process to dehull the seeds – meaning that it is not necessary to remove the seed skin. In a first operation, the seeds are flattened, followed by an operation in which the flakes are dried. The objective of the seed pretreatment is to make the oil as accessible as possible to the alcohol, without, however, causing too much modification to its mechanical strength. This prevents the formation of a paste and fines, respectively prejudicial to the implementation of a continuous process and to the final purification of the esters produced. Moreover, the processed seed should allow easy passage of the reaction fluid (alcohol–basic catalyst mixture) according to a simple percolation phenomenon.

The seeds processing comprises a seed preheating operation, carried out before the flattening operation. This preheating operation gives the seed greater plasticity and therefore more effective crushing during flattening (gain in terms of contact surface, of alcohol percolation rate and therefore of extractive capacity of the alcohol). It is preferably carried out at a temperature of less than or equal to 100 °C. Fresh seeds are flattened on a mechanical flattener with flat or serrated rollers. The serrated rollers proved to be more efficient with such large seeds as Castor beans. The seeds thus flattened are dried, for example in a ventilated oven, which is thermoregulated or industrially in a continuous belt or rotary hot-air dryer. The drying time and the temperature are chosen to obtain a decrease in the moisture of the seeds to values of less than 2 wt%. The amount of water remaining in the seeds has to be minimized to avoid the formation of free fatty acids (and soaps) during the transesterification.

3.4.2 Extraction and Transesterification Steps

The seeds pretreated as described above are brought into contact with a light anhydrous alcohol and an alkaline catalyst under temperature and time conditions sufficient to allow the extraction and the transesterification of the vegetable oil, and producing a mixture comprising fatty acid esters and glycerol, and a seed cake. Methanol was used in the present work for several reasons:

- First of all, one of the targeted products is a methyl ester of castor oil, and this should be enough by itself.
- The process developed is a loop process, in which excess alcohol will be recycled, and anhydrous methanol is easier to obtain than anhydrous ethanol, for example (There is an azeotrope between water and ethanol, that requires more complex process design to produce anhydrous ethanol).
- It is well known that FAME is more easily obtained than FAEE, due to the higher reactivity of methanol.

However, ethanol could have been selected for other advantages:

- A large alcohol loop is generated in the process, and a large heat consumption results from the evaporation/distillation of the alcohol. In that sense ethanol would be a better choice than methanol, reducing energy consumption.
- Methanol traces could remain in the meal, and it is expected that it could denaturize the proteins more than ethanol would do, and generate some issues if the meal is used in animal feed.

Certainly this latter reason would be valuable for many other crops where the meal already has a high commercial value such as for soybean and rapeseed, for which it would be important to preserve the nutritional value. But in case of Castor, the meal is not edible currently, and the best it can happen for it, is that it finds some additional value in feed formulation, once detoxified.

With regard to the process, the basic catalyst used in the method can be chosen from a wide group including sodium and potassium hydroxide, alcoholic sodium or potassium hydroxide, sodium or potassium methoxide. In order to minimize the cost, and also due to of past experience, sodium hydroxide was selected here.

3.4.3 Reactive Crushing of Castor Seeds with Methanol Extraction Carried Out in 3 Stages

The experiments are carried out in fixed bed reactors but aim to mimic continuous processes (method carried out in a fixed-bed reactor). 500 g of fresh castor seeds were processed on a Henry flattener with flat rollers having a fixed gap of 0.05 mm. The flattened seeds are in the form of flakes 0.2 mm thick and 2 mm in diameter approximately. The flattened seeds were dried at 60 °C for 16 h, reaching a final water content of 1.3 wt%. In a thermoregulated fixed-bed percolation column, these flattened and dried seeds were brought into contact with a mixture of sodium hydroxide and methanol, containing 0.5 wt% of sodium hydroxide relative to the seed and having an alcohol/seed weight ratio of 1.15. The extraction and transesterification reactions are carried out at a temperature of 50 °C for 30 min. The bed is then drained for 15 min. Extraction and washing of the seeds are then carried out with methanol in three stages and in a countercurrent direction. The liquid phase obtained is separated by decantation in order to recover a lighter phase rich in esters and a dense phase glycerol-rich. The ester yield is 77.2%. The oil cake obtained is subjected to drying in an oven at 120 °C for 4 h. The defatted oil cake is relatively well depleted, with a residual fat content of 5.4% (determined in accordance with standard NF ISO 659). The tests carried out in order to assay the ricin and the CB-1A allergen in the oil cake show that the oil cake is detoxified (see Tables 3.3, 3.12, 3.13 and 3.14).

3.4.4 Results of Ricin Dosage Assaying of Ricin according to an ELISA ("CEA Method")

A batch of defatted flakes of castor seed (batch 1B) as prepared in 3.3.3 and the oil cake resulting from the reactive seed crushing method of test 09-E08 (detailed conditions in Table 3.4), preheated in an oven at 120 °C for 4 h, were evaluated. The

TABLE 3.3 Evaluation of Defatted Castor Seeds and Oil Cake.

Product Tested	Ricin (%/Crude Raw Material)	Level of Reduction Compared with the Defatted Flake (%)
Defatted castor seed flakes	1.8	–
Oil cake test **09-E08**	8.6×10^{-4}	99.95

TABLE 3.4 Effect of the Catalyst Content on the Processing.

Criterion	TEST 09-E08	TEST 09-E11	TEST 09-E12
Weight of flakes of castor seed of batch 1B, g	350	350	350
Weight of methanol, g	700	700	700
Weight of sodium hydroxide used, g	1.1	0.7	0.35
Methanol/flake weight ratio	2	2	2
% Catalyst relative to flakes	0.30	0.20	0.10
Methyl ester yield, %	87.1	86.3	57.5
% Fat in the defatted oil cake, % methyl ester equivalents	2.6	3.5	3.7
Loss of esters (calculated value),[a] %	10.3	10.2	38.8
Glycerine yield, %	161	153	203

[a]Loss of esters = [theoretical weight of esters] – [weight of esters produced] – [potential weight of esters in the reduced-fat oil cake].

results obtained are given in Table 3.3. It is noted that the ricin initially present in the defatted flake at a high content (1.8%) is virtually no longer present in the oil cake resulting from the reactive seed crushing method (very low content $<10^{-3}$%), i.e., more than 2000 times lower than initially.

3.4.5 Influence of the Amount of Reactive Crushing Basic Catalyst

Tests with a varying content of catalyst used in the Reactive Crushing reaction (sodium hydroxide) were carried out. Results are summarized in Table 3.4. It is observed that, for a catalyst content of 0.3 wt% (tests 09-E08), a maximum yield of methyl esters (87.1%) is obtained, whereas, with 0.1 wt% of catalyst, the yield obtained is lower (57.5%). The glycerine mass yields are all greater than 100%, indicating that the glycerine contains impurities: water, salts, soaps, polar fat, and carbohydrates.

The purified esters obtained in the various tests were analyzed (Table 3.5).

For catalyst contents of 0.1 and 0.2 wt%, the esters produced have modest methyl ricinoleate contents (<75%) and, at the opposite, high contents of unwanted by-products such as monoglycerides, diglycerides and triglycerides. In the presence of 0.3 wt% of catalyst, the methyl ricinoleate content is high (>90%) and the glycerides are very correctly converted: 1.2% of monoglycerides and a total absence of diglycerides and triglycerides. In the presence of 0.3 wt% of catalyst, the esters obtained have the lowest residual acidity. Finally, the robustness of the method is to be noticed, since this method allows to directly transesterify, with an appreciable yield, oils of seeds having high acidities (>2 mg KOH/g) that are incompatible with the conventional methods.

3.4.6 Influence of the Amount of Alcohol

Tests were carried out with varying alcohol contents (Table 3.6). In these tests, the amount of flakes was reduced, rather than varying the amount of methanol, in order

TABLE 3.5 Effect of the Catalyst Content on the Quality of the Methyl Esters Recovered.

Criterion	TEST 09-E08	TEST 09-E011	TEST 09-E12
% Catalyst relative to flake	*0.3*	*0.2*	*0.1*
Acid number (mg KOH/g)	0.46	0.52	0.60
MeC16 content (%)	0.85	0.75	0.50
MeC18:2 content (%)	2.90	3.02	1.92
MeC18:1 content (%)	3.46	2.09	1.08
MeC18:0 content (%)	0.96	0.71	0.73
Methyl ricinoleate content (%)	90.65	73.16	49.80
Monoglyceride content (%)	1.19	16.14	12.22
Diglyceride content (%)	nd	0.80	4.28
Triglyceride content (%)	nd	0.38	28.77

nd: not detected.

TABLE 3.6 Effect of the Amount of Methanol Relative to the Amount of Seed Flakes.

Criterion	TEST 09-E25	TEST 09-E08
Weight of flakes of castor seed of batch 1B, g	280	350
Weight of methanol used in the reactive crushing step, g	700	700
Weight of sodium hydroxide used, g	0.84	1.1
Methanol/flake weight ratio	2.5	2.0
% Catalyst relative to flake	0.30	0.30
Weight of methanol used during the flake washing steps (in 5 rounds), g	1400	1750
Methyl ester yield, %	88.7	87.1
% Fat in the defatted oil cake, % methyl ester equivalents	4.3	2.6
Loss of esters (calculated value),%	7.0	10.3
Glycerine yield, %	141.3	161

to keep the liquid flow constant. The amount of alcohol used has no significant effect on the overall yield of methyl esters.

The results given in Table 3.7 hereinafter show that the final acidity of the esters is improved when more alcohol is used.

3.4.7 Influence of the Reaction Cycle Length

A test was carried out by doubling the contact time of the alcoholic and catalytic solution over the bed of seed flakes, i.e., by doubling the reaction time (Table 3.8).

TABLE 3.7 Effect of the Amount of Methanol on the Quality of the Methyl Esters.

Criterion	TEST 09-E25	TEST 09-E08
Methanol/flake weight ratio	*2.5*	*2.0*
Acid number (mg KOH/g)	0.23	0.46
MeC16 content (%)	1.01	0.85
MeC18:2 content (%)	3.98	2.90
MeC18:1 content (%)	3.14	3.46
MeC18:0 content (%)	0.85	0.96
Methyl ricinoleate content (%)	89.19	90.65
Monoglyceride content (%)	1.82	1.19
Diglyceride content (%)	0.00	nd
Triglyceride content (%)	0.00	nd

TABLE 3.8 Effect of the Reaction Time on the Processability.

Criterion	TEST 09-E26	TEST 09-E08
Weight of flakes of castor seed of batch 1B, g	350	350
Weight of methanol, g	700	700
Weight of sodium hydroxide used, g	1.1	1.1
Reaction cycle duration (contact time), min	**60**	**30**
Methanol/flake weight ratio	2.0	2.0
% Catalyst relative to flake	0.3	0.3
Methyl ester yield, %	90.4	87.1
% Fat in the reduced-fat oil cake, % methyl ester equivalents	4.5	2.6
Loss of esters (calculated value),%	5.1	10.3
Glycerine yield, %	137	161

Increasing the contact time increases the overall yield of methyl esters. Likewise, the glycerine yield decreases as the contact time increases, indicating a smaller presence of impurities in the glycerine.

From the qualitative point of view, the results in Table 3.9 show that increasing the contact time improves the residual acidity of the final esters but does not contribute to improving the monoglyceride content. The reaction is limited by the equilibrium between glycerol and esters on one side and methanol and oil on the other. Once the equilibrium is reached, increasing the contact time does not affect much the overall yield. Here, there is one more parameter, which is the cake, which retains some of the oil and which is not available for the reaction straight away, and which adsorb some of the catalysts (sodium hydroxide). In a process improvement, part of the glycerine produced could be removed during the reactive extraction process, thereby shifting further the equilibrium.

TABLE 3.9 Effect of the Reaction Time on the Product Quality.

Criterion	TEST 09-E25	TEST 09-E08
Reaction cycle duration (contact time), min	**60**	**30**
Acid number (mg KOH/g)	0.29	0.46
MeC16 content (%)	0.71	0.85
MeC18:2 content (%)	3.31	2.90
MeC18:1 content (%)	2.90	3.46
MeC18:0 content (%)	0.86	0.96
Methyl ricinoleate content (%)	90.34	90.65
Monoglyceride content (%)	1.82	1.19
Diglyceride content (%)	0.05	nd
Triglyceride content (%)	0.00	nd

3.4.8 Method for the Reactive Seed Crushing of Castor Seeds Including a Step of Neutralizing the Free Fatty Acids by Esterification

A variant of the method was tested by adding a step of neutralizing esterification of the liquid phases collected. This step comprises the following operations:

- The liquid phases of the method are collected and then combined.
- A "dry extract" measurement of these phases is carried out by evaporating off the methanol present in an aliquot of the liquid phases. The residual weight of product gives the "dry extract" content according to the following formula: *"dry extract" content = 100 × [residual weight after methanol evaporation]/weight of the liquid phases before evaporation.*
- Varying amounts of 2.5N sulfuric acid are added to the pool of liquid phases.
- The acidified liquid phase is then brought to 50 °C for 15 min.
- The conventional unitary steps of purifying the esters (methanol evaporation, glycerol separation, washing of esters, and drying) are then carried out.

A not very acidic seed (batch 3) was used in the context of these tests. In view of the results in Table 3.10, it appears that, in the presence of a not very acidic seed, and without the neutralizing esterification step, the ester yield is greater than 90%. Under the same conditions, the glycerine yield falls to a value of 124%, indicating that fewer soaps have formed as a result of the partial saponification of the catalyst. The values in italics display a considerable uncertainty (±20%) related to carrying out tests on liquid-phase aliquots.

From the qualitative point of view, the effects of the neutralizing esterification step, presented in Table 3.11 hereinafter, show that:

- The residual acidity of the final esters tends to slightly increase with the amount of sulfuric acid used during the neutralizing esterification step.

TABLE 3.10 Effect of the Amount of Acid Used during the Post-Esterification Step.

Criterion	TEST 09-E30-1	TEST 09-E30-2	TEST 09-E30-4	TEST 09-E30-3
Weight of flakes of castor seed of *batch 3*, g			350	
Weight of methanol, g			700	
Weight of sodium hydroxide used, g			1.1	
Reaction cycle duration (contact time), min			30	
Reaction time (°C)			50	
Methanol/flake weight ratio			2	
% Catalyst relative to flake			0.30	
Neutralizing esterification				
% H_2SO_4 versus dry extract content	0.0	0.1	0.3	0.4
2.5N H_2SO_4, mL	0.0	1.6	4.8	6.4
Methyl ester yield, %	96.0	*93 ± 20*	*70 ± 20*	*70 ± 20*
% Fat in the reduced-fat oil cake, % methyl ester equivalents	4.0	4.0	4.0	4.0
Loss of esters (calculated value), %	0	nd	nd	nd
Glycerin yield, %	124	*90 ± 20*	*90 ± 20*	*90 ± 20*

TABLE 3.11 Effect of the Post-Esterification Step on Product Quality.

Criterion	TEST 09-E30-1	TEST 09-E30-2	TEST 09-E30-4	TEST 09-E30-3
% H_2SO_4 versus dry extract content	*0.0*	*0.1*	*0.3*	*0.4*
Acid number (mg KOH/g)	0.14	0.17	0.49	0.63
MeC16 content (%)	0.97	0.79	0.84	0.86
MeC18:2 content (%)	3.76	3.37	3.52	3.55
MeC18:1 content (%)	3.76	3.37	3.52	3.55
MeC18:0 content (%)	1.31	1.30	1.20	1.36
Methyl ricinoleate content (%)	87.65	89.52	90.55	90.32
Monoglyceride content (%)	2.11	1.59	0.33	0.32
Diglyceride content (%)	0.44	0.04	0.03	0.03
Triglyceride content (%)	0.00	0.00	0.00	0.00

- On the other hand, the purity with respect to ricinoleate (target product) tends to increase with the amount of sulfuric acid used during the neutralizing esterification step. This result is concomitant with a decrease in unwanted monoglycerides.
- An optimum corresponding to the best compromise between, on one hand, the yield and the purity with respect to methyl ricinoleate and, on the other hand, the residual acidity of the esters, lies in the range of 0.1–0.3% of sulfuric acid used during the neutralizing esterification step.

3.4.9 Effect of the Final Drying Temperature on the Detoxification of Castor Meals

Results reported in Table 3.12 indicate that a reactive methanol extraction process (methanol plus NaOH) plays an important part on the detoxification of castor meal (Table 3.1). More than 90% of ricin activity was lost after processing with methanol and 0.3–0.5% of NaOH and drying at 100–120 °C for 4 h. However, only less than 50% ricin activity was lost if the final drying of the meal is done at 50 °C for 4 h using the same condition, suggesting the thermal instability of ricin in the presence of residual methanol and sodium hydroxide. Similar results were obtained from cell-based assay and CFT assay besides the experiment 10-E54-100, where less activity was detected by cell-based assay (33%) than by CFT assay (64%). The effects of hexane co-solvent process and the non-reactive methanol extraction process on the detoxification of castor meal were relatively smaller.

3.4.10 Evaluation of the Acute Toxicity and of the Sensitizing Capacity of the Castor Seed Flakes and Oil Cakes Produced during the Method

3.4.10.1 Evaluation of the sensitizing capacity A batch of defatted castor seed flakes as prepared according to section 3.3.3 and the oil cake resulting from the Reactive Crushing method of test 09-E08, preheated in an oven at 120 °C for 4 h, were evaluated. From the results given in Table 3.13, it results that the oil cake resulting from the Reactive Crushing method according to the invention no longer exhibits any sensitizing capacity, unlike the defatted flake which is highly sensitizing.

3.4.10.2 Evaluation of Acute Toxicity These results show that the administration of defatted castor seed flakes to rats causes 100% mortality at the dose of 2000 mg/kg, from the second day. On the other hand, the administration of the oil cake resulting from the reactive seed crushing method causes no mortality of the rats and no clinical sign at the dose of 2000 mg/kg for 8 days. Consequently, the oil cake is well detoxified at the end of the method. A weight gain is even noted at the highest dosage.

3.5 DISCUSSION

The main results are gathered in Table 3.12. The main conclusions of the new reactive seed crushing process from the study are:

1. *Impact of process conditions in reactive seed crushing*: When samples 1 (defatted seed from lot 1) and 6 (reactive seed crushing with methanol), are compared, it can be seen that the ricin reduction due to reactive seed crushing is around 50 times in the CEA measurement and 20 in the USDA. We know that ricin is not thermally decomposed at a temperature of 50 °C (sample 6), which corresponds to the defatted seed. Nevertheless, we also observe that a seed meal

TABLE 3.12 Ricin Quantification and Toxicity Measurement by CEA and USDA. (2 Independent Labs).

Samples Reference	N°	Lot Number of Castor Seeds	Process Type	NaOH (% vs flakes)	Drying Temp(°C)/ Duration (h)	CEA	USDA	
						Immunoaffinity – Ricin (%)	Tox CFT Assay (%)[a]	Tox Cell based (%)[a]
Defatted seed 09-TD-20[1]	1	1	–	–	20/48	1.8	100	97
Defatted seed 09-TD2-20	2	2	–	–	20/48	21[b]	100	99
08E12-120[2]	3	1	Batch reactive with methanol	0.5	120/4	0.01	9	9
09E17-120[2]	4	1	Reactive methanol	0.3	120/4[c]	2.2×10^{-4}	3	5
09E17-100	5	1	Reactive methanol	0.3	100/4	8.6×10^{-4}	5	5
09E29-50	6	1	Reactive methanol	0.3	50/4	0.034	52	50
09E30-100	7	2	Reactive methanol	0.3	120/4	5.4×10^{-4}	8	6
09-CMA1-TD Indian cake after solvent extraction	8	–	–	0.0	Unknown	0.76	81	72
09-E09-100	9	1	Reactive 50Meth/50Hex	0.3	100/4	0.06	39	47
10-E53-100	10	2	Non reactive 50Meth/50Hex	0.0	100/4	1.86	94	99
10-E54-100	11	2	Non reactive methanol	0.0	100/4	0.14	33	64
11-CMA2-TD20 Madagascar cake	12	–	–	0.0	20/48	0.19	91	94
11-E01-100 Madagascar cake	13	–	Reactive methanol	0.3	100/4	0.026	26	25
11-E03-100 Madagascar cake	14	–	Non reactive methanol	0.0	100/4	0.07	46	32

[1]Oral toxicity (300 < DL50 < 2000 mg/kg) and skin sensitizing (at 10%).

[2]Oral toxicity (DL50 > 2000 mg/kg) and not skin sensitizing (at 10%).

[a]Cell-based assay using 500 ng/mL of total protein – data represents the mean of 3 replicates.

[b]CEA considers that the value is over estimated due to a molecule that would interfere with the method (false positive).

[c]+Final drying at 100 °C, 4 h.

Tests 09E17, 09E29, and 09E30 have been made in the same conditions than the reference test 09E08, which also give the best ester quality.

TABLE 3.13 Allergenicity Tests.

Product Tested	Main Test	
	Maximum Test Dose (%)	Main Observations
Defatted castor seed flakes	10[a]	Lymphocyte hyperproliferation attributed to contact hypersensitivity Conclusion: product is sensitizing
Oil cake, test 09-E08	10	No lymphocyte hyperproliferation was noted Conclusion: product is not sensitizing

[a]Dose defined during the preliminary test following the death of 2 animals out of 3 after application of the product formulated at 10% and 25%. The surviving third animal did not, on the other hand, trigger any ear irritation at the dose of 10% (dose retained for the main test).

TABLE 3.14 Evaluation of Acute Toxicity.

Product	Daily Dose (mg/kg)	Main Observations	LD50
Defatted castor seed flake	50	No mortality on day 8 (D8) No clinical sign No histological abnormality No weight gain	LD50 > 50
	300	No morality on D8 No clinical sign No histological abnormality Slight weight gain	LD50 > 300
	2000	100% mortality on day 2 (D2) (piloerection preliminary to death)	300 < LD50 < 2000
Oil cake Test 09-E08	50	No clinical sign from D2 to D8	LD50 > 50
	300	No clinical sign from D2 to D8	LD50 > 300
	2000	No clinical sign from D2 to D8 Slight weight gain	LD50 > 2000

produced by the conventional process – sample 8, seed meal obtained from Indian cake produced after solvent extraction – is richer in ricin (CEA test) and more toxic (USDA test) than a seed meal obtained after reactive seed crushing. This observation shows that even at these low temperature methanol and/or sodium hydroxide are efficient to reduce the ricin content.

2. *Impact of final drying temperature of seed meal*: When samples 6 and 7, respectively dried at 50 and 120 °C, are compared it can be seen that the residual toxicity is around 6–8% at 120 °C, much lower than that at 50 °C, which is around 50–52% in the cell-free and cell-based tests performed at the USDA laboratory. When the ratio of concentration measured by the immunoreactive test of CEA is around 60. This result confirms that ricin is sensitive

to temperature (at above 100 °C). The result also illustrates the detoxifying effect of NaOH and/or methanol (used in reactive seed crushing) at higher temperature.

3. *Impact of hexane in a co-solvent reactive seed crushing process (samples 5 vs 9)*: The USDA test shows a difference of toxicity that is not favorable to the use of hexane (40–50% toxicity vs 5% for the reactive seed crushing with methanol only). Methanol in the process has a denaturing effect on proteins, and here less methanol is engaged in the co-solvent process – mixture of methanol and hexane – but also another hypothesis is that the hexane limits the access of methanol and sodium hydroxide to the proteins. NaOH effect might also be limited in the presence of hexane.
4. *Impact of non reactive seed crushing process with methanol (samples 11 vs 7 and 5)*: The effect of NaOH seems important for the final toxicity of the seed meal, with a ratio of 5–10 in favor of the reactive process. Methanol alone (without sodium hydroxide) has a positive effect to reduce the ricin content compared to hexane, but the effect is enhanced in the presence of the catalyst. The denaturing effect of methanol on proteins – and here the effect might not be limited to ricin – versus the non-reactive seed crushing in the presence of a hexane–methanol process (samples 11 vs 10, with a difference of toxicity in a ratio of 1.5:2).
5. *Impact of the reactive and non reactive crushing processes on a fat seed meal (samples 12–14)*: On a meal that was obtained after pressing in Madagascar (here there is an issue with the quality of the sample which had to travel to France and probably had time to hydrolyze since the Free Fatty Acid content was fairly high) the detoxifying effect is significantly reduced compared to a full seed (toxicity of 25% vs 5%). For a non-reactive process using methanol only on a meal, the decrease of toxicity is 46% to 36% versus 33% to 64% when done on the seed. Since the meal was containing a large amount of free fatty acids, it is quite probable that a large amount of NaOH was consumed by the free fatty acids (formation of soaps) and was not available for detoxification – in line with the lower transesterification observed. However, even in this case a reduction of the ricin content is observed compared to a conventional Indian castor cake.

Both ricin quantification methods (so-called CEA and USDA methodologies) confirmed the significant reduction in ricin content in a seed meal processed through the reactive seed crushing process. Ricin is made of two parts (A and B) which are linked by a S—S bond. Methanol and/or sodium hydroxide contributes through a chemical reaction, probably S—S bond splitting to the reduction of the toxicity. The combination of a basic catalyst and methanol, in the presence of the S—S bond, probably leads to a reaction at the sulfur bond breaking the ricin molecule in two parts and leading to its inactivation. The reactive seed crushing process can be used either in a dedicated unit that would only produce castor oil methyl ester, or in a combined unit that would produce oil through cold or hot crushing, and methyl ester by reactive processing on the fat seed meal. Several other combinations can also be investigated, in order to keep flexibility to the plant which would then serve different markets with oil and esters.

Castor meal is currently used as a natural fertilizer in the producing countries. Even if it cannot be used as animal feed, its current value in India is about one fifth to one third of the soybean meal value. It is also sold in European countries as a natural fertilizer for organic farming. However it still contains the toxic compounds. Most of the castor production is still done by manual harvesting, since the racemes are not all mature at the same time, and this suits countries with cheap labour. Using a non-edible oilseed meal as a cheap fertilizer makes sense for the countries where castor is grown and where there are no need for animal feed. As for many non-edible crops like Jatropha, there is an image in the public that this is a crop appropriate for "marginal land" where nothing else would grow; survive would be a more appropriate word. In marginal lands leading usually to marginal yields, fertilization is a must to reach industrial production. If the meal would not be used, chemical fertilizers would be necessary, and it is not sure that they could access the remote locations where the crops are being grown. So the use of castor meal should not be banned as fertilizer. However, the meal still contains compounds that are toxic for humans and animals (Table 3.15).

The present paper relates to a method for the reactive seed crushing of castor seeds which, starting from specifically processed castor seeds in the presence of a light alcohol and a basic catalyst, makes it possible to carry out the crushing and the reaction for transesterification of the triglycerides present in the castor oil in a single step, simultaneously producing a defatted cake, glycerol and esters of fatty acids, in particular ricinoleic acid. Said esters are intended mainly for the production of 11-aminoundecanoic acid, the monomer of Rilsan® PA11, which is a polyamide with exceptional physical properties, developed by Arkema, but could also be used to

TABLE 3.15 Seed Meal Value for Export from India, US $/t.

Year	Castor Meal (Bulk) Ex Kandla (US $/t)
2003	30
2004	
2005	67
2006	55
2007	81
2008	74
2009	75
2010	85
2011	114
2012	93
2013	97

Source: The Solvent Extractors' Association of India.

produce biodiesel or aviation fuel. In addition, with the reactive crushing method it becomes possible to obtain a completely detoxified and deallergenized cake in which both the toxin and the allergen have been inactivated, reducing the risks to the health of the individuals who handle them.

The reactive seed crushing method allows going directly from the seed to the fatty acid esters, in particular ricinoleic acid methyl ester, while avoiding the steps of crushing, refining and purification and the production of by-products: meaning reduction of CAPEX. These esters are also suitable for the production of biofuels, and the method can be applied to several other toxic crops such as Jatropha. However, Jatropha needs a slightly different procedure since the oil composition, but also the toxic compounds, and the seed texture are not the same than for castor.[13]

It is well-known to prepare fatty acid methyl esters (FAME) from oilseeds in two steps, namely a step of oil extraction, with or without solvent extraction, and a step of transesterification of this oil in the presence of alcohol and of catalyst, producing an ester phase and a glycerine phase. The concept of reactive seed crushing has been described in the patent literature[14–16] aiming to develop a new technology to produce biodiesel that would avoid the costly, and sometime monopoly, seed crushing and oil refining operations. However, in these earlier works conventional edible crops were addressed, meaning rapeseed, soybean, sunflower, etc. In addition, ethanol was often the alcohol of choice. There might be several reasons for the choice of ethanol, which is a more expensive alcohol than methanol, but the main one was to produce a 100% renewable fuel.

The use of alcohols as solvents for the extraction step, has been widely investigated, since the industry is pushed by new regulations to substitute the hexane that it has been used for a long time. When the goal is to produce an ester, an alcohol as a solvent might be a good choice since an alcohol/vegetable oil mixture would be directly produced. For castor oil, this combination has also been described in the patent literature.[17] According to the process described, the oil is extracted with an anhydrous alcohol and the miscella produced is subjected to a transesterification step.

Castor oil comprises glyceryl triricinoleate as the main component. Ricinoleic acid is a hydroxylated fatty acid. No other known natural oil contains such a high proportion of hydroxylated fatty acids. It is this characteristic on the glycerides composition which distinguishes castor oil from any other vegetable oils and fats and it is this composition which gives it its notable physical and chemical properties. Castor oil, which is a non-drying oil, has the highest viscosity number and density of all the natural oils. These properties are due in particular to the hydrogen bonds formed between the hydroxyl groups. Moreover, methanol is much more soluble in castor oil, but unfortunately so is glycerol. It is these properties which give castor oil a very particular behavior in chemical reactions and which mean that the methods for transesterification of "ordinary" vegetable oils that are known cannot quite simply be transposed to castor oil (N.B., the term "ordinary oil" is intended to mean non-hydroxylated oils extracted from oleaginous plants, such as sunflower, rape, soy, peanut, olive tree, sesame, safflower, coconut, and palm). There are several reports on processes for the transesterification of castor oil, even using heterogeneous catalysis,[18] although a low temperature process is preferable because of the reactivity of the hydroxyl group on ricinoleic acid.

An alternative process, for obtaining FAME for biodiesel application has been described mainly in the patent literature. A method, using a batch process (stirred bed reaction), has been described using castor seeds.[19] This method comprises, a step of prior processing of the seeds, during which the hulls are removed from the seeds and the seeds are then cleaned using a vibrating sieve, and the clean seeds are dried so as to reduce their moisture content to less than 0.5% by weight. The dried seeds are fed into a reactor with anhydrous alcohol; this heterogeneous mixture is converted into a homogeneous suspension by means of an agitator. It is only at this moment that a basic catalyst is fed into the reactor. This reaction mixture is then heated for 30–90 min at a temperature of 30–78 °C, resulting in transesterification of the triglycerides to give esters, with a high conversion rate of between 98% and 100%. However, this method has the following drawbacks: (1) additional production cost linked to the use of dehulling equipment, (2) generation of solid by-products with very low added value (hulls), (3) poor mechanical strength of the oil cake, (4) formation of fines responsible for clogging of filters, and (5) in the case of castor seeds, toxic and allergenic risk due to the tendency of the dry fines to be dissipated in the ambient air. In addition, all our attempts to reproduce this process have led to failures.

The castor seed contains, in addition to a powerful toxin (ricin), toxic alkaloids such as ricinine[20] and ricinidine, and also a group of highly allergenic compounds (CB-1A) which, active at concentrations of about 1 ppm (whereas its concentration by weight is 6.1–9% in the de-oiled hulled seed and 0.09–4.2% in de-oiled commercial oil cakes[21,22]), can cause severe skin reactions and serious respiratory system irritations in humans. Ricin is destroyed at high temperature and can therefore be inactivated by heating the cake.[23] Ricinine and ricinidine, for their part, are present only in a very small amount and do not pose toxicity problems since castor oil cake would be only involved in limited amounts in the formulation of cattle feeds. Once properly detoxified, castor meal can be used for livestock.

An important factor affecting the ricin content is linked to cultivation practice. The ricin content depends on the genotype, but also seems to increase with the maturity of the crop. The ricin content in the seeds collected from the main stem, was lower than the ricin content from the first branches (second pick) in manual harvesting.[24] These observations are very important since it would mean that the seeds specifically developed for mechanical harvesting, where all the racemes have to be mature at the same time, might lead to lower ricin content. Another strategy could also be to increase population density, in order to force the crops to have a single raceme, and therefore decrease the average ricin content.

Consequently, the only critical point in the exploitation of the castor oil cake is the CB-1A allergen, which is capable of withstanding heat denaturation. Spies et al. have shown the allergen to be a protein–polysaccharide of relatively low molecular weight.[25] In this respect, many studies have been carried out in order to develop a method for denaturing this compound; most result in treatments that are effective, but the other proteins present in the oil cake are also degraded and the oil cake then loses its nutritive value.

Several processes have already been proposed to deallergenize and detoxify castor oil cakes.[26] For example, by means of a treatment based on a mixture of water and $Ca(OH)_2$ so as to obtain a pH of 9.5–12.5 in the oil cake, and then to heat the oil cake for 1 h at 100–120 °C. In order for this treatment to be effective, the $Ca(OH)_2$ treatment must be at least 8% by weight relative to the weight of the oil cake for a heating temperature at minimum of 100 °C and at least 2% for a temperature of 120 °C.[27] In addition to the fact that this method requires a long and expensive additional step, the addition of calcium contributes to stabilizing phytates, which are antinutritional compounds of the oil cakes, according to a complexation process involving calcium and phytic acid and resulting in the formation of insoluble stable salts which cannot be digested by the animal.[28,29]

It is therefore desirable to have a method for treating castor seeds which makes it possible to simply, rapidly and inexpensively inactivate, in addition to the ricin, the powerful allergen CB-1A, which will thus enable risk-free handling by humans and the use of castor oil cake in animal feed. This is particularly important for the economy of countries which are large producers of castor oil (India, China, Brazil), since, while castor oil has many industrial uses, castor oil cakes have not yet found a use on the industrial scale, in particular owing to the allergy problems mentioned above. Current methods of treatment of the allergen, lead to a destruction of some of the essential amino acids.[30,31] Most probably the chemical reactions involved destroying both ricin and the allergen, involving high pH and high temperature, lead not only to a destruction of the S—S bond of the ricin, but also affect all the sulfur containing amino acids such as methionine. This would be a very important issue for an edible seed cake, which would lose some commercial value in such a treatment, but in the case of a non-edible cake, the loss of some of the amino acids has to be balanced by the detoxification.

3.6 CONCLUSION

The experimental results of the case study have shown that the reactive seed crushing method can be applied to castor seeds having a degree of acidity of less than 2 mg KOH/g, the method comprising the following steps:

- A seed pretreatment step in which, the seeds are flattened and dried.
- A reactive extraction in which a light anhydrous alcohol and an alkaline catalyst are contacted with the flakes under temperature and time conditions sufficient to allow the simultaneous extraction and transesterification of the vegetable oil, producing a mixture of fatty acid esters and glycerol, and an oil cake.

According to the conditions used, the method can directly produce a detoxified oil cake. The process developed can be implemented continuously on the industrial scale, for example by means of a continuously operating, moving belt reactor–extractor (of

De Smet or Crown extractor type); of a rotary filter; or of a centrifuge. The reactive seed crushing is carried out with methanol in a counter-current with respect to the oil cake, on several consecutive stages.

The method for the reactive seed crushing has many advantages:

- The pre-treatment of the seeds increases the contact surface for better percolation of the alcohol–catalyst mixture and therefore better extraction of the lipids and their subsequent conversion to esters.
- No prior impregnation of the processed seeds is necessary.
- The ester fraction obtained from the mixture comprising fatty acid esters and glycerol is particularly suitable for the production of ll-aminoundecanoic acid, a monomer used in the synthesis of polyamide-11.

Starting from whole seeds:

- Reduces the formation of fines, making the subsequent filtration steps easier, and limiting the allergenic and toxic risks since the dry fines have a tendency to dissipate/disperse in the ambient air.
- Maintains a good mechanical strength of the bed of flattened seeds (that will form the oil cake), this being a very advantageous property if it is desired to carry out the reaction in a continuous mode.

Reactive seed crushing provides a method for treating castor seeds which limits the number of seed treatment steps and the handling of the oil cake, with a view to a continuous industrial application aimed at producing ricinoleic acid methyl esters, and which allows to destroy "in planta" the toxin and the allergen present in the castor seeds. The other advantage of the method compared with the conventional methods lies in the small amounts of water used. The operations for refining the crude oil, for example, use very large amounts of water. This water saving is a major asset in the context of the development of this technology in developing countries and, to a lesser extent, in rich countries since water is becoming an increasingly expensive commodity.

The process described herein works well with castor seeds, and should be very efficient with many other crops rich in hydroxy fatty acids. In order to understand the limitations of the process, some tests were carried out using physical mixtures of castor seeds and other seeds like soybean and safflower. These tests were designed to simulate what could happen if Genetically Modified Crops, producing some ricinoleic acid, were to be processed. Some other tests (not reported here) were also done using Lesquerella seed and seeds containing no hydroxyfatty acids such as Jatropha.

ACKNOWLEDGMENTS

Acknowledgments are addressed to ADEME, the French Environment and Energy Management Agency for financial support of the RICHARB project which contributed to these results. The author also wishes to acknowledge the contribution of

Dr. Xiaohua He (USDA) for ricin measurements which have been discussed in more details elsewhere[32] and to Dr. A. Piccirilli and J. Magne (Valagro) who carried out the process development in collaboration with Arkema.

REFERENCES

[1] M.L. Wang, J.B. Morris, B. Tonnis, D. Pinnow, J. Davis, P. Raymer, G.A. Pederson, *J. Agric. Food Chem.* **2011**, *59*, 9250–9256.

[2] S.M. Bradberry, K.J. Dickers, P. Rice, G.D. Griffiths, J.A. Vale, *Toxicol. Rev.* **2003**, *22*, 65–70.

[3] Anonymous – The International Castor Oil Association, **1989**. The processing of Castor Meal for Detoxification and Deallergenation", Technical Bulletin number 1.

[4] J. Audi, M. Belson, M. Patel, J. Schier, J. Osterloch, *J. Am. Med. Assoc.* **2005**, *294*, 2342–2351.

[5] D.J. Barnes, S.B. Brian, A.B. Dwaine, *J. Ind. Crop Prod.* **2009**, *29*, 509–515.

[6] A.S. De Oliverira, J.M.S. Campos, M.R.C. Oliveira, A.F. Brito, S.C.V. Fihlo, *Anim. Feed Sci. Technol.* **2010**, *158*, 15–28.

[7] S. Roels, V. Coopman, P. Vanhaelen, J. Cordonnier, *J. Vet. Diagn. Invest.* **2010**, *22*, 466–468.

[8] I.H. Hong, T.E. Kwon, S.K. Lee, J.K. Park, M.R. Ki, S.I. Park, K.S. Jeong, *Exp. Toxicol. Pathol.* **2011**, *63*, 113–117.

[9] Düngemittelverordnung -DüMv-, Attachment 2, Nr. 7.1.5. Available online: http://www.gesetzeim-internet.de/bundesrecht/d_mv_2008/gesamt.pdf (accessed on 04.08.2011).

[10] F. Becher, E. Duriez, H. Volland, J.C. Tabet, E. Ezan, *Anal. Chem.* **2007**, *79*, 659–665.

[11] J.R. Spies, H.S. Bernton, H. Stevens, *Agric. Food Chem.* **1962**, *10*, 140–145.

[12] J.P. Lepoittevin, Revue Française d'Allergologie et d'Immunologie Clinique, *French Rev. Allerg. Clin. Immunol.* **2008**, *48*, 120–122.

[13] J.-L. Dubois, J. Magne, A. Piccirilli, WO2011/092430, (**2011**).

[14] C. Lacaze, Z. Mouloungui, J. Leyris, L. Rigal, A. Gaset, F. Silvestre, FR2747128, (**1997**).

[15] L.C. Hoang, EP1119600B1, (**2001**).

[16] M.J. Haas, WO03085070 (**2003**).

[17] S. Bhatnagar, IN153357 (**2010**).

[18] G. Hillion, J.J. Rouxel, FR2794768 (**2000**).

[19] C.N. Khalil, L.C. Ferreira Leite, US7112229; WO2005/014765 (**2005**).

[20] J.W. Hinkson, C.A. Elliger, G. Fuller, *J. Am. Oil Chem. Soc.* **1972**, *49*, 196–199.

[21] S. Anandan, G.K. Anil Kumar, J. Ghosh, K.S. Ramachandra, *Anim. Feed Sci. Technol.* **2005**, 120, 159–168

[22] B. Kim, *Food. Biotechnol.* **1995**, *4*, 93–97.

[23] Anonymous, Ricin (from *Ricinus communis*) an undesirable substances in animal feed-Scientific Opinion of the Panel on Contaminants in the Food Chain (Question N°EFSA-Q-2003-062, Adopted on 10 June 2008), *EFSA J.* **2008**, *726*, 1–38.

[24] C.V.S. Srinivas, G. Nagaraj, *J. Oil Technol. Assoc. Ind.* **2000**, *32*, 21–23.

[25] J.R. Spies, J.R. Coulson, H. Stevens, *J. Am. Chem. Soc.* **1944**, *66*, 1798–1799.

[26] K.H. Rao, *J. Food Sci. Technol.* **1970**, *7*, 77–82.

[27] E.J. Coulson, J.R. Spies, P. Wells, US3101 266 (**1963**).

[28] A. Mikic, V. Perić, V. Đorđević, M. Srebrić, V. Mihailović, *Biotechnol. Anim. Husb.* **2009**, *25*, 1181–1188.

[29] A.S. Ekop, I.B. Obot, E.N. Ikpatt, *E-J. Chem.* **2008**, *5*, 736–741.

[30] A.C. Mottola, B. Mackey, H.G. Walker, G.O. Kohler, *J. Am. Oil Chem. Soc.* **1972**, *49*, 662–664.

[31] B.K. Kim, *Food Sci. Biotechnol.* **2006**, *15*, 441–446.

[32] J.-L. Dubois, A. Piccirilli, J. Magne, X. He, *Ind. Crop Prod.* **2013**, *43*, 194–199.

4

AVANTIUM CHEMICALS: THE HIGH POTENTIAL FOR THE LEVULINIC PRODUCT TREE

JAN C. VAN DER WAAL AND ED DE JONG
Avantium Chemicals, Amsterdam, The Netherlands

4.1 INTRODUCTION

Levulinic acid (LA in short) is readily available from sugars and known to mankind long before the onset of the petrochemical revolution. The earliest account of levulinic acid dates from 1875 by Freiherr, Grote, and Tollens,[1] who used both sulfuric acid

Industrial Biorenewables: A Practical Viewpoint, First Edition. Edited by Pablo Domínguez de María.

and hydrochloric acid to dehydrate a wide variety of sugars such as glucose, fructose, sucrose, and starches. In the middle of 20th century a lot of research has been done on synthesis and on properties of levulinic acid and products derived thereof.[2] Recently LA was identified by Werpy[3] and revised by Bozell[4] as one of the high-potential renewable biomolecules.

When starting from C6 sugars, the formation of levulinic acid is thought to proceed via hydroxymethylfurfural (HMF) as the intermediate. Recently van Putten[5] reviewed the HMF chemistry extensively and concluded from the vast literature examined that HMF formation is difficult for aldoses, that is, glucose, mannose, and galactose, and proceeds readily from ketoses, that is, fructose. The low yields from glucose-based sugars and high prices for fructose-based feedstocks, combined with the rise of cheap petro-derived alternatives, severely hindered commercialization of LA after the Second World War.[2] In recent years, a new route has emerged starting from C5 sugars (i.e., xylose, arabinose) with furfuryl alcohol as the intermediate.[6] Especially in the presence of alcohols the isomerization of furfuryl alcohol has been shown to proceed smoothly with the corresponding levulinate esters (alkyl levulinates) as the products.[7] Both routes will be discussed in more detail later.

Since the rise of the petrochemical industries severely hindered commercialization of LA in the past and availability of lignocellulosic C5 and C6 sugar sources – either as agricultural waste stream, that is, straw, bagasse, rice husks, etc., or from direct lignocellulosic production, that is, forestry due to a leveling off of paper demand – has become more readily and cheaply available, it is not surprising that the attention to LA as renewable feedstock chemical has reemerged. Yet LA is not an end product in itself, and further conversion is still required. This was first highlighted by Werpy[3] and later revised by Bozell[4] and has since led to a substantial rise in companies that either offer LA or build on levulinates as chemical building blocks to create a wide array of products. In this chapter a full overview will be given into nowadays levulinic synthesis routes, its market players, and its versatility in many chemical outlets. All of this shows the high potential of levulinates in future applications.

4.1.1 Properties

The physical properties of levulinic acid and its lower alcohol ester derivatives have been collected and reported in Table 4.1. It is most likely that these will be the molecules that will serve as major building blocks in further conversions to useful chemicals.

4.1.2 The Levulinate Market

The acid-catalyzed dehydration of C6 sugars is long known to yield Levulinic acid, and the potential of the molecule has long been recognized, however the LA market seems to be relatively small. Ghorpade and Hanna[2] give reasons for the lack of commercialization such as the relatively high cost of raw materials combined with a low overall yields in combination with technology challenges in separation and purification around the Second World War. Remarkable though that a similar process for the

TABLE 4.1 Collected Physical Properties of Levulinic Acid and Methyl and Ethyl Levulinate.

	Levulinic Acid	Methyl Levulinate	Ethyl Levulinate
Trivial names	4-Oxovaleric acid, 4-oxopentanoic acid, β-acetylpropionic acid	Methyl 4-oxovalerate, methyl 4-oxopentanoate	Ethyl 4-oxovalerate, ethyl 4-oxopentanoate, ethyl ketovalerate, ethyl levulinate
Chemical formula	$C_5H_8O_3$	$C_6H_{10}O_3$	$C_7H_{12}O_3$
CAS#	123-76-2	624-45-3	539-88-8
Classification	GRAS	GRAS	Food-grade flavor ingredient, mildly irritating but classified as GRAS
Molecular mass (g/mol)	116.11	130.14	144.17
Appearance	White solid	Clear colorless liquid with mild caramel-like aroma	Clear colorless to pale yellow liquid with apple odor
Relative density (g/cm^3)	1.340 at 20 °C	1.051 at 20 °C	1.016 at 25 °C
Melting point (°C)	30–34	−24	
Boiling point (°C)	245–246	193–195	206.2[8]
Flash point (°C)	137[9]	72[10]	91[8]
Freely soluble in	Water, ethanol, diethyl ether, chloroform	Soluble in water and miscible with most organic solvents	Water, alcohol
p*K*a	4.78[9]		
Heat of formation ΔH_f (kJ/mol)	−2417 [9]	−2369.6[a]	−2408.2[a]
Heat of vaporization $\Delta_{vap}H_m$ (kJ/mol)	74.4[11]	50.4 and 51.1[11]	51.6 and 58.3[11]

[a]Assuming ΔH_f for water (l), methanol (l), and ethanol (l) of −285.8, −238.4, and −277.0 kJ/mol, respectively.

production of furfural, that is, the acid-catalyzed dehydration of C5 sugars to furfural with similar low yields and downstream purification issues, is a large commercial process with current production of around 300 kt/year.[12]

In the last decade the production volume of LA has been estimated by several authors. Moens[13] estimated in 2002 that around 1 kt/year is produced by DSM but solely through a petrochemical-based route with a typical price for the LA around 5 US$/lb. Currently, most of the levulinic acid is produced in China. Companies active in this area include Hebei Langfang Triple Well Chemicals Co., Ltd.; Hebei Shijiazhuang Worldwide Furfural & Furfuryl Alcohol Furan Resin Co., Ltd.; Jiangsu Yancheng China Flavor Chemicals Co., Ltd.; Shijiazhuang Pharmaceutical Group

Ouyi Pharmaceutical Co., Ltd.; and Shanghai Apple Flavor & Fragrance Co., Ltd. These producers typically produce chemicals in batch-operated plants. No information on the actual process employed was found. However, since furfural is cheaply and readily available and even produced by some of the companies themselves, we expect that processes using furfuryl alcohol are the main production routes used. The development of bio-based routes via the BioFine process is mentioned as a potential way to increase volumes to 180 kt/year. In a recent review of the BioFine process, Hayes[14] estimates the Levulinic acid market at 0.5 kt/year at a price of 5 US$/kg in 2008. Biomass-derived levulinic acid has been produced in a pilot plant located in Glens Falls, New York, operating at 1 t/day dry mass of feedstock per day using the BioFine process. But Hayes mentions that this pilot plant was only operational for several test run periods since 1996. Hayes[14] also reports the start-up of a larger pilot plant in Caserta, Italy, with a processing capability of 50 dry tons of feedstock per day with levulinic acid and ethyl levulinate as the main products. To the best knowledge of the authors this plant became operational in 2005 and operated until 2014. Recently the plant at Casserta, Italy has been aqcuired by GFBiochemicals.[17] In a recent market study by Grand View Research,[15] the LA production in 2012 is estimated at 2.5 kt at a price of 5 US$/kg. Although it is hard to have clear numbers on the total market size, the consensus seems to be that only minor amounts of levulinic acid are produced via petrochemical routes with current market price around 5 US$/kg.

Recently four parties have announced plans for levulinic acid and levulinate production plants. Segetis announced that it has successfully started its pilot plant facility in Golden Valley, Minnesota, demonstrating the viability of its proprietary process to convert biomass to levulinic acid. The pilot plant has a nameplate capacity of 80 metric ton/year and currently uses corn sugar as feedstock although the company said its process has broad capability and is intended to be used to explore many other feedstocks as Segetis' technology develops.[16] Segetis[18] has additionally announced plant plans in Laskin Energy Park in Hoyt Lakes, Minnesota, for the production of 10 kt/year. However, at this moment in time it is unclear if they can secure sufficient money to continue with operation and proceed with plans announced. Most of the produced levulinic acid will be for captive use since Segetis will convert most of the levulinics produced to new classes of chemicals (see succeeding text for more details). Therefore no price target has been given yet. Segetis has in the past stated that levulinics should become available at 0.5 US$/kg. GFBiochemicals has announced plans to expand there production capacity further to 50 kt/year.[17]

Avantium Chemicals[19] announced its plans for a 50 kt/year plant for their furanics platform chemicals. In the latter case, the plant will produce mostly methoxymethylfurfural (MMF), which is further converted to furandicarboxylic acid (FDCA) as one of the two bio-based building blocks for application in polyethylene furanoate (PEF) polymer, with methyl levulinate (ML) as a substantial side product. But Avantium Chemicals has stated that their furanics technology is flexible and that plants can be designed to predominantly produce methyl levulinate. Avantium's pilot plant at the Chemelot site in Geleen, the Netherlands, produces several ton of methyl levulinate, which is currently used for application development by third parties.

The development of a novel technology for the cost-effective production of a new biopolymer PEF by Avantium is a good example of the requirements needed for successful introduction of new biomass-derived chemicals. Avantium started out as a spin-off of Royal Dutch Shell in 2000 and specialized in screening and development of new catalysts and catalytic processes using their advanced high-throughput platform to accelerate the development of the new catalytic processes needed to be able to produce these new biochemicals in commercially viable processes. The strength of the approach taken by Avantium is not only the increase in experimental power but to a large extent use of conceptual process design early at even the earliest research stages.

DSM[20] has announced plans for producing adipic acid starting from levulinic acid. The basis for levulinic acid production is very similar to the BioFine technology. Though the recent start-up of GFBiochemicals by DSM employees may indicate a shift in plans.[17] Several patents have recently emerged that indicate that the Levulinic acid route is chosen to produce bio-based adipic acid.[21–23] Interestingly, in the same presentation Janssen[20] mentions fermentative routes to levulinic acid using sugarcane in Brazil as an alternative production option.

4.2 LEVULINIC PRODUCTION ROUTES

The synthesis of levulinic acid is industrially possible via several routes based on the feedstock employed (Table 4.1). As already mentioned earlier, the currently employed industrial process uses petrochemical-derived maleic anhydride. The emerging bio-based technologies, however, can either use C6 sugars or C5 sugars as the starting point. Important targets are the development of a selective dehydration without side reactions, especially the reduction of the tarry humins/char co-products, and the development of new, preferably heterogeneous catalysts to replace the traditional inorganic strong acids such as H_2SO_4, HCl, and H_3PO_4 or alternatively improve existing catalyst-based systems. Werpy and Peterson[3] identified the tolerance to inhibitory elements or components of biomass-based feedstocks (sugar streams) and the development of robust catalysts as the major technical hurdles that needed to be addressed for successful commercialization of the bio-based processes.

Since the identification of levulinic acid by the DoE, only very few companies have been openly active on levulinic acid and derived products. NBioFine focused on the production of levulinic acid from cheap feedstocks such as lignocellulosic and/or municipal waste streams. This approach is now further developed by GFBiochemicals. Segetis intended to use levulinic acid as basis of their product portfolio. Another company active in the field is Avantium Chemicals, which reported methyl levulinate as platform chemical that their furanics platform was capable of producing at commercial and economic scales (Table 4.2).[28]

TABLE 4.2 Levulinic Acid and Levulinate Production Routes.

Feedstock	Process	Key Intermediate	Product	References
Petrochemical routes				
Maleic acid esters	DSM	Acetylsuccinates	Levulinic acid	[24]
C6 sugar routes				
Glucose	Fermentation	Pyruvate + acetaldehyde	Levulinic acid	[25]
	Acid dehydration	HMF	Levulinic acid	
Fructose	Acid dehydration	HMF	Levulinic acid	
	Acid dehydration in methanol	MMF	Methyl levulinate	
D-glucosamine, *N*-Ac-D-glucosamine	Acid dehydration	HMF	Levulinic acid	[26]
C5 sugar routes				
Xylose (Quaker Oats to furfural [12])	Dehydration, hydrogenation, hydrolysis	Furfuryl alcohol	Levulinic acid	[6]
	Dehydration, hydrogenation, hydrolysis, etherification	Furfuryl alcohol + ethanol	Ethyl levulinate	[7]
Polymeric sources				
Paper sludge	DSM	Glucose	Levulinic acid	[22]
Lignocellulose	BioFine	Glucose and xylose	Levulinic acid	[14,17,27]
Starch	GFBiochemicals	Glucose	Levulinic acid	[17]
Chitin			Levulinic acid	[26]

4.2.1 Maleic Anhydride Route

It is possible to produce levulinic acid from petrochemical intermediates. Moens[13] reports that DSM used a maleic anhydride-based route at their production site in Linz[29] (Figure 4.1). In this route maleic acid is first converted to its diester, usually the diethyl ester. The diester subsequently undergoes an acylation of the activated double bond to form the acetylsuccinate. This reaction proceeds via a free-radical addition mechanism as detailed by Patrick, and yields up to 81% have been reported.[30] In the last step the acetylsuccinate undergoes a selective decarboxylation using either acid[31–35] or basic catalysts in water.[31] In the same step the remaining ester functionality is typically hydrolyzed by acids or saponified by bases. The best yields are reported for HCl ranging from 90% to 92%[32,33] and *p*-toluenesulfonic acid of 94%[36] as the catalyst. The overall process has been performed at 3 ton/day scale[30] by DSM, Linz, Austria.

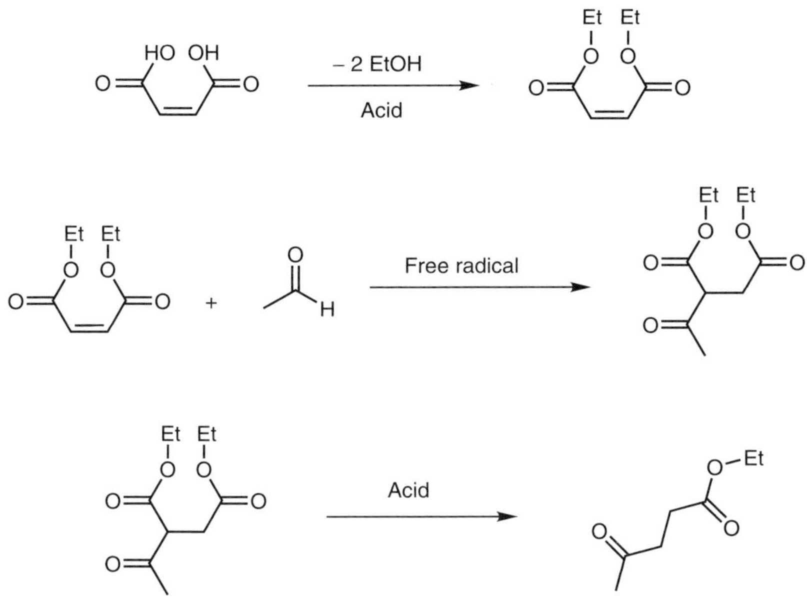

Figure 4.1 Maleic anhydride route to levulinic acid.

4.2.2 C6 Sugar-Based Routes

The most common route to levulinic acid is from C6 sugars; in particular glucose and fructose are reported as feedstocks. The first report on the synthesis and isolation of LA was by Freiherr et al. in 1875, though it is expected that in the HCl-catalyzed degradation of sucrose reported by Mulder in 1840, levulinic acid was also present as product.[36] In was not until Frost and Kurt showed in 1951 that LA could also be obtained from cellulosic feedstocks that low-cost routes seems to become within reach. Theoretically one could obtain levulinic acid in a 67.8 wt% yield from C6 sugars, with the remaining being formic acid and water (Figure 4.2). In practice the reported yields are significantly lower. Chalid summarizes such reported yields for a wide range of starting materials.[37] Best results reported for glucose is 41.4 wt% though typically significantly lower yields are achieved; for the more expensive fructose better yields of 52 wt% are given. Interestingly, several sources of lignocellulosic biomass have also been used, with yields in levulinic acid ranging between 5 and 15 wt%, though one should realize that the typical content of C6 sugars in these feedstock is only 30–40 wt%.[38]

Recently Avantium Chemicals reported the formation of methyl levulinate in the acid-catalyzed dehydration of sugars – fructose in particular – using methanol as the solvent as well as the reactant.[39] In the examples given, MMF (Figure 4.2), the precursor to methyl levulinate (ML), is still present, though Balu[28] reported that the process is fully tunable toward ML.

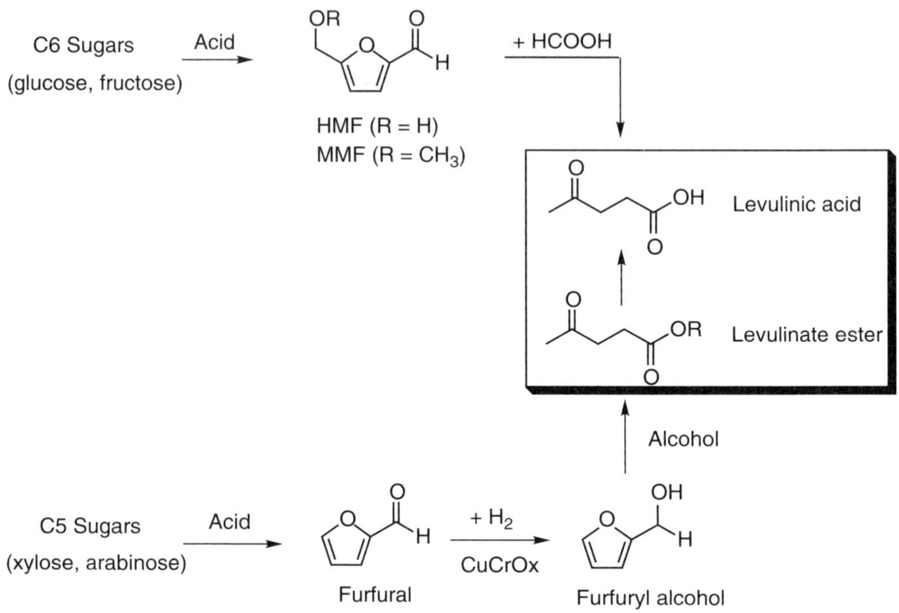

Figure 4.2 C6 and C5 sugar-based routes to levulinic acid.

There is a marked difference between glucose, including most lignocellulosic feedstocks, and the results achieved for fructose. The latter gives significantly higher amounts of levulinic products. This is generally explained by the mechanism of levulinic acid formation. The initial step is the formation of HMF in water, and RMF in the case of alcoholic solvents, by a triple dehydration step. HMF and to a lesser extent MMF are unstable molecules and easily condense, together with sugars and sugar degradation products, into black insoluble materials often referred to as humins.[38,40] In an extensive review on the HMF formation, van Putten et al.[5] shows that the temperature and reaction conditions required for the dehydration of glucose and other aldoses are significantly more severe than those for fructose, which is tentatively explained by the need for glucose to first isomerize to fructose via an enediol. This isomerization is a base-catalyzed reaction and thus proceeds very slowly under the often very acidic conditions used for the subsequent dehydration of the fructose. As a result of those harsher reaction conditions for aldoses, significantly higher amounts of humins are formed.

An alternative route using sugars is by a cascading enzymatic reaction. Zanghellini[25] claims a 6-step enzymatic process by converting pyruvate to C5 intermediates using an aldolase (Figure 4.3). No overall yield of glucose to levulinic acid is given, but the possibility to perform the enzymatic cascade in a living organism, bacterial or fungal, opens up the possibility of efficient fermentation processes.

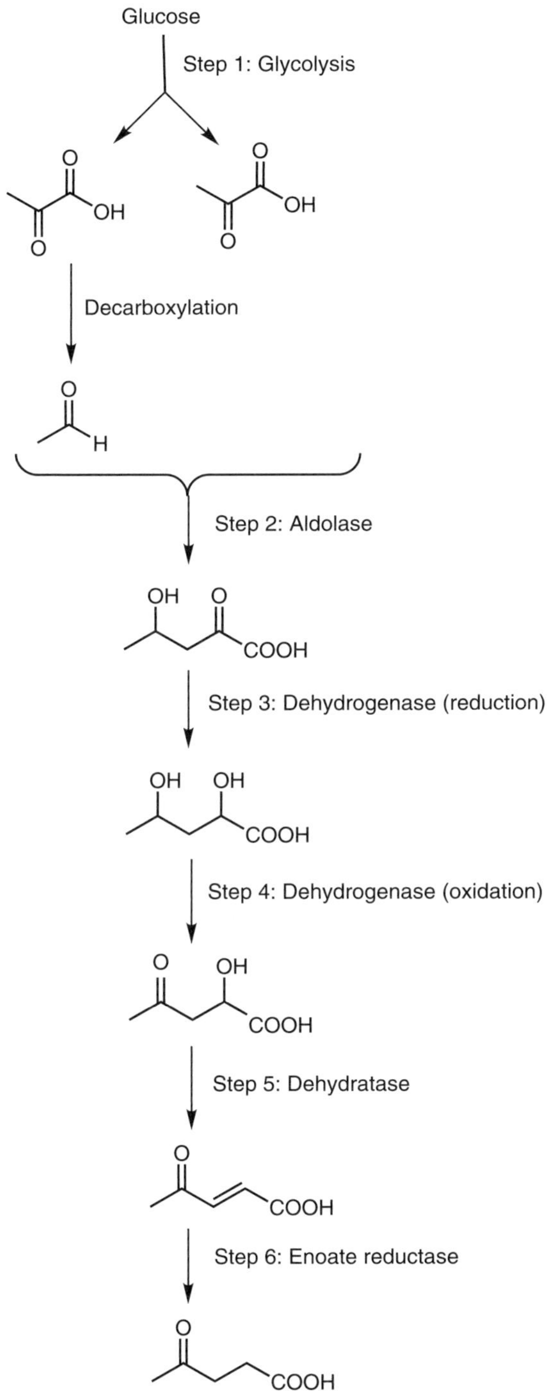

Figure 4.3 Enzymatic cascade for glucose to levulinic acid.

4.2.3 C5 Sugar Routes to Levulinic Acid

The use of cellulose in lignocellulosic biomass for levulinic acid production is long known (see preceding text). Yet cellulose only consists of 25–40% of the dry mass of biomass; thus inherently low overall yields will be obtained when using only the cellulosic part. Another significant part of lignocellulosic biomass is hemicellulose, which consists primarily of C5 and C6 sugars (its composition is very much dependent of the feedstock[41]) and C6 sugar acids. Conversion of the hemicellulose to furfural was developed by Quaker Oats in 1922 and described by Zeitsch[12] in great detail. Already in 1938 Leger and Hibbert reported the conversion of furfuryl alcohol to levulinic acid.[6] Otsuka et al.[42] described a process using furfuryl alcohol as the feedstock in the conversion either to levulinic acid or to levulinate esters.

The direct conversion of furfuryl alcohol yields levulinic acid. The overall process from biomass is a 3-step process consisting of an acid-catalyzed hydrolysis of hemicellulose dehydration and simultaneous dehydration of the C5 sugar and C6 sugar acids formed to furfural.[12,43] After gas-phase hydrogenation of furfural over copper chromite-based catalysts at 135 °C furfuryl alcohol is obtained. In the final step the furfuryl alcohol is hydrolysed in refluxing water using strong acid catalysts to the desired levulinic acid.[44,45] Yields in levulinic acid typically are in the 83–93% range.[38]

An alternative route that is currently investigated[46,47] employs alcohols rather than water as the hydrolysing agent. This gives the corresponding levulinate esters as the final product. Yields have been reported to be over 95% and typically over 98%. The advantage of this route is that the esters (see properties in Table 4.1) have a lower boiling point, thus allowing distillation as means of purification, and more importantly are less prone to give angelica lactone upon heating.

It is noted that both methods alone are currently less attractive due to the high cost of the possible feedstocks. Furfural prices typically range between 700$/t and 1500$/t and furfuryl alcohol is 1500$/t and higher. This would make the levulinic acid too expensive compared to a BioFine-based cellulose process. Hayes[14] noticed that the BioFine process could still be improved by incorporation of these C5 valorization route. Since the Quaker Oats process[12] degrades the cellulose to humins; such an approach can only work if the hemicellulose is converted to furfural while leaving the cellulose intact for conversion in a second step. It is likely to work in a biorefinery concept, which first separated the lignocelluloses in its major components. The BioFine process scheme is depicted in Figure 4.4.

4.2.4 Interconversion of Levulinic Acid and Its Esters

Since the levulinate esters are often the more preferred starting point for further conversions, an important reaction at this point is the interconversion of levulinic acid and levulinate esters. Like all esters this is an equilibrium that can be shifted by application of the appropriate conditions. Some attention must be taken because of the γ-carbonyl group as it is able to react as well. Under acidic conditions acetal formation with alcohols and intramolecular cyclization to angelica lactone may take place, while under basic conditions reverse aldol condensation may occur. Already

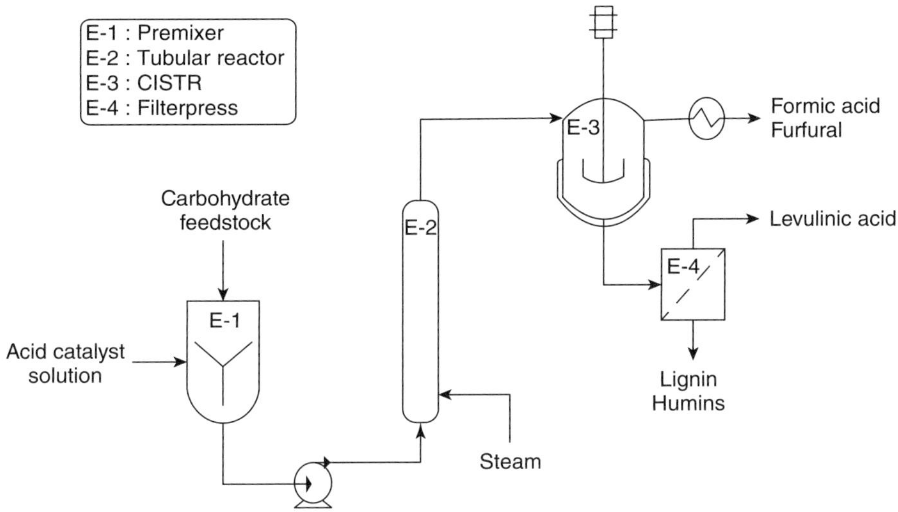

Figure 4.4 BioFine process scheme.[47] *Source*: Reproduced with permission of RUG/ Heeres/Girisuta.

in 1980 Hsu and Dwight[48] reported that the use of reactive distillation techniques make the hydrolysis of the esters a very easy and economically attractive route. Since alcohols – methanol and ethanol – have the lowest boiling point, they are selectively removed from the reaction mixture, thereby causing the equilibrium to be shifted to levulinic acid in quantitative yields simultaneously recovering the alcohols.

4.3 THE LEVULINIC ACID PRODUCT TREE

Currently levulinic acid is used as specialty chemical in pharmaceutical, cosmetic, and food applications. As already detailed earlier the market size for levulinic acid is very limited, and it is thus unlikely that these would be able to absorb large quantities of levulinic acid, which are anticipated to enter the marketplace. As such, further conversion of levulinic acid is essential to create new outlet markets. It is eminent that these new routes will need to be developed in parallel to the development of efficient levulinic biorefineries. One of the first to realize the potential of levulinic acid as building block for a wide range of chemicals were Werpy and Petersen[3] in 2004 and Bozell in 2010.[4] In Table 4.3 the combined list of identified molecules is given. In the consecutive paragraphs a selection of these target molecules and applications are discussed in detail. Figure 4.5 depicts relevant examples of levulinic acid derivatization.

4.3.1 Levulinic Acid Derivatives as Fuel Components

Soon after companies like BioFine announced that the levulinic acid would become available at very cost competitive prices when lignocellulosic- and cellulosic-based

TABLE 4.3 Reported Promising Target Molecules from Levulinic Acid.

Target Molecule	Source
Angelica lactone	[3,4,49,50]
1,4-Pentanediol	[3,4,51]
γ-Valerolactone	[3,4,51]
2-Methyltetrahydrofuran	[3,4,51]
Δ-Aminolevulinic acid	[3,4,51]
Diphenolic acid	[3,4,51]
β-Acetylacrylic acid	[3,4,51]
Levulinate esters	[3,4,17]
Acrylic acid	[3,4,51]
N-Alkyl- and N-aryl 2-methylpyrolidones	[17,51]
Glycerol ketals and polymers derived thereof	[17]
2-Methylene-γ-valerolactone	[17]
Succinic acid	[51]
Nonanone	[52]
Tetrahydrofuran	[51]
Dilevulinic acid and sebacic acid	[51]
Adipic acid and polymers derived thereof	[20]

municipal waste stream would be employed,[14] several groups investigated the potential of Levulinic acid as fuel component or additive.

It is clear that levulinic acid, which has low solubility in hydrocarbon liquids, would have to be derivatized to become a fuel. The first group of identified structures is those of the levulinate esters. These can easily be obtained from the levulinate or more preferably are directly formed in the production process starting from sugars, as done by Avantium, or from furfuryl alcohol (see preceding text). In order to still keep the fuel component fully 100% bio-based, it is clear that the alcohols used must also originate from biomass. At this moment mature bio-based technologies are in place for the production methanol, butanols, and especially ethanol. So far only industrial production routes for the methyl and ethyl levulinate esters as fuel have been reported by Maine BioProducts[51] and by Avantium.[19] In both cases the esters are reported to have good fuel properties as gasoline with only their slightly lower caloric value per kg of fuel as major drawback.[8]

To overcome low heating value and make application as biodiesel possible, Lange at Royal Dutch Shell introduced the concept of valeric fuels.[53] Here the levulinic acid is first hydrogenated to give *n*-pentanoic acid (valeric acid) and *n*-pentanol. Esterification of the valeric acid with the pentanol gives a low-oxygen-containing, straight-chain C10 ester with good biodiesel properties. Esterification of valeric acid with other bio-based glycols gives branched diesters with a relatively higher oxygen content with good biogasoline properties.

Another approach has been suggested by the TMFB consortium at the RWTH Aachen.[54] In a selection procedure they evaluated several bio-based molecules both

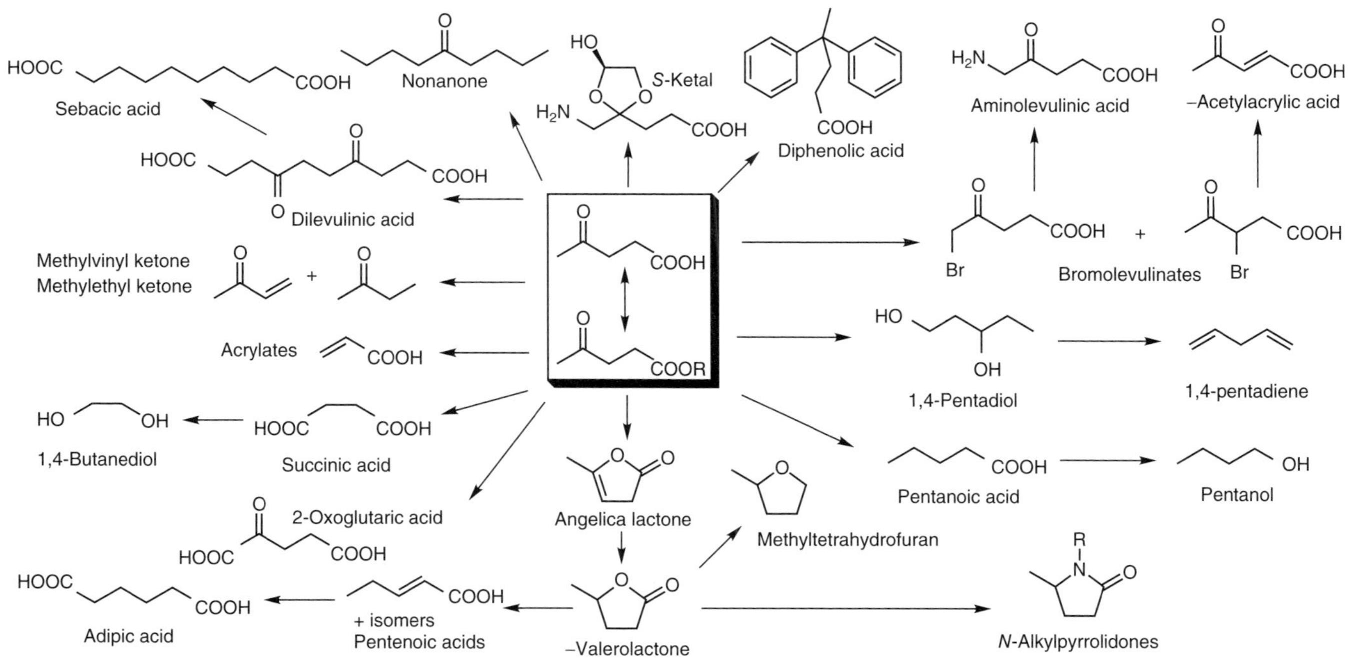

Figure 4.5 Chemical products derived from levulinic acid and its esters.

on fuel and engine properties and on economic production. One of the best candidates that they found was 2-methyltetrahydrofuran (MeTHF), another promising candidate being butyl levulinate. The economic advantage of this molecule originates from the two manufacturing routes. MeTHF can be derived by hydrogenation of furfural, a dehydration product from hemicellulosic sugars in plant materials, and it can also be derived from γ-valerolactone (GVL), a chemical feedstock component (see succeeding text for more details) that can be obtained from levulinic acid and thus from the cellulosic sugars in plant materials. Since the total content of hemicellulose and cellulose in plant materials typically is over 70%,[43] this means a highly efficient usage of widely available lignocellulosic biomass. A major disadvantage of MeTHF is the high amount of additional hydrogen needed.

Dumesic reports another group of interesting fuels that can be made via levulinic acid. Like in the case of the valeric acids from Shell, the levulinic acid is first hydrogenated to valeric acid to reduce the oxygen content. Here Dumesic points out a very efficient use of the formic acid that is coproduced in the formation of levulinic acid from C6 sugars. He realized that this is also a good donor of hydrogen and could thus be used to reduce by half the H_2 needed for obtaining the valeric acid. In order to further reduce the oxygen content, Dumesic uses the ketalization reaction between two valeric acid molecules to yield nonanone and CO_2.[55]

4.3.2 Levulinic Acid Derivatives as Chemical Feedstock Components

Levulinic acid is considered to be a versatile platform chemical to be further converted in solvents, plasticizers, monomers for polymers, etc. Examples of chemicals from LA and their potential application are illustrated in Figure 4.5. Since many of the chemical conversion routes proceed via esters of LA, it is not surprising that Methyl Levulinate can be considered as a feedstock with equal importance. As was already shown above, the interconversion between the two is facile and quantitative and will not be further discussed here.

Important chemicals that can be synthesized from both Levulinic acid and Methyl Levulinate are: γ-valerolactone, diphenolic acid (DPA), succinic acid, adipic acid, pentanoic acid, methyltetrahydrofuran, methyl ethyl ketone (MEK), and methyl vinyl ketone. Each of these applications are discussed in more detail later, with a particular focus on the advantages of using the esters as feedstock. The development of new chemicals from levulinic acid is, however, still in its infancy, in part because the price for levulinic acid at bulk scale production is still unclear. Another reason is the fact that many of the envisaged molecules are not yet commodity chemicals, in part because the economic competitiveness of drop-in molecules still has to be shown.

4.3.2.1 γ-Valerolactone (GVL) Many synthesis routes and catalysts are known to produce GVL with ML as feedstock. A direct synthesis of GVL from ML exists, but yield is only 9% product at a low conversion of 13.3%.[56] Several routes from LA have been described with two catalysts Pd/SiO_4[52] and $RuSn_4$.[57] These catalysts have a high conversion of 88% and selectivity of 96%. The high selectivity is relevant at high conversion that results in a small recycle flow and more importantly low angelica lactone formation, a known precursor to coke and thereby catalyst deactivation. Both catalysts operate at a temperature of 180 °C for a period of 10 h with the LA present as a liquid. Other catalysts identified for this conversion have less favorable combinations of conversion, selectivity, reaction temperature, pressure, and/or catalyst type. Examples of reported systems include both homogeneous and heterogeneous catalyst systems. Ru- tris(*m*-sulfonatophenyl)phosphine is a homogenic catalyst with good conversion and selectivity but requires an extra purification step to remove the catalyst for recycling.[58] Cu/Fe shows a relatively low selectivity compared to the other catalysts, and Ru/C could not compete to the two selected catalysts on overall yield per pass.[59]

In industrial practice conversion is typically less important compared to selectivity if the unreacted material can be recycled relatively easy. Since the boiling point of GVL is lower than that of levulinic acid, it can be distilled off with the unconverted levulinic acid recycled from the bottom stream without further purification steps needed. In this case the selectivity to other relatively volatile by-products – that is, pentanoic acid, methyltetrahydrofuran, and the already mentioned angelica lactone – is key in developing new catalysts systems. Angelica lactone is the inherent intramolecular lactonization product from levulinic acid. It is here that the less reactive methyl levulinate offers substantial advantages as the amount of free levulinic acid under catalytic conditions can be maintained low (Figure 4.6).

4.3.2.2 Diphenolic Acid (DPA) Bisphenol A (BPA) was first synthesized by the Russian chemist A.P. Dianin in 1891. This compound is synthesized by the condensation of acetone (hence the suffix A in the name) with two equivalents of phenol.[60] The reaction is catalyzed by a strong acid, such as HCl or a sulfonated polystyrene resin. Industrially, a large excess of phenol is used to ensure full condensation; the product mixture of the cumene process – acetone and phenol – may also be used as starting material.[61] The world production capacity of BPA was more than 2.2 million tons in 2009.[61]

The condensation reaction is not restricted to the use of acetone as the carbonyl source. The condensation of levulinic acid and levulinate esters with phenol has been described by several authors[62–66] to afford the resulting DPA as the product. In analogy with BPA, the key parameter in the synthesis is the selectivity of the two consecutive alkylations on both phenol rings. Similar to BPA only the p,p′-substituted isomer is of relevance, and the other two isomers, o,p′- and o,o′, cannot be recycled and are thus to be considered a waste stream. The first catalyst reported with good

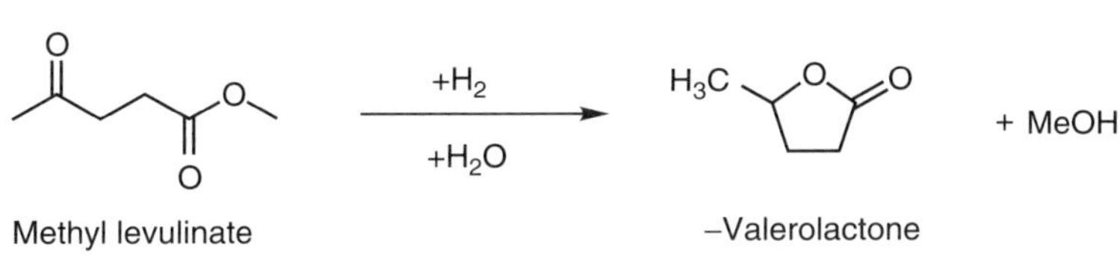

Figure 4.6 Reaction scheme of ML to GVL.

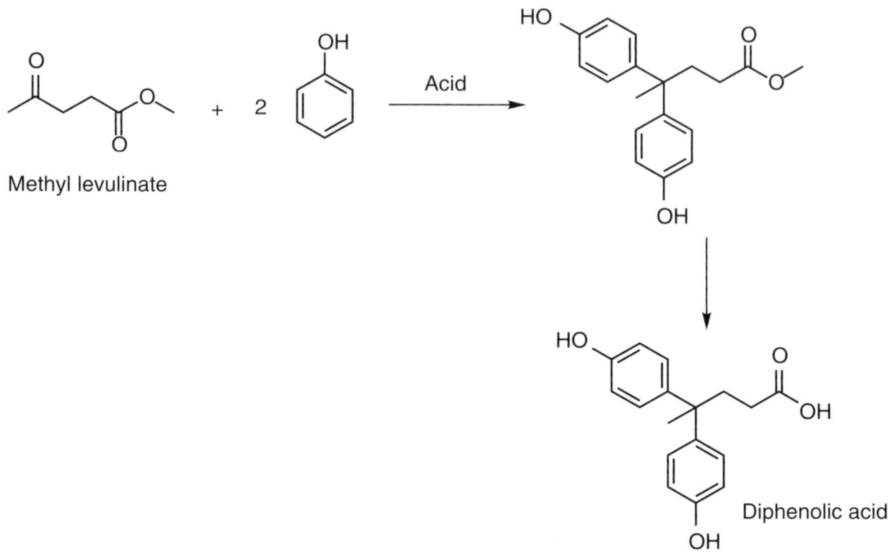

Figure 4.7 Reaction scheme of ML to DPA.

p,p′ selectivity is a Cs–HPA complex that has a conversion of 36% and a selectivity of 88% to p,p-DPA (overall yield 32%).[63] The second catalyst is a sulfonated hyperbranched poly(arylene oxindole) (SHPAO) catalyst, which is characterized by a conversion of 70% and a p,p′ selectivity of 72%[64] illustrated in Figure 4.7. Recently Chang[65] reported the use of HCl in combination with mercaptoacetic acid as highly selective catalyst system. A high selectivity of 97.1% was reported when using highly concentrated HCl and high phenol/LA ratios.

In all cases only the selectivity to the desired p,p′-isomer is reported and only little information is given on the exact nature of the other components present. This makes it difficult to estimate if these impurities are still the mono-p-hydroxyalkylated intermediate that could still be recycled or whether it is any of the unwanted ortho-substituted mono- or diphenols that need to be removed and disposed. It is clear that further research and process optimization are still required.

4.3.2.3 Pentanoic Acid (PA) Another interesting product from levulinics would be PA, also known as valeric acid. The direct synthesis from levulinic acid is complicated by the fact that under the reaction conditions needed for the complete ketone hydrogenation, the levulinic acid readily forms the angelica lactone. This is a highly reactive molecule and either hydrogenates to the stable GVL or is easily polymerized, which tends to deactivate the catalyst. As a consequence no direct chemical interesting process was found. However, indirect and costly routes for synthesizing PA via GVL have been reported with good overall conversion and selectivity. The production of GVL as described earlier, and direct conversion of GVL to the desired

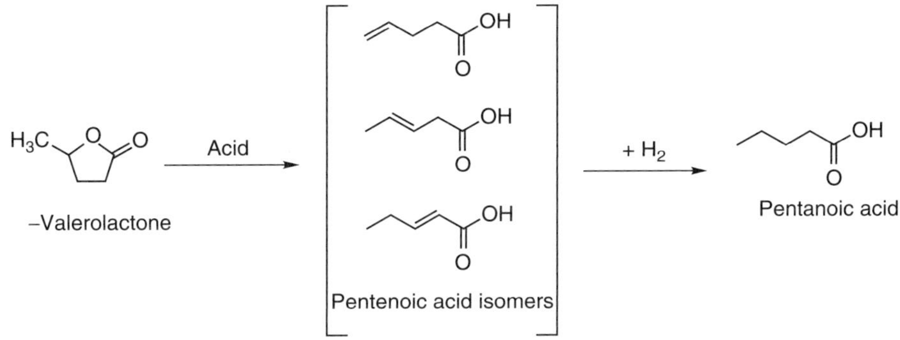

Figure 4.8 Reaction scheme of GVL to PA.

pentanoic acid, via pentenoic acid as intermediate, has been described by Chalid[38] using a Pd/niobia silica catalyst at 300 °C as illustrated in Figure 4.8.

4.3.2.4 2-Methyl Tetrahydrofuran (MeTHF) The production of MeTHF from levulinic acid is typically by hydrogenation of GVL; in fact multiple articles describe MeTHF as a minor by-product in the synthesis of GVL.[59,60] However, since the aim of these experiments was to reduce the selectivity of MeTHF to a minimum level, they do not report highly selective catalysts. The most common approach to MeTHF is by further hydrogenation of GVL. Direct conversion starting from LA has also been reported. The earliest examples were developed by PNNL as described by Moens.[13] A one-step process with several lactone and diol intermediates is mentioned without great detail. The group of Luque[67] describes a continuous hydrogenation of LA to MeTHF based on the Cu nanoparticle-based catalyst system, earlier described by Pravin.[68] This reaction was performed at 265 °C, under 25 bar H_2, with a conversion of 100% and a selectivity of 89% illustrated in Figure 4.9. The catalyst used for the reaction is a nanoparticle Ni–Cu/SiO_2 and had no significant activity loss after 320 hours on stream.

4.3.2.5 Succinic Acid The conversion of levulinates into succinic acid is interesting. Succinic acid has received considerable attention and several companies have build fermentation plants totaling up to 267 kt/year in 2016 [69] and announced plans up to expand to 700 kt/year by 2020.[70] The price range for succinic acid at the end

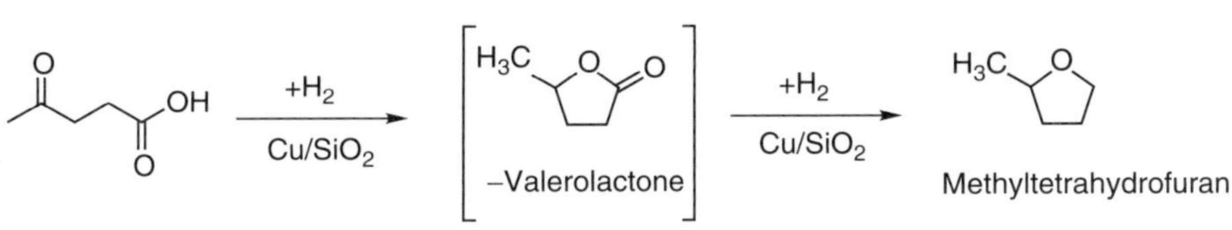

Figure 4.9 Reaction scheme of ML to MeTHF.

of 2013 is 2866–2932 US$/t,[71] but price erosion is expected once all the anticipated additional capacity has come online. In this respect, Markets and Markets expects the global succinic acid market to be worth $486.7 million by 2019,[72] which would mean that at full projected capacity, prices may drop to as low as 1850$/t.

It has long been known that LA can be converted to succinic acid in the presence of strong bases.[14] This method consumes stoichiometric amounts of base, usually NaOH, and in addition requires additional H_2SO_4 to neutralize the product mixture and liberate the succinic acid. This method is not only expensive, but it also produces large quantities of Na_2SO_4 waste. The more obvious route would be to oxidize levulinates using oxygen. Several authors report different approaches. Podolean[73] uses Ru nanoparticles as an efficient catalyst with a selectivity to succinic acid of 96% at 46.1% conversion. Liu uses Mn salts as an efficient catalyst system and reports 78% yields in SA (Figure 4.10).[74]

An interesting possibility has recently been reported by ADM[75] using the well-known industrially applied CoMnBr system. Though yields of 18% reported are still low compared to those for the Ru nanoparticle and Mn salts discussed previously, it should be noted that the levulinic acid was present in small quantities next to HMF and that the reaction conditions were optimized for the latter conversion into FDCA.[76]

4.3.2.6 *Methyl Ethyl Ketone and Methyl Vinyl Ketone*

Another interesting route to drop-in chemicals is by decarbonylation to methyl vinyl ketone and decarboxylation to methyl ethyl ketone (Figure 4.5). The first is a reactive intermediate in the synthesis of biperiden, vinclozolin, and synthetic vitamin A,[76] the latter being commonly used as solvent. The first approaches were reported by Dumesic[77] in a process similar to many conversions of this type. The feedstock, that is, levulinic acid, is passed over a metal oxide catalyst at high temperatures. Here levulinic acid is volatized and passed over catalyst bed consisting of solid acid catalyst. The maximum yield in methyl vinyl ketone reported is 41% with less than 5% methyl ethyl ketone being observed. Interestingly the mechanism of methyl vinyl ketone formation is in part suggested to proceed via angelica lactone intermediates, and angelica lactone is even fed to the reactor in some examples. However, in the production of GVL the formation of angelica lactone is typically tried to be avoided due to the high tendency to polymerize and form coke on the catalysts.

An alternative method has recently reported via oxidative decarboxylation.[78] Copper oxide supported on cerium dioxide (CuO/CeO_2) and alumina (CuO/Al_2O_3) were prepared and used for the oxidation of LA. The experiments revealed that

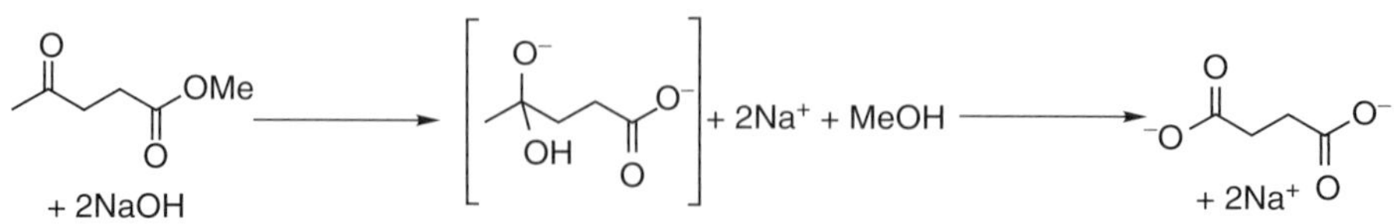

Figure 4.10 Base-assisted formation of SA from LA.

CuO/CeO_2 and CuO/Al_2O_3 can oxidize LA and get methyl vinyl ketone with a yield of 15.5% at 57.4% conversion under mild reactive conditions. In contrast pure CuO oxidizes LA under the same reaction conditions to give butanone (MEK) in 5% yield at low conversions of 24%.

4.3.2.7 δ-Aminolevulinic Acid Another possible outlet for levulinic acid is 5-amino-4-oxopentanoic acid (also δ-aminolevulinic acid (DALA)). DALA has plant fertilizer properties and is suggested to increase salt tolerance[79] and is also an herbicide. Currently DALA is produced by Cosmo Oil and targets a market of 200–300 million lb/year at a projected cost of \$2.00–3.00 lb^{-1}.[15] The production of DALA by Cosmo Oil is, however, directly from sugars in a fermentation process. In a chemical route from levulinic acid, Werpy suggested that levulinic acid-derived β-acetylacrylic acid could serve as an intermediate in the production of DALA. Although this route is possible as reported by Wynn and Corwin,[80] current research seems to focus on amination strategies involving 5-bromo levulinic acid. Using protected amines like phthalimide, hexamethylenetetramine, sodium formamide, and sodium azide, followed deprotection in presence of HCl, the amino group of DALA is protected against reactions with the ketone or carboxylic acid function. Key step is the selective bromination of levulinic acid to 5-bromo levulinic acid using Br_2. Specifically the regioselective control of the radical bromination to prevent the 2-bromo levulinic acid and over bromination to di-bromo-substituted products seems key. As study by Zavozin[81] showed that high selectivity to the desired 5-bromo levulinic acid can be obtained by protecting the ester function and using methanol as the solvent. Here an opportunity for methyl levulinate as feedstock is clearly at hand.

4.3.2.8 Adipic Acid DSM (the Netherlands) has announced their plans for the production of adipic acid starting from lignocellulosic biomass.[20,82,83] Many details are not yet disclosed, but the key step in this process appears to be the economic production of levulinic acid, followed by conversion in GVL and subsequent ring opening to pentenoic acid. In the final step a homogeneous Pd catalyst is used in the selective hydrocarboxylation with CO/H_2O with high selectivity at the C5 position. It is pointed out that the cleavage of the GVL into pentenoic acids is usually performed over mildly acidic solids and yields a mixture of all three isomers. Kinetically the 3- and 4-pentenoics are formed first from GVL, but over the acidic catalyst isomerization will occur, and considerable amounts of the thermodynamically most stable 2-pentenoic acid is form. Remarkable is that the applied homogeneous Pd catalyst can isomerize the pentenoics while selectively only inserting the CO at the C5 position. As the 4-pentenoic acid is removed from the equilibrium, the isomerization will assure that all of the three pentenoic acids are converted to the desired adipic acid.

4.3.2.9 Levulinic Acid-Based Ketals Segetis Inc. has been focussing on other derivatives of levulinic acid, for example, levulinic acid-based ketals, which have reached advanced stages of development and are technologically ready to move to advanced production levels. This novel class of ketals is formed by the reaction of diols/polyols such as glycerol with the ketone group of levulinic acid in the presence

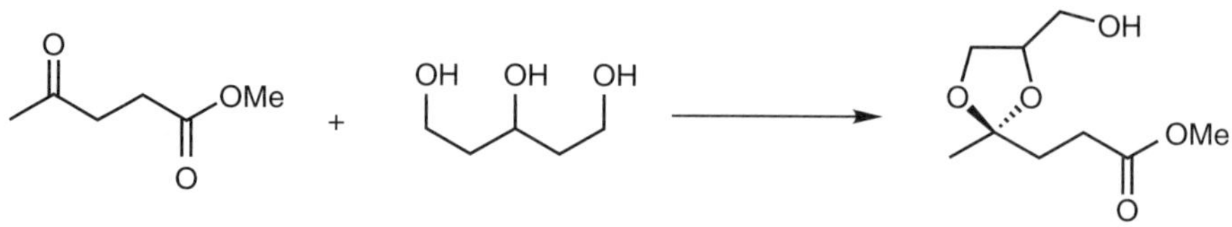

Figure 4.11 Reaction scheme of the reaction of LA and glycerol into ketals.

of a homogeneous mineral acid such as HCl (Figure 4.11). The new ketal products have excellent functionality and can potentially replace existing petroleum-based solvents, surfactants, and plasticizers.[84] The levulinic ketal esters can be readily extended with alcohols, esters, and amines using standard catalysts and reaction conditions to produce a wide range of bio-based ester technologies.

4.4 CONCLUSIONS AND OUTLOOK

As one of the earliest biochemicals produced and now with the onset of processes like BioFine, in operation by GFBiochemicals, and Avantium YXY process, it is expected that levulinates will be readily available as chemical building blocks. Several routes to new and to drop-in chemicals have been discussed. These are mostly based on levulinic acid, though from a catalytic and from a process point of view the use of methyl levulinate has several advantages. It is a liquid at room temperature and has a lower boiling point allowing gas-phase reactions, is not a weak organic acid, thus with less corrosion issues, and it does not readily form angelica lactone, which is a known coke precursor, to have more stable catalytic processes.

As is demonstrated levulinic acid and methyl levulinate are versatile bio-based chemical platform chemicals. The basic production technology has been developed and demonstrated by several companies, and is in operation by GFBiochemicals for levulinic acid and Avantium for methyl levulinate. It is thus to be expected that they both will be readily available in the future. Yet market introduction will require that both the levulinates are produced at economically attractive prices.[13] The direct use of levulinates as flavor and fragrance compounds represents only a small market volume. The subsequent conversion in other molecules is therefore needed and has been described in detail. The success of drop-in molecules will depend mostly on price competition with the existing petrochemical-based equivalents, where those molecules with new functionality will initially be driven by improved performance criteria. Given the wide range of possibilities demonstrated already, it will only be a matter of time before these products will hit the market.

The unique skills, experience, and equipment acquired in the development of a new production method for methyl levulinate from sugars will enable Avantium Technologies to develop novel and/or improved catalytic processes starting from an affordable and readily available methyl levulinate.

REFERENCES

[1] A. Freiherrn, V. Grote, B. Tollens, *Justus Liebigs Ann. Chem.* **1875**, *175*, 181–204.

[2] V. Ghorpade, M. Hanna, *Cereal* **1997**, 49–55.

[3] T. Werpy, G. Petersen, "Top Value Added Chemicals From Biomass", Pacific Northwest National Laboratory (PNNL) and the National Renewable Energy Laboratory (NREL), August 2004.

[4] J.J. Bozell, G.R. Petersen, *Green Chem.* **2010**, *12*, 539–554.

[5] R.-J. van Putten, J.C. van der Waal, Ed. de Jong, C.B. Rasrendra, H.J. Heeres, J.G. de Vries, *Chem. Rev.* **2013**, *113*, 1499–1597.

[6] F. Leger, H. Hibbert, *Can. J. Res., B: Chem. Sci.* **1938**, *16*, 68–69.

[7] A. Démolis, N. Essayem, F. Rataboul, *ACS Sustain. Chem. Eng.* **2014**, *2*, 1338–1352.

[8] http://www1.eere.energy.gov/vehiclesandfuels/pdfs/deer_2002/session4/2002_deer_mccormick.pdf, accessed 22 Jul 2014.

[9] http://www.chemicalbook.com/ProductMSDSDetailCB3213533_EN.htm, accessed 22 Jul 2014.

[10] http://www.chemicalbook.com/ChemicalProductProperty_US_CB9311358.aspx, accessed 22 Jul 2014.

[11] J.S. Chickos, W.E. Acree, "Enthalpies of Vaporization of Organic and Organometallic Compounds 1880–2002", http://www.umsl.edu/~chickosj/JSCPUBS/vap2003.pdf, accessed 23 Sep 2014.

[12] K.J. Zeitsch, "*The chemistry and technology of furfural and its many by-products*", 1st Ed.; Elsevier, Amsterdam, **2000**.

[13] L. Moens, "Advances in the chemistry and processing of beet and cane sugar", Proceedings of the 2002 Sugar Processing Research Conference held in New Orleans, Louisiana, USA, **2002**, 10–13, 26–41.

[14] D.J. Hayes, J. Ross, M.H.B. Hayes, S. Fitzpatrick, "The biofine process: production of levulinic acid, furfural and formic acid from lignocellulosic feedstocks", In "*Biorefineries-Industrial Processes and Products: Status Quo and Future Directions*" (Eds. B. Kamm, P.R. Gruber, M. Kamm). Wiley GmbH, **2008**.

[15] http://www.prweb.com/releases/Levulinic-Acid-Market/GrandViewResearch/prweb11705450.htm accessed 3 Apr 2014.

[16] http://greenchemicalsblog.com/2013/10/10/segetis-starts-levulinic-acid-pilot-production/, accessed 12 Jul 2014.

[17] GFBiochemicals. http://www.gfbiochemicals.com/company/

[18] http://greenchemicalsblog.com/2014/06/09/segetis-gets-325k-grant-plans-commercial-facility/, accessed 12 Jul 2014.

[19] K. Laird In "Heeft Avantium de Technologie van de Toekomst", *Kunststof magazine*, Nov. 2013, 10.

[20] M. Janssen, "Renewable Adipic Acid from Levulinic Acid", NCCC XV, Noordwijkerhout, The Netherlands, March 2015.

[21] R.F.M.J. Parton, M.P.W.M. Rijkers, J.A. Kroon, WO20120330040, EP2684875, **2012**.

[22] A. de Rijke, G.W.A. Hangx, R.F.M.J. Parton, B. Engendahl, WO/2014/087014, **2014**.

[23] A. de Rijke, G.W.A. Hangx, R.F.M.J. Parton, B. Engendahl, WO/2014/087013, **2014**.

[24] L. Farnleitner, H. Stueckler, H. Kaiser, E. Kloimstein, EP 401532 B1, **1990**.

[25] A.L. Zanghellini, WO 2012/030860 A1, **2012**.

[26] H. Hamburger, *Biochem. Zeit.* **1912**, *36*, 1–4.

[27] S.W. Fitzpatrick, US 5608105 A, **1995**.

[28] A.M. Balu, N. van der Puil, N. Kemeling, J.C. van der Waal, E. de Jong, "Insights in technical and economical aspects for sustainable production of methyl levulinate (ML) towards its commercialization", 247th ACS National Meeting and Exposition, Dallas, USA, March 2014.

[29] Brochure "Fine Chemicals program at a glance. Regular product range – Development products", DSM Fine Chemicals Austria, Version Q2, **2006**.

[30] T.M. Patrick, *J. Org. Chem.* **1952**, *17*, 1009–1016.

[31] A.Y. Sizov, L.A. Yanovskaya, V.A. Dombrovskii, *Izvestiya Akademii Nauk SSSR, Seriya Khimicheskaya*, **1990**, *2*, 474–475.

[32] M.G. Zalinyan, V.S. Arutyunyan, M.T. Dangyan, *Armyanskii Khimicheskii Zhurnal*, **1978**, *31*, 545–546.

[33] J. Rong, M.E. Nelson, B. Kusche, N.D. Priestley, *J. Nat. Prod.* **2010**, *73*, 2009–2012.

[34] C. Mentzer, D. Billet, *Bull. Soc. Chim. Fr.* **1945**, *12*, 292–295.

[35] A. Treibs, K. Hintermeier, *Chem. Ber.* **1954**, *87*, 1163–1166.

[36] J.J. Mulder, *J. Prakt. Chem.* **1840**, *21*, 229.

[37] M. Chalid, "Levulinic Acid as a Renewable Source for Novel Polymers", PhD Thesis, University of Groningen, the Netherlands, **2012**.

[38] I. van Zandvoort, Y. Wang, C.B. Rasrendra, E.R.H. van Eck, P.C.A. Bruijnincx, H.J. Heeres, B.M. Weckhuysen, *ChemSusChem* **2013**, *6*, 1745–1758.

[39] A.S. Dias, G.J.M. Gruter, R.-J. van Putten, WO2012091570, **2012**.

[40] J.-M. Pin, N. Guigo, A. Mija, L. Vincent, N. Sbirrazzuoli, J.C. van der Waal, E. de Jong, *ACS Sust. Chem. Eng.* **2014**, *2*, 2182–2190.

[41] E. de Jong, R.J.A. Gosselink, "Lignocellulose-based chemical products", In: "*Bioenergy Research: Advances and Applications*" (Eds. V.K. Gupta, C.P. Kubicek, J. Saddler, F. Xu, M.G. Tuohy). Elsevier, Amsterdam, The Netherlands, **2014**, 277–313.

[42] M. Otsuka, Y. Hirose, T. Kinoshita, T. Masawa, US 3752849. **1973**.

[43] M.-A. Bornik, L.W. Kroh, *J. Agric. Food Chem.* **2013**, *61*, 3494–3500.

[44] B.V. Timokhin, V.A. Baransky, G.D. Eliseeva, *Russ. Chem. Rev.* **1999**, *68*, 73–84.

[45] W.D. van de Graaf, J.P. Lange, WO 2007023173 A1, **2005**.

[46] U.M. Dzhemilev, R.I. Khusnutdinov, A.R. Baiguzina, A.A. Smirnov, R.R. Mukminov, RU 2319690 C1, **2006**.

[47] B. Girisuta, "Levulinic Acid from Lignocellulosic Biomass", PhD Thesis, University of Groningen, the Netherlands, **2007**.

[48] C.C. Hsu, W. Dwight, WO 80/02423 A1, **1980**.

[49] L. Reid, *Ind. Eng. Chem.* **1956**, *48*, 1331–1341.

[50] B. Kamm, M. Kamm, M. Schmidt, T. Hirth, M. Schulze, "Lignocellulose-based chemical products and product family trees", In "*Biorefineries-Industrial Processes and Products*" (Eds B. Kamm, P.R. Gruber, M. Kamm). Wiley GmbH, **2006**, *2*, 97–149.

[51] http://umaine.edu/mcspolicycenter/files/2009/08/MBBP_ECTA_Aug09.pdf, accessed 22 Jul 2014.

[52] K. Yan, T. Lafleur, G. Wu, J. Liao, C. Ceng, X. Xie, *Appl. Catal. A: Gen.* **2013**, *468*, 52–58.

[53] J.-P. Lange, R. Price, P.M. Ayoub, J. Louis, L. Petrus, L. Clarke, H. Gosselink, *Angew. Chem. Int. Ed.* **2010**, *49*, 4479–4482.

[54] RWTH Aachen University, "Tailor-Made Fuels from Biomass", Yearly Report **2009**.

[55] J.S. Serrano-Ruiz, D. Wang, J.A. Dumesic, *Green Chem.* **2010**, *12*, 574–577.

[56] X. Tang, H. Chen, L. Hu, W. Hao, Y. Sun, X. Zeng, L. Lin, S. Liu, *Appl. Catal. B: Environ.* **2014**, *147*, 827–834.

[57] S.G. Wettstein, J.Q. Bond, D.M.Alonso, H.N. Pham, A.K. Datye, J.A. Dumesic, *Appl. Catal. B: Environ.* **2012**, *117–118*, 321–329.

[58] M. Chalid, A.A. Broekhuis, H.J. Heeres, *J. Mol. Catal. A: Chem.* **2011**, *341*, 14–21.

[59] P.P. Upare, J.-M. Lee, D.W. Hwang, S.B. Halligudi, Y.K. Hwang, J.-S. Chang, *Fuel*, **2014**, *115*, 101–108.

[60] http://en.wikipedia.org/wiki/Bisphenol_A, and references therein, accessed 22 Jul 2014.

[61] "Experts demand European action on plastics chemical", Reuters, 22 June 2010.

[62] X. Yu, Y. Guo, K. Li, X. Yang, L. Xu, Y. Guo, J. Hu, *J. Mol. Catal. A: Chem.* **2008**, *290*, 44–53.

[63] S. van de Vyver, J. Geboers, S. Helsen, F. Yu, J. Thomas, M. Smet, W. Dehaen, B.F. Sels, *Chem. Commun.* **2012**, *48*, 3497–3499.

[64] C. Chang, X. Han, M. Xiuli, D. Xiaojian, S. Deng, P. Cen, School of Chemical Engineering, Zhengzhou University, Hangzhou, Peop. Rep. China. , *Huaxue Shijie* **2006**, *47*(5), 293–295.

[65] Y. Ide, N. Kagawa, M. Itakura, I. Imae, M. Sadakane, T. Sano, *ACS Appl. Mat. Interf.* **2012**, *4*, 2186–2191.

[66] P.N. Honkhambe, M.V. Biyani, N.S. Bhairamadgi, P.P. Wadgaonkar, M.M. Sallunkhe, *J. Appl. Pol. Sci.* **2010**, *117*, 2545–2552.

[67] J.M. Bermudez, J.A. Menéndez, A. Romero, E. Serrano, J. García-Martínez, R. Luque, *Green Chem.* **2013**, *15*, 2786–2792.

[68] P.P. Upare, J.M. Lee, Y. Kyu Hwang, D. Won Hwang, J.H. Lee, S.B. Halligudi, J. Soo Hwang, J.S. Chang, *ChemSusChem* **2011**, *4*, 1749–1752.

[69] www.biofuelsdigest.com.

[70] http://www.alliedmarketresearch.com/bio-succinic-acid-market, accessed 20 May 2014.

[71] P. Davies, Tecnon OrbiCHem, "Chemical Business Focus" Issue 003, accessed 28 Nov 2013.

[72] http://www.marketsandmarkets.com/PressReleases/succinic-acid.asp, accessed 20 May 2014.

[73] I. Podolean, V. Kuncser, N. Gheorghe, D. Macovei, V.I. Parvulescu, S.M. Coman, *Green Chem.* **2013**, *15*, 3077–3082.

[74] J. Liu, Z. Du, T. Lu, J. Xu, *ChemSusChem*, **2013**, *6*, 2255–2258.

[75] B. Subramaniam, X. Zuo, D.H. Busch, P. Venkitasubramaniam, WO 2013033081 A2, **2013**.

[76] H. Siegel, M. Eggersdorfer, "*"Ketones", Ullmann's Encyclopedia of Industrial Chemistry*", Wiley-VCH, Weinheim, **2005**.

[77] J.A. Dumesic, R.M. West, WO2011087962, **2011**.

[78] Y. Gong, L. Lin, B. Zhang, *Chin. J. Chem.* **2012**, *30*, 327–332.

[79] H. Kuramochi, M. Konnai, T. Tanaka, Y. Hotta, US 5661111 A, **1997**.

[80] R.W. Wynn, A. H. Corwin, *J. Org. Chem.* **1950**, *15*, 203–208.

[81] A.G. Zavozin, N.E. Kravchenko, N.V. Ignat'ev, S.G. Zlotin, *Tetrahedron Lett.* **2010**, *51*, 545–547.

[82] A.M.C.A. Castelijns, M.C.C. Janssen, H.W.L.M. Vaessen, US 20120329981A1, EP2537840, **2012**.

[83] S.M.A. De Wildeman, M. Schurmann, P.C. Raemakers-Franken, A.C. Trefzer, WO/2010/104391, **2010**.

[84] S. Selifonov, S.D. Rothstein, B.D. Mullen, US 8604223, **2013**.

5

C5LT: BIORENEWABLES AT C5 LIGNO TECHNOLOGIES AB

KAISA KARHUMAA AND VIOLETA SÀNCHEZ I NOGUÉ

C5 Ligno Technologies AB, Lund, Sweden

5.1 INTRODUCTION

A multitude of technical possibilities combined with the fact that yeast has been domesticated for human use for thousands of years make baker's yeast

Industrial Biorenewables: A Practical Viewpoint, First Edition. Edited by Pablo Domínguez de María.

Saccharomyces cerevisiae a well-suited microorganism for its use in biorenewable fuel and chemical production. Yeast is widely used as biocatalyst in many processes from food and beverage production to production of pharmaceuticals. Since the beginning of the development of genetic engineering methods, yeast has proved to be a good workhorse for both laboratory testing and industrial applications due to its ease of handling. It was one of the first organisms that were completely sequenced, which is a prerequisite for efficient genetic engineering work. The advantage of the whole genome sequence has made it one of the earliest organisms extensively used for genetic engineering, which has kept it at the forefront of the developments in molecular biology. Moving from academy to industry, baker's yeast is also one of the first examples where genetic engineering has been put into practice in industrial processes. It is naturally tolerant to many stress factors, which is why it is the chosen microorganism for fermentation of lignocellulosic feedstocks.

The use of lignocellulosic materials such as agricultural and forestry waste for biofuel production has got increasing attention during the latest decades and is now approaching a commercial reality. While ethanol has been commercially produced from starch (corn, potato, wheat grains) and sucrose (sugarcane), a sustainable and economically feasible production of biofuel requires the use of the whole-plant biomass for the industrial process. The second-generation ethanol production process includes the use of the hemicellulose part of the plant biomass, coming from by-product streams of agricultural or forestry processing or from energy crops specifically grown for the purpose of fuel production. Hemicellulose contains pentose sugars xylose and arabinose, which have to be efficiently fermented to create an economically sound ethanol production process. Although theoretically possible and despite thorough academic studies, the second-generation ethanol production has proved to be much more challenging than starch- or sucrose-based first-generation ethanol production. Although the knowledge from the first-generation ethanol production is a great advantage, the process methods cannot be as such transferred to the second-generation process, but specific and thorough process optimization is further required.

Production of ethanol from lignocellulosic raw materials is the renewable liquid biofuel process that is closest to commercialization in large scale at the moment. The second-generation ethanol production is reaching a level where there are more than 16 demonstration-scale facilities and 11 commercial-scale facilities globally that are near completion.[1] Some commercial operations, for example, Iogen,[2] have been running for some time, but as a whole the demonstration and commercial activities have seriously started only recently. Solving the technical issues in lignocellulosic ethanol production will largely function as a pilot case even for the production of other chemicals, since many similar problems are likely to appear in all lignocellulose-based microbial production processes. Effective usage of all sugars in the lignocellulosic hydrolysates is a prerequisite for any production process. Thus, a fully economically viable process for ethanol will pave way for the commercialization of other, more advanced, biofuels in the near future.

This chapter will describe the background and present situation for the development of yeast strains for the second-generation ethanol production, developed at C5 Ligno Technologies AB (C5LT).

5.2 LIGNOCELLULOSIC ETHANOL PRODUCTION: PROCESS

The second-generation ethanol production process consists of two major parts: (1) pretreatment and hydrolysis of the raw material to release the monomeric sugars to the liquid phase and (2) the fermentation part where a microorganism, most often yeast, is added to ferment the sugar to ethanol (Figure 5.1).

Numerous different methods exist for the pretreatment of the lignocellulosic biomass, of which the most well described are pretreatments with dilute acid and steam explosion, reviewed in the open literature.[3] Pretreatment is usually followed by enzymatic hydrolysis of the sugar polymers by using a commercial enzyme mixture of cellulases and hemicellulases, such as glucanases and xylanases. The resultant liquid is subsequently subjected to fermentation (separate hydrolysis and fermentation (SHF)). In another process concept, simultaneous saccharification and fermentation (SSF),[4] the hydrolysis and fermentation steps are combined so that the enzymes and fermenting microorganism are added at the same time, and the sugar that is released by the enzymes is continuously fermented (Figure 5.1). The advantage of this method is the reduced sugar concentration in the fermentation liquid, which relieves product inhibition on the hydrolyzing enzymes. In addition, the low glucose concentrations in SSF may also improve pentose fermentation by yeast. The details of this phenomenon will be discussed later in this chapter. To our knowledge, most of the precommercial facilities do not yet apply the SSF method. Rather, different variants of an SHF process are usually applied.

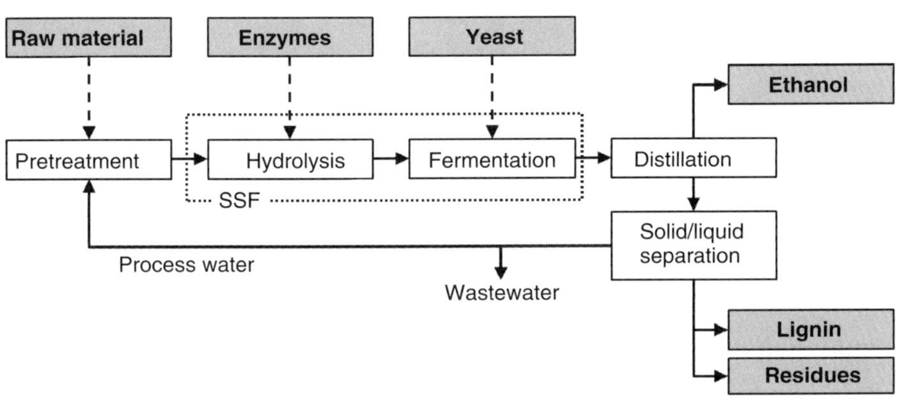

Figure 5.1 Configurations used for the conversion of biomass to ethanol. The dotted line indicates SSF configuration.

5.2.1 Lignocellulosic Ethanol Production: Hydrolysate Composition

Lignocellulosic materials contain cellulose and hemicellulose, and their relative amounts and composition vary with the type of raw material (recently reviewed in the literature[5]). In most materials glucan is the major polymer constituting 32–43% of the dry matter. Xylan and arabinan vary much more between different feedstocks. Of these two, xylan is more abundant, varying from 15% of the dry matter in salix to 26% in wheat straw. Corn stover contains about 22% xylan. Arabinan is much less prevalent, but even the 2–5% arabinan in feedstocks like switch grass and corn stover will become significant in the final optimization of the process.

After hydrolysis, the produced lignocellulosic hydrolysates contain variable amounts of the different sugar monomers depending on raw material and the hydrolysis method. The absolute amounts of a single type of sugar may vary from as low as below 10 g/L to almost 100 g/L in lignocellulosic hydrolysates produced with different methods from different raw materials. The variety of lignocellulosic raw materials, their composition, and the many different process designs for pretreatment and hydrolysis (Figure 5.2) give rise to lignocellulosic hydrolysates with widely varying composition and concentrations of fermentable sugars and fermentation inhibitors. The microbial metabolism differs for each sugar type, which has great importance for the design and selection of the appropriate fermenting microorganism. These aspects will be discussed later in this chapter.

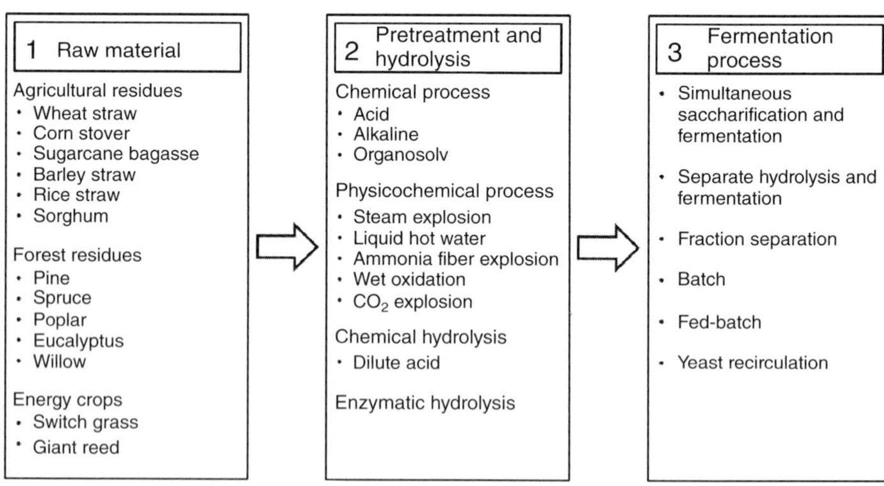

Figure 5.2 Variety of different raw materials, pretreatment and hydrolysis methods, and process configurations. As the various options each three categories can be combined in almost any combination, the number of resulting ethanol production processes is almost infinite. This generates a wide variety of different lignocellulosic hydrolysates and requirements for the fermenting microorganism.

5.2.2 Lignocellulosic Ethanol Production: Pentose Fermentation

In techno-economic evaluations of processes for lignocellulosic ethanol production, it becomes clear that to make the process economically profitable, the pentose fraction of the sugars must be efficiently fermented to ethanol.[6] Furthermore, the whole process economy of second-generation ethanol has to be strictly optimized. Today, however, neither the hydrolysis nor the fermentation steps of this process reach theoretical yields. As handling of the lignocellulosic feedstocks is costly and the value of the produced ethanol is low, there is very little room for inefficiency in the process configuration. Some sugar is lost in the pretreatment due to heating and in the hydrolysis step due to the inefficiency of the enzymes in certain conditions. Yet, since the ethanol concentration in the fermentation broth must be over 40 g/L to allow efficient distillation,[7] at least a total of 80–100 g/L of fermentable sugars have to be initially present in the fermentation liquid. Also in the fermentation step, theoretical yields of ethanol are difficult to reach, and this is why every improvement that can be made with regard to ethanol production is important. For pentoses especially the fermentation of xylose has a large impact on the overall ethanol yield as it is the more abundant of the two pentose sugars and can represent an important fraction of all the fermentable sugars present in the hydrolysate.

5.2.3 Lignocellulosic Ethanol Production: Fermenting Microorganisms

Several microorganisms are able to ferment both hexose and pentose sugars. There are bacteria, fungi, and yeasts that naturally consume xylose, but most of them are not robust enough to tolerate the toxic and acidic conditions in lignocellulosic hydrolysates. In comparisons between different fermenting microorganisms,[8] baker's yeast *S. cerevisiae* has been repeatedly found to be more tolerant to fermentation inhibitors than bacteria and some fungi. Although several robust pentose-consuming fungal species exist, they usually require oxygen for growth and consumption of pentoses.[9] The fermentation conditions in industrial bioreactors are anaerobic, because aeration is not an option for a large and low-margin process due to its cost. Thus, the microorganism of choice should not only tolerate anaerobiosis but also be able to ferment well without oxygen present. At the same time, the use of obligate anaerobic bacteria would be challenging due to the practical difficulties in handling them. For these reasons, baker's yeast *S. cerevisiae* is now considered the best option for the second-generation ethanol production, as it is already used for the first-generation ethanol production as well as in various other industrial applications. Since it has been "domesticated" and naturally selected specifically for the production of alcohol, it is already naturally adapted for some of the stress factors present in industrial fermentation processes.

The aforementioned advantages of high tolerance and robustness of baker's yeast outweigh the limitation that it does not naturally consume pentose sugars at reasonable rates. This has generated intense research on genetically engineering baker's yeast to enable efficient xylose fermentation, recently reviewed in the literature.[10] The topic has been studied since the late 1980s[11] and has served as a good example case in the research field of metabolic engineering[12] as it will be discussed later in this chapter. Currently, the xylose fermentation performance of genetically engineered *S. cerevisiae* is reaching commercially viable levels.

5.2.4 History of Metabolic Engineering for Pentose Fermentation

Metabolic engineering applies genetic engineering to introduce exogenous genes or to alter the levels of the endogenous metabolic enzymes to change the product or substrate range of a certain microorganism. What has become evident during the studies in this field is that it is not the mere presence of the enzymes necessary for a certain metabolic pathway but the extent of a metabolic flux, which determines the success of a metabolic engineering strategy. The metabolic flux from a substrate toward a target product depends on multitude of factors including the specific activities of the enzymes involved in the metabolic pathway, their catalytic properties such as affinities and catalytic rates, the thermodynamics of the reactions, and the concentrations of the intermediate metabolites that may vary depending on other metabolisms taking place at the same time. Just introducing the relevant genes is not necessarily enough to create sufficient metabolic fluxes for the production of a desired product, but in-depth careful optimization of the pathways is needed. This is why metabolic engineering strategies that in principle should work but in reality sometimes prove unsuccessful.

The need for metabolic engineering of *S. cerevisiae* for xylose fermentation became evident when in the search of a suitable microorganism during the 1980s, no microorganism was found that could produce ethanol from xylose, grow anaerobically, and tolerate the inhibitory conditions in lignocellulosic hydrolysates. The best solution was to make *S. cerevisiae* xylose fermenting, rather than improving xylose-consuming microorganisms under anaerobic and inhibitory conditions. In the initial studies with naturally xylose-fermenting yeasts such as *Scheffersomyces (Pichia) stipitis* and various *Candida* species, it was found that they metabolize xylose through the use of the enzymes xylose reductase (XR)[13] and xylitol dehydrogenase (XDH; Figure 5.3).[14] Upon the advent of genetic engineering techniques, trials were started with heterologous expression of XR and XDH in *S. cerevisiae*.[11b]

The first engineered xylose-consuming yeast strains consumed very little of the available xylose in the medium. Soon it was found that also the overexpression of the yeast's endogenous xylulokinase (XK) was necessary to direct the xylose metabolism toward central metabolism.[11a] The next major improvement step was made when the four enzymes in the nonoxidative pentose phosphate pathway (PPP), namely, transketolase (TKL), transaldolase (TAL), ribose-5-phosphate ketol-isomerase (RKI), and D-ribulose-5-phosphate 3-epimerase (RPE), were simultaneously overexpressed.[15] The full benefit of this engineering strategy did not become obvious until it was combined with multicopy overexpression of the XR and XDH.[16] It illustrates that identification of the rate-controlling metabolic steps in a pathway is not always straightforward and depends on the order in which the new modifications are introduced. Despite the improved aerobic xylose consumption

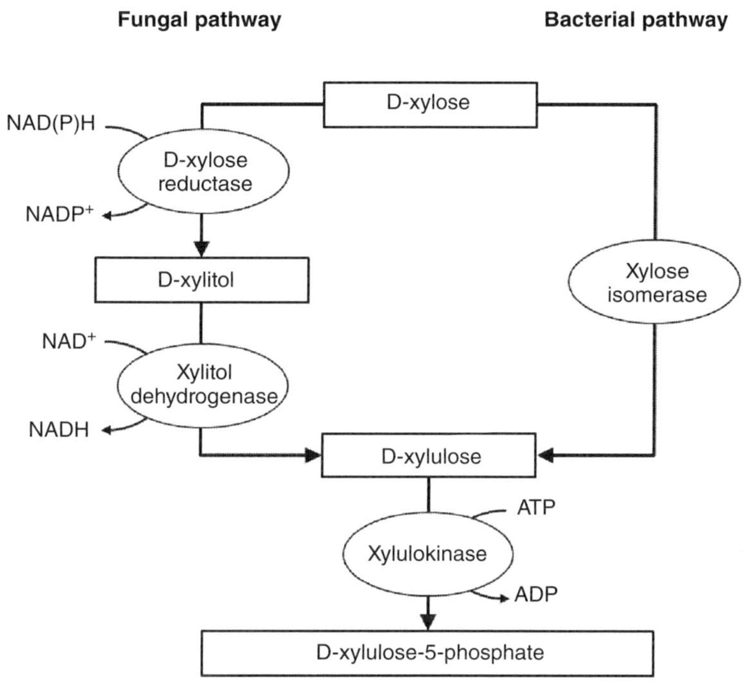

Figure 5.3 The initial xylose conversion pathways in fungi (XR–XDH) and bacteria (XI).

by these yeast strains, significant anaerobic growth on xylose was not yet reached. This was attributed to the mismatching cofactor preferences of the XR and XDH enzymes (Figure 5.3).

Another consequence of the cofactor mismatch between XR and XDH was that the first yeast strains expressing XR and XDH primarily produced xylitol from xylose in mineral laboratory medium. The native XR of *S. stipitis* uses both NADH and NADPH as cofactors, whereas the following enzyme XDH uses NAD^+ only. This becomes a problem especially in anaerobic conditions where the yeast cell has no means for regeneration of enough of NAD^+. Metabolic pathway modifications to force the yeast cell to make more NAD^+, and less NADPH available for XR, were unsuccessful due to the significant decrease in the overall xylose consumption rate.[17]

The obvious strategy to circumvent the formation of xylitol is to use the enzyme xylose isomerase (XI; Figure 5.3) and enzyme that is common in bacteria and plants. It proved difficult to find an XI enzyme that is sufficiently well expressed in yeast. After trials with several bacterial XIs,[18] a suitable XI enzyme was found in the fungus *Piromyces* sp.[19] Subsequently, several other XI enzymes have been expressed in baker's yeast.[20] The initial results obtained with functionally expressed XI proved that the previous observation of the presence of an enzyme activity is not enough for an effective metabolic pathway, but rather that several factors need to be coupled to allow sufficient metabolic flux in reality. As for the case of XR and XDH, expression of XI only was not sufficient for xylose fermentation, but also overexpression of XK and the PPP enzymes were necessary to reach reasonable xylose consumption rates. Still, some xylitol is formed even in XI-carrying strains due to activities of unspecific aldoreductases in *S. cerevisiae*.

As the early xylose-fermenting strains only grew on xylose under aerobic conditions, reaching anaerobic growth on xylose was long considered a target point to show that xylose consumption that is fast enough for efficient ethanolic fermentation had been reached. Indeed, it was later shown in laboratory conditions that anaerobic growth correlates well with the specific ethanol productivity.[5]

Extensive evolutionary engineering programs were undertaken to reach anaerobic growth on xylose with yeast strains expressing XI.[21] In evolutionary engineering, cells are grown repeatedly in strongly selective environment for hundreds of generations, usually taking months of laboratory work time. The natural mutation rate of cell replication generates mutations, and the best growing mutants are enriched. This very efficient, but time-consuming, strain improvement strategy was used to generate the first anaerobically growing yeast strains carrying XI.[21,22] Later investigations indicated that it was partially xylose transport but most of all the increased number of the XI gene copies that were incorporated to the yeast chromosomal DNA.[21c] The early XI-expressing strains carried the gene on a multicopy plasmid, and subsequent evolution had apparently caused integration of parts of the plasmid also in the genomic DNA. This suggests that a very high number of copies of the XI gene are required to support efficient anaerobic growth and ethanolic fermentation of xylose.

Another approach to achieve anaerobic growth was to genetically modify the XR or XDH proteins to change their cofactor preferences.[23] After comparisons between XR protein sequences from different sources, mutations were made in candidate sites that lie in the cofactor-binding region. Thus, mutation K270R resulted in a decreased affinity for NADPH, which probably increases the use of NADH by this enzyme.[23] An even better result was achieved with the N272D mutation, which had higher NADH-using activity than the wild-type enzyme.[24] Also mutated XDH enzymes have been generated.[25] On this basis, *C5LT bases its xylose fermentation technology on mutated XR–XDH pathway enzymes.*

The advantage of the XR–XDH pathway is that only a limited number of gene copies are required for efficient xylose uptake and anaerobic growth on xylose. When compared in yeast strains with isogenic (i.e., genetically identical) backgrounds, the same expression levels of either XR–XDH or XI genes result in very different xylose fermentation rates.[26] When expressed from a plain multicopy plasmid with the same promoter as XR, the *Piromyces* sp. XI did not provide as fast xylose consumption rate as the XR-carrying strain in the same genetic setup. Only after evolutionary engineering of the XI-carrying strains, fast xylose consumption rates were achieved, which was primarily the result of multiplication of the XI gene to sufficiently high copy numbers.

The need for tailored yeast strains for different combinations of lignocellulosic material and process design calls for a system that allows fast and reproducible construction of xylose-fermenting yeast strains in various strain backgrounds and with individual genetic modifications. In this context, the need for evolutionary engineering significantly reduces the number of strains that can be developed in a certain time frame. C5LT has therefore chosen to use the mutated XR–XDH strategy to optimize the cofactor usage, since it allows quick development of xylose-fermenting strains with minimal genetic engineering and well-defined genetic modifications.

5.3 C5LT GENE PACKAGE TECHNOLOGY

Since the possible combinations for different raw materials, pretreatments, and process conditions are almost infinite, it is unlikely that a single yeast strain will perform equally well in all ethanol production processes worldwide. Rather, every process will require a yeast strain that is tolerant to the lignocellulosic raw material at hand and fulfills the requirements of the process configuration.

C5LT is basing its genetic engineering technology on an easily repeatable transformation system so that we not only generate a collection of C5-fermenting strains based on different strain backgrounds but also can modify any existing or newly isolated *S. cerevisiae* strain into a C5-fermenting one (Figure 5.4).

The system is based on a collection of DNA cassettes, gene packages, all of which can be transformed into industrial yeast strains in different combinations. As the XR–XDH system does not require evolution for effective xylose fermentation, final industrial strains can be screened directly. Our technology has proved to be reproducible and to generate genetically stable strains. It is demonstrated in different strain backgrounds, reaching similar levels of xylose consumption rates, despite the strains having individual characteristics with respect to tolerance and other physiological factors. The introduction of the necessary genes renders the yeast strains effectively xylose fermenting and able to grow on xylose under anaerobic conditions. Since the system does not require evolution, this can be done without jeopardizing the other capabilities of the specific yeast strain. Importantly, such traits may easily be lost during extended evolutionary engineering programs. This is why the rational and fast C5LT gene package technology is optimal for modifying yeast strains with industrially interesting properties.

Furthermore, the C5LT genetic engineering technology is not limited to ethanol production. Also, yeast strains developed to produce chemicals other than ethanol can quickly be transformed to utilize xylose.

5.3.1 From Laboratory to Industrial Conditions

When developing xylose-fermenting yeast strains, it is crucial to test strain performance in actual industrial lignocellulosic substrates as they contain numerous compounds that are inhibitory to yeast. Although laboratory media are more defined, working only under laboratory conditions may be thus misleading. As an example of this, much work was dedicated to reduce the xylitol formation in xylose fermentation. But, ironically, when yeast was finally used in real industrial lignocellulosic hydrolysate, very little xylitol formation was actually observed even when the

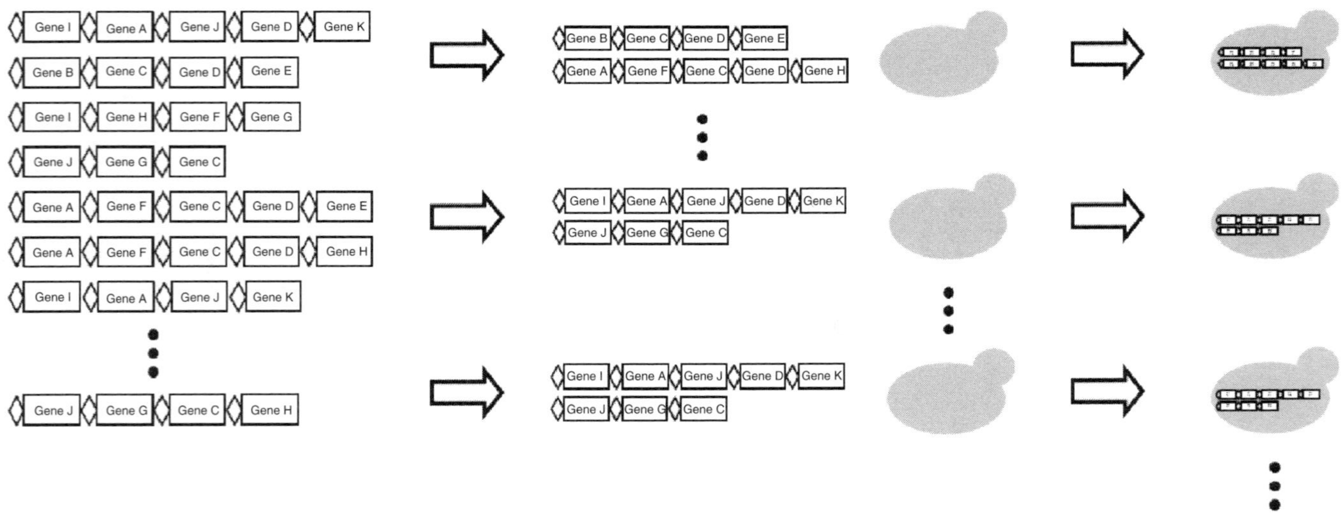

Figure 5.4 Schematic illustration of the C5LT gene package technology to generate xylose-fermenting yeast strains.

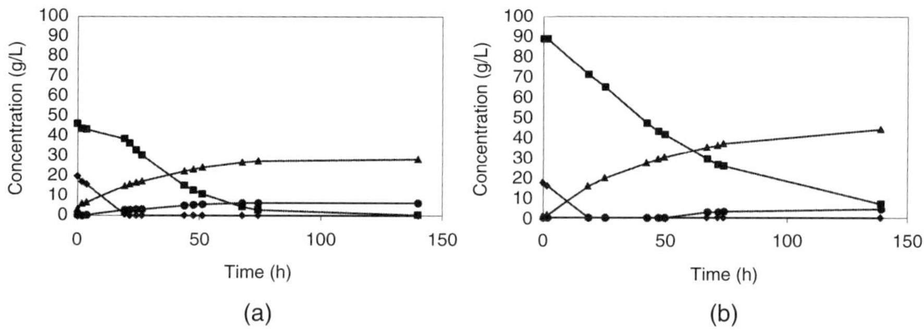

Figure 5.5 Anaerobic fermentation of mineral medium (a) and lignocellulosic hydrolysate (b) using the same C5LT strain. Xylose – squares; glucose – diamonds; ethanol – triangles; xylitol – circles.

yeast carried the native NADPH-dependent XR enzyme.[26,27] The phenomenon is explained by the reduction of lignocellulose-derived inhibitors, such as furfural and 5-hydroxymethylfurfural (HMF), to their less inhibitory alcohols by enzymes similar to those that convert xylose to xylitol.[28] It seems to affect the intracellular cofactor levels so that the availability of NAD^+ increases, which improves xylitol conversion to xylulose by XDH. The same effect can be achieved by addition of an external electron acceptor, such as acetoin.[28] In Figure 5.5 the fermentation performances of the same xylose-fermenting yeast strain in mineral laboratory medium (a) and in lignocellulosic hydrolysate (b) are shown. Although about 10% of xylose is converted to xylitol in mineral medium, virtually no xylitol is formed in lignocellulosic hydrolysate. Thus, relying exclusively on tests made in laboratory conditions would point toward the need for further engineering to diminish xylitol formation, whereas in the real process conditions, this is actually not a problem.

Another remarkable example of how laboratory testing may lead to wrong conclusions is when strains are compared in mineral medium. When comparing strains in mineral medium, they may seemingly have similar xylose consumption rates, but the differences between the strains only become obvious when fermentation tests are run in lignocellulosic hydrolysates. This demonstrates the importance of performing strain development work in the conditions that are actually present in the industrial processes.

5.3.2 Lignocellulosic Hydrolysates: Inhibitors

In addition to sugars, the lignocellulosic hydrolysates contain complex mixtures of other compounds, many of which are inhibitory to microbial fermentation. When the lignocellulosic material is heated during pretreatment, some of the sugars are dehydrated to furans such as furfural (from pentoses) and HMF (from hexoses), which are toxic to most microorganisms. Moreover, various compounds formed from the lignin fraction may also be inhibitory. Further, when the hemicellulose is hydrolyzed to release the monomeric sugars, acetic acid is formed by the deacetylation of this

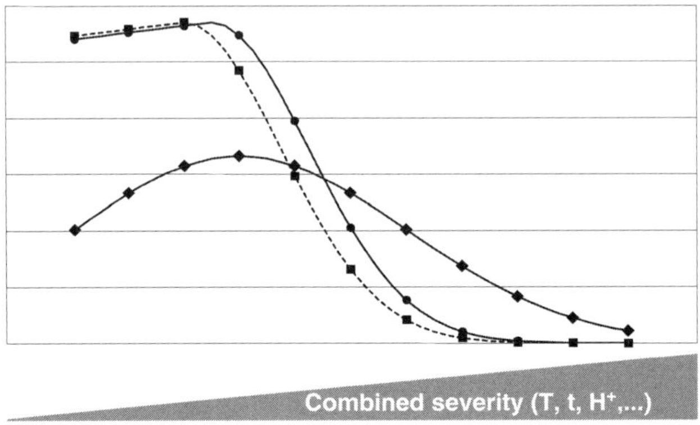

Figure 5.6 Sugar yield (diamonds), C6 fermentability (circles), and estimated C5 fermentability (squares) as function of the combined severity values. Data retrieved from [29].

fraction. More acids are formed if the lignocellulosic hydrolysate containing furfural and HMF is further heated, since these compounds are degraded into formic and levulinic acids, which are even more potent inhibitors of microorganisms than acetic acid.[29] Consequently, the toxicity and acidity of the pretreated and hydrolyzed lignocellulosic material present a strong limitation on the choice of the fermenting microorganism.

In process design for ethanol production from lignocellulosic hydrolysates, the most challenging balance is the one between recovering the highest possible amount of monomeric sugar from the raw material and at the same time maintaining the fermentability of the resultant hydrolysate. During pretreatment, monomeric sugars are released mainly from hemicellulose. After this, in most cases an enzymatic hydrolysis is performed with cellulases and other similar enzymes, which releases monomeric sugars from the cellulose. During the pretreatment, various parameters, like temperature and time, determine how much sugar, acetic acid, and inhibitors like furfural and HMF are released. Generally, longer residence times and higher temperatures increase the formation of inhibitory furans from the monomeric sugars. Thus sugar is not only lost from the fermentation process, but also the fermentability of the mixture is decreased as these compounds are toxic to yeast.

Larsson et al.[29] have shown how the severity value (i.e., combination of the process parameters) of the pretreatment affects sugar yields on one hand and fermentability of them on the other hand (Figure 5.6). At higher severity values, more sugar is released, but the maximum fermentability, measured as the maximum ethanol yield, may be achieved at lower severity than where maximum sugar yield is reached. This effect is likely to be even more pronounced for pentose fermentation, which is generally known to be more inhibited by lignocellulosic degradation compounds. In Figure 5.6 a schematic estimate of the situation for pentose sugars is shown.

The higher sensitivity of xylose fermentation to inhibitors and especially acidic conditions may result in situations where a process that is optimized for fermentation of hexose sugars works poorly when switched to pentose-fermenting yeast. Thus, fermentation processes have to be separately optimized for second-generation ethanol production. It is very tempting to expect that pentose-fermenting yeast just can be handled like any other yeast, but much better results can be expected if efforts are put on optimizing the fermentation conditions in close collaboration with process design.

5.3.3 Different Yeast Strains Suit Different Processes

Although the yeast *S. cerevisiae* is collectively called baker's yeast, there are numerous different strains of this yeast that have different properties with regard to, for example, growth and fermentative capacity. Different yeast strains are used for baking, brewing, and winemaking, as well as for the production of spirits or fuel ethanol. In, for example, wine- or beermaking, different yeasts are used in order to produce different taste characteristics, which is a result of slight differences in the metabolic routes to flavor compounds. Similarly, tolerance to inhibitors and stress factors present in lignocellulosic hydrolysates requires metabolic routes in yeast, and these routes are different depending on the type of inhibitor compound. As lignocellulosic hydrolysates from different raw materials contain their individual combination of inhibitors, different yeast strains are needed for optimal fermentation of each hydrolysate. The process configuration sets further constraints to the choice of strain, for example, by determining how much the yeast is supposed to grow and under which conditions or, reversely, how much carbon should be used for yeast growth at the expense of ethanol production.

Yeast strains have been compared for tolerance in lignocellulosic hydrolysates in several studies.[30] Tolerance of yeast toward inhibitors can also be measured separately for each inhibitor, but it is apparent from our practical experience that there are synergistic effects when all inhibitors are present at the same time. This is why tolerance is best assessed in the actual lignocellulosic hydrolysates. In a screening by Almeida et al.,[30a] 13 different strains were compared for tolerance against three different lignocellulosic hydrolysates in a test based on anaerobic growth. Most strains had a widely varying performance in the different hydrolysates, but there were also strains that were always among the best performing in all of the tested hydrolysates. Although growth in the presence of lignocellulosic hydrolysate is essential for the possibility to ferment it, other parameters, such as flocculation, acid tolerance,[31] and thermotolerance,[32] may also be important for the choice of yeast strain. *The C5LT gene package technology allows quick genetic modification of any yeast strain to a xylose-utilizing one so that individually tailored strains can be designed for each combination of raw material, pretreatment, and process configuration.*

5.3.4 Fermentation of Lignocellulosic Hydrolysates by C5LT Yeast Strains

At C5LT, fermentation of lignocellulosic hydrolysates is routinely assessed in 1 L batch fermentation to evaluate the fermentation performance of yeast strains. Reversely, it is also used to evaluate the fermentability of hydrolysates with a strain with known xylose fermentation capabilities.

Here we present fermentation results with C5LT's industrial xylose-fermenting yeast strains in lignocellulosic hydrolysates originating from corn stover, bagasse,

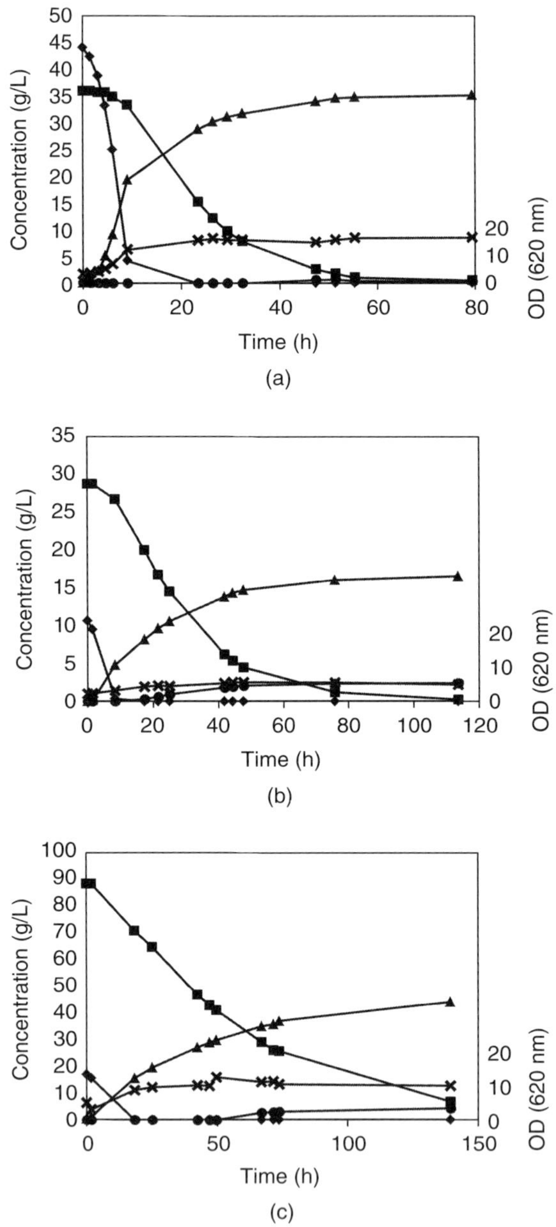

Figure 5.7 Anaerobic batch fermentation of corn stover (a), wheat straw (b), and bagasse (c) using three different C5LT strains. Xylose – squares; glucose – diamonds; ethanol – triangles; biomass – crosses; xylitol – circles.

and wheat straw (Figure 5.7). We have selected from our strain collection the best performing strains for each of the hydrolysates. Since there is a strong incentive to keep the yeast concentration low, we run our fermentations inoculated with 1–2 g/L yeast cell dry weight. The higher concentration of 2 g/L is used for hydrolysates with high sugar concentrations, in this case bagasse, which had over 100 g/L of total sugar (Figure 5.7c). The effect of yeast concentration on fermentation will be further discussed later in this chapter. Under these conditions ethanol is produced with yields of 81–86% of theoretical, corresponding to yield of 0.41–0.44 g ethanol per gram total sugar in the hydrolysate. Small amounts of xylitol were produced due to relatively low inhibitor concentrations. With C5LT yeast strains, the fermentation of about 35 g/L xylose, together with the other sugars, is complete within 60–80 h when using the aforementioned yeast concentrations. In this case, fermentations were inoculated with untreated yeast cells grown in rich laboratory medium (YPD), for the sake of reproducibility. By pregrowing cells in tolerance-inducing conditions, even faster fermentation can be achieved.[33]

5.3.4.1 Fermentation Parameters As yeast developers, we have long asked from process developers whether it is the ethanol yield or the ethanol productivity that is the most important target parameter for lignocellulosic fermentation. The optimal situation with maximal yield and maximal productivity seems to be difficult to reach in practice, and the two parameters often seem to be mutually exclusive. High yields of ethanol, with regard to the sugar present in the raw material, are easier to reach when the process is run until sugar consumption is complete. However the cost of running the plant has to be balanced against the gain in ethanol produced, especially when the pentose fermentation sometimes proceeds slowly at the end.

High productivity, that is, short fermentation times, is best reached with a yeast strain that ferments xylose fast in the presence of inhibitors, but also process design can be applied to reduce fermentation time. For example, ethanol productivity can be increased by using higher yeast concentrations or designing the process so that yeast is allowed to grow. By keeping the yeast concentration high, more fermentative capacity is present in the reactor, and fermentation will proceed faster. However, the correlation between cell concentration and fermentation efficiency is not necessarily linear. There is also a connection between yeast concentration and tolerance to inhibitors, as higher yeast cell mass faster detoxifies the inhibitors in the hydrolysates. Therefore a thorough analysis is required for each process regarding the optimal yeast concentration that is required, because, as stated earlier, the yeast production cost is high considering the low margins in ethanol production. There is also to possibility to add a yeast propagation step to the process configuration, which may lower the cost of yeast production. Figure 5.8 shows lignocellulosic fermentation starting with 2 g/L (dry weight) cells, compared with 5 g/L yeast dry weight concentration. Fermentation proceeds about twice as fast when the yeast concentration is doubled.

For comparing the performance of yeast strains from various sources, it would be interesting to have general fermentation parameters that could easily be compared. It would also aid in setting a common “goal” value for satisfactory yeast performance. Since the public data often are incomplete and originate from different conditions, no

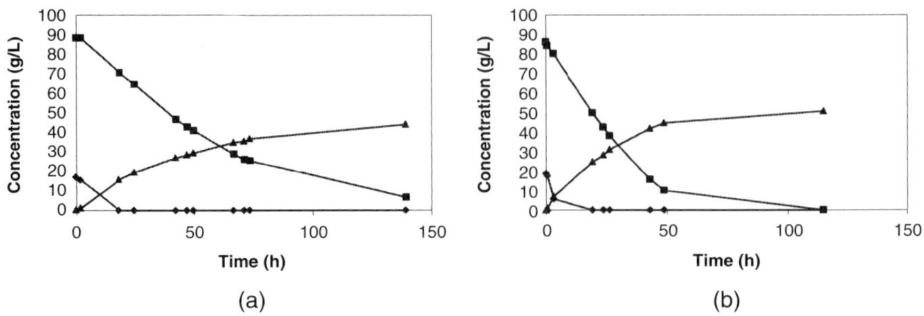

Figure 5.8 Anaerobic batch fermentation of lignocellulosic hydrolysate using 2 g/L (a) or 5 g/L (b) of initial yeast cell dry weight concentration. Xylose – squares; glucose – diamonds; ethanol – triangles.

such benchmark values are publicly available. However when interpreting data from different sources, yield is not a very informative measure as crucial process parameters are seldom reported, although they may have large effect on the yield values. Yeast concentration is an example of such a crucial parameter. In laboratory conditions, specific productivity, that is, g ethanol/h/g cell dry weight, is considered the most reliable parameter when comparing different experiments because the cell concentration is then taken into account. In lignocellulosic hydrolysates, however, the concentration of cells is hard if not impossible to measure due to the viscosity and color of the material. Volumetric productivity (g ethanol/h/l) is frequently reported, but it does not include the yeast concentration, which has immense effect on both fermentation and process economies. As it is known how much yeast is initially added from the beginning, we prefer to use this yeast concentration as a parameter to relate fermentation performance to ethanol. Productivity as a general variable is not suitable for lignocellulosic hydrolysates, since different hydrolysates contain different amounts of different sugars, and the fermentation time in different process configurations varies.

5.4 FERMENTATION OF LIGNOCELLULOSIC HYDROLYSATES: REMAINING CHALLENGES

The main challenge in lignocellulosic fermentation is that xylose fermentation is slower than glucose fermentation for all currently known yeast strains. Furthermore, with the early xylose-fermenting strains, xylose fermentation remained incomplete regardless of fermentation time. For the latest ones involving more robust strains, complete depletion of xylose is rather a question of how long the fermentation is allowed to proceed.

The heterogeneity of the substrate, that is, the mix of different sugars that all need to be efficiently fermented by the same microorganism, is another challenge. The molecular regulation of yeast metabolism is such that not all sugars can be consumed at the same time. This depends partially on steric hindrances when different sugars are to be transported into the cell through the same transporters but is also due to different levels of regulatory processes in the cells. The fact that galactose is only consumed after glucose is depleted is usually not a problem as the amount of galactose in most lignocellulosic hydrolysates is very low. On the contrary the xylose fraction is often significant, and the sequential consumption increases the fermentation time. A large number of studies focus on cofermentation of hexoses and pentoses, and the question is, is it necessary? As long as all sugars are fermented within a target time frame, it should not be important in which order the sugars are consumed.

The ratio of glucose and xylose concentrations in the hydrolysate may also affect fermentation behavior. Although a certain degree of cofermentation of glucose and xylose is often seen, the general pattern is that xylose is consumed when glucose concentration is low or depleted. Yet, sometimes a small amount of glucose actually improves xylose fermentation: in the SSF fermentation setup, glucose is constantly released by the function of the hydrolyzing enzymes, and the constant presence of low level of glucose improves ethanol formation and decreases xylitol formation in the early XR–XDH-carrying strains.[34] Furthermore, hydrolysates with low amounts of glucose often appear more toxic, since detoxification of the inhibitory compounds is more efficient during the glucose consumption phase.

As discussed previously, the most effective process for the generation of monomeric sugars does not always produce the most fermentable hydrolysate. Therefore the importance of evaluating hydrolysis processes based on the yield of ethanol rather than yield of sugar cannot be stressed too much. This implies that efficient xylose fermentation must be developed not only from the fermentation perspective but also from the pretreatment and hydrolysis perspective.

5.5 CONCLUSIONS

Baker's yeast *S. cerevisiae* is the microorganism of choice not only for the second-generation ethanol production but also for the production of other chemicals from renewable lignocellulose. The development in genetic engineering of yeast to allow xylose fermentation is now reaching a level that a commercial application in real industrial conditions is becoming a reality. C5LT has developed a series of industrial xylose-fermenting yeast strains with good fermentation performance in lignocellulosic hydrolysates. The C5LT technology is based on interchangeable gene packages that can easily be transformed into any *S. cerevisiae* strain. The technology is fast and reproducible and does not require lengthy evolution protocols for successful yeast strain development. It constitutes a "plug-in" system for *S. cerevisiae* strains from any variety of environments, such as strains domesticated in specific ethanol plants, as well as strains engineered for the production of chemicals other than ethanol. Furthermore, it particularly allows for the introduction of new gene modifications as they are discovered. Regardless of the end product to be produced from lignocellulosic materials, it is essential for any process that the raw material must be efficiently converted. Without effective uptake and metabolism of all sugars present in the lignocellulosic hydrolysate, it is impossible to reach high product yields. The C5LT gene package technology allows the development of tailored yeast strains that respond to these requirements.

ACKNOWLEDGMENTS

We thank Prof. emerita Bärbel Hahn-Hägerdal for critically reading this manuscript.

REFERENCES

[1] Cellulosic Biofuels Industry Progress report 2012–2013, Advanced Ethanol Council.

[2] www.iogen.ca.

[3] P. Alvira, E. Tomás-Pejó, M. Ballesteros, M.J. Negro, *Bioresour. Technol.* **2010**, *101*(13), 4851–4861.

[4] W.F. Gauss, S. Suzuki, M. Takagi, **1976**, US3,990,944.

[5] J.R. Almeida, D. Runquist, V. Sànchez i Nogué, G. Lidén, M.F. Gorwa-Grauslund, *Biotechnol. J.* **2011**, *6*(3), 286–299.

[6] (a) S. Macrelli, M. Galbe, O. Wallberg, *Biotechnol. Biofuels* **2014**, *7*(1),26; (b) P. Sassner, G. Zacchi, *Biotechnol. Biofuels* **2008**, *1*(1),4.

[7] A. Wingren, Ethanol from Softwood - Techno-Economic Evaluation for Development of the Enzymatic Process. PhD thesis, Lund University, **2005**.

[8] (a) L. Olsson, B. Hahn-Hägerdal, *Process Biochem.* **1993**, *28*(4), 249–257; (b) M.W. Lau, C. Gunawan, V. Balan, B.E. Dale, *Biotechnol. Biofuels* **2010**, *27*(3), 11.

[9] R. Maleszka, H. Schneider, *Can. J. Microbiol.* **1982**, *28*(3), 360–363.

[10] S.R. Kim, Y.C. Park, Y.S. Jin, J.H. Seo, *Biotechnol. Adv.* **2013**, *31*(6), 851–861.

[11] (a) S.F. Chang, N.W. Ho, *Appl. Biochem. Biotechnol.* **1988**, *17*, 313–318; (b) P. Kötter, R. Amore, C.P. Hollenberg, M. Ciriacy, *Curr. Genet.* **1990**, *18*(6), 493–500.

[12] G. Stephanopoulos, J.J. Vallino, *Science* **1991**, *252*(5013), 1675–1681.

[13] C. Verduyn, R. Van Kleef, J. Frank, H. Schreuder, J.P. Van Dijken, W.A. Scheffers, *Biochem. J.* **1985**, *226*(3), 669–677.

[14] M. Rizzi, K. Harwart, N.A. Bui-Than, H. Dellweg, *J. Ferment. Bioeng.* **1989**, *67*(1), 25–30.

[15] (a) B. Johansson, B. Hahn-Hägerdal, *FEMS Yeast Res.* **2002**, *2*(3), 277–282; (b) K. Karhumaa, B. Hahn-Hägerdal, M.F. Gorwa-Grauslund, *Yeast* **2005**, *22*(5), 359–368.

[16] K. Karhumaa, R. Fromanger, B. Hahn-Hägerdal, M.F. Gorwa-Grauslund, *Appl. Microbiol. Biotechnol.* **2007**, *73*(5), 1039–1046.

[17] M. Jeppsson, B. Johansson, B. Hahn-Hägerdal, M.F. Gorwa-Grauslund, *Appl. Environ. Microbiol.* **2002**, *68*(4), 1604–1609.

[18] (a) K.A. Briggs, W.E. Lancashire, B.S. Hartley, *EMBO J.* **1984**, *3*(3), 611–616; (b) A.V. Sarthy, B.L. McConaughy, Z. Lobo, J.A. Sundstrom, C.E. Furlong, B.D. Hall, *Appl. Environ. Microbiol.* **1987**, *53*(9), 1996–2000; (c) M. Walfridsson, X. Bao, M. Anderlund, G. Lilius, L. Bülow, B. Hahn-Hägerdal, *Appl. Environ. Microbiol.* **1996**, *62*(12), 4648–4651.

[19] M. Kuyper, H.R. Harhangi, A.K. Stave, A.A. Winkler, M.S. Jetten, W.T. de Laat, J.J. den Ridder, H.J. Op den Camp, J.P. van Dijken, J.T. Pronk, *FEMS Yeast Res.* **2003**, *4*(1), 69–78.

[20] (a) A. Madhavan, S. Tamalampudi, K. Ushida, D. Kanai, S. Katahira, A. Srivastava, H. Fukuda, V.S. Bisaria, A. Kondo, *Appl. Microbiol. Biotechnol.* **2009**, *82*(6), 1067–1078; (b) D. Brat, E. Boles, B. Wiedemann, *Appl. Environ. Microbiol.* **2009**, *75*(8), 2304–2311; (c) L. de Figueiredo Vilela, V.M. de Mello, V.C. Reis, E.P. Bon, F.A. Gonçalves Torres, B.C. Neves, E.C. Eleutherio, *Bioresour. Technol.* **2013**, *128*, 792–796; (d)R.E. Hector, B.S. Dien, M.A. Cotta, J.A. Mertens, *Biotechnol. Biofuels* **2013**, *6*(1), 84.

[21] (a) M. Kuyper, A.A. Winkler, J.P. van Dijken, J.T. Pronk, *FEMS Yeast Res.* **2004**, *4*(6), 655–664; (b) L. Diao, Y. Liu, F. Qian, J. Yang, Y. Jiang, S. Yang, *BMC Biotechnol.* **2013**, *13*, 110; (c) H. Zhou, J.S. Cheng, B.L. Wang, G.R. Fink, G. Stephanopoulos, *Metab. Eng.* **2012**, *14*(6), 611–622.

[22] M. Kuyper, M.J. Toirkens, J.A. Diderich, A.A. Winkler, J.P. van Dijken, J.T. Pronk, *FEMS Yeast Res.* **2005**, *5*(10), 925–934.

[23] (a) Y. Zhang, H. Lee, *FEMS Microbiol. Lett.* **1997**, *147*(2), 227–232; (b) M. Jeppsson, O. Bengtsson, K. Franke, H. Lee, B. Hahn-Hägerdal, M.F. Gorwa-Grauslund, *Biotechnol. Bioeng.* **2006**, *93*(4), 665–673.

[24] D. Runquist, B. Hahn-Hägerdal, M. Bettiga. *Appl. Environ. Microbiol.* **2010**, *76*(23), 7796-7802.

[25] S. Krahulec, M. Klimacek, B. Nidetzky, *Biotechnol. J.* **2009**, *4*(5), 684–694.

[26] K. Karhumaa, R. Garcia Sanchez, B. Hahn-Hägerdal, M.F. Gorwa-Grauslund, *Microb. Cell Fact.* **2007**, *6*, 5.

[27] Y. Lu, R. Warner, M. Sedlak, N. Ho, N.S. Mosier, *Biotechnol. Prog.* **2009**, *25*(2), 349–356.

[28] C.F. Wahlbom, B. Hahn-Hägerdal, *Biotechnol. Bioeng.* **2002**, *78*(2), 172–178.

[29] S. Larsson, E. Palmqvist, B. Hahn-Hägerdal, C. Tengborg, K. Stenberg, G. Zacchi, N.-O. Nilvebrant, *Enzyme Microb. Technol.* **1999**, *24*, 151–159.

[30] (a) J.R. Almeida, K. Karhumaa, O. Bengtsson, M.F. Gorwa-Grauslund, *Bioresour. Technol.* **2009**, *100*(14), 3674–3677; (b) E. Albers, C. Larsson, *J. Ind. Microbiol. Biotechnol.* **2009**, *36*(8), 1085–1091; (c) T. Brandberg, C.J. Franzén, L. Gustafsson, *J. Biosci. Bioeng.* **2004**, *98*(2), 122–125.

[31] A. Garay-Arroyo, A.A. Covarrubias, I. Clark, I. Niño, G. Gosset, A. Martinez, *Appl. Microbiol. Biotechnol.* **2004**, *63*(6), 734–741.

[32] V. Wallace-Salinas, M.F. Gorwa-Grauslund, *Biotechnol. Biofuels* **2013**, *6*(1), 151.

[33] V. Sànchez i Nogué, V. Narayanan, M.F. Gorwa-Grauslund, *Appl. Microbiol. Biotechnol.* **2013**, *97*(16), 7517–7525.

[34] K. Öhgren, O. Bengtsson, M.F. Gorwa-Grauslund, M. Galbe, B. Hahn-Hägerdal, s G. Zacchi, *J. Biotechnol.* **2006**, *126*(4), 488–498.

6

Cepsa: TOWARDS THE INTEGRATION OF VEGETABLE OILS AND LIGNOCELLULOSIC BIOMASS INTO CONVENTIONAL PETROLEUM REFINERY PROCESSING UNITS

MARIA FÉ ELÍA, OLALLA DE LA TORRE, RAFAEL LARRAZ, AND JUANA FRONTELA

Cepsa Research Centre, Alcalá de Henares, Madrid, Spain

Industrial Biorenewables: A Practical Viewpoint, First Edition. Edited by Pablo Domínguez de María.

6.1 ABOUT Cepsa

The *Compañía Española de Petróleos S.A.U.*, Cepsa (Spanish Petroleum Company) is an integrated, energy service company that is active in the entire petroleum value chain. Cepsa performs commercial activities in the field of exploration and production of crude petroleum, in areas of refinery and subsequent petrochemical processes, in sectors like gas and electricity, and it is also present in the distribution and commercialization of all its products. In August 2011, the International Petroleum Investment Company (IPIC), an investment group established by the Abu Dhabi government in 1984 – which had been already present as stakeholder of Cepsa over the last 26 years – acquired 100% of Cepsa (Figure 6.1).

Cepsa has operated in the energy service sector since 1929, when it was constituted as the first privately owned petroleum company of Spain. It has traditionally been a dynamic company with an innovating spirit that – across its broad history – has always put emphasis on research activities applied to production processes, as well as to the development of novel innovative products. Currently, Cepsa has a worldwide international presence in Europe, America, Africa, and Asia (Figure 6.2) and has over 10,500 employees.

The overall distillation capacity of Cepsa is ca. 27 MT/y, that is, more than 35% of total distillation capacity in Spain.[1] Cepsa's refining operations are concentrated in three refineries focused on delivering fuels and petrochemical commodities, as well as a fourth plant producing asphalts:

- Gibraltar-San Roque Refinery, in operation since 1967, and distilling 12 MT/y.
- La Rábida Refinery, in operation since 1969, distillation capacity of 10 MT/y.
- Tenerife Refinery, in operation since 1930, distilling 4.5 MT/y.
- ASESA Refinery (50% co-owned), distilling 1 MT/y.

The three first refineries – delivering fuels – are oriented to the optimized production of gas oil for its use as automotive fuels in heating and for agricultural purposes (Figure 6.3).

Figure 6.1 International Petroleum Investment Company (IPIC). (*See color plate section for the color representation of this figure.*)

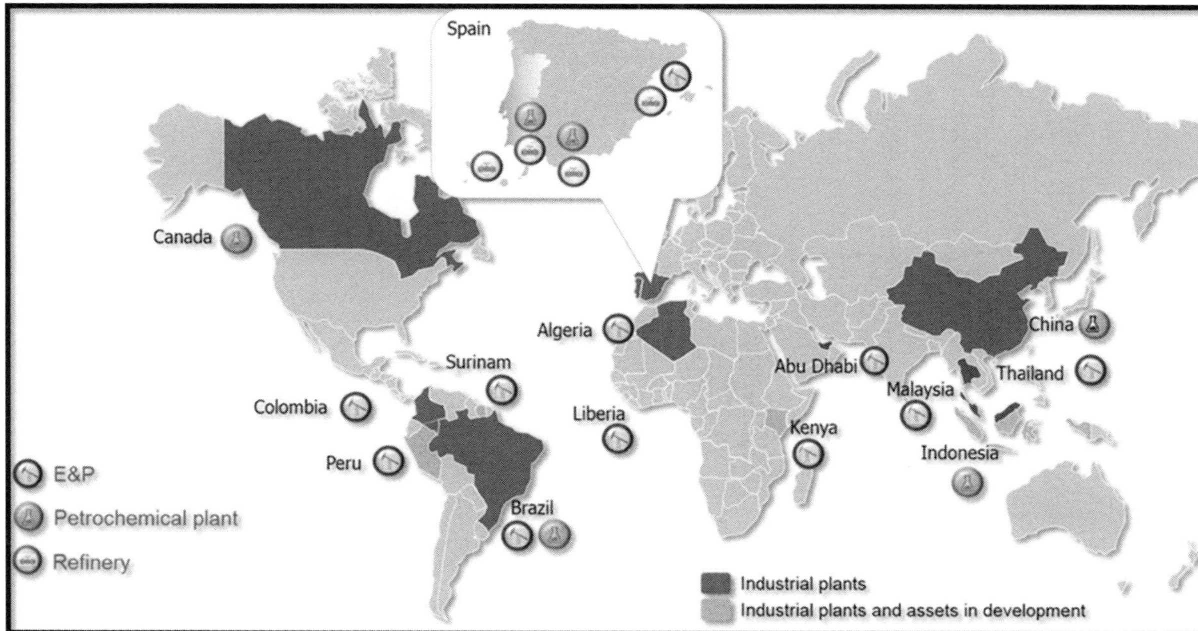

Figure 6.2 Current presence of Cepsa worldwide.

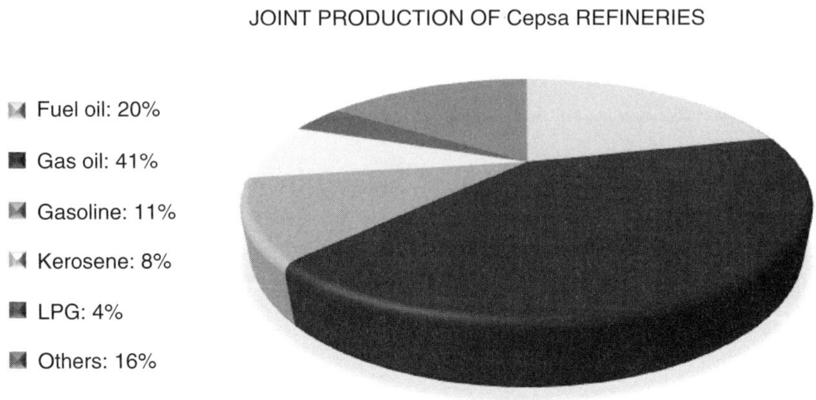

Figure 6.3 Product distribution portfolio of Cepsa refineries.

6.1.1 Our Strategy

Cepsa is strongly committed to meeting four core important aspects as part of its strategy as an energy service company:

- International projection. Continuous improvement, efficiency, and responsibility are the three basic pillars of all Cepsa activities at the international arena.
- Efficient and sustainable energy. We are committed with society to provide secure and sustainable energy and thus contributing to the development of regions where operating plants are, committed with the respect to Human Rights and ethical conduct, and making efforts to minimize the impact of our activity in the environment. Cepsa has implemented a health and safety working strategy – based on OHSAS 18001:2007 – aiming at eliminating or minimizing associated risks to employees or third parties in all working operations worldwide. It is noteworthy that Cepsa invests annually ca. 100,000 man-hours in personal training for work incident prevention.
- Innovative spirit and technical excellence in its activities. The combination of experience, knowledge, and innovation maintains Cepsa at the highest competitive standards in all of its activities. In this area, research is, for Cepsa, a powerful tool for value generation, to enhance competitiveness, and to stimulate (economic) growth, all of them contributing to an improved quality product as well as to the technological development and capacity of the research group.
- Customer satisfaction. The highest product and service quality at Cepsa are the core competences to keep our clients always satisfied with our products.

Furthermore, Cepsa is convinced that its capacity to generate value is closely linked to its aptitude to understand the perspectives and visions of the society. In this respect, Cepsa has been growing with the objective of anticipating customer's

Figure 6.4 Cepsa R&D&I center.

needs, making use of top technology and implementing a broad commercial network worldwide. Moreover, Cepsa is strongly committed with the environmental protection, safety of people and operations, as well as with the quality of the products that are delivered. Since 1975 – when the R&D&I center of Cepsa was launched – the company has converted research into a tool for value generation, competitiveness, and sustainable development, contributing to optimize production processes and improve product quality and technological competences of the company (Figure 6.4).

6.1.2 Cepsa and Renewable Energies

In the current challenging global environment, with the steady increase of the population, the progressive depletion of fossil resources, petroleum price increase, and environmental concerns – such as climate change and air quality – Cepsa is committed to contribute to tackle these challenges by developing novel biomass-based fuels that can be integrated in its business scheme.

However, despite the hurdles associated to fossil-based resources, the European economy is largely dependent on them as carbon and energy sources. Therefore, the European space is vulnerable to (political) insecurities and/or to supply constraints of needed petroleum-based resources, as well as to market fluctuations and volatility of these raw materials. Thus, to maintain (and even enhance) its competitiveness and environmentally aligned growth, the EU must evolve to a low-carbon emission society – in which industries make an efficient use of resources – and must foster the provision of bioproducts and bioenergy within the bioeconomy concept.

On the basis of these reasons, Cepsa has decided to introduce renewable raw materials in its production processes to deliver biofuels. As stated previously, fossil-based fuels (derived from petroleum, natural gas, coal) contribute significantly to industrial growth and technological development. Besides being a crucial energy source for transportation, most of the used chemical materials contain carbon in their structure – namely, fuels, fibers, pharmaceuticals, or other materials – and thus they derived from fossil resources.

All associated problems to the handling of fossil raw materials[2,3] have triggered several governmental actions and initiatives. Among them, several regulations have been established, aiming at gradually replacing such fossil resources by renewable raw materials and thus decreasing the greenhouse effect. According to the directive 2009/28/EC from European Parliament and of the Council on April 23, 2009 – dealing with stimulating the use of renewable energy – a 20% improvement in energetic efficiency must be reached by 2020. 20% of that must proceed from renewable biomass resources, while for the transportation sector, this must account to 10%.[4] The refereed renewable energies covered in such directive are biofuels and bioliquids, biogas, biomass, tidal energy, wind energy, geothermic, and others: hydroelectric, as well as municipal and industrial wastes, EDAR sludge, and solar energy.

Within this described context of more restrictive regulations and increasing social and environmental awareness, Cepsa – fully aware of these environmental problems – established as a core objective of the research in the development of sustainable and environmentally safe fuels. On this basis, Cepsa has been incorporating biofuels in its fuels since 2000. The addition of renewable-based biofuels to energy sources has allowed the company to contribute with the reduction of greenhouse gas emissions, while at the same time diversify the primary energy consumption by decreasing the dependency of fossil-based fuels.

The production of petroleum-derived liquid fuels and organic chemicals starts with the distillation of crude petroleum, leading to different fractions according to the molecular weight. In this process gases are separated, as well as small fractions of solid materials such as waxes and asphalts. The subsequent manufacturing of the obtained fractions demand additional operations, such as sulfur and nitrogen abatement in diesel fuels, or chemical processes to convert hydrocarbons into more structured materials with high added value (Figure 6.5).[3]

Regardless of the automotive fuel – diesel or gasoline – the incorporation of a renewable-based fuel is done after the production of the mineral fuel at the refinery, and thus a blend of fossil-based and renewable fuel is achieved. Nevertheless the type of incorporated biofuel varies from diesel to gasoline. With respect to gasoline, Cepsa incorporates renewable-derived ETBE. In this way the quality of the fuel is improved and enables its transportation through pipelines in a safer and more efficient distribution manner. In concrete, Cepsa has in operation two chemical plants to produce renewable bioETBE, with an annual consumption of 60,000 t of bioethanol. On the other hand, for diesel fuel, the EN-590 regulation establishes that a maximum of a 7% volume in biodiesel can be incorporated.[5] Biodiesel is the most broadly used biofuel in Europe, and it is produced from vegetable oils through a chemical transesterification, which leads to a liquid mixture of fatty acid methyl esters (FAME).[6] Currently, to fulfill regulations and normative, Cepsa purchases the required biodiesel to external companies and in-house blends such biodiesel with the obtained products in the refineries of the company. In detail, Cepsa has in operation two associated plants to its refineries, with a production capacity of 450 MT/y of biodiesel.

Thus, by introducing bioETBE and biodiesel in fuels produced in its refineries, Cepsa covers the current demand under the specifications required by legislation, and what is more important, introduces products with a strong commitment with the environment and customers. In this respect, in 2014 Cepsa incorporated 160 MT of bioETBE in gasolines and 177 MT of biodiesel in gas oil. Likewise, 500 t of B-10, B-15, B-20, and B-30 were labeled *on spec* with European fuel quality standards.

6.1.3 Biofuels at Cepsa: Research and Development

The current use of fossil-based raw materials is more economically efficient than using renewable-based resources. Currently, the cost of renewable resources implies, in some occasions, up to 75% of the overall costs of the process. Likewise, the applied technology for converting petrochemical raw materials into chemicals is largely developed and optimized and is basically different from those routes required to transform vegetable-based raw materials with potential industrial use. The difference lies in the different composition of both fossil-based and renewable-based raw materials.

Crude petroleum is a complex mixture of hundreds of highly hydrophobic hydrocarbon compounds, mainly formed by carbon and hydrogen. Some other elements such as oxygen, nitrogen, and sulfur or phosphorous, can be present in some of the molecules as well. The composition comprises mainly linear hydrocarbons – mostly saturated polymers of units – (CH_2), from 4 to 30 units, as well as some aromatic,

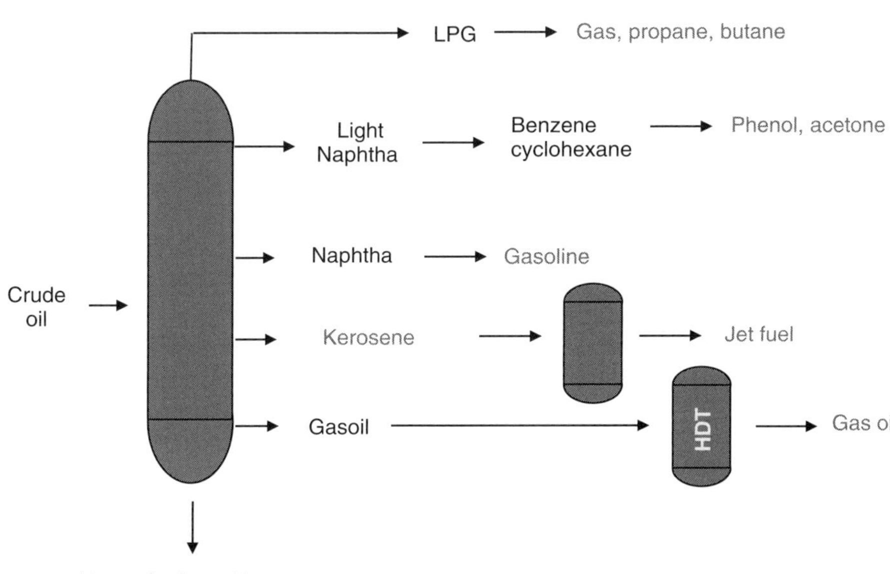

Figure 6.5 Simplified flowchart of a refinery.

naphthenic compounds. In summary, gasoline is a mixture of short-chain hydrocarbons (4–8 atoms), whereas diesel consists of larger chains of hydrocarbons (12–20 atoms). Likewise, mineral jet fuels are a more specific mixture of hydrocarbons, free of oxygen-containing compounds, and with a higher quality – due to stronger regulations – compared to automotive or other used fuels for burning purposes. It is actually a mixture of linear, isomerized alkanes, cycloalkanes, and aromatic compounds within the range of C8–C16.[4]

Conversely, vegetable-derived raw materials present a high oxygen content, and most of the obtained compounds are of a chain length that is not adaptable to the conventional fuel ranges commonly used in transportation sector. Furthermore, biomass is mostly formed by solid materials bearing a large chemical variety in terms of function and structure. Thus, one of the main challenges when working with biomass as raw material is to develop strategies to depolymerize the original material, while at the same time keeping its intrinsic functional structure intact for its valorization.[7]

Based on the previously described differences among fossil-based and vegetable-derived raw materials, the current technologies for mineral processing – largely mature, from which Cepsa has an ample knowledge – must be adapted to process renewable raw materials. Given this, why has Cepsa, with a successful fossil fuels based business, invested in biofuels and renewable raw materials?

On the basis of the earlier reported arguments, for a company which business is based in crude petroleum as raw material, the incorporation of renewable feedstocks – currently not economically competitive – might represent rather a threat. Its incorporation in petroleum units would force to undertake operational risks, due to the already mentioned differences between renewable raw materials and mineral-based petroleum. On the other hand, such incorporation would decrease the sales and profitability of mineral-based products currently commercialized by the company. Nevertheless, despite these reasons, Cepsa has an alternative future vision and has then strategically prioritized the incorporation in its production schemes the use of renewable raw materials. Albeit not envisaged in a short-term perspective, the integration of renewable feedstock's in refineries can become competitive, leading to environmental advantages, for example, in the reduction of CO_2 emissions. In another line, the incorporation of renewable materials would improve the quality of the products that are currently being produced in refineries, achieved by reduction of sulfur, aromatics, and olefins, while simultaneously improving the properties of conventional fuels, such as combustion in gasolines or lubricity in diesel.

Based on this strong commitment for innovation and continuous improvement, the company is tackling the challenge of adapting its technology to the characteristics of the novel renewable feedstocks, while expanding the know-how of novel processes within the company at the same time. Herein, the R&D&I Cepsa center is playing a key role. As main target, the Research Center aims at aligning the investigated processes with the global strategy of the company – organized through dedicated Business Units to exploration and production, refining, and petrochemistry – for the optimization of all processing units, as well as for the quality improvement of products delivered. To this end, there is a close relationship between the Research Center and each of the company Business Units, enabling the optimization of these processes and products, maximizing the efficiency of the operational units, and decreasing costs. Nevertheless, a substantial part of the funds of the Research Center are allocated to innovation, fostering nonmarketable processes at short term. In this area, renewable feedstock is having a key role. Based on the viability studies of each of these developed processes, decisions are made on whether industrial scale-up and implementation to company refineries will take place.

6.1.4 From the Laboratory to the Industry

As an energy source company, our main challenge is the incorporation of raw materials whose compositions differ significantly from mineral-derived petroleum. This must be done using the present implemented technology, to obtain biofuels with an equivalent composition to the mineral ones, and/or biobased additives that improve the quality of the automotive fuel. As stated earlier, apart from being present in all phases of the petroleum value chain processing, Cepsa has also considered as essential the exploitation of different ways of biofuel production. Thus, from the Research Center several initiatives with these aims have been fostered, starting from renewable raw materials – of different origin – with special focus on those proceeding from second and third generation (not competing with food production). In this way, over the last years several projects focusing on vegetable oils (Section 6.2) and lignocellulosic biomass (Section 6.3) have been conducted. In the following sections some remarkable results will be discussed.

6.2 VEGETABLE OILS

The use of vegetable oils as renewable raw material represents a viable and efficient alternative to current fossil-based fuels. Thus, besides contributing to diminishing CO_2 emissions – having a neutral carbon cycle – their H/C ratio is high, and their composition is similar to that of conventional fuels. In this respect, as long as the use of edible vegetable oils and food-producing lands are avoided, the use of vegetable oils as renewable raw materials is a viable alternative for biofuel production.[8]

Vegetable oils and animal fats are mainly formed by fatty acid triglycerides, resulting from the esterification of glycerol and three fatty acids with an even number of carbon atoms. Regardless of their origin, these materials have analogous chemical structures and similar compositions.[9] The biodiesel that is presently being incorporated in conventional fuels is obtained from vegetable oils through transesterification of the fatty acid triglycerides with methanol or ethanol (Figure 6.6).[7]

However, there are many drawbacks associated both to the production process of biodiesel, as well as related to the final product achieved. With regard to problems associated to the transesterification step, economic and environmental hurdles may be pinpointed. They are mainly caused by the formation of glycerol during the process – a by-product with low value and difficult commercialization – as well as to the used catalysts. On the other hand, other issues are related to the final product (FAME)

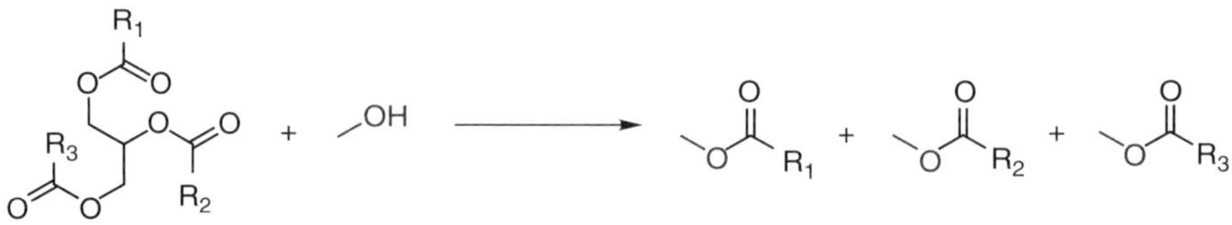

Figure 6.6 Transesterification reaction of triglycerides to form FAME.

properties, which may present some incompatibilities to some regulations – diesel specifications – that must be fulfilled with regard to the Norm EN-590.[5] Furthermore, FAME are not very stable toward oxidations, due to the presence of unsaturated fatty acid chains, and thus lead to obstructions in filters and formation of deposits in the injectors, poor cold properties of the final fuel, and the generation of higher emissions of NO_x than those of conventional mineral diesels. Due to this, only small proportions of FAME (up to 7%) can be blended with mineral-based diesel fuel. Last but not least, if FAME is from edible seeds, it is characterized as first-generation biodiesel, which is another strong constrain for its use as fuel.[7]

Cepsa, in the quest for alternatives to FAME production from vegetable oils, has been undertaking several research projects, supported by public funding provided by the *Ministerio de Economía y Competitividad Español* (MINECO), through its *Centro de Desarrollo Tecnológico* (CDTI). In the following sections, a detailed description of some of these projects will be given.

6.2.1 Industrial Research for the Conversion of Vegetable Oils into High-Quality and Ecological Fuels (IDI-2009-0349)

Due to the previously described significant challenges that FAME may present – as diesel component – since 2005 Cepsa is fostering the development of novel production processes of biofuel, starting from more sustainable vegetable oils, to deliver a final product with better stability and sustainability than transesterification-based biodiesel. The new process is based on the full hydrogenation of triglycerides of vegetable oils to afford hydrotreated vegetable oil (HVO).

Within *the Research Center at Cepsa*, the first pilot plant tests were conducted, reproducing hydrotreating units of company's refineries, using the same commercial catalysts and under conventional operation conditions. The goal of such hydrotreating units is the reduction of sulfur and nitrogen in the gas oil upon addition of hydrogen at high pressure. In this case, by incorporating vegetable oils to the hydrotreating units – achieving a full hydrogenation of the triglycerides – a paraffinic product is obtained, which can be then blended with the pool of diesel obtained from conventional mineral-based resources.

In the hydrogenation mechanism of vegetable oils – using hydrotreating catalysts – at least four main reaction types are involved: the hydrogenation of C=C bonds of fatty acid chains, the hydrodeoxygenation of carbonyl groups (HDO), and the decarboxylation and decarbonylation of such groups (Figure 6.7).[10]

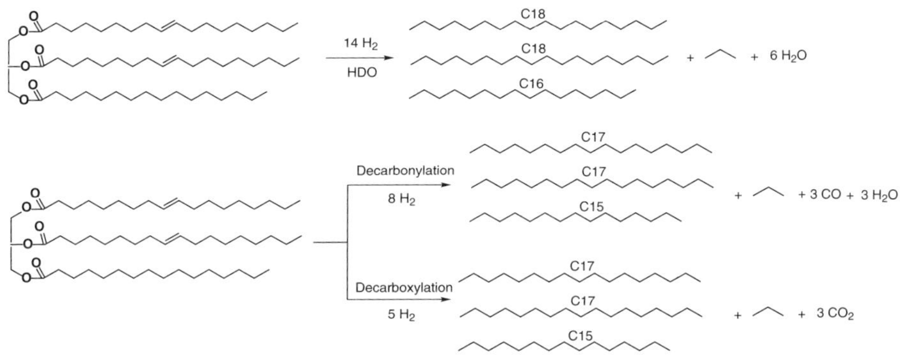

Figure 6.7 Possible reaction pathways in the hydrotreating of vegetable oil. *Source*: [9].

In the hydrodeoxygenation reactions fatty acids are converted into hydrocarbon linear chains with a number of atoms equivalent to the respective fatty acid. Conversely, in decarboxylation and decarbonylation processes, produced paraffin presents (obviously) one carbon atom less than their original fatty acids. Therefore, through a hydrotreating process of vegetable oils, it is possible to obtain paraffin with odd or even carbon atom numbers, which will vary depending on the original triglyceride profile. In a hydrotreating conventional process, the ratio between the decarboxylation and decarbonylation (together) compared to the hydrodeoxygenation is 65:35.

The first experiments performed at the Research Center of Cepsa, using as feedstock soybean, sunflower, rapeseed, and used frying oils (UFO), gave hints on how the final product was changing depending on the oil. Moreover, experiments showed the possible technical bottlenecks that might occur through the hydrotreating of such raw materials in conventional hydrotreating units (Figure 6.8).

In these cases, raw materials were hydrotreated at moderate pressures (55 bar) and employing commercial hydrodesulfurization catalysts. In all cases the diluted vegetable oil was coprocessed with straight run gas oil (SRGO), and it was observed that in all cases the achieved product resulted a mixture of *n*-paraffin adaptable to the desired diesel range. Thus, they could be blended with the conventional diesel pool (Figure 6.9).

Nevertheless, during the hydrotreating process of these raw materials, a number of important aspects were observed as well, which are of importance when the scale-up of the process to a refinery plant is considered. The hydrotreating of vegetable oils is a more *exothermic* process than the hydrodesulphurization of conventional diesel. On the other hand – as it was already predicted from the reaction schemes (Figure 6.7) – the hydrogen consumption required for a full conversion of the vegetable oil is higher than the needed for the hydrotreating of conventional diesel. In Figure 6.10 pictures of the original raw material and the subsequent hydrotreated products are shown.

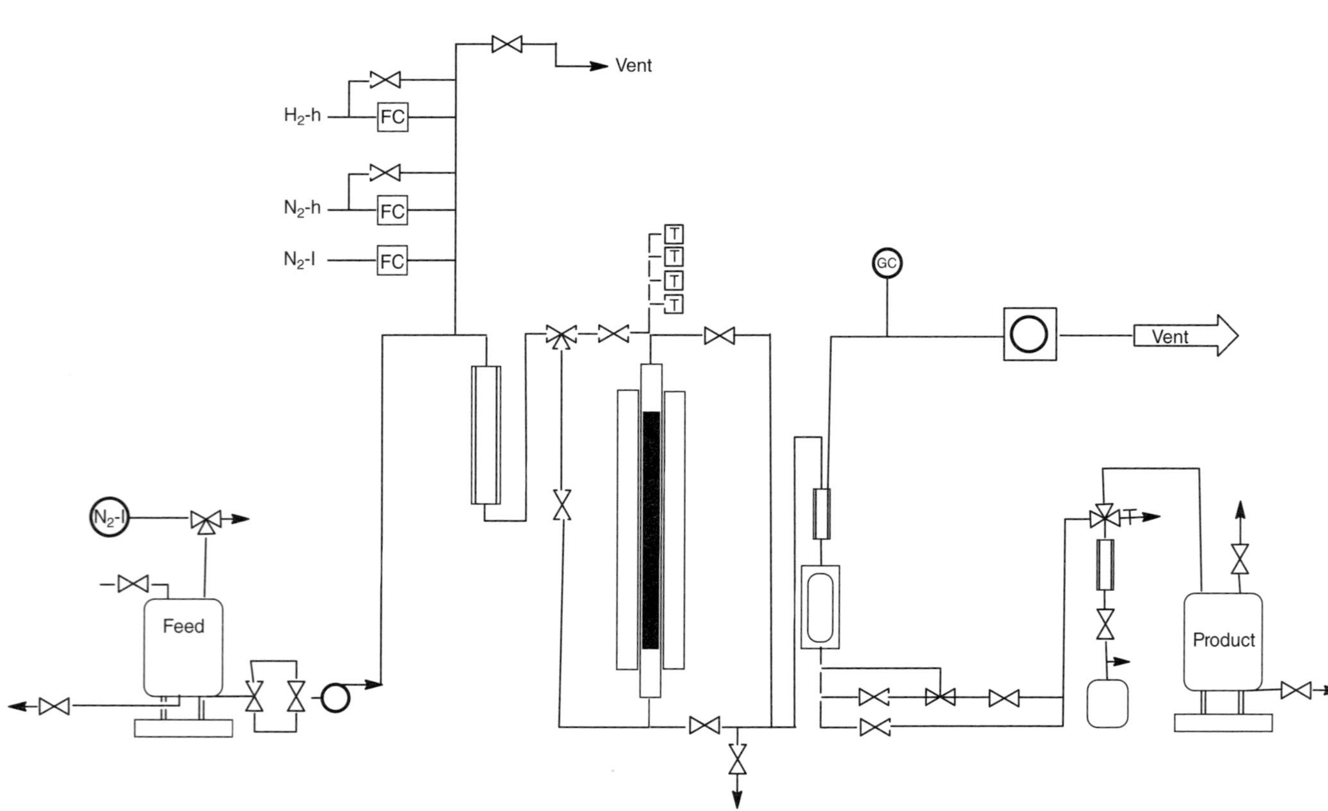

Figure 6.8 Simplified scheme of the pilot plant used in the hydrotreating of vegetable oils.

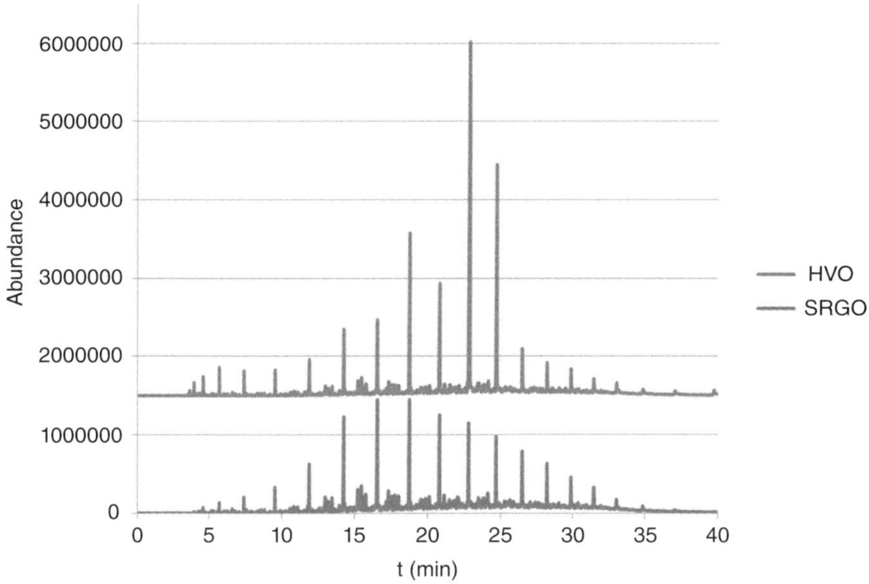

Figure 6.9 Gas Chromatography analysis of mineral Diesel (dark gray) and the one obtained by coprocessing of mineral and vegetable oil (light gray).

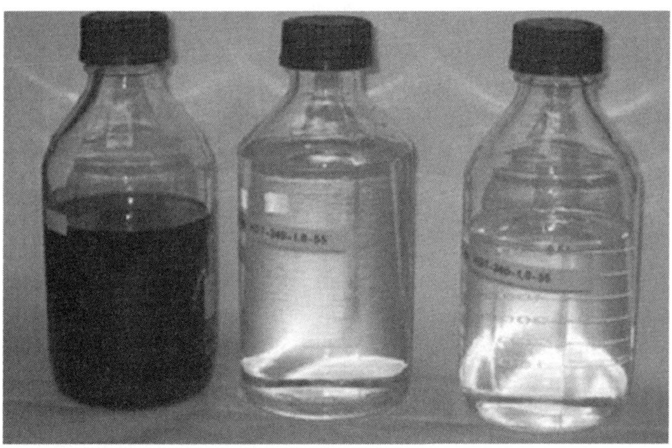

Figure 6.10 Used frying oil and two hydrotreated products. (*See color plate section for the color representation of this figure.*)

To further assess the actual viability of the process – before considering its further scale-up – all obtained products were characterized and evaluated according to Norm EN-590.[5] In all cases, and under the established operating conditions, the blend of Staright Run Gasoil and HVO displays better properties than those observed for hydrodesulfurated diesel. Table 6.1 summarizes data of two products obtained with hydrotreated sunflower oil and UFO at 40% blend in a SRGO. The density

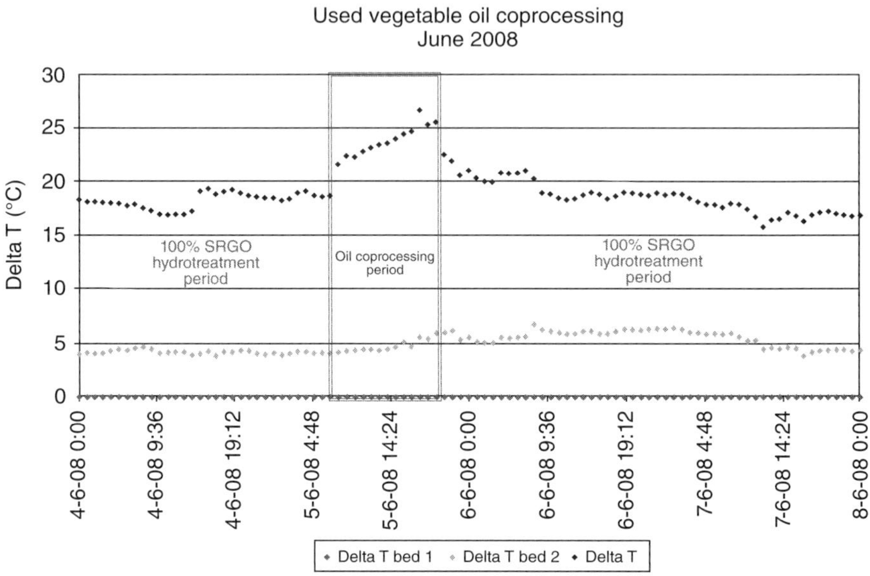

Figure 6.11 Observed exothermic behavior through the catalytic bed during the hydrotreating of 100% of SRGO and during the coprocessing of used frying oil in the refinery in Tenerife.

improvement is larger when oil is incorporated to the mixture, and the cetane number increases with the vegetable oil proportion in feed (Table 6.1).

The previously described promising results obtained in these preliminary studies at the Research Center of Cepsa confirmed the technical viability of the process and triggered Cepsa to conduct several in-house industrial tests in hydrotreating units with moderate severity in the refineries of La Rábida (Huelva, Spain) and Tenerife (Canary Is., Spain). In this respect, in the refinery of La Rábida, the coprocessing of vegetable oils within a hydrodesulphurization commercial unit was conducted in Spain for the first time. The test took place in 2007 using a commercial catalyst of Co–Mo type and using nonrefined soybean oil in concentrations ranging from 5 to 16 wt%. Subsequently, and given the success of the first test, other evaluations were conducted as well. On the other hand, in the refinery of Tenerife, in 2008 the first Spanish commercial test of coprocessing Used Frying Oil was performed. The commercial unit in which the experiment was carried out was a hydrodesulfurization unit for automotive diesel and was loaded with a Ni–Mo-type catalyst. Until 6 wt% of UFO was coprocessed (Figure 6.11).

All tests carried out at industrial level at the hydrodesulphurization units within Cepsa refineries enabled the identification of bottlenecks of the operation, as well as the potential technical viability of the process. Herein, the drawn conclusions in the industrial tests confirmed the preliminary results obtained in the pilot plant at the Research Center. The *exothermic behavior* was observed during the vegetable oil coprocessing, reaching 41 °C when 16 wt% of oil was loaded. This effect is mainly due to the presence of oxygenated, more reactive compounds than sulfur-containing

TABLE 6.1 Properties of HVO Achieved in Pilot Plant in the Coprocessing of SRGO with Used Frying Oil (UFO) and Sunflower Oil.

		100% SRGO	40% UFO 60% SRGO		40% Sunflower Oil 60% SRGO	
	WABT[a]	360	362	340	360	
	LHSV[b] (h^{-1})	2	1	1	2	
	Method					*Specification*
S (ppm)	UNE-EN-ISO 20486		7	35		Max. 10
Density (kg/m^3)	UNE-EN-ISO 121855	829.3	809.9	812.8	809.0	0.8200–0.8450
Aromatics (%vol)	UNE-EN-12916					
Mono-		22.5	12.1	13.8	12.8	
Di-		1.8	1.0	0.8	1.2	
Tri-		0.5	0.2	0.2	0.3	
PAH's[c]		2.3	1.2	1	1.5	Max. 11
Cloud point (°C)	UNE-EN-ISO-23015	−6	5	6	6	6(s)–0(w)
CFPP[d] (°C)	EN 116	**−6**	**−2**	−2	1	0(s)−10(w)
Cetane index CCI	EN ISO 4264	59	68	67	68	Min. 46
CCI-4V[e]		63	74	73	76	
Avg. CCI[f]		61	71	71	72	

[a]Weighted average bed temperature.
[b]Liquid hourly space velocity.
[c]Poly aromatic hydrocarbons.
[d]Cold filter plugging point.
[e]Cetane index four values.
[f]Cetane index average.

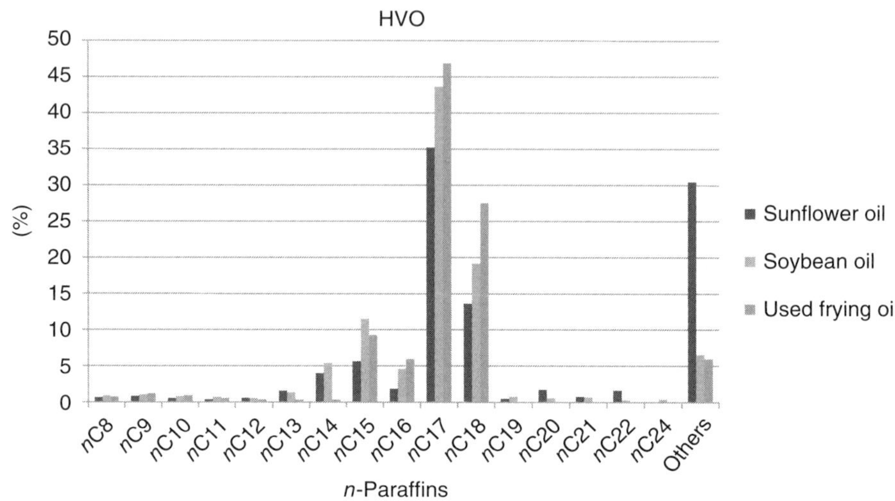

Figure 6.12 Proportion profiles of *n*-paraffin obtained when hydrotreating sunflower, soybean, and used frying oils in operational units at moderate pressure.

species of a mineral crude oil, as well as to a *higher consumption of* H_2, when the vegetable raw material is introduced.[11,12]

The obtained product after hydrotreating of vegetable oils is formed by *n*-paraffin, the same components that are forming the mineral diesel and with optimized properties to those of products delivered by incorporating FAME. Full conversions were reached in all cases, and the presence of nonconverted carbonylic groups in the hydrotreated product was not analytically detected. These experiments were actually *the first industrial tests conducted with vegetable oils in Spain* that positioned Cepsa at the forefront of the research for HVOs production from vegetable oils and UFO. In Figure 6.12 the proportion profiles of the *n*-paraffin obtained with the different vegetable oils are depicted.

As a drawback of the reported technology, CO and CO_2 were detected in the recycling gas. The presence of those compounds is not common in conventional hydrodesulfurization of diesel and are actually responsible of inhibiting the activity of the hydrodesulphurization catalyst. It was observed, however, that the deactivation effect was reversible, and the initial catalytic activity was recovered when the vegetable oil was not fed to the unit. As stated previously, the outstanding properties of the final product obtained after hydrotreating were largely better than those observed when FAME were incorporated. Thus these hurdles did not demotivate Cepsa to continue with the project. In this line, studies went on at the R&D level, focusing on which operative conditions would prioritize the hydrodeoxygenation route (HDO) against the decarboxylation and decarbonylation. The rationale behind that research strategy is that the presence of CO/CO_2 is caused by those decarboxylation and decarbonylation steps, and thus they must be minimized along the process. That may enhance the proportion of *n*-paraffin with an even number of carbons, with a beneficial impact in the volumetric yield to biofuel of the reaction.

Following this line, again several previous studies in pilot plant were conducted – using sunflower oil as feedstock – and together with the assistance of chemical processes simulators and theoretical calculations, it was concluded that *a higher pressure did prioritize the HDO process* over the decarboxylation and decarbonylation. After these experiments in pilot plant, the third industrial test in the refinery La Rábida was conducted in 2009 in a high-pressure unit (>100 bar). Again, it was an exothermic process and a higher consumption of H_2 was measured when compared to the coprocessing of mineral crude oil. Notably, in this case the minimization of CO_x gases was confirmed, and *no inhibition of the hydrodesulphurization process (HDS) was observed.* Again, Cepsa was positioned as a pioneering company in the realization of an industrial test of coprocessing vegetable oils in a commercial high-pressure hydrofinishing unit.

The cooperation between the Research Center of Cepsa and the refineries of the company enabled the successful development of this project and has allowed Cepsa to incorporate HVO into the pool of automotive gas oil. These biofuels must cope with the European specifications of quality established by the Norm EN-590.[5]

In addition to the previously discussed work – which it has actually been converted now into a novel research line at Cepsa – it is necessary to evaluate the operating impact that the coprocessed vegetable oil may have in process units that are typically operating with mineral-based crude fractions. In other words, the influence in the metallurgy of the units as well as the deactivation of the hydrotreating catalysts need to be assessed in detail.

Currently, Cepsa produces regularly HVO in several of its hydrodesulphurization units for automotive diesel. This project represents a clear example of research, development, and innovation, starting at laboratory scale, and evolve to end up in an implemented process at industrial scale.

6.2.2 Microalgae Production of Energetically Valuable Lipids with Industrial CO_2 (IDI-2011-1086)

Triggered by the excellent results obtained in the coprocessing of vegetable oils with SRGO in the hydrotreating units (see Section 6.2.1), the possibility of expanding the raw material types for the hydrocarbon production was considered. To that end, a novel research line was started, mainly using microalgae-based oils to be converted into third-generation biofuels. Algae are aquatic organisms with a rapid growth, able to store lipids within their cells, and showing an oil productivity largely superior to these of other vegetable crops. Given that algae cultivation is not competing with agricultural lands used for edible crops, there is no conflict with food production, representing a clear advantage over other vegetable oils. In another line, for its growth algae need CO_2 and water. This aspect would enable Cepsa to recycle its own gas effluents and industrial nutrient-enriched wastewater generated in the refineries.

Figure 6.13 Cepsa partners in the project ALGINDCO2.

To tackle the project "*Microalgae production of energetically valuable lipids with industrial* CO_2" (ALGINDCO2) to produce biofuels, Cepsa acquired a new funding package of MINECO (through CDTI). The project was executed in the years 2011–2012, and several Spanish companies and academic groups with remarkable scientific experience in this sector were involved as well, aiming at developing a global biofuel production project from the starting phase – algae cultivation – to the technical characterization of the end biofuel achieved.

The partners of Cepsa for this project were the University of Cádiz, the University of Huelva (through the group BITAL – *Biotecnología de Algas*), and the company BioOils (Figure 6.13).

The project was divided in several steps, each one coordinated and developed by the most specialized partner.

The BITAL group was responsible for the selection of microalgae, as well as for the cultivation of them. They were collected from the wastewater treatment plants from the refinery of Huelva and other aquifers of the surroundings (Figure 6.14). The algae able to adapt rapidly to biological and ecological conditions of the environment were selected based on their growing rate as well as on the content of fatty acids produced. The selected microalgae were of genus *Scenedesmus* and *Chlorella*. Actually, the isolated *Chlorella* strain had not been registered in the taxonomical collections yet, and therefore its identification is presently being carried out.

As necessary nutrients for the growth of microalgae, Cepsa provided the CO_2 from its effluents in the refinery of La Rábida, and the water from the wastewater treatment plant of the same refinery. After growing the selected strains in raceways (300 L), an experimental plant for microalgae production was built, a horizontal photobioreactor (PBR) with a total capacity of $5\,m^3$, and a follow-up and valorization of obtained products were performed (Figure 6.15).

The University of Cádiz was responsible for the development and optimization of an extraction method for the lipids stored within the selected algae. Extractive methods using different solvents or supercritical CO_2 were evaluated.

BioOils shared its broad knowledge in areas like the characterization of the extracted lipids and the chemical conversion routes to produce FAME (transesterification).

Finally, *Cepsa* contributed from its ample knowledge acquired in the conversion of vegetable oils via hydrotreating (previous section),[13] as well as its experience in the development of competitive industrial processes. In this respect, Cepsa was the responsible for the conversion of algae-based lipids to biofuels, and the project was undertaken at the Research Center of Cepsa. In Figure 6.16 the raw material and the hydrotreated algae oil (HAO) are depicted.

Figure 6.14 Places close to the refinery of Huelva, where microalgae were isolated (indicated by numbers).

(a) (b)

Figure 6.15 Two of the used systems for the project. (a) Raceway (300 L). (b) Horizontal photobioreactor (PBR) of 5 m^3.

The coprocessing of algae oils under diesel hydrotreating conditions led to a Hydrotreated Algae Oil, which displayed excellent chemical properties, raising the potential to incorporate it to the company diesel pool. The *n*-paraffin profile of the obtained HAO is shown in Figure 6.17.

The good results obtained in every step of the project, together with the fact that the final profile of paraffins after the hydrotreating step was within the fuel range, triggered Cepsa to tackle the technological challenge of implementing the process in-house at industrial scale. This is obviously an ambitious goal, which would enable Cepsa to contribute to the future development of this industrial sector and would be

Figure 6.16 Algae-based oil and HAO.

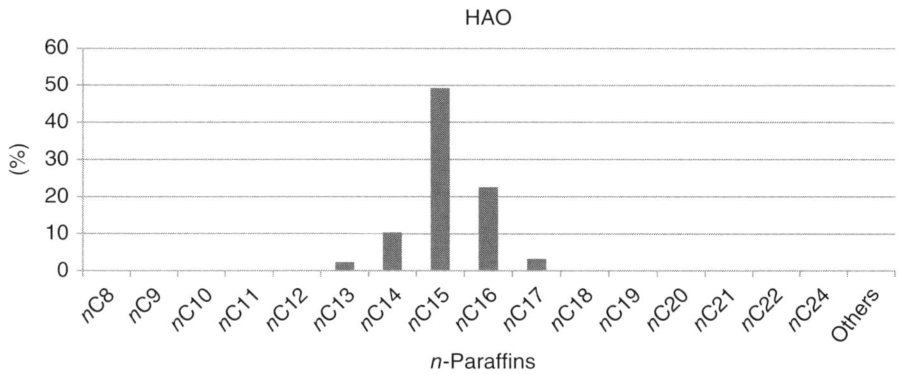

Figure 6.17 Profile of *n*-paraffin of HAO under diesel hydrotreating conditions.

aligned, at the same time, with the Directive 2009/28/EC of the European Parliament, and the Council of 23 of April of 2009, which fosters the use of energy from renewable resources.[4]

Currently, the bottleneck is to have large quantities of algae oil and at a competitive price. The overall approach implies a great effort in R&D, with the goal of improving operating costs (OPEX) of the overall process, from algae to the oil. The overall scheme is depicted in Figure 6.18. The investigations reported in this section have been recently compiled in an edited book.[14]

6.2.3 A Novel Process for the Production of Biokerosene Compatible with Refinery-Based Processes (IDI-2012-0462)

All the aforementioned concerns about the future availability of petroleum – together with an increasing energetic demand – as well as the environmental problems associated to the use of mineral-based crude can be also extended to the aviation sector.

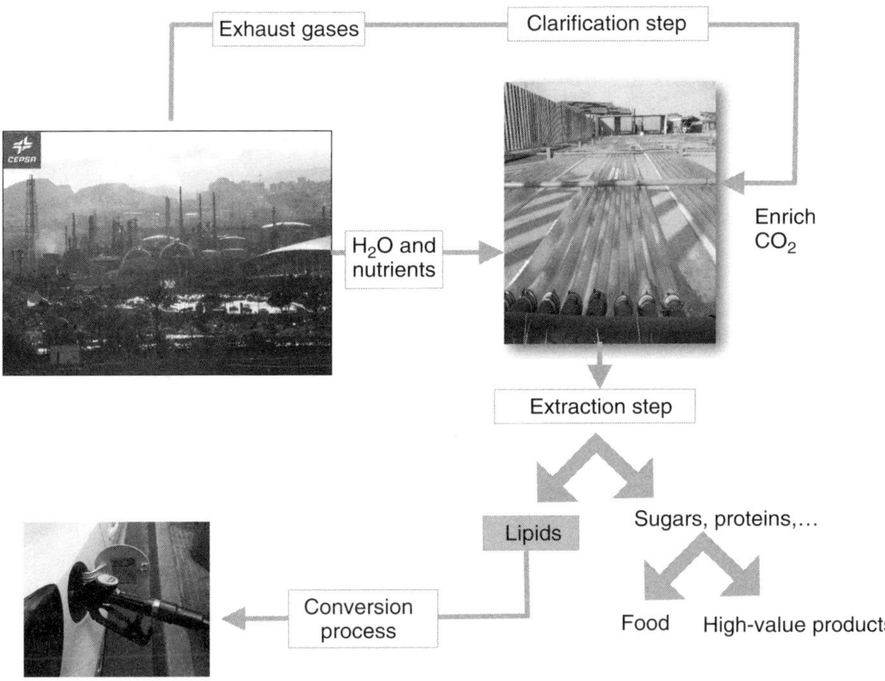

Figure 6.18 Simplified scheme of the production of biofuel from algae oil.

This sector consumes annually ca. 1.5–1.7 billions of barrels of fossil fuel Jet-A and accounts for 2% of the generated CO_2 by the human activity.[13,15] Nevertheless, it is expected that these figures will increase on a short term basis – due to the strong increasing demand in the sector – with increments of up to 4.5% (annually) in flights and passengers until 2050.[16] To that expected increase – which will bring a higher fuel consumption and CO_2 emissions – one must add the associated costs of such fuel, which represent 25–40% of overall costs of aviation companies.[16] Conclusively, any optimization in the field would have a great impact, both at economic and at environmental level. With these considerations in mind, and with the aim of reducing the environmental impact without compromising the quality of the fuel, the aviation sector has put forth either the reduction of CO_2 emissions up to 50% by 2050 or the substitution of 6% of the mineral-derived fuel by biofuels – coming from renewable resources – by 2020.[16]

The obtained quality of the HVOs (described in previous sections), and the reluctance of the aviation sector for using renewable raw materials, stimulated Cepsa to continue its studies of conversion of vegetable oils from different origins, in this case for the production of biojet fuel. Yet, the incorporation of biofuel in the current jet fuels present more challenges than its incorporation in automotive fuels. Actually, in the aviation sector there is no a viable alternative to mineral fuels, whereas in the automotive sector one may envisage gas or electricity as further possible future options.

To that drawback, it has to be added that not all existent liquid fuels can be employed for jet fuels due to security reasons clearly stated in restrictive normative and specifications. Therefore, alternative jet fuels must fulfill the approved operational and technical specifications, and behave in an analogous way to current jet fuels (*drop-in fuels*). In other words, novel biojet fuels must keep properties of fluidity, viscosity, or properties at cold temperatures and low pressures in which aviation is routinely operating and described in the Norm ASTM D7566.[13]

Cepsa forms part of the Spanish initiative for the production and consumption of biokerosene for the aviation sector, and within this context it has developed the project "A novel process for the production of biokerosene compatible with refinery-based processes," funded by MINECO (through CDTI). The main goal of that project is the production of jet biofuels within a conventional refinery, that is, compatible with current operations, in a sustainable way and with a minimal impact for the environment. The project was executed at the Cepsa Research Center. The goal was to establish two chemical conversion routes of vegetable oils to afford biokerosene. Both processes were based on the hydroprocessing of the renewable raw material, and depending on the composition of the raw material, one or the other route would be preferred to maximize the biokerosene production. Obviously, in the two processes the previous knowledge turned out to be a solid asset, especially the previously described hydrotreatment of vegetable oils and algae lipids to produce biolinear paraffin.

Previous studies confirmed that the hydrotreatment of vegetable oils leads to paraffin comprising a mixture of the same number of carbon atoms and one carbon atom less that the fatty acid composing the original triglycerides. Each oil exhibits a different fatty acid triglyceride profile, and therefore studies must be conducted individually to evaluate its viability and compatibility with jet fuels. For instance, it was already observed that for diesel fuel applications, those oils containing fatty acids of 18 carbon atoms are the proper ones.

One of the main differences between mineral jet fuel and the experimental fuel that might be obtained from vegetable oils is the content in aromatics, compounds that cannot be found from renewable sources. Thus, given that the novel synthetic fuels must also be adapted to the required standards (in terms of cold flow properties), they must be blended with conventional fuels at a maximum proportion of 50% (v/v)[16–18] according to the same Norm ASTM D7566. Another drawback related to the end product composition is the expected deficient behavior of a HVO at low temperatures – given its high content in paraffin – what would certainly hamper its final use as jet fuel. These poor cold flow properties cause the formation of waxes and crystals when the fuel is at low temperatures (normal ones at the aviation industry).

Considering these challenges, the first envisaged alternative for the conversion of vegetable oils was a two-step process. The first step is the paraffin formation starting from the *hydrotreating* of the vegetable oils was conducted. Subsequently, the obtained paraffin must be subjected to a deparaffinization process to improve its cold flow properties through isomerization of the paraffin.[10,18–22] The second alternative studied was the *hydrocracking* of vegetable oils. In the hydrocracking units at refineries, a severe hydrotreating process proceeds, in which heavy fractions are converted selectively into lighter ones, formed by higher-value products such as naphtha,

Figure 6.19 Coconut, camelina, and UFO. (*See color plate section for the color representation of this figure.*)

jet fuel, kerosene, and diesel, which could not be achieved through other processes. Therefore, different from hydrotreating process, in the hydrocracking the boiling point and the density of the fraction are reduced, and thus it is an adequate process for the conversion of vegetable oils into biokerosene.

At the Research Center of Cepsa, the first tests with three prototypical oils – having different fatty acid chain lengths – were conducted. Moreover, the difference between refined and nonrefined oils in the hydroprocessing was evaluated as well. Refined coconut oil, crude camelina oil, and UFO (not refined) were selected (Figure 6.19). Likewise, the same commercial catalysts – used at Cepsa refineries for hydrotreating and hydrocracking – were employed. With regard to differences with refined oil, the two other oils presented a higher content in metals and free fatty acids. These aspects will need to be considered in the hydroprocessing, as it was concluded from previous processes as well.

The different triglyceride profile will determine the obtained paraffin after the typical hydroprocessing. The treatment was conducted at moderate pressures and using commercial catalysts of Co–Mo type (for camelina and UFO) and Ni–Mo (for coconut oil). With each vegetable oil a different range of paraffin were obtained. As it can be expected – according to the initial composition of each oil – the refined coconut oil afforded paraffin with lower number of carbon atoms (37% *n*C11, 9% *n*C12, 15% *n*C13, and 4% *n*C14) than the obtained with used crude vegetable oil (9% *n*C15, 6% *n*C16, 47% *n*C17, and 28% *n*C18) or crude camelina oil (6% *n*C15, 3% *n*C16, 20% *n*C17, 11% *n*C18, 9% *n*C19, and 5% *n*C20). Table 6.2 summarizes the theoretical composition of the oils in fatty acids.

All obtained products through a conventional hydrotreating process are heavier than the theoretical range of kerosene (C9–C13), albeit starting from refined coconut oil, a paraffin distribution closer to conventional Jet-A1 is obtained (Figure 6.20).

TABLE 6.2 Theoretical Composition of Fatty Acids of Crude Camelina Oil, Refined Coconut Oil, and Nonrefined Used Frying Oil.

	8:0 (wt%)	10:0 (wt%)	12:0 (wt%)	14:0 (wt%)	16:0 (wt%)	18:0 (wt%)	18:1 (wt%)	18:2 (wt%)	18:3 (wt%)	20:0 (wt%)	20:1 (wt%)	22:1 (wt%)
Camelina oil					8	3	17	24	32		12	3
Used frying oil			0.5	2	19	8.8	49	16	2	1		
Coconut oil	8	7	46	18	9	3	7	2				

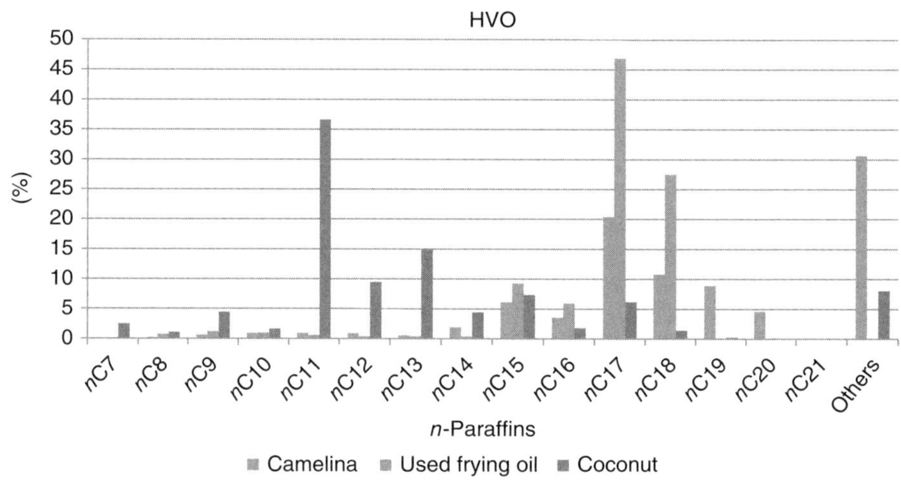

Figure 6.20 Profile of *n*-paraffin in HVO obtained at 55 bar, LHSV = 2 h^{-1} and 372 °C for camelina, 375 °C for refined coconut oil, and 366 °C for nonrefined UFO.

To adjust the product to the required *on-spec* range, the HVOs were subjected to catalytic dewaxing employing Pt-SAPO-11 – a catalyst developed at the Research Center of Cepsa – at three different temperatures (330, 340, and 350 °C). After that processing, the chain lengths of all products were actually reduced. Yet, for HVOs produced by camelina oil and UFO, the paraffin content of 17 and 18 carbon atoms was still too high (Figure 6.21) to target biokerosene requirements.

It was concluded that to afford biokerosene it is necessary *a hydrotreating process followed by catalytic dewaxing step, and only when the number of carbon atoms of paraffin* is similar to that of coconut oil, that is, *mostly of 12 carbon atoms*. After this combined two-step process, the coconut oil renders a product within the kerosene range and displaying excellent cold flow properties. It was actually confirmed that the upper pour point and the cloud point of the isomerized product were, as expected, better than those of HVO (Figure 6.22).

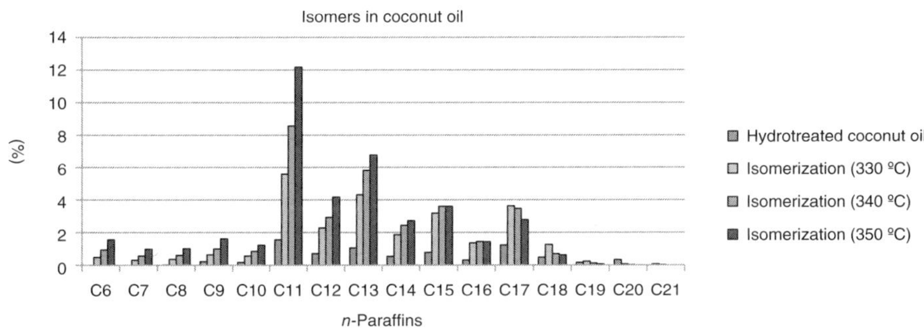

Figure 6.21 Isomer distribution after isomerization of HVO derived from refined coconut oil between 330 and 350 °C at $P = 30$ bar and LHSV = 2.7 h^{-1}.

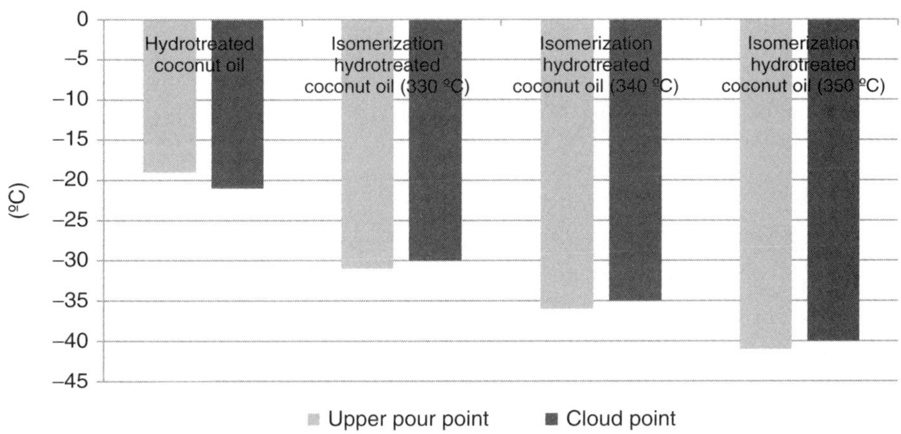

Figure 6.22 Cold flow properties of coconut oil HVO isomerized.

To maximize the kerosene fraction in HVO from camelina and UFO – formed by triglycerides with fatty acids of chain lengths superior to 16 carbon atoms – the second alternative, the hydrocracking, was considered. To evaluate the effectivity of the process, several experiments in pilot plant were conducted under typical mild hydrocracking conditions, that is, between 350 and 390 °C, and employing a commercial hydrocracking catalyst of type Ni-W/Al_2O_3-SiO_2. Herein, when hydrocracking catalysts were employed, a mixture of compounds was achieved. After a fractionation of that mixture, a product with kerosene range was obtained (40% yield for UFO and 70% for camelina oil) (Table 6.3).

These obtained final products, besides, display outstanding cold flow properties to be used as aviation fuels and actually reproduce the chemical composition of mineral-based fuels, both in molecular weight and in the distribution and functionality (Figure 6.23).

TABLE 6.3 Paraffins, Aromatics, and Alkanes Yield in Hydrocracking Products of UFO (55 bar, LHSV = 1.1 h^{-1}).

	Paraffins (%)	Aromatics (%)	Others (%)
350 °C	35	11	54
370 °C	30	12	58
390 °C	26	20	54

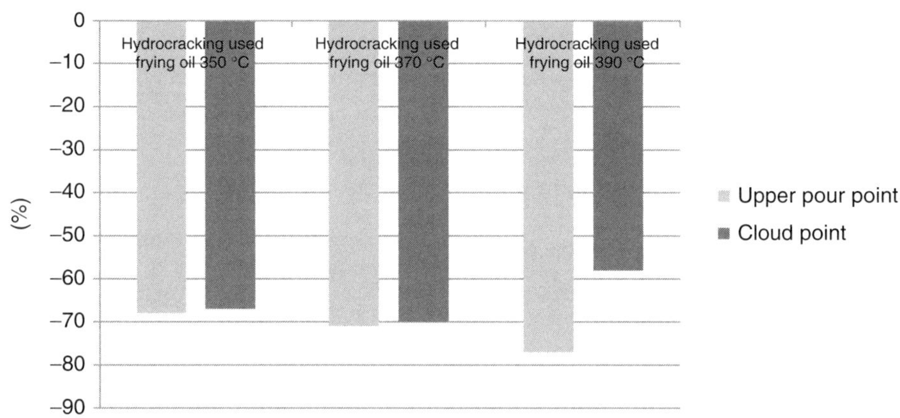

Figure 6.23 Cold flow properties of kerosene fraction obtained after hydrocracking process using crude frying oil as raw material.

After these results, two environmentally and economically sustainable routes for the production of a novel biofuel alternative to Jet-A mineral-based fuel have been developed at the Research Center of Cepsa. It is environmentally sustainable because it is using renewable raw materials. And it is economically sustainable because it makes use of in-house currently employed catalysts of Cepsa refineries in the same infrastructure that currently is being used at hydrocracking units for mineral-based fractions. Each process is adequate to the triglyceride profile (fatty acids) of the starting material, and thus a large flexibility may be achieved: depending on the raw material – subjected to actual availability depending on the region or the season – the chemical processing can be straightforwardly adapted.

Through a *hydrotreating process, it is possible to maximize the kerosene fraction of oils with fatty acid chain lengths of the range C8–C16*. Conversely, a hydrocracking catalyst – under mild hydrocracking operation conditions – enables the diminution of fatty acid chain lengths toward the desired fraction in the case of oils containing fatty acids of C16–C20 range. Thus, a final product with outstanding cold flow properties and high performance is obtained. In Figure 6.24, the alternatives are depicted.

6.2.4 Concluding Remarks: Vegetable Oil for Refineries

The herein reported projects dealing with vegetable oils as raw materials (Section 6.2) have permitted to broaden the classic refinery perspective depicted in Figure 6.25.

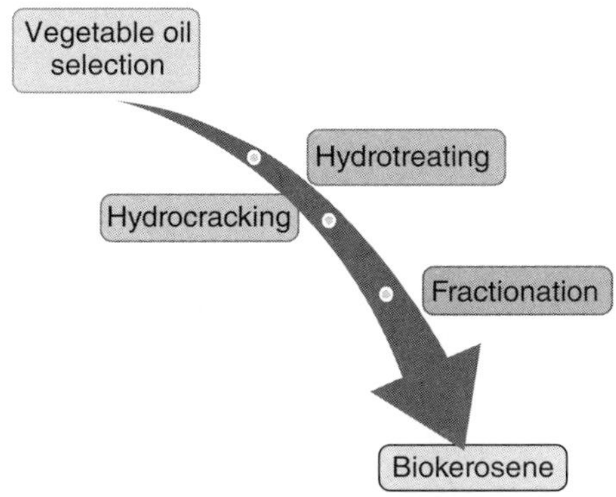

Figure 6.24 Alternatives for the production of biokerosene using vegetable oils as raw material.

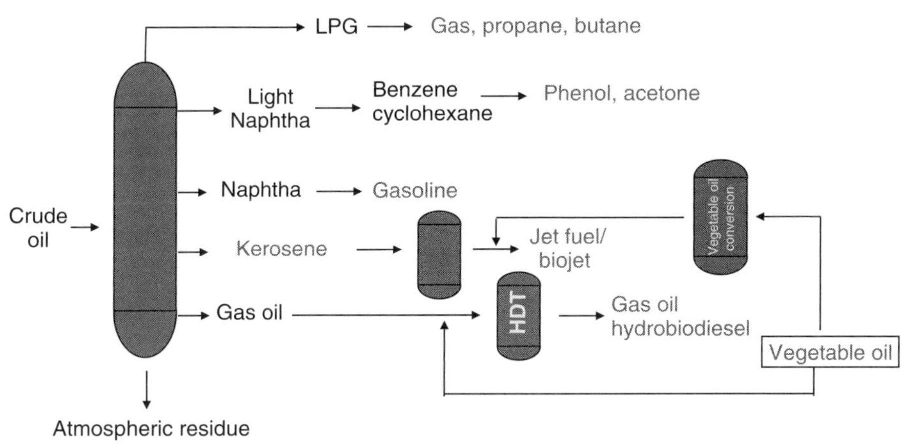

Figure 6.25 Envisaging the integration of vegetable oils as raw materials for a conventional petroleum refinery.

Thus, this new model enables the incorporation of vegetable raw materials in the production pipeline, as an alternative to mineral crude oils. The synthetized biofuels are not only chemically analogous to the conventional ones but also they present even better performances than the original fossil-based ones in certain cases. The novel model for a refinery is envisaged in Figure 6.25.

6.3 LIGNOCELLULOSIC BIOMASS

In addition to the vegetable-based components that Cepsa is incorporating to automotive gas oils – which are still under investigation and development – as well as to the other ongoing research lines on lipid and vegetable oils conversions (Section 6.2), Cepsa considers that lignocellulosic biomass is yet another raw material source – different from vegetable oils – that can be used for its conversion

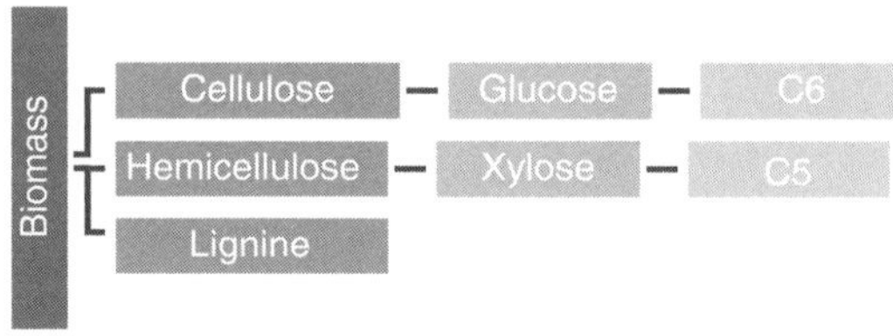

Figure 6.26 Main components of lignocellulosic biomass.

into biofuels or bioproducts. The incorporation of this raw material type will allow the company to continue fulfilling the Directive 2009/28/EC.[4] Lignocellulosic biomass is a very interesting fraction to be employed as renewable feedstock, as it is the most abundant in the earth, cheap, and with a rapid growth. Chemically it is a complex fibrous material forming the structure of cell walls in plants, and it is constituted by three main biopolymers: cellulose, hemicellulose, and lignin.[7,23] The carbohydrates that are forming these biopolymers represent ca. 75% of renewable biomass and constitute the natural raw material from which it is possible to produce organic compounds in an economic and industrially viable manner, for its further use as food, paper, heat, wood derivatives, or fuels. The biofuels obtained from these raw material are regarded as second-generation biofuels, as they are derived from agricultural or forest residues or from nonedible crops.

The main part of carbohydrates present in polysaccharide fractions of lignocellulose – hemicellulose and cellulose – are C5 and C6 sugars (Figure 6.26). As it was in the case of vegetable oils (Section 6.2), to be used as fuels, it is necessary to decrease the oxygen content, and this is typically carried out via different chemical routes.[24,25] In 2013, Cepsa entered in the quest for lignocellulosic-based biofuels, thus broadening its efforts for integrating the "bioconcept" within its productive schemes.

6.3.1 Renewable Diesel as Alternative to Conventional Gas Oil (DRAGO) (IDI-2013-0457)

At the Cepsa Research Center, an ongoing project is focusing on hemicellulose as raw material. The two-year project (again funded by MINECO through CDTI) was successfully accomplished, and due to this, the research line is still open and Cepsa is presently dedicating own resources to the R&D within this field. Hemicellulose is a polysaccharide mainly formed by pentose sugars and some hexoses. Pentoses are a source of furfural and its derivatives (Figure 6.27), which constitute the raw material for the project "*Renewable diesel as alternative to conventional gas-oil (DRAGO).*"

Presently paper industries are converting the fibrous biomass fraction – mainly composed by cellulose – into cellulosic pulp, which is subsequently processed to produce paper. To that end, a previous separation of the biomass components is conducted, that is, cellulose is separated from hemicellulose and lignin, which are often incinerated as heating source. Thus, the development of the Cepsa project on furfural

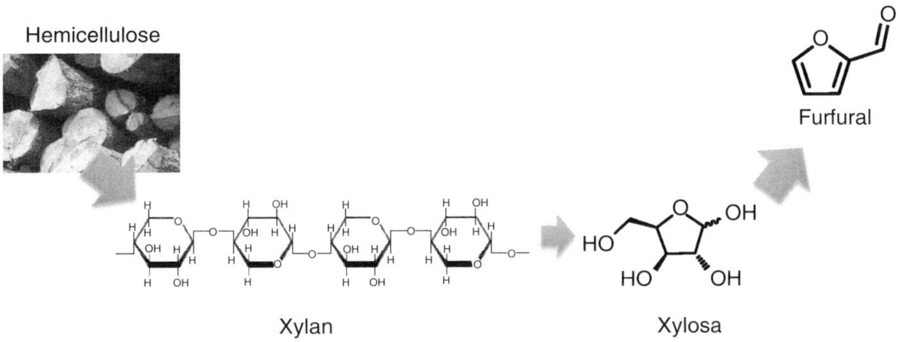

Figure 6.27 Furfural production from hemicellulose.

will enable the valorization of a fraction that is currently incinerated or discarded in other sectors (Figure 6.28).

The main objective of DRAGO project is the biofuels production – or a bioadditive for diesel pool – starting from hemicellulose.[26,27] Again, for the development of this project, the cooperation with two Spanish scientific centers was crucial:

The University of La Laguna (Tenerife) has carried out in-depth studies of agricultural and forestry residues of a Spanish region based on its availability and content in hemicellulose (Figure 6.29). At the same time, the development and optimization of methods for the extraction of hemicellulose were performed.

The Instituto de Tecnología Química (ITQ, CSIC Valencia) has been the responsible of the development of a process to convert hemicellulose into furfural and 2-methylfuran, the two platform chemicals of interest. In this area, the ITQ developed a process for the production of highly added value products from biomass residues, and this line is the basis of the DRAGO project.[28,29]

In the *Cepsa Research Center*, the conversion of the furan-based molecules into biofuels was performed. The process comprised two steps of alkylation and hydrodeoxygenation. In concrete, a mixture of highly ramified alkanes is obtained. The Research Center has also been responsible for the characterization of the obtained biofuel, as well as with the analysis of the properties of the product obtained as biobooster (Figure 6.30).

From all analytic studies conducted by the Research Center of Cepsa, it was concluded that the final product displays excellent properties to be used as diesel additive, or to be blended with the pool of mineral-based diesel, aiming at improving the properties of the fossil fuel. If the improvement of the properties described in Norm EN-590 is large, the addition of the synthetized biobooster would enable the blending of diesel fuel with other heavy fractions or with poor quality streams. Table 6.4 summarizes the properties of the synthetized biobooster.

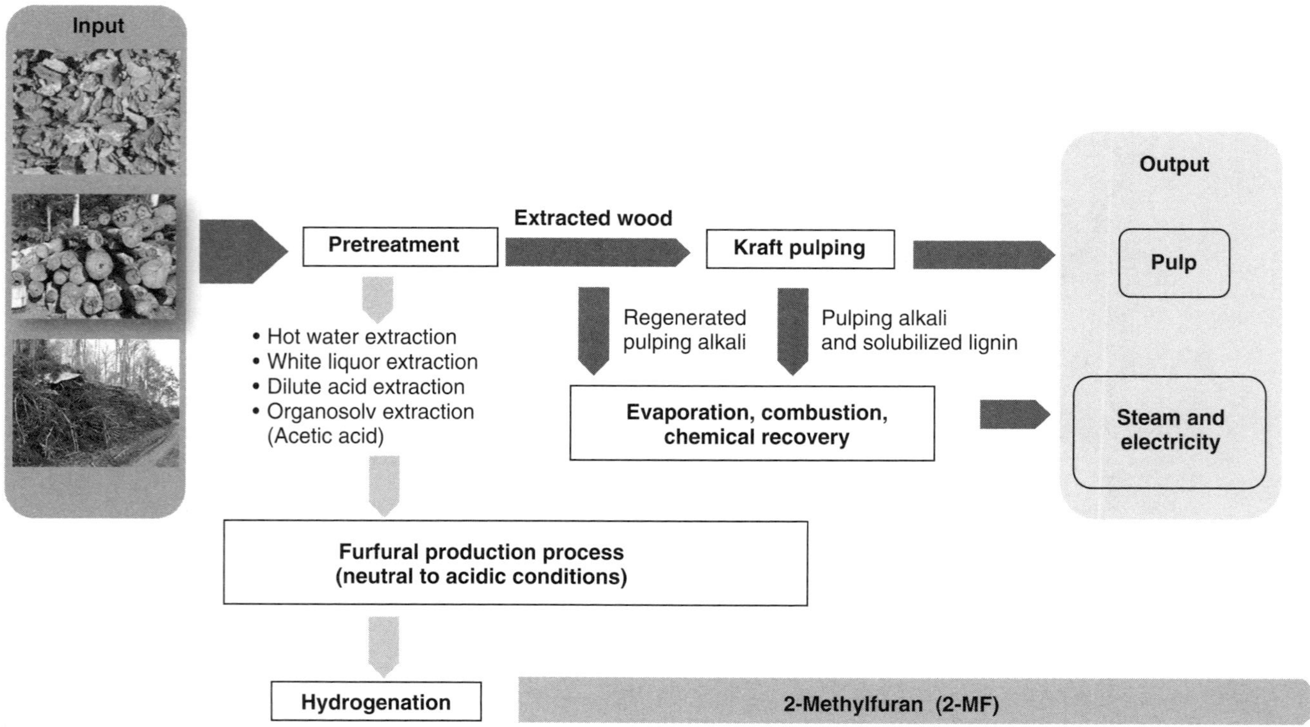

Figure 6.28 Integration of DRAGO project within a pulp and paper industry.

Figure 6.29 Pine residues used by the University of La Laguna in Tenerife (Canary Islands, Spain).

To confirm these biobooster quality in a blend, several mixtures of direct distillation diesel and light cycle oil (LCO) were prepared as low-quality stream fraction, and different loadings of the synthetized biobooster were added. The maximum limit of addition of LCO in diesel fractions – and in absence of the biobased component – is given by the specification of the density of the final mixture according to the Norm EN-590.[5] It is only possible to blend 2% of LCO with diesel to fulfill with the

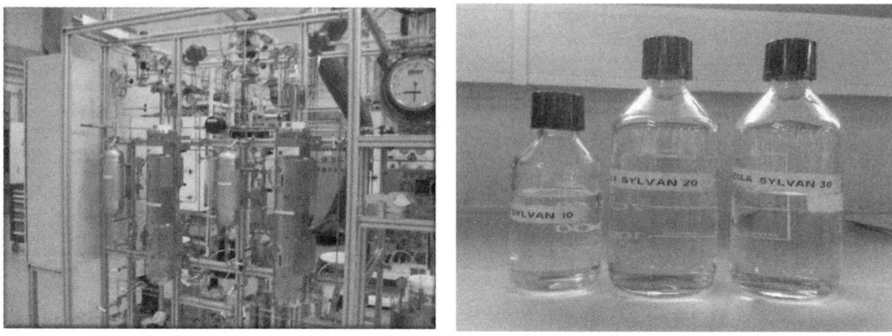

Figure 6.30 Pilot plant for hydrodeoxygenation of furan-based molecules and diesel fuel obtained after the process.

TABLE 6.4 Properties of Biobooster Synthetized at DRAGO Project.

	Method	Biobooster	Specification
Density (kg/m^3)	UNE-EN ISO 12185	0.78337	0.820–0.840
S (ppm)	UNE-EN ISO 20486	<1	10
Número cetano	UNE-EN-ISO 5165	72.2	min 51
Upper pour point (°C)	ASTM D-97	−75	–
Cloud point (°C)	ASTM D-2500	−80.6	
Flash point (°C)	UNE-EN-ISO 2719	80.1	>55
Kinematic viscosity 40 °C (mm^2/sg)	UNE-EN-ISO 3104-96	1.915	2.0–4.5
POFF (°C)	UNE-EN 116-98	−48	Winter <−10°
Lubricity (μm)	UNE-EN-ISO 12156-1-07	419	<460

maximum limit of density specified in the Norm. However, it is possible to blend up to 10% of LCO if 12% of the biobooster is added as well (Figure 6.31).

Based on the results observed in the Cepsa Research Center, the optimization of the yields of every step of the DRAGO project would allow the production of a second-generation bioproduct that would enhance the flexibility of the diesel pool and different from FAME and HVO. Furthermore, it would allow the addition of high-density crude fractions or low-value streams in higher proportion without being *off spec* with the current normative for automotive diesels.

6.4 CONCLUDING REMARKS

Along this chapter, the activities of Cepsa in the field of biofuels have been presented. Different case studies, ranging from the use of vegetable oils to produce diesel additives or HVO products to activities in lignocellulosic biomass, have been in-depth

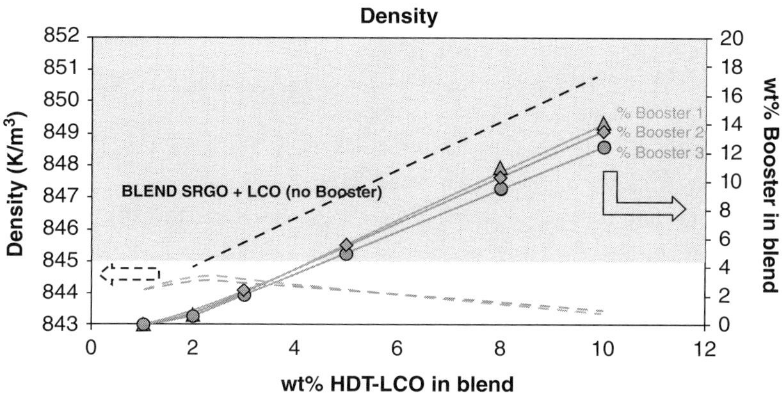

Figure 6.31 Density of the blend of SRGO and LCO with the synthetized biobooster of the DRAGO project.

discussed. Processes have been assessed both at pilot plant and at industrial scale, within operational units of Cepsa refineries. Once processes have been optimized – for example, type of catalyst, reactor, conditions, actual chosen process, etc. – it has been found that many of the synthetized products can be actually blended with different conventional mineral-based fuels to afford products with optimized performances and with better environmental prognosis. Given the commitment of the company with the sustainability and renewable energies, Cepsa has found in other companies and in Spanish research centers the proper cooperation to tackle every envisaged challenge, broaden its knowledge, and sharing its experience and know-how in the field of fossil-based fuels.

REFERENCES

[1] 2013 CORES Report. http://www.cores.es/en/publicaciones/annual-statistical-report-es/annual-statistical-report-2013, accessed Mar 2015.

[2] S. Bezergianni, A. Dimitriadis, L.P. Chrysikou, *Fuel* **2014**, *118*, 300–307.

[3] B. Kamm, P. R. Gruber, M. Kamm, S. Kromus, "*Biorefineries-Industrial Processes and Products, Status Quo and Future Directions*", *Vol. 1,* (Eds. B. Kamm, P.R. Gruber, M. Kamm). VCH, Weinheim, **2006**, 3–32.

[4] Directive 2009/28/EC of the European Parliament and of the Council of 23 April 2009 on the promotion of the use of energy from renewable sources and amending and subsequently repealing Directives 2001/77/EC and 2003/30 EC, *Official Journal of the European Union,* http://eur-lex.europa.eu/legal-content/EN/ALL/?uri=CELEX%3A32009L0028.

[5] EN-590 Standard: European Diesel Fuels Specifications, https://en.wikipedia.org/wiki/EN_590.

[6] S. Bezergianni, S. Voutetakis, A. Kalogianni, *Ind. Eng. Chem. Res.* **2009**, *48*, 8402–8406.

[7] L. Petrus, M.A. Noordermeer, *Green Chem.* **2006**, *8*, 861–867.

[8] A. Corma, M. Renz, *Catal. Rev.* **2012**, *25*(2), 9–18.

[9] B. Veriansyah, J.Y. Han, S.K. Kim, S.A. Hong, Y.J. Kim, J.S. Lim, Y.W. Shu, S.G. Oh, J. Kim, *Fuel* **2012**, *94*, 578–585.

[10] G. Huber, P. O'Connor, A. Corma, *Appl. Catal. Gen.* **2007**, *329*, 120–129.

[11] M. Al-Sabawi, J. Chen, *Energ. Fuel.* **2012**, *26*, 5373–5399.

[12] Confidential Information. "Hidrotratamiento de Aceites Vegetales" **2012** Cepsa Affidavit.

[13] www.obsa.org.

[14] E. Forján, C. Vilchez, J.M. Vega, "*Biotecnología de Microalgas*", Cepsa. ISBN: 978-84-617-2314-0. Madrid, Spain. **2014**.

[15] www.enviro.aero/biofuels.

[16] www.iata.org.

[17] J.K. Satyarthi, C. Thota, G.D.T. Shastry, P.S. Viswanathan, *Petrol. Technol. Quart.* **2013**, *Q2*, 75–79.

[18] R. Egeberg, N. Michaelsen, L. Skyum, P. Zeuthen, *Petrol. Technol. Quart.* **2010**, *Q2*, 101.

[19] P. Zeuthen, H. Topsoe, H. Rasmussen, *Hydrocarbon Eng.* **2012**, 29–35.

[20] S.J. Reaume, N. Ellis, *Energies* **2013**, *6*, 619–633.

[21] T. Kasza, J. Hancsók, *Hung. J. Ind. Chem.* **2011**, *39*, 121–126.

[22] R. Szynkarczuk, M. Robinson, L. Huve, **2014**. www.digitalrefining.com/article/1000927, accessed Mar 2015.

[23] J.P. Lange, Biofuels, *Bioprod. Bioref.* **2007**, *1*, 39–48.

[24] D.J. Hayes, S. Fitzpatrick, M.H.B. Hayes, J.R.H. Ross, "*Biorefineries-Industrial Processes and Products, Status Quo and Future Directions,*" *Vol. 1,* (Eds. B. Kamm, P.R. Gruber, M. Kamm). VCH, Weinheim, **2008**, 139–161.

[25] F.W. Lichtenthaler, "*Biorefineries-Industrial Processes and Products, Status Quo and Future Directions,*" *Vol. 2,* (Eds. B. Kamm, P.R. Gruber, M. Kamm). VCH, Weinheim, **2008**, 2–59.

[26] A. Corma, O. de la Torre, M. Renz, *ChemSusChem* **2011**, *4*, 1574–1577.

[27] A. Corma, O. de la Torre, M. Renz, *Energ. Environ. Sci.* **2012**, *5*, 6328–6344.

[28] A. Corma, O. de la Torre, M. Renz, N. Villandier, *Angew. Chem. Int. Ed.* **2011**, *50*, 2375–2378.

[29] A. Corma, M. Renz, O. de la Torre, WO 2011/157876 (A1) **2010**.

7

DuPont : BIORENEWABLES AT E.I. DU PONT DE NEMOURS & CO

MICHAEL A. SALTZBERG, ARMANDO M. BYRNE, ETHEL N. JACKSON, EDWARD S. MILLER JR., MARK J. NELSON, BJORN D. TYREUS, AND QUINN ZHU

E.I. DuPont de Nemours & Co, Wilmington, DE, USA

Industrial Biorenewables: A Practical Viewpoint, First Edition. Edited by Pablo Domínguez de María.

This chapter describes the strategy and activities of DuPont in the field of industrial biorenewables, covering both business-related and technical issues and developments. The topics to be covered include:

1. A short history of the evolution of DuPont's business portfolio that has led it to its current strong position in industrial biorenewables.
2. A discussion of DuPont's philosophy of innovation.
3. A description of DuPont's current commercial businesses and publicly disclosed development programs in industrial biorenewables.
4. A case history of how a new biomonomer was produced through metabolic engineering and fermentation (Bio-PDO®), along with development of the related downstream polymer (Sorona® triexta) to create a fast-growing and substantial business.
5. A case history of the creation of a new premium category of "harmoniously raised" farmed salmon (Verlasso®) based on a development of a sustainable alternative to fish oil through the metabolic engineering of a yeast to produce the essential omega-3 nutrient eicosapentaenoic acid (EPA).
6. A discussion of future directions for DuPont's efforts in industrial biorenewables.

7.1 DuPont HISTORY AND STRATEGIC PRIORITIES

The history of the DuPont company has been well documented.[1,2] E.I. du Pont de Nemours & Co. was founded in 1802 in Wilmington, DE, United States, by French immigrant Éleuthère Irénée du Pont to manufacture black powder, an important explosive. Over the next century or so, the DuPont company continued to invest in research and in manufacturing capacity for black powder and other explosives to become one of the world's leading suppliers of these critical materials. Some of the keys to success in the company's first century of existence were the successful establishment of safety as a core value of the company, as well as a commitment to substantial and consistent investment in relevant new areas of science and engineering.

Beginning in the early twentieth century, DuPont entered the chemicals and materials businesses through both internal investment and acquisition. DuPont's first forays into the market were renewable materials based on extensions from its cellulose-based explosives business, including cellophane films, nitrocellulose plastics and lacquers, and viscose (rayon) fibers. DuPont was also a leader in the petrochemical revolution of the first half of the twentieth century, inventing and introducing a wide variety of materials such as nylon polyamides for fibers and engineering plastics, Teflon® fluoropolymers, Neoprene® synthetic rubber, and Lucite® and Orlon® acrylics for coatings, windows, and fibers. Following the Second World War, DuPont's portfolio of materials continued to broaden and deepen to include products such as Dacron® and Mylar® polyesters for films and fibers,

Delrin® polyacetals, and Butacite polyvinyl butyral interlayers for safety glass windshields.

In the 1960s and 1970s, other advanced and specialty materials were invented and added, including Lycra® spandex elastomeric fibers, Kevlar® and Nomex® aromatic crystalline polymers for applications such as body armor and fire protection apparel, Tyvek® spunbonded nonwoven fabrics, and Riston® photoresist for patterning of printed circuit boards. Throughout the 1970s and 1980s, DuPont focused on expanding into additional high-value advanced materials for applications such as electronics (e.g., Kapton® polyimides for flexible circuits) and printing (e.g., Cromalin® proofing systems). It was also during this time that DuPont made the decision to begin investing in long-term research in the emerging area of industrial biotechnology.

Since the 1990s, DuPont has focused on becoming a sustainable, science-based growth company and has thus undergone an extensive transformation of its business portfolio. The Conoco® energy business along with the traditional nylon and polyester fiber businesses was divested, and large investments were made in acquiring the Pioneer® seeds business and expanding DuPont's presence in production agriculture. In 2011, DuPont acquired Danisco, a company with strong businesses in nutrition and health and in industrial enzymes, which, combined with DuPont's existing businesses in these segments, made the company a world leader in both areas. DuPont has since also divested its performance coatings business and announced its intent to separate the performance chemicals businesses (Figure 7.1).

The drive behind these portfolio changes has been to focus the company on three strategic priorities that are driven by innovation: agriculture and nutrition, biobased industrials, and advanced materials. The new DuPont company is a world leader in

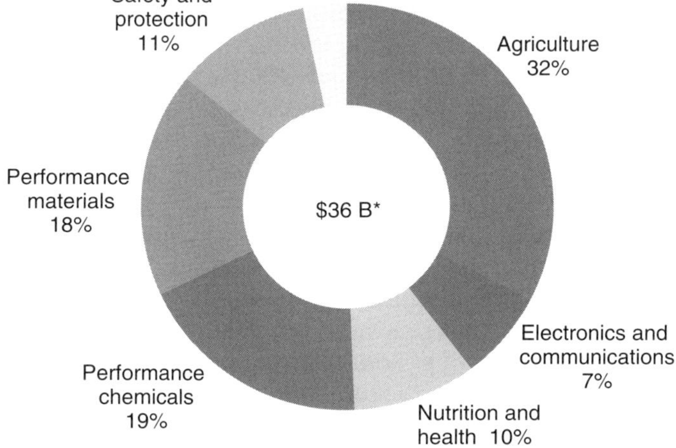

Figure 7.1 DuPont's 2013 sales by business unit.

all three of these areas, with the strategy to drive new business growth in each area and to find synergies between them. One of the most important of these trends is to transform the advanced materials businesses through the use of industrial biotechnology and chemistry to enable the movement from petro-based to renewably based supply chains. DuPont has broad and deep understanding of the customers, market trends, and applications of advanced materials based on its long history of innovation and business success in that space. This position provides the company with a strong platform of capabilities and credibility to help drive new renewable processes and materials to market based on industrial biotechnology.

7.2 DuPont's INNOVATION PHILOSOPHY

DuPont's Chief Science and Technology Officer, Douglas Muzyka, recently described the company's innovation philosophy,[3] which includes three major components:

1. *Market-driven innovation.* The close connection of research and development to customers and market-driven insights.
2. *Integrated science.* The application of multiple and sometimes diverse scientific and engineering disciplines to find effective and practical solutions to complex and difficult problems.
3. *Inclusive innovation.* Partnerships with other organizations, for example, customers, academia, suppliers, government laboratories, and start-up companies to accelerate progress and come to solutions that any one organization could not find alone.

7.2.1 Market-Driven Innovation

As mentioned previously, DuPont's long history and strong market presence in advanced materials provides close connections with customers and therefore the ability to glean important insights about their requirements for both end-use properties and processing parameters. The development of either new renewable materials or new processes to manufacture existing materials from renewable feedstocks requires understanding of the impact of key materials properties and impurity profiles on the behavior of such materials in customer manufacturing processes, as well as deep insight into the value chain economics. These considerations must guide decisions on which projects to pursue and which technical approaches to take to the development. For example, DuPont's intimate knowledge of the end-use requirements of demanding under-the-hood applications for the automotive market, combined with an understanding of the important processing characteristics required to make complex parts from engineering polymers, enabled the development of Zytel® RS renewable nylon resins. These materials have received several awards from automotive industry organizations for successfully commercialized parts, for example, for renewably sourced polyamide fuel lines for biodiesel.[4]

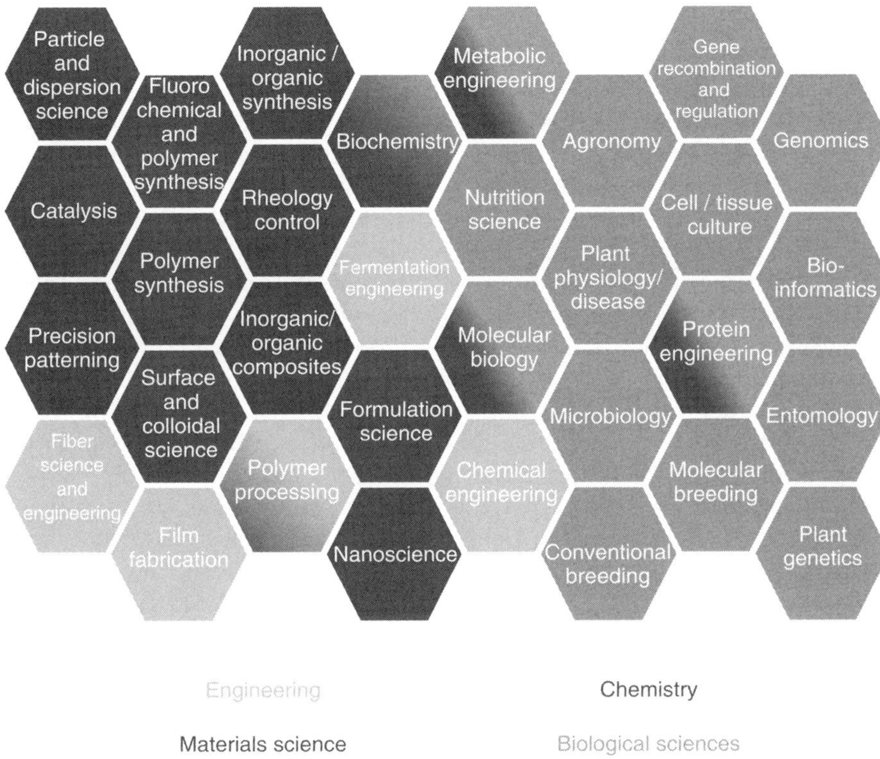

Figure 7.2 DuPont's scientific competencies. This breadth and depth in a variety of disciplines enables successful development of industrial biorenewables. (*See color plate section for the color representation of this figure.*)

7.2.2 Integrated Science

DuPont has a unique breadth of scientific competencies that range from agronomy to chemistry to materials science and engineering, as shown in the following (Figure 7.2).

DuPont believes that the successful development of new materials based on renewable resources requires this broad set of capabilities. For example, development of the technology behind DuPont's cellulosic ethanol plant in Nevada, IA, United States, required a wide variety of disciplines, including agronomy to understand how corn stover could be harvested sustainably from nearby corn fields, chemistry to develop an effective pretreatment process to breakdown the stover biomass, protein engineering and application technology to develop the enzymes required to saccharify the cellulose and hemicellulose in the biomass, metabolic and fermentation engineering to develop a novel microbe to ferment both the C5 and C6 sugars to ethanol, and the project engineering to manage the construction and start-up of a large and complex plant based on new-to-the-world technology.

7.2.3 Inclusive Innovation

To complement this breadth of scientific competencies available within the company, DuPont believes that collaboration is critical to the success of the innovation process. The development of the Bio-PDO organism and process – described more fully in the next section – is an excellent example of this "inclusive innovation" approach. In the early 2000s, DuPont worked with Genencor International (now part of DuPont following the Danisco acquisition) to develop the *Escherichia coli* organism that metabolizes glucose directly into 1,3-propanediol. Genencor's expertise in protein engineering of enzymes and metabolic engineering was complementary to that available at DuPont, and the two research organizations were able to work closely to do the groundbreaking work required to reach the challenging rate, titer, and yield targets. Once the organism was developed, DuPont collaborated with Tate & Lyle to scale up and further develop the fermentation process. Tate & Lyle has extensive experience with large-scale commercial fermentations, while the Bio-PDO program was DuPont's first foray into this area. Once the scale-up of the new Bio-PDO technology was demonstrated, the two companies formed a joint venture (JV) to build the commercial plant in Loudon, TN, United States, adjacent to a corn wet mill owned by Tate & Lyle so that the JV would have access to a reliable and low-cost source of glucose, the most critical raw material. The inclusive innovation approach helped make this complex and difficult development a reality, and it remains a cornerstone of DuPont's strategy.

7.3 DuPont's INDUSTRIAL BIORENEWABLE PORTFOLIO 2013

DuPont today is a world leader in industrial biorenewables, with sales of more than $1.4 billion in 2012. The majority of these sales are through the industrial biosciences business unit, with some renewable materials also being sold in business units such as performance polymers and nutrition and health. The main product lines and projects can be organized into three categories:

1. Enzymes. Catalytic proteins used at relatively low concentrations as processing aids for a wide variety of applications. The enzymes are produced in a host organism in large-scale fermentation processes and then provided to customers in a variety of solid and liquid formats.
2. Biomaterials. Chemicals and polymers made from renewable resources. The current commercial product line consists of Bio-PDO and several high-performance polymers, which are partially or completely renewable as described in the following.
3. Biofuels. Renewable fuels made from biomass or sugars through fermentation. Current programs include cellulosic ethanol and biobutanol.

The enzyme businesses were acquired as part of the Danisco transaction in 2011. The Genencor enzyme division of Danisco was built over several decades through both internal development and acquisition. The capabilities to engineer and produce customized proteins through fermentation in a "cell factory" to meet the specific market needs is the core capability of this business, along with the applications development to make the products work in diverse customer processes. The commercial enzyme businesses today serve the following segments:

1. Animal nutrition. DuPont develops and sells a variety of enzymes that aid primarily swine, poultry, and cattle to more effectively digest and utilize the nutrients in their feed. For example, Phyzyme® XP is a phytase feed enzyme designed specifically to improve the availability of phosphorus, calcium, energy, and amino acids from feed ingredients. DuPont also sells complementary animal nutrition products such as probiotics and betaine and recently introduced yeast-based omega-3s for salmon feed, which will be discussed more fully later in the chapter.
2. Grain processing. DuPont enzymes are used extensively in the processing of carbohydrates to value-added products such as sweeteners and fuels. For example, DuPont Spezyme® alpha-amylase catalyzes the breakdown of starches to fermentable sugars, critical to the efficient production of ethanol. Other enzymes aid the liquefaction, saccharification, and isomerization of starches and sugars into the form required by the customer. DuPont also offers some complementary products to this market such as Fermasure® XL, a fermentation additive that rapidly attacks and decreases harmful bacteria in processes such as ethanol production.
3. Home and personal care. The products in this portfolio are used in detergents and related products to catalyze cleaning of a variety of stains, including proteases for removal of proteins and amylases for removal of starches. There are emerging applications of enzymes and other proteins in personal care as well.
4. Food. DuPont enzymes are used to improve product quality, shelf life, and/or process cost of food in applications including bakery, brewing, and dairy.
5. Specialties. Industrial applications such as textile processing and pulp and paper also benefit from the use of DuPont enzymes to reduce the usage of harsh chemicals or create desirable effects, for example, "stone-washed" denim.

DuPont's current portfolio of biomaterials is led by Bio-PDO and Sorona, which will be described more fully in the next section. The largest application for Bio-PDO is as a key monomer in Sorona triexta polymer, which finds its major applications in carpet and textile fibers. Bio-PDO is also used in direct applications such as personal care and sustainable deicing fluids under the brands Zemea® and Susterra® sold by the DuPont Tate & Lyle Bioproducts JV. DuPont also sells renewable engineering polymers such as Zytel RS nylon resins that are based on sebacic acid derived from castor beans and Hytrel® RS thermoplastic elastomers containing PTMEG derived from cellulosic biomass. DuPont soy polymers provide 100% renewable alternatives for industrial applications such as paper coatings and inks and are derived from manufacturing processes associated with DuPont's Solae® soy protein business.

In biofuels, DuPont has two major ongoing programs. The first, as mentioned earlier, is the cellulosic ethanol program. At the time of the writing of this chapter (Spring 2014), the first large-scale (30 million gallons/annum) commercial plant utilizing this technology and corn stover as the feedstock was under construction in Nevada, IA, United States. This project followed the successful demonstration of the technology at a pilot plant in Vonore, TN, United States. Some of the most critical aspects of this program include:

1. Development of a cost-effective and efficient corn stover collection process that also meets agronomic targets for the farmers involved.
2. An environmentally sustainable, low-cost pretreatment process utilizing a synergistic combination of enzymes to hydrolyze and saccharify the cellulose and hemicellulose to render the sugars fermentable.
3. Development of a novel ethanologen based on *Zymomonas mobilis* that could efficiently transform both the C6 and C5 sugars to ethanol in the "dirty" environment of the stover hydrolysate.

The biobutanol development program is being executed by Butamax®, a joint venture between DuPont and BP, which represents another good example of the inclusive innovation approach. The goal of this joint venture is to develop a process to make isobutanol from corn through fermentation. Biobutanol would provide superior fuel value and compatibility with the current fuel and vehicle infrastructure as compared to the incumbent biofuel, ethanol.

7.4 CASE HISTORY #1: Bio-PDO AND SORONA

Polyester materials have been well known since the 1930s. They are made through a condensation reaction from a diacid, most commonly terephthalic acid (TPA), and a glycol where the hydroxyl groups are on either end of the molecule (e.g., ethylene glycol or 1,4-butanediol). Polyethylene terephthalate (PET) is made from ethylene glycol and TPA and is the most ubiquitous polyester, used widely for rigid and flexible packaging, fibers for textiles and nonwovens, and engineering polymers. The next most common polyester is polybutylene terephthalate (PBT) made from 1,4-butanediol and TPA, used primarily in engineering polymer applications.

It has been known for some time that polytrimethylene terephthalate (PTT) polymer made from 1,3-propanediol (PDO) and TPA had properties that were distinct from both PET and PBT and were particularly interesting for fiber applications.[5] PTT was never a success commercially due to the high cost of producing PDO from petroleum-based feedstocks using traditional chemical routes. Most reaction routes favor production of 1,2-propanediol (propylene glycol (PPG)). PPG is used widely in solvent and chemical applications but is not useful as a monomer in polyester applications.

In the mid-1990s, DuPont began to reconsider the use of PTT, particularly for carpet fiber. The polymer offers a number of advantages for this market, including an almost unique combination of softness, crush resistance, and inherent stain resistance.[6] With the success of products like Stainmaster® nylon carpets that provided consumers with enhanced stain protection based on surface treatment of the fibers with fluorosurfactants or other chemicals, there was a building consumer demand for a new fiber that could provide even better properties, including permanent and "built-in" stain resistance (stain resistance that is inherent to the polymer rather than conferred by a surface treatment). For textiles, PTT offers softness as well as some stretch-and-recovery properties, which enhance the comfort of fabrics. As applications for Lycra and other spandex fibers grew in the market, there was an increasing consumer demand for comfort fabrics. Polyesters are also relatively easy to spin into fibers and to dye a wide variety of colors, especially as compared to nylons and spandex, so there was also demand from carpet and textile fiber spinning firms for a material that processed as easily as a polyester but that could offer properties that could compete with nylon or other high-value fiber materials. With the relatively low cost of TPA, the other key monomer in PTT, the major roadblock to the commercial success of PTT was the lack of an economically viable manufacturing process for PDO.

In the 1990s, DuPont scientists began to explore the possibility of using a fermentation approach to manufacture PDO. A primary consideration in the development was that the new process must offer sufficiently low production cost to enable PTT to compete in the market with the incumbent materials. An added advantage would be the ability to market a biobased PDO and PTT polymer based on its improved environmental footprint, although at the time the program was initiated, most consumers did not state a preference for environmentally friendly materials nor a willingness to pay a premium for them. This trend has evolved over the past fifteen years or so, as consumers are now more willing to consider environmental performance in their purchase decisions, but even today Sorona PTT polymer is marketed first based on its performance proposition and second on its environmental benefits.

7.4.1 Bio-PDO Biocatalyst Development

Examples of bacteria that produce PDO naturally as a result of anaerobic growth on glycerol can be found in several genera, including *Clostridia*, *Lactobacillus*, and *Klebsiella*. These organisms catalyze a difficult chemical dehydration of glycerol to 3-hydroxypropionaldehyde with subsequent NADH-dependent reduction to PDO allowing growth by preserving cellular NADH/NAD^+.[7] Although a fermentation process based on glycerol had been well studied and advocated, the cost of glycerol was viewed as a significant challenge to commercialization.[8,9] A more desirable process would incorporate a microorganism that had the ability to convert a cost-effective carbon source such as carbohydrates or sugars directly to PDO. An ideal example was glucose, which had an acceptable cost structure and where the ability to produce glycerol from glucose by fermentation was well established.[10] The conversion of glucose to PDO, however, was not known to exist in nature, and so it was realized that a glucose-based process for the production of PDO would require the construction of a novel metabolic pathway with genes from disparate microorganisms.[11]

Before embarking on the construction of a novel PDO-producing microorganism, it was important to establish quantitative and measurable targets that could link research objectives with industrial performance requirements. Metrics commonly used in the production of chemicals by fermentation include product yields, production rates, and product titers. Values for each of these metrics for the envisioned commercial PDO process were calculated based on a combination of theoretical biochemical pathway analyses together with an early design of the unit operations surrounding fed-batch fermentation, product separation, and refining. Each individual metric had specific issues that needed to be considered as part of the design process. For example, *yield* requires a balance to be struck between the demands of PDO production with those of cellular growth and maintenance, as well as the elimination of unnecessary by-products. *Volumetric productivity* is the product of the biomass concentration in the fermenter and the specific productivity of that biomass. It was necessary to grow most of the biomass during the course of the production fermentation, again requiring management of the metabolic competition between the PDO production pathway and pathways necessary for growth within the cell. A high *titer* was required in order to minimize the cost associated with recovering PDO from the aqueous fermentation broth, but it was unknown at the outset of the project just how toxic PDO would be to the production organism at that titer and what biological or engineering solutions would be required to mitigate that toxicity.

The conversion of glucose to PDO requires the combination of two natural pathways: glucose to glycerol and glycerol to PDO. The best natural pathways for the production of glycerol were known to exist in yeast. *Saccharomyces cerevisiae* produces glycerol from the glycolytic intermediate dihydroxyacetone 3-phosphate using two enzymes: glycerol-3-phosphate dehydrogenase (G3PDH) and glycerol-3-phosphate phosphatase (G3PP). This pathway utilizes enzymes with high substrate specificity and catalytic efficiency, is energetically favorable, and ends with a functionally irreversible step. A natural pathway for the production of PDO from glycerol was known to exist in the bacterium *Klebsiella pneumoniae* and involve two enzymes: glycerol dehydratase and 1,3-propanediol dehydrogenase (Figure 7.3).

To construct a single microorganism to produce PDO, the decision was made to utilize heterologous pathway genes expressed in the bacterium *E. coli* to avoid issues involving natural regulation of the pathways. At the time, *E. coli* was the most completely studied organism with a well-understood metabolism and physiology and was genetically tractable with a rich set of genetic tools. There was also a strong base of industrial experience in large-scale fermentations and production. To build the pathway from glucose to glycerol, genes encoding G3PDH[12] and G3PP[13] (*DAR1* and *GPP2*, respectively) were cloned from *S. cerevisiae* and introduced into *E. coli* K12 on a plasmid, a multicopy extrachromosomal element. Enzyme expression levels were varied by influencing the rates of transcription and translation and by the relative positioning of the *DAR1* and *GPP2* genes on the plasmid to achieve optimal conversion of glucose to glycerol. *E. coli* is capable of allowing glycerol to reenter central carbon metabolism rather than be used to make PDO, so elimination of these catabolic reactions was necessary. Therefore, the genes encoding glycerol kinase (*glpK*) and glycerol dehydrogenase (*gldA*) were deleted from the *E. coli* chromosome. Early in

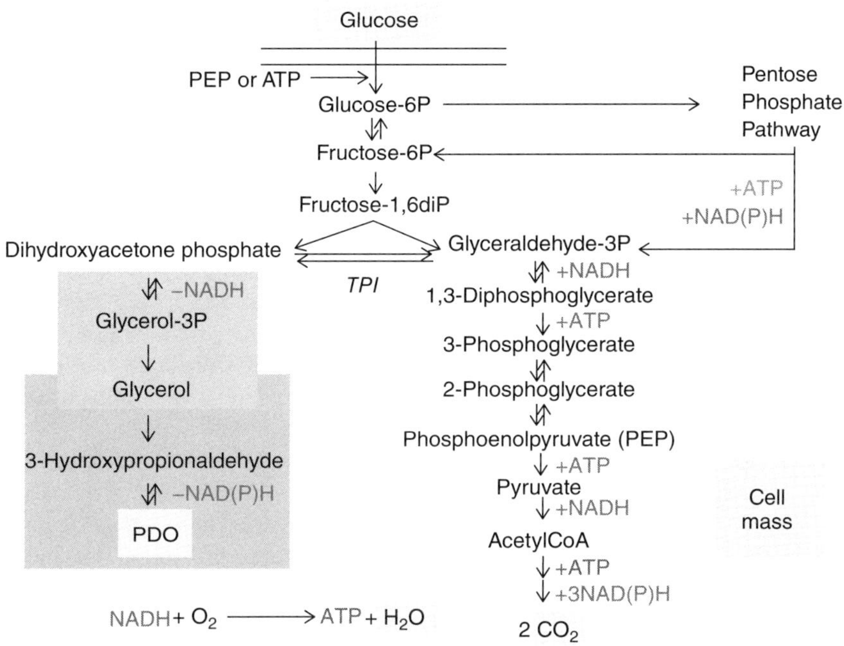

Figure 7.3 Metabolic pathways for the production of PDO and biomass. Reactions inserted from *Saccharomyces* are in the blue background, and those inserted from *Klebsiella* in the green background. Initial and final products are in yellow backgrounds. Important redox and energy cofactors are in red.

strain development the triosephosphate isomerase gene (*tpi*) was deleted, forcing a 50:50 split of carbon flow, to demonstrate the ability to channel carbon into the glycerol production and away from the tricarboxylic acid cycle. Although this created an artificial ceiling on carbon yield at 50%, it proved a useful tool in developing the glycerol/PDO pathway without dealing with concerns surrounding the cell's ability to remodel metabolism in favor of growth. This genetic modification would later be removed by returning *tpi* in subsequent strains to increase the carbon yield to greater than 50% by controlling the balance between PDO production and central metabolism.

Another limitation to carbon yield was identified in glucose transport. Glucose transport into *E. coli* takes place via the phosphorylation of glucose by the phosphotransferase system (PTS), which requires an equivalent of phosphoenolpyruvate (PEP). PEP is primarily generated through the enolase reaction in lower glycolysis. Each glucose molecule provides carbon for one glycerol molecule from the introduced glycerol pathway and for one PEP molecule by way of lower glycolysis; thus glycerol carbon yield is limited to 50%. The team realized that this constraint might be overcome by removing the PEP requirement for glucose transport. This was achieved by preventing glucose uptake and phosphorylation by the PTS through deletion of the *ptsH*, *ptsI*, and *crr* genes.[14] Glucose transport was then restored with an ATP-dependent phosphorylation system encoded by the *E. coli* galactose permease (GalP) and glucokinase (Glk) genes.[15] Their levels of expression were increased by replacing their native promoters with a synthetic, constitutive *trc* promoter.[16] Further improvements in glucose uptake were achieved through selection of faster-growing *E. coli* cells in a 10 L fermenter.

To surpass the yield ceiling imposed by the *tpi* deletion, it was necessary to restore triosephosphate isomerase activity while downregulating the flux of carbon toward lower glycolysis by reducing the expression of glyceraldehyde 3-phosphate dehydrogenase (GapA). This was achieved through a combinatorial approach, replacing the natural *gapA* promoter with promoter variants that decreased expression while altering expression levels of the glycerol pathway enzymes present on a low-copy plasmid. A strain with reduced *gapA* expression in combination with a glycerol pathway plasmid produced glycerol from glucose at a yield of 0.54 mol/mol.

Conversion of glycerol to PDO is carried out in two enzymatic steps. Glycerol dehydratase (*dhaB1*, *dhaB2*, *dhaB3*) enables the conversion of glycerol to 3-hydroxypropionaldehyde, and 1,3-propanediol dehydrogenase (*dhaT*) catalyzes the reduction of the aldehyde to PDO. The genes encoding both enzymes are organized into two operons in *K. pneumoniae*, along with four other genes – *dhaX*, *orfW*, *orfX*, and *orfY*.[17,18] Together these genes are part of the *dha* regulon, which is present on a contiguous fragment of DNA.[17,19,20] An anticipated problem with the dehydratase, a multisubunit ($\alpha_2\beta_2\gamma_2$) B_{12}-dependent enzyme, was that it undergoes catalytic inactivation during reaction with glycerol, with $k_{cat}:k_{inact} \sim 10^5:1$.[21,22] PDO was also shown to cause dehydratase inactivation. The mechanism of inactivation involves the modification of the coenzyme B_{12} moiety leading to $B_{12\text{-inact}}$, which remains tightly bound to the dehydratase. Reactivation of glycerol-inactivated dehydratase requires the addition of a reactivation complex comprising two proteins, DhaX and OrfX,[18,23,24] which catalyzes the release of bound $B_{12\text{-inact}}$ in an ATP-dependent fashion.[25] Reactivation allows the apoenzyme to bind coenzyme B_{12} once again and catalysis to resume. In contrast to the dehydratase reactivase genes, the genes *orfY* and *orfW*, which encode a coenzyme B_{12} regenerating function, were not required for efficient PDO production. Unlike natural PDO-producing organisms, *E. coli* does not synthesize coenzyme B_{12} *de novo* and so it was understood that B_{12} would either be provided by adding the B_{12} biosynthetic pathway to the production organism or by addition of vitamin B_{12}, a precursor to coenzyme B_{12} to the production tank. The technically more feasible route of adding vitamin B_{12} was selected; however, cost estimates suggested that replenishing inactivated glycerol dehydratase with coenzyme B_{12} in commercial-scale fermentation would require an unaffordable amount of vitamin B_{12}. To decrease the rate of inactivation, a directed evolution project was undertaken, resulting in a set of variant dehydratases with $k_{cat}:k_{inact}$ approximately sevenfold higher than the native enzyme.[26] One of these was used in the production organism.

With the genes for conversion of dihydroxyacetone 3-phosphate to PDO in hand, the pathway was assembled on a two-plasmid system in *E. coli*. The resulting strain produced PDO from glucose at a carbon yield and titer of 0.28 mol/mol and

68 g/L, respectively. The fermentation of this strain on glucose proved robust up to the point where PDO production stopped after reaching a maximum. It was observed, however, that glucose consumption either continued with significant formation of glycerol and cell mass or it abruptly ceased. Investigations showed the cause was neither inactivation of glycerol dehydratase by PDO leading to glycerol accumulation nor loss of 1,3-propanediol dehydrogenase leading to toxic accumulation of 3-hydroxypropionaldehyde. An effort was launched to explore alternative oxidoreductases and dehydratases, assuming that the ones in use were insufficient for an unknown reason. To assist in the screening of alternative oxidoreductases, a plasmid variant carrying the *Klebsiella dha* regulon was constructed without the 1,3-propanediol oxidoreductase (*dhaT*). It was expected that a production strain carrying this plasmid variant would not have the ability to convert 3-hydropionaldehyde to PDO in the presence of vitamin B_{12}. Surprisingly, a fermentation with this strain in the presence of vitamin B_{12} resulted in robust PDO production with a 66% increase in titer.[27] Further investigation revealed the existence of an *E. coli* gene, *yqhD*, which encoded the 3-hydroxypropionadehyde-reducing activity and was induced in the absence of the *Klebsiella* DhaT. Noteworthy was the discovery that YqhD utilizes NADPH as a cofactor, unlike DhaT that utilizes NADH. It is likely that the differences in the cofactor reduced/oxidized ratios *in vivo* contribute to the significantly higher titer.

With the identification of the appropriate primary glycerol and PDO pathway enzymes, attention was directed toward the construction of a single plasmid system that would allow the optimization of expression of each enzyme while achieving enhanced plasmid stability. The pathway genes were organized into three separate operons – the glycerol operon (*DAR1* and *GPP2*) and two operons incorporating the PDO pathway genes. Each operon was driven by separate promoters and transcriptionally isolated by appropriate transcriptional terminators.[28] The PDO pathway operons allowed for the optimal glycerol dehydratase/reactivase ratio to be achieved. Additionally the antibiotic marker present on the plasmid was eliminated and replaced with the *E. coli serA* gene, an essential gene necessary for serine biosynthesis to complement the deletion of the *serA* gene in the production strain genome.

Additional genetic modifications were introduced in the *E. coli* production host to improve yield further. For example, acetic acid accumulation had been observed in fermentations with glucose excess. To address this, the levels of PEP carboxylase (*ppc*) were increased to improve TCA cycle efficiency and thereby decrease acetate accumulation. The methylglyoxal synthase gene (*mgsA*) was deleted to avoid an alternate route of DHAP entry into central metabolism and potential toxic intermediates.[29] Once both modifications had been introduced into the production strain, glucose-limited feed in fermentation was no longer needed to control by-product synthesis, substantial recovery of volumetric rate was achieved, and carbon yield was significantly increased.

Expression of *yqhD* was increased to raise the capacity of the PDO pathway, while expression of the native cobalamin adenosyltransferase (*btuR*) was increased to complement elimination of *orfW* from the expression plasmid. The resulting

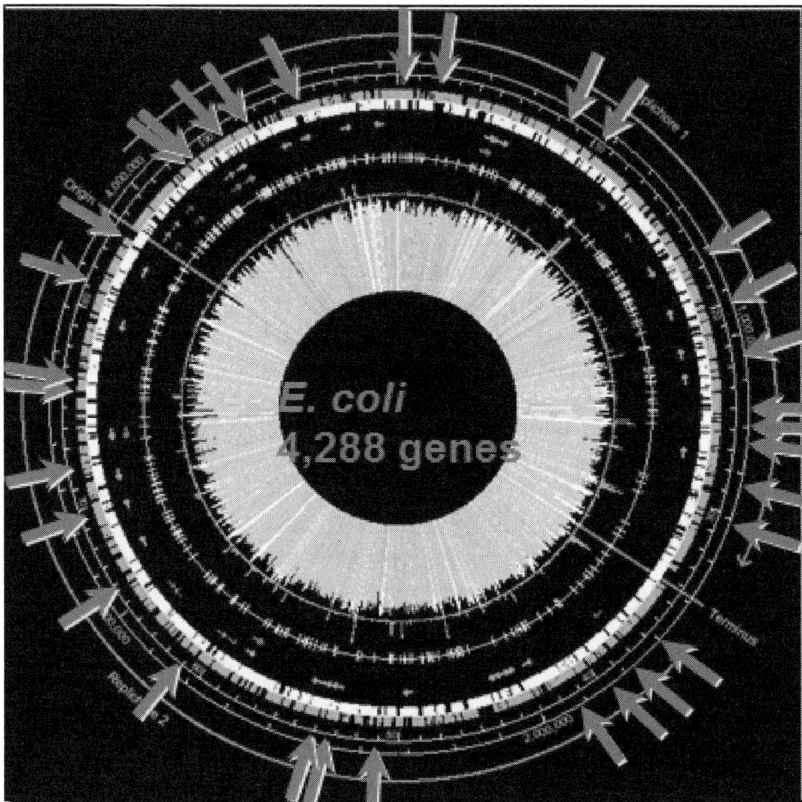

Figure 7.4 Alterations made in the *E. coli* genome in the course of making the PDO biocatalyst. Changes were made to glucose transport, glycerol metabolism, glycolysis (Entner–Doudoroff, pentose phosphate, and tricarboxylic acid cycles), respiration, amino acid biosynthesis pathways, anaplerotic reactions, and global regulators. Not shown – the plasmid containing the pathway genes. (*See color plate section for the color representation of this figure.*)

strain provided higher PDO carbon yield at constant total carbon yield. Finally, four genes known or predicted to be involved in carboxylic acid production were deleted from the production organism: *pta* (a putative subunit of phosphotransacetylase) and *ackA* (acetate kinase), both with the potential to reduce acetate production, and *aldA* and *aldB*, putative aldehyde dehydrogenases acting on 3-HPA to make 3-hydroxypropionic acid. The resulting strain together with the final pathway plasmid provided the high yield, rate, and titer of 0.9 mol/mol, 3.5 g/L/h, and 135 g/L, respectively (Figure 7.4).

Subsequent improvements to PDO-producing strains included adding the ability to utilize sucrose by expression of a sucrose transporter gene, a sucrose hydrolase, and a fructose kinase gene to enable efficient conversion of both glucose and fructose products of sucrose hydrolysis to the corresponding sugar phosphates that can enter glycolysis.

7.4.2 Bio-PDO Process Development and Commercialization

The Bio-PDO and Sorona products were developed simultaneously to improve speed to market and thus gain commercial advantage. Prior to the availability of Bio-PDO, Sorona polymer was produced using petro-based PDO to begin the downstream development process. Bio-PDO material made during the piloting phase was then used in small-scale Sorona polymer trials to study intrinsic polymer material properties and quality attributes. Polymer processing performance, quality, and consistency were checked periodically on a 100 lb/h continuous polymerization line. The semiworks process produced enough Sorona polymer material to enable testing of key downstream polymer spinning, dyeing, and weaving processes.

A conservative design strategy was implemented in the development of this "first-of-its-kind" Bio-PDO process to balance the risk of the untested fermentation process and biocatalyst. The challenge for process development was to link the units together in such a way as to recover a water-soluble product at 10–15% concentration from an aqueous fermentation broth and deliver exceptional finished product purity (>99.95%), exceeding that for food-grade materials, in a cost-effective manner befitting commodity product economics. The opposing goals of high product purity and low production costs necessitated the selection of mild reaction and purification conditions to minimize generation of unwanted side products that contribute to color and UV absorbance in the finished product.

The commercial Bio-PDO process was developed with three principal steps: fermentation, separation, and refining. Each had its own challenges, and the individual elements had to work together to create a commercially viable process:

- *Fermentation.* The fermentation process produces Bio-PDO at 10–15% concentration in an aqueous medium that contains biocatalyst, mineral salts, proteins, ammonia-containing organic acid salts, glycerol, 1,2,4-butanetriol, and unreacted sugars. The aerobic fermentation is operated as a batch process that contains two large intermediate storage tanks to provide the transition from the batch upstream fermentation section to the continuous downstream purification section. The fermentation area contains four 600,000 L bubble column fermenters fueled by four 60,000 L bubble column seed fermenters. D-glucose (as 95% dextrose syrup) is delivered to the Bio-PDO process from the 125,000 bushel/day Tate & Lyle corn wet mill. Prior to making the transition from the fermentation area to the intermediate storage tanks and downstream purification process, the end-of-cycle fermentation contents are subjected to high temperatures for several minutes in an external heat exchanger to deactivate the microorganisms.
- *Separation.* A three-step filtration scheme consisting of microfiltration, ultrafiltration, and nanofiltration is the first step used to remove cellular biomass, particulates, and high molecular weight contaminants following the heat kill of the

finished fermentation contents. The heat-deactivated biomass is first removed in a crossflow microfiltration process. The concentrated permeate that passes through the membranes is collected in a header system. The filtrate is washed with water and subjected to further crossflow filtration (diafiltration) to achieve a higher overall recovery of Bio-PDO. The permeate streams from the two steps are combined and subjected to an ultrafiltration step to remove high molecular weight contaminants so the subsequent nanofiltration step will be efficient and effective. The Bio-PDO-containing fermentation broth exiting the ultrafiltration step is essentially free of insoluble material.

Nanofiltration affords rejection of disaccharides, higher molecular weight oligomers, and UV absorbers and fractional rejection of ionic components. Smaller molecular components such as glycerol, 1,2,4-butanetriol, and Bio-PDO pass through the membranes completely. Similar to the aforementioned filtration steps, both concentration and diafiltration stages are used in the nanofiltration step.

Following an evaporation step, ion exchange is used to remove the weak ions and UV absorbers from the Bio-PDO stream. The ion exchange columns are arranged in a CACA configuration: strong acid cation, weak base anion; strong acid cation, weak base anion, followed by a mixed bed composed of a mixture of strong acid cation and strong base anion. Evaporation using a mechanical vapor recompression system reduces the moisture content in the feed from 90% to 20%. Evaporation is done under vacuum so as to minimize the evaporation time and temperature and thus reduce potential for formation of UV-absorbing compounds caused by high-temperature reaction of residual sugar, glycerol, 1,2,4-butanetriol with Bio-PDO, and ammonia-containing compounds.

- *Refining.* The refining area comprises a vacuum distillation train consisting of four packed distillation columns interrupted by a hydrogenation step between the second and third columns. The crude feed leaving the mixed bed polish step is sent to the first of the four distillation columns, a packed distillation column, which removes water. A falling film reboiler is used to minimize the holdup time in the reboiler and therefore undesirable side reactions in the bottom of the column. The bottom stream from column 1 containing Bio-PDO is pumped to column 2, another packed column that removes the heavy impurities, mainly glycerol and unreacted sugars, and any heavy impurities originating from side reactions in the bottom of column 1 and the evaporator. The overhead stream from column 2, which contains Bio-PDO, is sent to a catalytic hydrogenation reactor system. Nickel-catalyzed hydrogenation of components containing double bonds is critical to meet the required color specifications of the final product, primarily reducing yellowing in the final polymer. Following chemical reduction, the product stream is subjected to the final two stages of distillation. The objective of column 3 is to distill off water and further light impurities arising from column 1 bottoms, side reactions in column 2, and products from the hydrogenation reactor. The bottoms of column 3 are sent column 4, which provides >99.98% Bio-PDO product.

7.4.3 Key Learnings from the Bio-PDO Process Development

1. *Integrated science and engineering are required* to make a commercial metabolic engineering project a success. It was critical to have process engineering input at the beginning of the project to set metrics and guide program development and process chemists and engineers at the end of the program to develop the commercial process. Throughout the entire program, the biocatalyst development was carried out by molecular biologists, biochemists, and fermentation scientists with consultation with process chemists. Late in the project it was necessary to have continued contributions from biochemists and molecular biologists to address challenges that arose during piloting (Figure 7.5).
2. When pursuing groundbreaking approaches, *many parallel paths must be explored* within a strong team environment. New process development programs are risky and expensive. In order to mitigate the risk of not reaching success in a commercially reasonable time period, parallel paths must be taken to any critical objective. In the end, one such path will be selected to be part of the overall solution, potentially leaving the talented people who worked on the other paths feeling unfulfilled. It is therefore important to foster a culture of team success rather than individual success in such highly complex research programs.
3. *Progress is discontinuous.* As in all breakthrough research programs, progress in metabolic engineering come in fits and starts. Frequently the team became stuck on a particular rate or yield number for an extended period, followed by a sudden jump in performance. This inconsistency and unpredictability of progress requires great patience and understanding on the part of the management team. It is critical to have cross-functional senior management

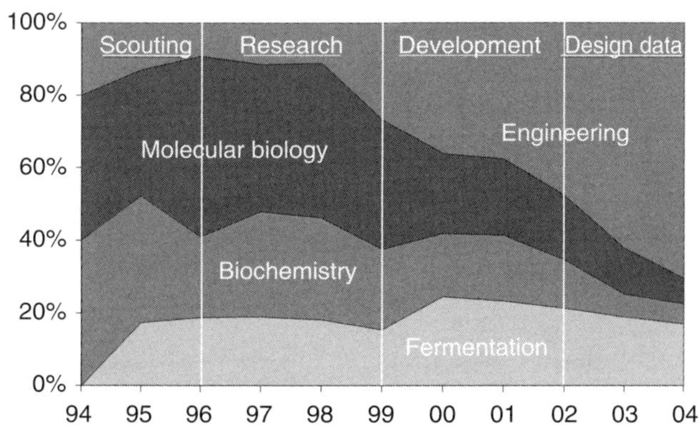

Figure 7.5 Distribution of skills of project team members year by year over the course of the Bio-PDO™ project.

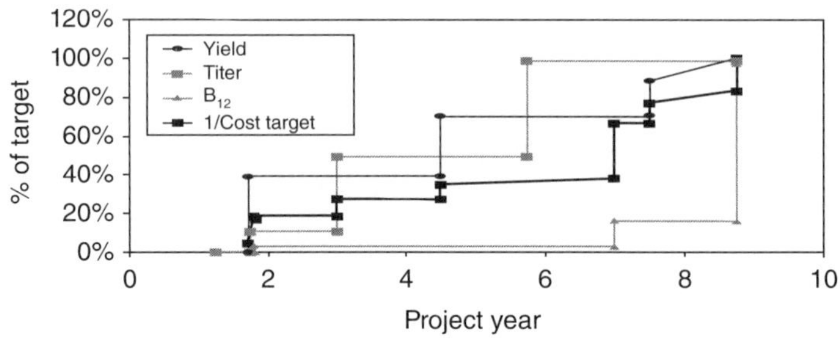

Figure 7.6 Progress toward key objectives year by year over the course of the project.

support and engagement in the program, with a disciplined review and risk-management process (Figure 7.6).

7.4.4 Commercial Results for Bio-PDO and Sorona

Commercial production of Bio-PDO began in October 2006 at the Loudon, TN, plant. The capacity of the plant was originally about 100 million pounds (45 million kg) per year based on a capital investment of about \$100MM. The plant has since been expanded to about 130 million pounds (60 million kg) per year based on the success of Sorona in the market place. The manufacturing process has achieved all of the technical targets originally envisioned. From an environmental perspective, the new process to make Bio-PDO consumes up to 40% less energy and reduces greenhouse gas emissions by more than 40% as compared to 1,3-PDO and PPG made using traditional petroleum-based processes.[30] The Sorona polymer made using this Bio-PDO contains 37% annually renewable plant-based ingredients by weight (28% biobased carbon).[31] The production of Sorona made with Bio-PDO uses 30% less nonrenewable energy and reduces greenhouse gas emissions by 63% compared to the production of an equal amount of nylon 6, the main competitor for Sorona in carpet applications. These data are derived from a peer-reviewed life cycle analysis (LCA) of Bio-PDO/Sorona production (Peer reviewer: Prof. Konrad Saur, Five Winds International), while the nylon 6,6 and nylon 6 LCA data are from Plastics Europe.[32]

The use of Bio-PDO enables the customers of DuPont Tate & Lyle Bioproducts to manufacture and market products with improved performance. These products also contain significant biobased content, an attribute that many customers and communities are beginning to demand. The importance of the Bio-PDO development has also been well recognized by the technical community. For example, in 2003, the US Environmental Protection Agency presented DuPont with its "Presidential Green Chemistry Award" for the company's research leading to the development of the 1,3-propanediol process, and in 2007, the American Chemical Society awarded scientists from DuPont and Tate & Lyle the "Heroes of Chemistry" award for development of the commercial process.

DuPont had to create the downstream demand for Bio-PDO by building the market for its main application, Sorona polymer. To build demand sufficiently far ahead of the start-up of the first Bio-PDO plant, DuPont introduced Sorona made from petro-based PDO in 2000. The cost of producing this market development material was well above the target cost for the final Sorona process. However, making commercial quantities of Sorona available was critical to getting customers to qualify and adopt this new polymer in their fiber spinning and downstream carpet and textile applications. In general, any new biomaterial – one offering a new composition, not just a new manufacturing process – will face a similar "chicken-and-egg" hurdle to market adoption. Without investment in a large-scale production plant, the production cost will remain high and customers will be reluctant to adopt the new material and drive demand volume. Without sufficient demand from customers, the manufacturer is reluctant to invest in the large-scale plant required to reach the target production costs. It therefore requires a large and long-term investment by the biomaterial manufacturer to make market development quantities available, normally at a loss, until the demand can support the investment in a large-scale plant. DuPont's corporate culture and history of introducing breakthrough new materials that took time to be accepted in the market and the long-term commitment to investment in industrial biotechnology supported the decision by its management to make this critical investment in Sorona.

Another critical factor to the success of Sorona is DuPont's close partnership with customers such as Mohawk Industries, a leader in the North American carpet market.[33] Technical teams at the two companies worked closely together to optimize the performance of Sorona in Mohawk's fiber spinning operations and to ensure that the required end-use performance was realized. Just as important, the teams worked together to create value for the Sorona brand in the carpet market through investment in advertising and other marketing programs. Mohawk's commitment to making this ongoing marketing investment and their strong presence, credibility, and relationships in the carpet market have been critical to the success of Sorona. In addition, the marketing efforts of this new polymer/fiber/carpet family leveraged the historical brand equity of both companies in the carpet market, aiding adoption of the new product. The end result has been that Sorona carpet is the most successful launch of a new product in the carpet market in over forty years. The product line has grown rapidly during a time of relative stagnation in the overall North American carpet market by taking share from incumbent materials such as nylon.

The technical and business success of Bio-PDO and Sorona required outstanding research and development, excellence in operations, and a sustained, effective marketing effort. DuPont brought many capabilities to the project, but the capabilities of partners Genencor, Tate & Lyle, and Mohawk Industries were also critical. The program is thus an excellent example of DuPont's three-pronged innovation approach:

1. *Market-driven innovation.* Understanding and recognizing the need for a new fiber material in the carpet and textile markets with differentiated performance that could be served by Sorona.
2. *Integrated science.* Creation of a new cost-effective manufacturing process for Bio-PDO including metabolic engineering and fermentation science to create

the *E. coli* organism, chemistry and engineering to separate the Bio-PDO from the fermentation broth, combined with materials science and engineering to develop Sorona polymer and its applications.

3. *Inclusive innovation.* The partnerships with Genencor to develop the Bio-PDO production organism; Tate & Lyle to develop the fermentation process, secure the feedstock, and build the commercial Bio-PDO plant; and Mohawk to develop the carpet applications and market the final product.

7.5 CASE HISTORY #2: DEVELOPMENT OF YEAST-BASED OMEGA-3S FOR VERLASSO HARMONIOUSLY RAISED SALMON

The long-chain omega-3 fatty acids EPA and docosahexaenoic acid (DHA) are essential nutrients for humans and for many commercially important fish and livestock. As essential nutrients, they are not produced by the body and must be obtained through the diet. The human health benefits of the long-chain omega-3s have been well documented.[34–36] There are two recommended sources for adding omega-3s to the human diet: (1) eating cold-water fish such as salmon or anchovies that are naturally rich in omega-3s and (2) taking omega-3-rich oils as nutritional supplements or pharmaceuticals.

The predominant source of omega-3s for both human nutrition and for the production of farmed salmonids – salmon, trout, and related species – is fish oil, which is sourced from the capture and processing of wild pelagic fish such as anchovies, menhaden, and jack mackerel. These smaller fish obtain their omega-3s by feeding on certain unicellular algae found in plankton. The supply of fish oil is essentially capped now, as the amount of pelagic fish that can be captured is highly regulated by the governments of the countries that control the key fisheries in the North Atlantic and South Pacific oceans. Strict enforcement of these regulations is critical to protect the health of threatened marine environments.

Traditionally, the majority of fish oil produced has been used in a crude form to feed salmon and other animals, with a relatively lower fraction of the fish oil being purified and concentrated for human nutrition applications. Farmed salmon production has grown strongly over the past twenty years, as reported by the Food and Agriculture Organization of the United Nations.[37]

The fraction of fish oil used for human nutrition has also been increasing substantially over the past decade or so, driven by growth in demand for the higher-value uses in supplements and pharmaceuticals. This dynamic reduces the availability of fish oil for raising salmonids. Typically, farmed salmon are raised on a diet relatively rich in fish oil, requiring the processing of 3–4 kg of wild pelagic fish to raise 1 kg of farmed salmon.[38] In 2002, it was estimated that roughly 60% of all the fish oil produced globally was used to raise farmed salmonids.[39]

The ratio of wild fish that must be processed to provide for the production of the farmed salmon, also known as the "fish in/fish out ratio" or "feeder fish dependency ratio" (FFDR), has been a serious concern to key environmental organizations that seek to protect the health of the oceans, such as Monterey Bay Aquarium's Seafood

Watch.[40] Seafood Watch rates the sustainability of wild-caught and farm-raised seafood so that consumers can make informed choices on the environmental impact of their seafood purchases. This organization has always rated farmed salmon as "Avoid," its lowest rating for sustainability, in great part due to the unfavorable FFDR for salmon aquaculture. Other important nonprofit organizations that focus on the health of the marine environment, such as the Environmental Defense Fund, have raised similar concerns.[41]

It is clear that further growth in the demand for long-chain omega-3s required by the human nutrition and aquaculture markets cannot be satisfied by expanding the exploitation of the oceans. A small fraction of the omega-3 market that requires pure DHA in infant formula and other specialty human nutrition applications is currently supplied from an algae-based fermentation process first commercialized by Martek Biosciences Corporation (recently acquired by DSM). The development of algal DHA technology was recently reviewed.[42] These fermentation-based products have not been used extensively in aquaculture applications.

In 2002, DuPont chartered a program to seek a new sustainable route to long-chain omega-3s. The next section provides a detailed description of this technical program, beginning with the selection of a suitable production host, followed by the complex metabolic engineering required to provide an efficient pathway to the synthesis of EPA, the dynamic modeling used to provide in-depth process understanding, and finally the scale-up of a commercially viable fermentation process. An overview of the metabolic engineering process can be found in these two articles.[43,44]

7.5.1 Selection of *Yarrowia lipolytica* as the Production Host

Several factors such as cell metabolism for fatty acid production and lipid accumulation, ease of metabolic engineering, fermentation performance, scale-up capability, operational safety, and product stewardship have to be considered when selecting a production host for any metabolic engineering program. After analyzing several organisms, the unconventional yeast *Yarrowia lipolytica* was selected as the host to produce omega-3 fatty acids by metabolic engineering.

Y. lipolytica is routinely isolated from cheeses, sausages, or natural environments like oil fields. It is an obligate aerobic yeast that can use hydrocarbons such as glucose, fructose, glycerol, and various fats as carbon sources. Most strains are haploid.[45] Depending on growth conditions, *Y. lipolytica* cells differentiate into yeast, pseudomycelium, and true mycelial forms (www.genolevures.org/yali.html). Some *Y. lipolytica* strains are oleaginous organisms that can accumulate up to more than 30% dry cell weight (DCW) as storage triglycerides (TAG) in oil bodies under the condition of nitrogen starvation and glucose excess. The lipid from glucose-grown cells is comprised mainly of TAG in which oleic acid (C18:1n-9) and linoleic acid (LA, C18:2n-6) are the two major fatty acids.

Y. lipolytica is one of the most studied unconventional yeasts.[46] A complete genome sequence of strain CLIB122 has been published (www.genolevures.org/yali.html). It has six chromosomes with a total of about 20 Mb DNA that encodes about 6,500 genes. There is no extrachromosomal plasmid existing in the wild-type

Y. lipolytica strain. Genetic transformation occurs when exogenous DNA integrates into the genome by homologous and nonhomologous recombination. *Y. lipolytica* has also been used as a model system for studying hydrophobic substrate utilization, peroxisome biogenesis, lipid metabolism, and biolipid production.[46,47]

It is relatively easy to develop *Y. lipolytica* auxotrophic mutants. Transformants can be selected by complementation of auxotrophic mutations and do not require the use of antibiotic resistance genes as selectable markers. The auxotrophic markers most commonly used are the *lys5 gene mutant* defective in saccharopine dehydrogenase (GenBank Accession No. M34929), the *leu2* gene mutant defective in beta-isopropylmalate dehydrogenase (GenBank Accession No. AF260230), and the *ura3 gene mutant* defective in orotidine 5′-monophosphate decarboxylase (GenBank Accession No. AJ306421). The counterselection system of the URA3 gene and 5-fluoroorotic acid (5-FOA), which selects for the *ura* phenotype, allows multiple rounds of integration of targeted genes into the *Y. lipolytica* genome and thereby introduction of many copies of desired genes.[48] There have been extensive studies on the genetics, molecular biology, and physiology of *Y. lipolytica* accumulated over the past thirty years, which have stimulated many scientists to use it as a model system for scientific research and industrial applications.

Y. lipolytica has an established history of robust fermentation performance and safe use in commercial-scale processes. The cultured cell density (DCW) can reach more than 100 g biomass/L with carbohydrates or fatty acids as the sole carbon source. *Y. lipolytica* was first produced commercially by British Petroleum as a source of single cell protein (SCP) for animal feeds under the trade name of *Toprina*® using cheap and abundant *n*-paraffins as the sole carbon source. There was no associated toxicity or carcinogenicity detected with this product.[49] *Y. lipolytica* was used to produce a gamma-decalactone flavoring agent from the alkyl ricinoleate derivative of castor oil. *Y. lipolytica* was also used as "generally recognized as safe (GRAS)" organism for commercial production of food-grade citric acid (US Food and Drug Administration list of microbial-derived ingredients approved for use in food; Title 21, Part 173, Section 165). Wild-type *Y. lipolytica* strains produce a mixture of citric and isocitric acids at consumed substrate yields of up to 130%.

Over 40 different *Y. lipolytica* strains from various public depositories all over the world were collected and screened for their fermentation performances and ability to accumulate omega-3 fatty acids such as EPA and DHA (C22:6n-3) when fed various substrates. After a careful execution of statistically designed experiments, the strain American Type Culture Collection (ATCC) #20362 was selected after it was demonstrated to achieve our fermentation performance targets: DCW greater than 100 g/L, lipid content greater than 30% of the DCW, and lipid productivity greater than 1 g/L/h. The ATCC #20362 strain was then selected for pathway engineering. The genome sequence of strain ATCC #20362 is >99% identical to the genome sequence of strain CLIB122 (www.genolevures.org/yali.html).

7.5.2 Natural Aerobic Pathways for Omega-3 EPA Production

Like other yeast, wild-type *Y. lipolytica* does not synthesize omega-3 fatty acids. The fatty acid profile of the wild-type strain ATCC#20362[43,48] showed that it can synthesize the EPA precursor LA (C18:2n-6). In other organisms, there are two different pathways to synthesize EPA from LA, the anaerobic polyketide synthase pathway[50] or an aerobic desaturase[51] and elongase pathway. The aerobic pathway (Figure 7.7) can be further classified into a delta (Δ)-6 desaturase pathway (the Δ-6 pathway, found in algae, mosses, fungi, and others) or a Δ-9 elongase and Δ-8 desaturase[52] pathway (the Δ-9 pathway, found in euglenoids).

In the "Δ-6 pathway" the first step utilizes Δ-6-desaturase to convert LA and alpha-linolenic acid (ALA, 18:3n-3) to gamma-linolenic acid (GLA, 18:3n-6) and stearidonic acid (STA, 18:4n-3), respectively; the second step is then use of the $C_{18/20}$-elongase to convert GLA and STA to dihomo-gamma-linoleic acid (DGLA, 20:3n-6) and eicosatetraenoic acid (ETA, 20:4n-3), respectively. In the "Δ-9 pathway" the first step is the Δ-9 elongase to convert LA and ALA to eicosadienoic acid (EDA, 20:2n-6) and eicosatrienoic acid (ETrA: 20:3n-3), respectively; the second step is then the Δ-8-desaturase to convert EDA and EtrA to DGLA and ETA, respectively.

To date, no Δ-15-desaturase has been identified to convert LA to ALA with 100% efficiency. Transformed cells with a heterologous Δ-15-desaturase gene will always contain both LA and ALA. Therefore, the "Δ-6 pathway" and the "Δ-9 pathway" can simultaneously use both LA and ALA as primary substrates if there is a functional Δ-15-desaturase or any omega-3 desaturase that can convert LA to ALA inside the cell. Most Δ-17-desaturases identified so far not only have high activity to convert arachidonic acid (ARA, 20:4n-6) to EPA but also have weak omega-3 desaturase activity to convert LA to ALA, GLA to STA, EDA to ETrA, and DGLA to ETA.

All of these desaturation and elongation enzymes carry out their reactions in the endoplasmic reticulum membrane.[51] Desaturation occurs on the acyl moiety of phospholipid (PL), while elongation occurs on the acyl moiety of the acyl-CoAs. The acyl-CoA pool is limited, leading to lower conversion efficiency by elongases. Therefore, elongation in both the Δ-6 and Δ-9 pathways is the rate-limiting step. It was thus concluded that the introduction of either the Δ-6 or Δ-9 pathway genes into the wild-type strain should allow the production of EPA through desaturation and elongation of the native fatty acid species.

7.5.3 Development of the Genetic Tools Required for Metabolic Engineering

Before 2012, the *Y. lipolytica* transformation system[53] had been established and a low-copy replication plasmid had been developed.[54] This system had been used as a host for efficient secretion of expressed heterologous proteins[55] or as a system to study certain biological functions. The genetic elements necessary for metabolic engineering, such as promoters and terminators, had limited availability.[56–59] In order to develop *Y. lipolytica* as a host for omega-3 fatty acid production by metabolic engineering, the DuPont team carried out the following four work streams simultaneously:

1. Characterization of a series of promoters
2. Characterization of genes encoding enzymes for omega-3 fatty acid production

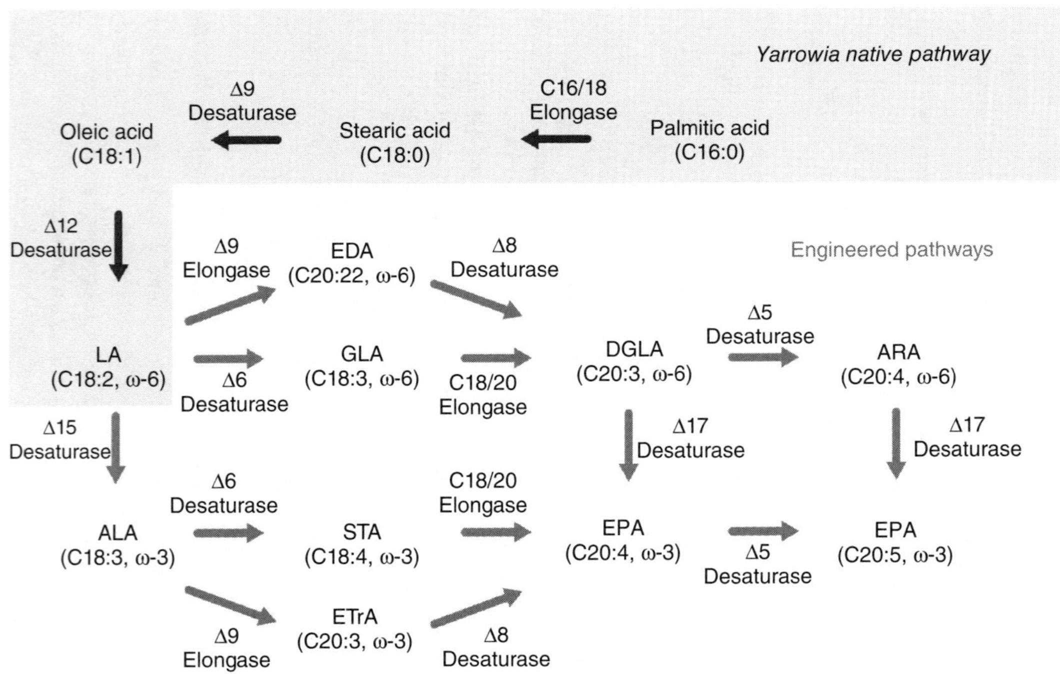

Figure 7.7 Schematic diagram of aerobic pathways for EPA biosynthesis.

3. Strain construction and methods for further strain improvement
4. Development of fermentation process

7.5.3.1 Characterization of a Series of Promoters According to the RNA expression profile of the wild-type strain under growth and/or nitrogen starvation conditions, several promoters from *Y. lipolytica* were isolated by genome-walking and PCR techniques.[60–62] The promoters that were isolated are from the genes encoding an export protein (*EXP*, YALI0C12034g), fructose 1,6-bisphosphate aldolase (*FBA1*; YALI0E26004g), glycerol-3-phosphate-*O*-acyltransferase (*GPAT*, YALI0C00187g), phosphoglycerate mutase (*GPM1*; YALI0B02728g), glycerol-3-phosphate dehydrogenase (*GPD1*; YALI0B02948g), and an ammonium transporter (*YAT1*, YALI0_E27181g). All these promoter activities were compared with the promoter of the translation elongation factor (TEF) gene of *Y. lipolytica*[56] by quantitative fluorometric assays of the β-glucuronidase (GUS) reporter[63] driven by individual promoter. The results showed that the *FBA1* promoter was the strongest among the six promoters. The *GPM1* promoter was as strong as the *TEF* promoter; the *GPD1* promoter was 2.5 times stronger than *GPM1* promoter; and the *FBA1* promoter activity was 5.5 and 2.2 times stronger than the *GPD1* and *GPM1* promoters, respectively.

In the N-terminal coding region (169 bp) of the *FBA1* gene, there is a 102 bp intron located between amino acids 20 and 21. Fusion of the *FBA1* promoter plus the N-terminal coding region covering the first 23 amino acids and the intron ($FBA1_{in}$) with the GUS reporter gene resulted in GUS activity about five times greater than the FBA promoter alone. The 169 bp covering the N-terminal coding region of the *FBA1* gene also enhanced the activity of other promoters when chimeric promoters were constructed. These data demonstrated that there was a transcriptional enhancer located in the intron of the *FBA1* gene.[61]

The YAT1 promoter has a unique feature – it has almost no activity under normal growth conditions, but under nitrogen-limiting conditions, its activity increases approximately 35-fold.[61] This promoter is useful for directing the expression of omega-3 biosynthetic genes since the lipid synthesis and accumulation is induced by nitrogen limitation. The relative strength of these promoters in limited nitrogen conditions was determined by quantitative GUS assays, and they are as follows: FBA_{in} > YAT1 > FBA > GPD, EXP > GPAT > GPM = TEF.

There are about 6500 genes in *Y. lipolytica*. The promoter of each gene has unique properties. A diverse promoter library was also generated through random mutagenesis using a specific promoter as a template.[64] Studies of promoter characteristics are required before they can be utilized to maximize the expression of targeted genes in certain conditions.

7.5.3.2 Characterization of Genes Encoding Enzymes for Omega-3 Fatty Acid Production *The Δ-6 pathway.* To find the best genes covering the Δ-6 pathway, all the wild-type genes identified by Ross Davison of Abbott Laboratories[58,59] in *Y. lipolytica* were screened. There are two Δ-6-desaturase genes from *Mortierella alpina* and *Saprolegnia diclina*; two $C_{18/20}$-elongase genes from *M. alpina* and

Thraustochytrium aureum; three Δ-5-desaturase genes from *Isochrysis galbana*, *M. alpina*, and *S. diclina*; and one Δ-17-desaturase gene from *S. diclina*.[48,58,59] Since all these genes were from other organisms, the codon usages of most genes were not preferred in *Y. lipolytica*. To achieve high levels of expression, all of the preferred target genes were designed and screened according to the codon usage pattern of *Y. lipolytica*. The consensus sequence (5′-ACC*ATG*G-3′) around the "ATG" translation initiation codon was also incorporated, and the GC content was controlled to about 54% (typical for genes from *Y. lipolytica*). It was also discovered that substrate conversion was increased in almost all of the codon-optimized genes except the Δ-5-desaturase gene derived from *M. alpina*. The improved substrate conversion efficiency of these "codon-optimized" genes was hypothesized to result from more efficient translation of their encoded mRNAs in *Y. lipolytica*. In the commercial strains, the genes introduced into *Y. lipolytica* for strain construction are all "codon optimized," and each of the synthetic genes has an *Nco*I site around its translation initiation site and a *Not*I site after its stop codon. The building block-like structure of these genes facilitates assembly of expression constructs with different configurations.

By increasing the number of copies of the Δ-6 pathway genes and "pushing" the carbon flux into the engineered pathway by amplification of the $C_{16/18}$-elongase gene of *M. alpina*[65] to convert the palmitic acid (C16:0) into stearic acid (C18:0) and the Δ-12-desaturase gene of *Fusarium moniliforme*[66] to convert the oleic acid to LA, a strain Y9027[47] was generated to produce EPA at about 40% total fatty acid methyl ester (FAME). This strain contained 19 copies of ten different heterologous genes that are integrated into its genome. The second highest fatty acid in the lipids was GLA, at about 21% FAME. The *in vivo* substrate conversion catalyzed by the Δ-6-, Δ-5-, and Δ-17-desaturases was about 86%, 90% and 97%, respectively, indicating that these enzymes functioned well in strain Y2097.[48] In contrast, the substrate conversion catalyzed by four copies of the $C_{18/20}$-elongase was only about 69% in that strain, suggesting that the $C_{18/20}$-elongation reaction was a bottleneck in the pathway. The GLA produced is the product of Δ-6-desaturase and the substrate of $C_{18/20}$-elongase. The Δ-6-desaturase introduces a double bond into the LA acyl chain esterified to a PL backbone; the $C_{18/20}$-elongase catalyzes the condensation of a malonyl group to GLA acyl chain esterified to CoA.[51,67] Thus, to reduce the amount of GLA in the EPA lipids, it was required to identify not only an efficient $C_{18/20}$-elongase but also active acyl exchange between the PL and CoA pools in the ER membrane.

To reduce the GLA and increase the EPA titer in the lipid phase, genes were separately amplified encoding glycerol-3-phosphate acyltransferase (*GPAT*), lysophosphatidic acid acyltransferase (*LPAAT*), diacylglycerol acyltransferase (*DGAT1* and *DGAT2*), lysophosphatidylcholine acyltransferase (*LPCAT*), choline phosphotransferase (*CPT1*), acyl-CoA synthase (*ACS*), phospholipase C (*PCL1*), and phospholipase D (*SPO22*). Although these modifications each improved the GLA conversion efficiency up to 15%, the GLA could not be completely eliminated by any of these modifications.[48] The fatty acid profile of the lipids from the engineered strains with Δ-6 pathway did not meet our target for EPA content in the accumulated lipid. To avoid GLA accumulation in the lipids, the decision was made to pursue the Δ-9 pathway for the commercial strain.

The Δ-9 pathway. To engineer *Y. lipolytica* to produce EPA through the Δ-9 pathway, genes are required to encode Δ-9 elongase, Δ-8 desaturase, Δ-5 desaturase, and Δ-17 desaturase. These genes were isolated for enzymes in the Δ-9 pathway and also other genes involved in lipid biosynthesis, fatty acid transport, and by-product reduction from several microorganisms such as algae, euglenoids and fungi (Table 7.1).[65,66,68–97]

7.5.3.3 Strain Construction and Methods for Further Strain Improvement To construct an EPA production strain utilizing the Δ-9 pathway, a series of strategies were employed. First, an efficient EPA biosynthetic pathway was built by using strong promoters such as EXP1, FBAINm, GPAT, GPD, and YAT. All heterologous genes were codon optimized, and multiple copies of structural genes for each step were introduced. To increase the Δ-8-desaturase activity, three Δ-9-elongase and Δ-8-desaturase bifunctional fusion genes[71] were also constructed; the Δ-8-desaturase activity in these fusion enzymes was increased almost 100% as compared with Δ-8-desaturase alone. Second, the carbon flux into the engineered Δ-9 pathway was "pushed" and "pulled" by overexpression of the $C_{16/18}$-elongase gene, the Δ-12-desaturase gene, and multiple copies of Δ-17-desaturase genes, respectively. Third, beta-oxidation was reduced by the knockout of *PEX* genes[43,72] such as *PEX10* that is involved in peroxisome biogenesis and matrix protein importation. Fourth, fatty acid transport was controlled by fine regulation of different acyltransferases to increase fatty acid flux for EPA production and lipid accumulation.

At each step of strain construction, gas chromatography[98] was used to screen many different transformants. The insertion of a DNA fragment into the *Y. lipolytica* genome occurs predominantly by nonhomologous end joining,[99] so that each transformation generated a library of transformants with the heterologous genes integrated at different location in the genome, resulting in diverse performance. One beneficial trait found through this screening was that the knockout of the *PEX10* gene[43,72] increased EPA titer in lipids more than twofold over other isolates from the same transformation. In these strains with *pex10*Δ, the activity of beta-oxidation is greatly reduced. There were no normal peroxisomes found inside the cells, only unidentified membrane-like structures that might be the deformed nonfunctional peroxisomes. To knock out undesired genes in the genome of *Y. lipolytica*, approaches were also developed to increase homologous recombination efficiency, which have been used for strain construction and improvement.

By combining the previous strategies, a Gen I strain Y4305[43] was generated that contained 30 copies of 9 different genes and produced EPA at 56.6% FAME, without GLA as an intermediate. The total EPA produced was about 15% DCW. The lipid produced by strain Y4305 had a unique and healthy fatty acid profile that was extremely low in saturated fatty acids and only small levels of intermediates. This is due to balanced expression of various genes in the Δ-9 pathway with strong "pushing" and "pulling" for EPA biosynthesis and accumulation. Like strain Y2097 using the Δ-6

TABLE 7.1 Isolated Genes for EPA Biosynthesis.

Genes	Source of Gene
Δ-5-desaturase	*Euglena gracilis* *Euglena anabaena* *Eutreptiella* sp. CCMP626
Δ-8-desaturase	*Euglena gracilis* *Euglena anabaena* *Eutreptiella* sp. CCMP389 *Eutreptiella* sp. CCMP1491
Δ-9-desaturase	*Yarrowia lipolytica*
Δ-12-desaturase	*Fusarium moniliforme* *Yarrowia lipolytica*
Δ-15-desaturase	*Fusarium moniliforme*
Δ-17-desaturase	*Pythium aphanidermatum* *Phytophthora ramorum* *Phytophthora sojae*
Δ-9-elongase	*Euglena gracilis* *Euglena anabaena* *Eutreptiella* sp. CCMP389
Synthetic Δ-9-elongase and Δ-8-desaturase bifunctional enzymes	Δ-9-elongase and Δ-8-desaturase mutant of *E. gracilis* Δ-9-elongase of *Eutreptiella* sp. CCMP389 and Δ-8-desaturase mutant of *E. gracilis* Δ-9-elongase and Δ-8-desaturase mutant of *E. anabaena*
$C_{14/16}$-elongase	*Yarrowia lipolytica*
$C_{16/18}$-elongase	*Mortierella alpina* *Yarrowia lipolytica*
Acyl-CoA:1-acyl-glycerol-3-phosphate acyltransferase (LPAAT)	*Mortierella alpina* *Yarrowia lipolytica*
Acyl-CoA:lysophosphatidylcholine acyltransferase (LPCAT)	*Yarrowia lipolytica*
ATP citrate lyase	*Mortierella alpina* *Yarrowia lipolytica*
Choline-phosphate cytidylyltransferase (PCT)	*Yarrowia lipolytica*
Diacylglycerol acyltransferases (DGAT1 and DGAT2)	*Yarrowia lipolytica*
Diacylglycerol cholinephosphotransferase (CPT)	*Yarrowia lipolytica*
Malic enzyme	*Yarrowia lipolytica* *Mortierella alpina*
Malonyl-CoA synthetase (MCS)	*Rhizobium leguminosarum bv. viciae 3841*
Phospholipid:diacylglycerol acyltransferase (PDAT)	*Yarrowia lipolytica*
Yeast AP-1-like transcription factor (YAP1)	*Yarrowia lipolytica*

pathway, the substrate conversion efficiency of the introduced desaturases was still significantly higher than the elongases in strain Y4305. However, the selection of the Δ-9 pathway ensured that the rate-limiting elongation was the first step of the engineered pathway. Accumulation of intermediates was therefore kept to a minimum, in contrast to cells engineered with the Δ-6 pathway where the first step was not rate limiting and accumulation of GLA becomes significant.

After construction of strains that produced a high proportion of lipid as EPA, the major focus became increasing the total lipid content in the engineered cells. To reduce the manufacturing cost for EPA production, the Gen II strain Z1978[44] was generated via twenty-four steps of genetic modifications. It contained 35 copies of 17 different genes and produced EPA at >58% FAME. The total EPA content in strain Z1978 was about 20% DCW. The fatty acid profile of strain Z1978 was similar to that of strain Y4305, with extremely low saturated fatty acid content. It also had only small amounts of intermediates and no GLA. The major improvement of strain Z1978 over strain Y4305 lies in the lipid content that was increased from about 30% DCW in Y4305 strain to >38% in strain Z1978.

Based on strain Z1978, strain Gen III HP Z5567[44] was generated by six more steps of genetic modification. Strain Z5567 contained forty-one copies of nineteen different genes. It produced EPA at about 50% FAME but with a total lipid content of more than 50% DCW. Therefore, strain Z5567 produced EPA at about 25% DCW, which was an improvement of 25% and 67% over strain Z1978 and Y4305, respectively. All of these strains are stable for at least 50 generations, which has allowed consistent EPA production at commercial scale.

7.5.3.4 Omega-3 Yeast Fermentation Process Development Fermentation process development was initiated from the very start of the program for three major reasons: (1) to support strain engineering by screening under controlled conditions, (2) to optimize fermentation conditions for each promising new strain developed, and (3) to be in a position for scale-up to generate sufficient amounts of product suitable for downstream process development, safety studies, and efficacy studies. Figure 7.8 gives a pictorial description of the fermentation support given to the research project. Part a) represents a cell and symbolizes the genetic engineering described previously. New, promising strains that are products of the genetic engineering were first tested in shake flask experiments (Part b). Since production of omega-3 oils is a two-stage process, cells were first grown in flasks to a specified cell density after which they were deprived of nitrogen for growth and fed a carbohydrate solution. Under those conditions lipid accumulates in the cells. At the end of the production stage, the amount of lipid in the cells was determined along with its composition.

Strains that show good results in flasks were taken one step further to lab-scale fermenters (1–2 L). In these, the dissolved oxygen and pH can be controlled more consistently than is possible in flasks and the fermenter results are thus more representative of a full-scale process. The strains that perform exceptionally well at lab scale qualified to be scaled up to pilot scale and beyond (Figure 7.8d).

The rate-limiting step in the process shown in Figure 7.8 was the use of shake flasks to identify promising strains. The flasks are relatively labor intensive to prepare

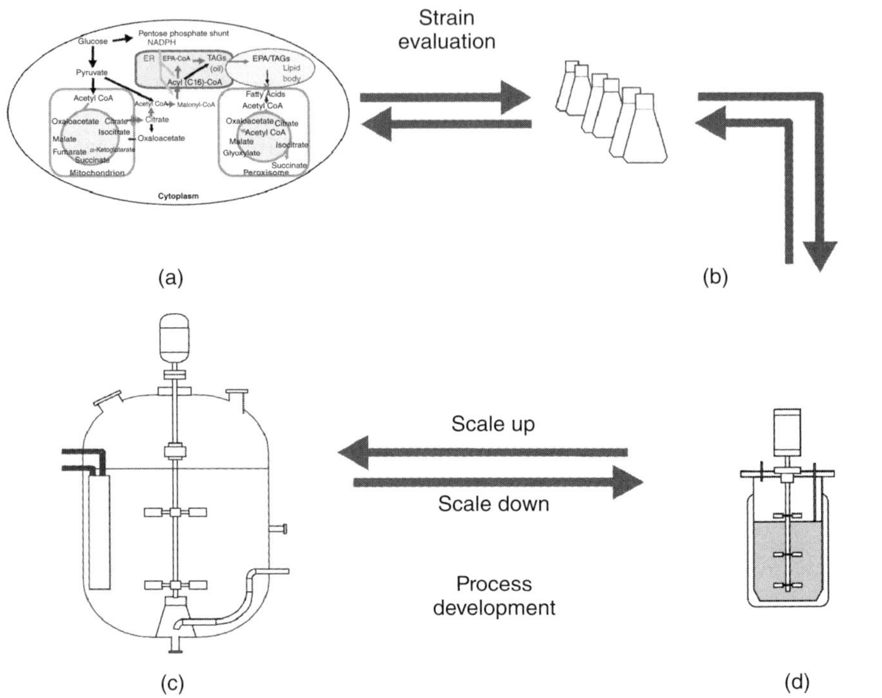

Figure 7.8 Workflow of strain and fermentation process development.

and require fairly bulky shakers to be processed. In addition, both pH and dissolved oxygen concentration typically vary during the experiment since neither is directly controlled. As a complement to the flask evaluations, it was found for the omega-3 project that microfermenters provided an elegant solution to the limitations of the flasks.[100] A microfermenter is a small vessel holding only a few milliliters of solution where the solution is held under controlled temperature, dissolved oxygen, and pH conditions. There are several different manufacturers of microfermenters. In our work, we used a unit with 24 individual microfermenters, each having its own strain and distinct processing conditions.

Figure 7.9 illustrates the way that the microfermenters were used both for strain screening and process development. The bar graph in Figure 7.9 shows the individual results from different strains subjected to the same fermentation conditions. Allowing for a standard strain to be run each time, eleven to twenty-three different new strains can be tested in each run depending on whether duplicates are used or not. This significantly increased throughput and reduces workload compared to flask experiments.

The lower graph of Figure 7.9 shows how the microfermenters allowed for media and condition optimization for a given strain. In this case the fermenters were inoculated with the same strain but the media and conditions were varied for the different units during the same run. The results obtained in the microfermenters were then validated by experiments at larger scale. Figure 7.9 shows that in the case of many

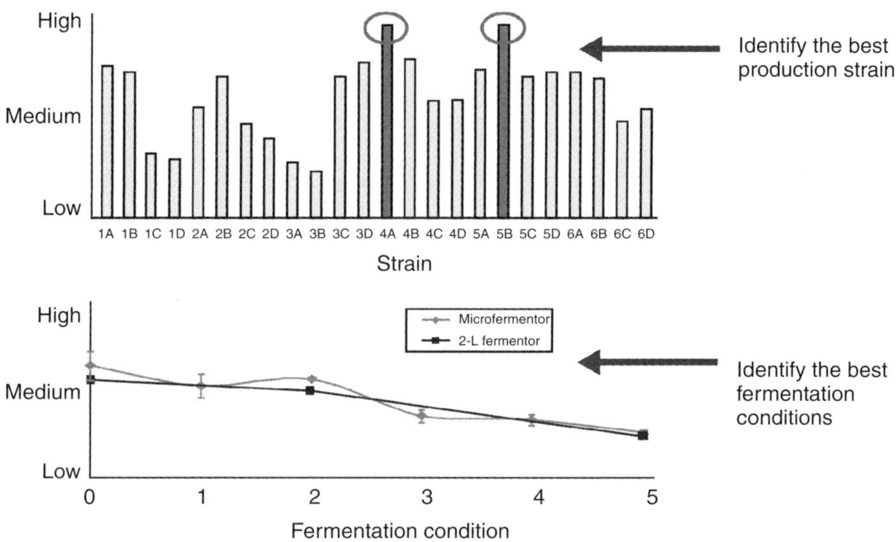

Figure 7.9 Microfermenters can be used for strain screening (bar graph on top) and for optimization of fermentation conditions (scatter plot below).

omega-3 strains, the endpoint performance of the microfermenter was comparable to endpoint results from the lab-scale fermenters, demonstrating the predictive power of this experimental approach.

While the microfermenter experiments provided guidance on optimization of the process conditions for a given strain, they were not sufficient for exploring the complete time history of a fermentation run or to generate sufficient information to build a dynamic model. For that, the lab-scale fermenters with their 1–2 L working volume were required. These were sufficiently large that samples could be taken frequently enough to provide a complete record of the transient behavior of the fed-batch fermentation, allowing design of optimized control strategies with respect to temperature, pH, and dissolved oxygen. Figure 7.10 shows a schematic of a lab-scale fermenter and how it was operated to generate high levels of lipid and omega-3. In the first stage of the fermentation, the fermenter was inoculated with the strain under study. While air was sparged into the fermenter and rapid stirring was applied, the cells grew exponentially on the carbohydrate substrate and the nitrogen provided with the base. At a specified time the ammonia base was switched to KOH to avoid further nitrogen supply and create the aforementioned nitrogen starvation conditions. Under these conditions, cell growth stopped and the cells shifted over to accumulating lipids while still using the carbohydrate supplied.

A critical aspect of rapid process development and scale-up in the omega-3 project was the use of predictive, dynamic simulation models. These models were built from first principles and included kinetic models of growth and product generation along with expressions for substrate, nitrogen, and oxygen consumption. The models were matched to the experimental data from the lab-scale fermenters. The models could

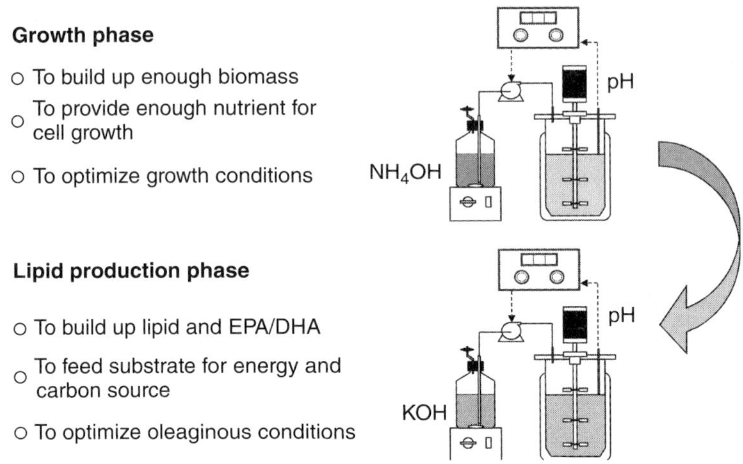

Figure 7.10 Omega-3 fermentation as a two-stage, fed-batch process developed primarily in lab-scale fermenters. During the first stage, biomass is grown to a specified cell density. During the second stage, lipid containing EPA/DHA is accumulated in the prepared biomass.

predict cell growth, DCW, oxygen consumption rate, dissolved oxygen tension, and CO_2 evolution rate, variables that also were measured during the fermentation. More importantly, the models could also predict lipid accumulation rates in the cell and rate of EPA accumulation in the lipid as a function of various process conditions. These analytical measurements were otherwise not available until several days after the run. The models were thus able to guide the experimentation and predict quality parameters before and during the run. Figure 7.11 shows an example of the predictive capability of the models with respect to one of the key product quality parameters, %EPA in the lipid. In the graph to the left, the model was used to predict the amount of EPA in the lipid when a particular process condition was changed. The model predicted severe retardation of EPA accumulation under certain conditions. To verify the results, experiments were run at lab scale with these conditions imposed. As seen in the rightmost graph, the actual behavior of the EPA in the lipid was, at least qualitatively, close to that predicted by the dynamic model. This feature was very important when it came to scaling up from lab scale to pilot scale and beyond, especially if the adverse condition was likely to be more accentuated in the large scale.

The last steps in any fermentation process development are scale-up to pilot-scale and then to commercial-scale assets. At the time when the omega-3 process was ready to scale up, an outside toll manufacturer was used to provide the required equipment and expertise to run the experiments. CP Kelco in San Diego, CA, United States, was selected as our scale-up facility. Here it was possible to test the omega-3 fermentation process at larger scale and also gather a wealth of information about the process and its dynamic behavior. The agitation and air handling features of the pilot plant were incorporated into our dynamic models so the mass transfer characteristics as functions of superficial gas velocity and agitation power could be studied for

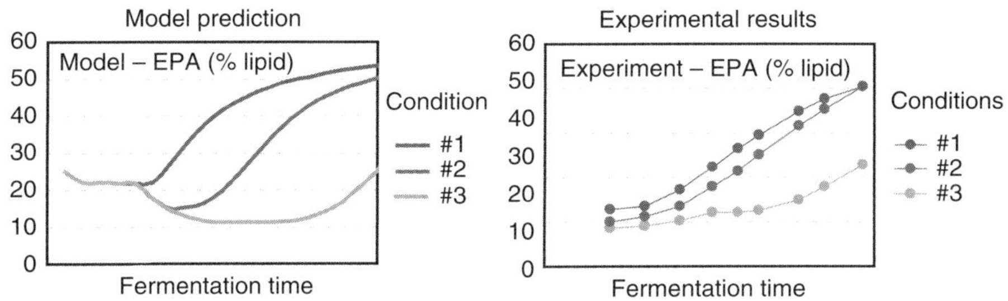

Figure 7.11 Example use of a predictive fermentation model.

each run. These features became quite important as the process was scaled up to commercial-sized fermenters where a common scale-up criterion is to keep the mass transfer characteristics constant.

As indicated in Figure 7.8, the path from strain engineering to commercial scale was not always unidirectional. For some strains, problems were encountered when the "standard protocol" was applied at that scale. Since the cost of a fermentation run increases quite dramatically with scale, the approach was always to go back to the next lower scale to elucidate the problem. These are called "scale-down" experiments. For example, at the pilot scale it was found that a strain had good quality attributes in the middle of the run but seems to lose these as the run progressed. By taking the strain back to the lab-scale fermenters, the process conditions that were believed to influence the quality attribute could be explored. Once these were identified, the protocol at the pilot scale was altered, and it was observed that the problem had been resolved. This new knowledge was always incorporated into the dynamic models such that they now could predict possible quality problems upon further scale-up of the process.

The benefits of highly predictive dynamic models became even more evident as the project moved from the pilot plant to another larger toll manufacturer for commercial production. The fermenters at this toll manufacturer were larger and the equipment different and they were not nearly as well instrumented as the pilot plant. It was now found that a dynamic model was an indispensable tool to interpret what was different at the commercial site relative to the pilot plant when the fermentation results were initially below expectation. The model filled in missing details from each initial run to supplement the relatively sparse set of process data collected by the instruments. With the information provided by the model, it was possible to make a set of corrections to the equipment and the operating procedures such that the project quickly got back on track, producing large batches of high-quality material.

The end result of this highly complex and challenging technical program was the development of a commercially viable, large-scale process to produce EPA in *Y. lipolytica*. At the time of the writing of this chapter, this product was in full commercial production at DuPont's fermentation facility in Cedar Rapids, IA, United States, providing a truly sustainable source of this critical long-chain omega-3 nutrient. This work was recognized as a "metabolic engineering tour de force" and "an example of the power of biotechnology in the push toward a sustainable society" in the August 2013 edition of *Nature Biotechnology*.[101]

7.5.4 Development of the Verlasso Salmon Business Proposition

To help develop the business proposition for the new omega-3 product, early on in the program DuPont business development personnel chartered several rounds of market research to understand the potential for using this new product to address the sustainability issues that face the salmon aquaculture industry. This research with North American consumers and retailers led DuPont to the conclusion that a reasonable fraction of consumers would be willing to pay a premium price for a farmed salmon raised in a more environmentally sustainable fashion than traditionally farmed salmon.

Consumers in North America are accustomed to paying a price premium for wild salmon, which is available seasonally from Alaska and other areas on the West Coast. Such premiums can range from ~50% to 300% or more of the typical price of farmed Atlantic salmon, dependent primarily on species (e.g., king salmon is generally more expensive than sockeye salmon) and availability of the wild salmon, which varies from one year to the next. The idea of a premium farmed salmon is also supported by the success of other similar categories of premium animal proteins such as free-range chicken and grass-fed beef. Based on the results of the market research, the DuPont marketing team concluded that a target price point for the new sustainable category of salmon around the low end of the wild salmon price point could be feasible.

DuPont first undertook some feeding studies with the world-renowned aquaculture institute Nofima to demonstrate the feasibility and safety of the new sustainable diet approach.[102] These studies showed that Atlantic salmon grow as well or better on the yeast-based diet as they do on traditional fish oil-based diets, with similar uptake of omega-3s. DuPont also worked with regulatory bodies in both the United States and Chile (the country that produces most of the farmed salmon sold in the Western Hemisphere) to obtain "GRAS" status in the United States and Servicio Agrícola y Ganadero (SAG) approval from the Chilean government for the omega-3 yeast as a feed ingredient. During this time period, DuPont approached Empresas AquaChile SA, one of the world leaders in salmon and tilapia aquaculture headquartered in Puerto Montt, Chile, to propose a partnership to create and market a new category of more sustainably raised salmon. AquaChile has been a leader in promoting sustainable aquaculture practices, and thus the partnership began in 2007.

AquaChile brought a deep knowledge of salmon agronomy, harvesting, logistics, and retailer relationships, all capabilities that were complementary to those available within DuPont. Technical teams from the two companies worked together and with AquaChile's key feed supplier Alitech (a JV of AquaChile and leading fish feed company Biomar Group) to develop the application of DuPont's omega-3 yeast in Atlantic salmon feed. This required several years of tests at the tank scale and then in ocean pens to optimize the feeding and raising strategies. AquaChile and DuPont also formed a unique partnership to market the salmon, which was branded "Verlasso harmoniously raised fish" to differentiate this new category from traditionally farmed salmon for consumers. As the development progressed, the "Verlasso Way" of raising salmon,[103] and thus the value proposition of the product, came to be defined by a holistic set of principles for raising salmon in harmony with nature, the most important of which are:

1. FFDR of 1.0 or less to reduce the amount of feeder fish used by ~75% versus traditional salmon diets (through utilization of the omega-3 yeast feed ingredient) to substantially reduce the pressure on ocean fish populations.
2. Pen densities for the fish that do not exceed 12 kg salmon/ton of water to allow the fish more room to swim and to reduce impact on the surrounding ocean.
3. No nets treated with copper-based antifouling compounds or use of organophosphates in order to protect the marine environment.

The salmon raised in this way also turned out to be lower in fat, with a taste that has been recognized as superior to traditionally farmed salmon by key influencers in the culinary community.[104] AquaChile introduced Verlasso salmon commercially in the United States in September 2011. As consumers and retailers have come to understand the Verlasso value proposition, sales have grown strongly. Verlasso was named a "Best New Food of 2011" by Seafood International Magazine and received the 2012 Seafood Summit Innovation Award[105] in recognition of the contributions of this breakthrough new product to the sustainability of the aquaculture industry. In August 2013, Seafood Watch announced that, after careful evaluation of Verlasso's farming practices, it was designating it as a "Good Alternative" in its rating system, the first and only ocean-raised farmed salmon to receive this designation.[106] With the endorsement of this highly influential environmental organization, Verlasso sales have accelerated.

7.5.5 Key Learnings from the Omega-3 Yeast/Verlasso Salmon Development

The omega-3 yeast/Verlasso program is another excellent example of DuPont's innovation approach as applied to industrial renewables. The success of the product required:

1. A long and complex metabolic engineering, fermentation, and scale-up program to invent a new, viable, and sustainable source of omega-3s.
2. Thorough applications work to develop the feeds and the feeding strategies and to obtain the required regulatory approvals for a new feed ingredient.
3. World-class brand development and marketing to design the positioning and sales approach for a new category of farm-raised salmon.
4. Broad and deep knowledge of salmon farming, harvesting, and logistics to get this highly perishable fresh product to market at the highest quality.

The DuPont principles of innovation were also well represented by the omega-3/Verlasso program:

1. *Market-driven innovation.* Realizing that the omega-3 technology breakthrough represented an opportunity to create and participate in the value of creating a whole new category of farmed salmon, not just an opportunity to make a more sustainable feed ingredient.
2. *Integrated science.* The metabolic engineering and related biological science required to create a yeast that could produce omega-3s coupled with the large-scale fermentation engineering to produce it economically and the agronomy expertise to prove out the application of the yeast in fish feed.
3. *Inclusive innovation.* The partnerships with AquaChile and their feed supplier Alitech were critical to driving this innovation all the way from the laboratory through the fish feed manufacturing and the fish farm to retailers and consumers.

7.6 FUTURE DIRECTIONS FOR DuPont IN INDUSTRIAL BIORENEWABLES

DuPont today has a substantial and profitable group of businesses in the area of industrial biorenewables and will continue to invest in extending the core offerings of these businesses. In addition, DuPont continues to invest more broadly in the area of industrial biorenewables to create fundamentally new products and businesses. As mentioned earlier, the strategy of the company is to focus on agriculture and nutrition, biobased industrials, and advanced materials, growing businesses within these spaces and finding new business opportunities by connecting them in novel ways. This investment takes the form of a rich portfolio of individual programs that are selected and resourced based on three key themes:

1. Broadening the applications for enzymes, peptides, and proteins.
2. Conversion of existing advanced materials businesses to renewable supply chains from fossil fuel-based supply chains and development of new renewable advanced materials for core DuPont markets.
3. Use of nonfood crops and agricultural by-products as feedstock to make value-added products and to create additional revenue opportunities for farmers.

7.6.1 Broadening Applications of Enzymes and Other Proteins

As described earlier, enzymes and related peptides or proteins are already used in a wide variety of applications ranging from food to detergents to textile processing and beyond. DuPont believes that these applications are just the beginning of the potential applications of these versatile and wide-ranging materials. Enzymes can be considered to replace harsh chemicals or other unit operations in a variety of industrial processes, formulated products of all sorts, or cleaning operations. Realizing this potential will require close connection with the market to understand the value of the applications versus the incumbent approaches, the ability to adapt the tools of enzyme engineering and production to create new functionalities, and the application knowledge and customer partnerships to drive these new products to market. An example of this emerging trend is the use of Accellerase® enzymes in cellulosic ethanol processes. These enzymes have been specifically designed to hydrolyze and saccharify recalcitrant cellulosic biomass and thus render it suitable for further processing. Another area where new uses of enzymes, peptides, and proteins are expected is in personal care, where today many products achieve a biological endpoint (e.g., acne reduction or hair dyeing) through the use of harsh chemicals. The use of more targeted enzyme or peptide approaches may improve both the efficacy and the safety of personal care products.[107]

7.6.2 Renewable Advanced Materials

The second major theme for DuPont's investment in industrial biorenewables is driving renewable advanced materials. DuPont has spent many years of investment building the market presence, technology expertise, and applications know-how to serve key materials markets such as automotive, electronics, construction, safety and protection, fibers, nonwovens, and others. The company also has a large stake in bringing new offerings and market opportunities to farmers, the single largest group of DuPont customers. It is therefore one of DuPont's key strategic priorities to develop new advantaged renewable routes to *existing* advanced materials as well as *new* renewable advanced materials offerings to serve its broad customer base. For example, in June 2013 the DuPont performance polymers business publicly stated its goal of converting at least half of its portfolio to renewable sources by 2020 as described in this interview on Plastics Today.[108] In some cases, like Bio-PDO/Sorona, these new routes will take the form of fermentation from corn-based dextrose to a new monomer that enables a new polymer. In other cases, such as Zytel RS, the renewable monomer (sebacic acid) will be derived chemically from an agricultural product (castor beans). In still other cases, such as DuPont soy polymers, the product will be a coproduct from an agricultural processing step (in this case the processing of soybeans) that can be adapted for industrial usage. In general, DuPont is not currently investing in renewable routes to commodity basic materials, because with the evolution of its advanced materials portfolio, these products are now not a good strategic fit for the company.

New renewable routes to existing materials offer the advantage that the product will "drop in" to existing customer processes as long as the impurity profile and overall quality are as good as or better than the incumbent petro-based material. This reduces or eliminates the market risk associated with introducing a new product but also caps the value of the product made by the new route at the current value of the petro-product plus any available premium for better environmental performance (such price premiums are generally small). This means that the technoeconomics of the new process must be substantially better than the average cost of manufacture of the incumbent processes, or it will be difficult to justify the investment and risk to invest in a new-to-the-world manufacturing process. On the other hand, development of new materials based on renewable routes offers the advantage of potentially bringing new desirable properties and therefore differentiated value propositions, as in Sorona for carpets and textiles, but also increases the market risk and required investment to get sales of the product up to economic scale. Here the challenge, as described earlier, is for the product to survive the "valley of death" faced by all new materials as they are first introduced and struggle to gain sufficient adoption to get to economic scale. DuPont is currently pursuing a biomaterials portfolio that contains both types of product developments (new renewable routes to existing materials and new materials from renewable routes) that will result in multiple new product introductions over the coming years.

"Second-Generation" Renewables." The final major theme for DuPont is the move from food crops to nonfood biomass and agricultural by-products as key feedstocks for both fuels and materials. The incumbent agricultural value chains have evolved over time to provide raw materials that are used for food, such as dextrose and vegetable oil, efficiently and effectively. It is therefore logical that many of the first forays into renewable materials were designed to utilize these incumbent value chains as feedstock, for example, corn-based ethanol for biofuels. In addition, many organisms that can be adapted for fermentation of key industrial products have also evolved to use dextrose or other C6 sugars as their substrates of choice. Both of these considerations often drive developers of industrial renewable products to the use of food ingredients as their key feedstock.

While food ingredients offer perhaps the easiest path to establishing a new supply chain for a renewable material, there are several challenges with this approach. The most important of these is the perception from many parts of the global community that food ingredients should not be used to make fuels or materials. This discussion, sometimes identified by the short-hand of the "food versus fuel debate," has been widely reported, for example, on the Deutsche Welle website.[109] There are many points of view on the idea of using agricultural crops for fuel and materials, but in any case this issue offers an important barrier to companies wishing to expand in the industrial renewables space. Perhaps the most elegant solution to this issue is to develop routes to fuels and materials that use nonedible by-products of the agricultural process, for example, corn stover, and/or use nonfood crops that can grow on marginal land or other areas where food crops cannot be practically grown, for example, switch grass or macroalgae (seaweed). This approach is commonly called "second-generation" biofuels or biomaterials.

As described earlier in the chapter, DuPont has invested heavily in the successful development of a second-generation process to convert corn stover into ethanol. DuPont believes that the tools developed in this program such as the economical collection of the agricultural by-product, the processes and enzymes to convert recalcitrant biomass into a fermentable substrate, the metabolic approaches that allow an organism to metabolize both C5 and C6 sugars, and the genetic tools to perform economic fermentations on an impure hydrolysate will be more generally applicable to a development of routes to a variety of materials and fuels. The industrial biotechnology industry is pursuing a number of approaches to using nonfood biomass as feedstock, and in line with this trend, some of the key programs in DuPont's current portfolio are *second-generation* approaches of one sort or another. DuPont's goal over time, as the development of the requisite technology tools allows, is to base all of its industrial biorenewable programs and products on nonfood biomass and/or agricultural by-products. This evolution is likely to occur over more than a decade.

7.7 SUMMARY

Over a history that spans more than two centuries, DuPont has invested in research, applications development, marketing, and sales to bring an unequaled stream of innovative and entirely new product categories to market. The company today continues this tradition with its focus on three strategic priorities: agriculture and nutrition, biobased industrials, and advanced materials. Industrial biorenewables is an important part of this strategy, providing the opportunity to create new businesses based on the rapid advances in biological sciences, including some large opportunities that connect the company's agriculture and nutrition businesses with its advanced materials businesses.

The efforts to develop new products and manufacturing routes to existing products at the company are guided by a three-pronged innovation philosophy: (1) market-driven innovation, (2) integrated science, and (3) inclusive innovation. Two case histories were presented to illustrate DuPont's innovation approach. The first was the development of Bio-PDO and Sorona. In this case, a biorenewable route enabled the key Bio-PDO monomer to be manufactured at lower cost with substantially lower environmental impact than the incumbent petro-based routes. This in turn enabled the introduction of a polymer with differentiated properties that has been a success in the market place. The second case history was on the development of a yeast-based source of omega-3s to replace fish oil derived from wild pelagic fish and the creation of a new category of premium farmed salmon based on this innovation.

Finally, DuPont's three major themes for future investment in industrial biorenewables were discussed. These include (1) new applications for enzymes, peptides, and proteins, (2) introduction of advanced materials and fuels based on renewable supply chain routes, and (3) moving toward a second generation of industrial biorenewables that are based on nonfood biomass and agricultural by-products. Industrial biorenewables are central to DuPont's strategy as a science company embarking on its third century of existence, and the company plans to continue to invest in this space to maintain its leadership position.

REFERENCES

[1] A. Kinnane, *DuPont: From the Banks of the Brandywine to Miracles of Science*, E.I. du Pont de Nemours and Company, Wilmington, DE, **2002**.

[2] DuPont company website "Innovation Starts Here", http://www2.dupont.com/Phoenix_Heritage/en_US/index.html, **2014**.

[3] D. Muzyka, *Vision 2025: How to Succeed in the Global Chemistry Enterprise*, Chapter 2, **2014**, 15–23, http://pubs.acs.org/isbn/9780841229389.

[4] SPE 41st Annual Innovation Awards, "Renewably Sourced PA for Biodiesel Fuel Lines", http://www.speautomotive.com/Awards%20Modules/2011Awards/Environmental/Biodiesel%20Fuel%20Lines.pdf, **2011**.

[5] M. Ward and M.A. Wilding, *J. Poly. Sci. Poly. Phys.*, **1976**, *14*, 263–274.

[6] J.V. Kurian, *J. Pol. Environ.*, **2005**, *13(2)*, 159–167.

[7] R. Daniel, T. A. Bobik, G. Gottschalk, *FEMS Microb. Rev.,* **1998**, *22*, 553–566.

[8] W.-D. Deckwer, *FEMS Microb. Rev.*, **1995**, *16*, 143–149.

[9] A.-P. Zeng, H. Biebl, *Adv. Biochem. Eng./Biotech.,* **2002**, *74*, 239–259.

[10] H.-J. Rehm, In "*Biotechnology*", *Vol. 6* (Eds. H.-J. Rehm, G. Reed). VCH, Weinheim, Germany, **1996**, 205–227.

[11] A.A. Gatenby, A. L. Haynie, V. Nagarajan, R. V. Nair, C. E. Nakamura, M. S. Payne, S. K. Picataggio, M. DiaZ-Torres, A.K.-H. Hsu, R.D. Lareau, D.E. Trimbur, G.M. Whited (E.I. du Pont de Nemours and Co., Genencor International, Inc.), WO 98/21339, **1998**.

[12] K. Larsson, R. Ansell, P. Eriksson, L. Adler, *Molecular Microbiol.*, **1993**, *10*, 1101–1111.

[13] J. Norbeck, A-K. Pahlman, N. Akhtar, A. Blomberg, L. Adler, *J. Biol. Chem.*, **1996**, *271*, 13875–13881.

[14] N. Floras, J. Xiao, A. Berry, F. Bolivar, F. Valle, *Nature Biotechnol.* **1996**, *14*, 620–623.

[15] M. A. Cervin, P. Soucaille, F. Valle. (Genencor International, Inc.) US Patent 7,371,558, **2008**.

[16] J. Brosius, M. Ertle, J. Storella, *J. Biol. Chem.* **1985**, *260*, 3539–3541.

[17] M. Seyfried, R. Daniel, G. Gottschalk, *J. Bacteriol.* **1996**, *178*, 5793–5796.

[18] T. Tobimatsu, H. Kajiura, M. Yunoki, M. Azuma, T. Toraya, *J. Bacteriol.* **1999**, *181*, 4110–4113.

[19] G.A. Sprenger, B.A. Hammer, E.A. Johnson, E.C.C. Lin, *J. Gen. Microbiol.* **1989**, *135*, 1255–1262.

[20] I-T. Tong, H. H. Liao, D. C. Cameron, *Appl. Environ. Microbiol.* **1991**, *57*, 3541–3546.

[21] T. Toraya, T. Shirakashi, T. Kosuga, S. Fukur, *Biochem. Biophys. Res. Commun.* **1976**, *69*, 475–480.

[22] W. W. Bachovchin, J. Robert, G. Eagar, K. W. Moore, J. H. Richards, *Biochemistry* **1977**, *16*, 1082–1092.

[23] L.V. Johnson, E. Pechonick, S.D. Park, G.D. Havemann, N.A. Leal, T.A. Bobik, *J. Bacteriol.* **2001**, *183*, 1577–1584.

[24] D.-I. Liao, L. Reiss, I. Turner, G. Dotson, *Structure*, **2003**, *11*, 109–119.

[25] S. Honda, T. Toraya, S. Fukui, *J. Bacteriol.* **1980**, *143*, 1458–1465.

[26] K.J. Gibson, X.-S. Tang, D.-I. Liao, (E. I. du Pont de Nemours & Co.) WO 2004/056963, **2004**.

[27] M.H. Emptage, S. Haynie, L.A. Laffend, J.P. Pucci, G. Whited, US Patent 6,136,976, **2003**.

[28] M.S. Payne, S.K. Picataggio, D.E. Trimbur, A.K.-H. Hsu, F. Valle, R. Nair, P. Soucaille, (E. I. du Pont de Nemours & Co., Genencor International, Inc.) WO 2003/089621, **2003**.

[29] M.M. Zhu, F.A. Skraiy, D.C. Cameron, *Metab. Eng.* **2001**, *3*, 218–225.

[30] LCA results and methodology described on the DuPont Tate & Lyle Bioproducts, LLC website, http://www.duponttateandlyle.com/life_cycle.

[31] DuPont Sorona® website, http://www.dupont.com/products-and-services/fabrics-fibers-nonwovens/fibers/brands/dupont-sorona/articles/biobased-products.html.

[32] Plastics Europe Eco-Profiles: http://www.plasticseurope.org/plasticssustainability/eco-profiles/browse-by-list.aspx.

[33] Press release, "Mohawk and DuPont Partner to Introduce Most Revolutionary Carpet Fiber Technology in 20 Years": http://www2.dupont.com/Media_Center/en_US/news_releases/2005/article20051104.html.

[34] R.J. Deckelbaum and C. Torrejon, *J. Nutr.* **2012**, *142*, 587S–591S.

[35] H.E. Bays et al., *Am. J. Cardiol.* **2011**, *108*, 682–690.

[36] P.J. Gillies, W.S. Harris, P.M. Kris-Etherton, *Curr. Atheroscler. Rep.* **2011**, *13*, 467–473.

[37] Fisheries and Aquaculture Dept. of the FAO website: http://www.fao.org/fishery/species/2929/en.

[38] A.G.J. Tacon, M. Metian, *Aquaculture* **2008**, *285*, 146–158.

[39] A. Jackson, *Int. Aquafeed*, **2012**, Sep/Oct 18–21.

[40] Seafood Watch website: http://www.seafoodwatch.org/cr/seafoodwatch.aspx.

[41] Environmental Defense Fund rating of salmon sustainability: http://seafood.edf.org/salmon.

[42] W. Barclay, C. Weaver, J. Metz, In "*Single Cell Oils*" (Eds. Z. Cohen, C. Ratledge). AOCS Press, Champaign, IL, **2005**, 36–73.

[43] Z. Xue et al, *Nature Biotech.* **2013**, *31*, 734–740.

[44] S-Y Hong, J. Seip, P.L. Sharpe, Z. Xue, N.S. Yadav, H. Zhang, Q.Q. Zhu, US Patent Application 20120052537, "Recombinant microbial host cells for high eicosapentaenoic acid production", **2011**.

[45] G. Barth, C. Gaillardin, *FEMS Microbiol. Rev.* **1997**, *19*, 219–237.

[46] J.-M. Nicaud, *Yeast* **2012**, *29*, 409–418.

[47] M. Tai, G. Stephanopolous, *Metab. Eng.* **2013**, *15*, 1–9.

[48] Q. Zhu et al., "Metabolic engineering of an oleaginous yeast for the production of Omega-3 fatty acids", In "*Single Cell Oils*" (Z. Cohen, C. Ratledge). 2nd Edition. AOCS Press, Urbana, Illinois, **2010**, 51–73.

[49] C. Ratledge, "*Single Cell Oils for the 21st Century*", AOCS Press, Champaign, Illinois, **2005**, 1–20.

[50] J.G. Metz et al., *Science* **2001**, *293*, 290–293.

[51] D. Meesapyodusk, X. Qiu, *Lipids* **2012**, *47*, 227–237.

[52] J.G. Wallis and J. Browse, *Arch. Biochem. Biophys.* **1999**, *365*, 307–316.

[53] D.C. Chen, J.M. Beckerich, C. Gaillardin, *Appl. Microbiol. Biotechnol.* **1997**, *48*, 232–235.

[54] P. Fournier et al, *Proc. Natl. Acad. Sci. U. S. A.*, **1993**, *90*, 4912–4916.

[55] L.S. Davidow et al., *J. Bacteriol.*, **1987**, *169*, 4621–4629.

[56] S. Muller, T. Sandal, P. Kamp-Hansen, H. Dalboge, *Yeast*, **1998**, *14*, 1267–1283.

[57] T. Juretzek, H.J. Wang, J.M. Nicaud, S. Mauersberger, G. Barth, *Biotech. Bioprocess Eng.* **2000**, 5, 320–326.

[58] D.S. Knutzon, J.M. Thurmond, Y.S. Huang, S. Chaudhary, E.G. Bobik Jr, G.M. Chan, S.J. Kirchner, P. Mukerji, *J. Biol. Chem.* **1998**, *273*, 29360–29366.

[59] S.K. Picataggio, N.S. Yadav, Q.Q. Zhu, US Patent 7,238,482, "Production of polyunsaturated fatty acids in oleaginous yeasts", **2007**.

[60] Z. Xue, Q.Q. Zhu, US Patent 7,264,949, "Glycerol-3-phosphate o-acyltransferase promoter for gene expression in oleaginous yeast", **2007**.

[61] S.Y. Hong, J. Seip, D. Walters-Pollak, R. Rupert, R. Jackson, Z. Xue, Q. Zhu, *Yeast*, **2011**, *29*, 59–72.

[62] Z. Xue, Q.Q. Zhu, US Patent 8,323,960, "Ammonium transporter promoter for gene expression in oleaginous yeast", **2012**.

[63] R.A. Jefferson, T.A. Kavanagh, M.W. Bevan, *EMBO J.* **1987**, *6*, 3901–3907.

[64] H. Alper, C. Fischer, E. Nevoight, G. Stephanopolous, *Proc. Natl. Acad. Sci. U. S. A.*, **2005**, *102*, 12678–12683.

[65] D.J. Macool, Z. Xue, Q.Q. Zhu, US Patent 7,470,532, **2008**.

[66] N.S. Yadav, Q. Zhu, H. Zhang, US Patent 7,504,259, **2009**.

[67] F. Domergue, A. Abbadi, C. Ott, T.K. Zank, U. Zahringer, E. Heinz, *J. Biol. Chem.*, **2003**, *278*, 35115–35126.

[68] H.G. Damude, H. Zhang, L. Farall, K.G. Ripp, J.F. Tomb, D. Hollerbach, N.S. Yadav, *Proc. Natl. Acad. Sci. U. S. A.* **2006**, *103*, 9446–9451.

[69] H.G. Damude, Q.Q. Zhu, US Patent 7,256,033, **2007**.

[70] Z. Xue, Q.Q. Zhu, US Patent 7,465,793, **2008**.

[71] H.G. Damude, A.J. Kinney, K.G. Ripp, Q.Q. Zhu, US Patent Application 20080254191, **2008**.

[72] S.P. Hong, P.L. Sharpe, Z. Xue, N.S. Yadav, Q.Q. Zhu, US Patent Application 20090117253, **2009**.

[73] N.S. Yadav, H. Zhang, Q.Q. Zhu, US Patent Application 20100317882, **2009**.

[74] H.G. Damude, D.M. Pollak, Z. Xue, Q.Q. Zhu, US Patent 7,678,560, **2010**.

[75] H.G. Damude, D.M. Pollak, Z. Xue, Q.Q. Zhu, US Patent 7,695,950, **2010**.

[76] N.S. Yadav, H. Zhang, US Patent 7,749,703, **2010**.

[77] H.G. Damude, Q.Q. Zhu, US Patent 7,790,156, **2010**.

[78] H.G. Damude, US Patent 7,794,701, **2010**.

[79] Q.Q. Zhu, Z. Xue, US Patent Application 20100159558, **2010**.

[80] H.G. Damude, Q.Q. Zhu, Z. Xue, US Patent 7,863,502, **2011**.

[81] H.G. Damude, Z. Xue, N.S. Yadav, Q.Q. Zhu, US Patent 7,879,591, **2011**.

[82] H.G. Damude et al., US Patent 7,932,077, **2011**.

[83] H.G. Damude, Q.Q. Zhu, US Patent 7,943,365, **2011**.

[84] H. Damude, Q. Zhu, US Patent 7,943,823, **2011**.

[85] H.G. Damude, H. He, D.-I. Liao, Q. Zhu, US Patent 8,026,089, **2011**.

[86] H.G. Damude, B. McGonigle, Z. Xue, Q.Q. Zhu, US Patent 8,048,653, **2011**.

[87] H.G. Damude, B. McGonigle, Q. Zhu, Z. Xue, US Patent 8,049,062, **2011**.

[88] H.G. Damude, Q.Q. Zhu, US Patent 8,049,070, **2011**.

[89] H.G. Damude, Q.Q. Zhu, US Patent 8,058,517, **2011**.

[90] H.G. Damude, Q.Q. Zhu, US Patent 8,119,860, **2012**.

[91] H.G. Damude, Q.Q. Zhu, US Patent 8,188, 338, **2012**.

[92] Q.Q. Zhu , D.M.W. Pollak, US Patent 8,268,598, **2012**.

[93] H.G. Damude, Q.Q. Zhu, US Patent 8,298,797, **2012**.

[94] Z. Xue, N.S. Yadav, Q.Q. Zhu, US Patent 8,323,935, **2012**.

[95] H. Zhang, H.G. Damude, N.S. Yadav, *Yeast*, **2012**, *29*, 25–38.

[96] D.W. Pollak et al., *Lipids*, **2012**, *47*, 913–926.

[97] Z. Xue et al., *Appl. Microbiol. Biotechnol.* **2013**, *97*, 1973–1985.

[98] E.B. Cahoon, K.G. Ripp, S.E. Hall, A.L. Kinney, *J. Biol. Chem.* **2001**, *276*, 2637–2643.

[99] E. Weterings, D.J. Chen, *Cell Res.* **2008**, *18*, 114–124.

[100] D. Xie, In "*Methods in Molecular Biology*", Humana Press, New York, NY, **2012**, 217–231.

[101] J.P. Wynn, *Nature Biotechnol.* **2013**, *8*, 716–717.

[102] B. Hatlen et al., *Aquaculture* **2012**, *364–365*, 39–47.

[103] Verlasso® Website: http://www.verlasso.com/our-story/ and associated web pages.

[104] S. Gugino, *Wine Spectator* **2013**, 2013.

[105] Press release on DuPont website, "Verlasso® Receives 2012 SeaWeb Innovation Award": http://www2.dupont.com/media/en-us/news-events/september/verlasso-receives-award.html.

[106] The Seafood Watch® Seafood Report on Verlasso® salmon can be found at http://www.seafoodwatch.org/cr/cr_seafoodwatch/content/media/MBA_SeafoodWatch_VerlassoFarmedSalmonReport.pdf.

[107] R. Lad, ed., "*Biotechnology in Personal Care*", Taylor and Francis Group, New York, **2006**.

[108] D. Smock, "DuPont Plans Major Shift to Biobased Feedstocks", Plastics Today website, June 24, 2013, http://www.plasticstoday.com/articles/dupont-plans-major-shift-biobased-feedstocks0624201301.

[109] C. Hasselbach, H. Jeppesen, article on Deutsche Welle website "Food or Fuel Debate Leads to EU Biofuel Changes": http://www.dw.de/food-or-fuel-debate-leads-to-eu-biofuel-changes/a-16313695.

8

EVONIK: BIOECONOMY AND BIOBASED PRODUCTS

HENRIKE GEBHARDT[1], PETER NAGLER[2], STEFAN BUCHHOLZ[2], STEFAN CORNELISSEN[1], EDDA SCHULZE[1], AND ACHIM MARX[1]

[1]*Evonik AG, Essen, Germany*
[2]*Evonik AG, Marl, Germany*

Industrial Biorenewables: A Practical Viewpoint, First Edition. Edited by Pablo Domínguez de María.

8.1 INTRODUCTION

8.1.1 Bioeconomy at Evonik

Evonik, the creative industrial group from Germany, is one of the world leaders in specialty chemicals. Its activities focus on the key megatrends health, nutrition, resource efficiency, and globalization. Among others, biorenewables may offer innovative solutions to meet the challenges of these megatrends. Biorenewables are part of the general concept of the bioeconomy as it is defined by the European Commission in its key publication "Innovating for sustainable growth: a bioeconomy for Europe," as follows: Bioeconomy "encompasses the production of renewable biological resources and the conversion of these resources and waste streams into value-added products such as feed, food, biobased products and bioenergy."[1] Morevoer, in the publications "Policy Strategy Bioeconomy," "Biorefineries Roadmap," and the "National Research Strategy BioEconomy 2030," the German government illustrates how it supports the shift from a petroleum-based society to a sustainable biobased economy and defines bioeconomy as "all sectors of economy and service that produce, use, process or trade renewable resources like plants, animals and microorganisms and their products."[2]

Applying these two definitions to Evonik, bioeconomy-related products fall under four different categories. They are chemicals and materials that are biobased and at the same time bioprocessed by microbial fermentation (1). This applies to amino acids such as Biolys®, ThreAMINO®, and TrypAMINO®, which are instrumental in lowering the amount of resources required for the production of healthy, nutritious food for a growing world population. Further products that comply with the definition of bioeconomy are chemicals produced from a biobased feedstock by traditional catalysis, without the use of biotechnology (2) such as the polyamide VESTAMID® Terra. Additionally, we include biodegradable products (3) into our bioeconomy product portfolio because biodegradability contributes to the performance and/or sustainability of the product. An example is the biodegradable surfactant REWOCARE® BDS 15, which functions as rinse aid in automatic dishwash as well as other hard surface cleaning products.

Furthermore, Evonik offers a range of products that may itself not comply with these three attributes but are indispensable to enable value creation in the bioeconomy (4). In this latter area, catalysts for several efficient conversions of biobased raw materials, for example, for processing dicarboxylic acids such as succinic acid to alcohols, for the hydrogenation of sugars, for the manufacture of fatty amines, and for the production of biodiesel from vegetable oils are part of Evonik's catalyst portfolio. For instance, an application that contributes to the overall sustainability of biogas production is the membrane SEPURAN® Green. This membrane removes carbon dioxide very efficiently form biogas so that the remaining methane can be fed into the natural gas grid or can be compressed to replace liquefied natural gas.

Figure 8.1 gives a comprehensive overview of Evonik products for the bioeconomy and a selection has been published online recently (www.evonik.com/bioeconomy).

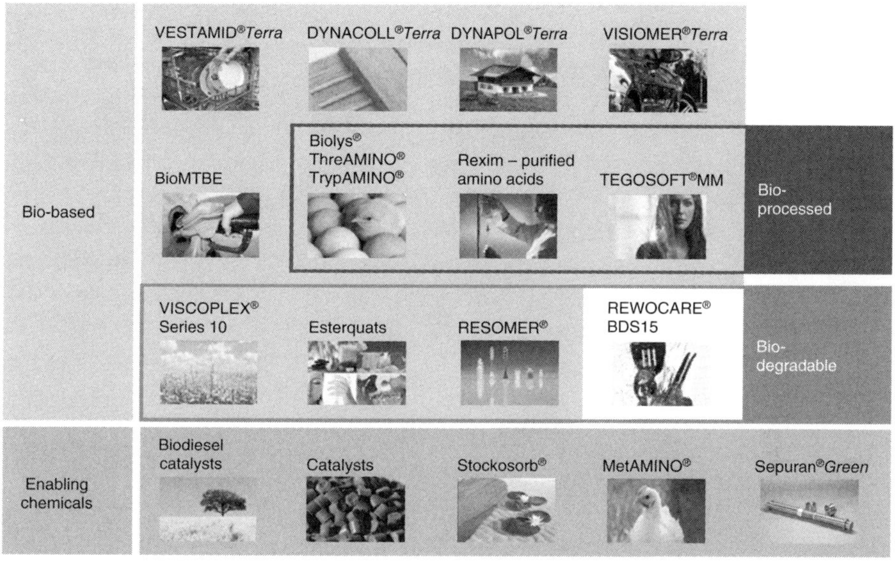

Figure 8.1 Following our definition, products contribute to the bioeconomy if they are biobased, bioprocessed, biodegradable, or if they are indispensable to enable value creation in the bioeconomy.

In any case, bioeconomy market activities of Evonik aim at higher sustainability when compared to established processes and raw materials. Bioeconomy is a key pillar of the sustainability strategy at Evonik – an outline of this can be found in the Corporate Responsibility Report 2012.[3] Sustainability comprises the three dimensions society, environment, and economics (Figure 8.2).

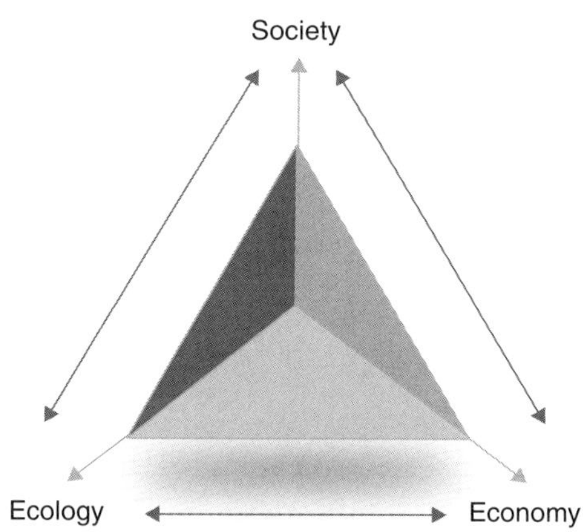

Figure 8.2 Each nation, institution, and company need to have a sustainability or CR strategy for keeping the balance of all three sustainability aspects in order to keep mankind's well-being on earth. *(See color plate section for the color representation of this figure.)*

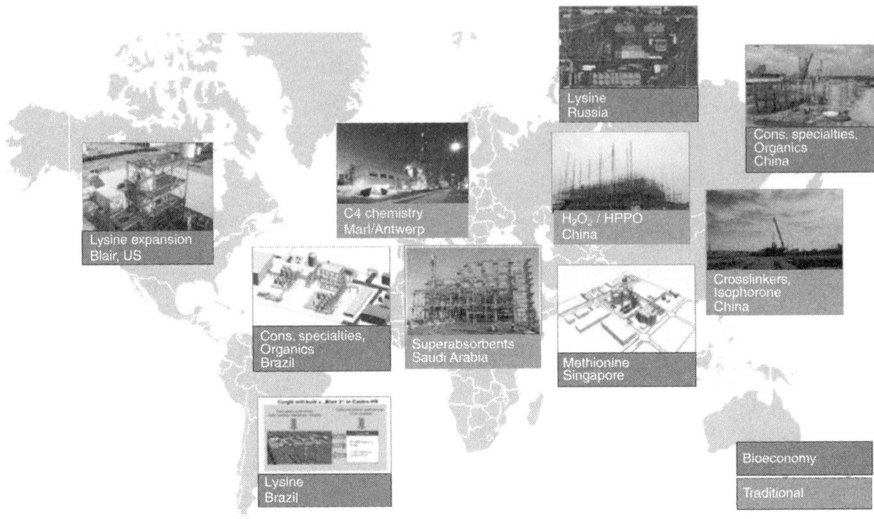

Figure 8.3 Lighthouse investment projects at Evonik.

Bioeconomy is becoming increasingly important for investments in high-growth chemical megatrends. Figure 8.3 presents lighthouse investment projects from Evonik. Out of the ten examples, six are related to bioeconomy. For the remaining four examples, biobased feedstock and/or bioprocesses are currently not advantageous. More research and development is needed to prove the potential of biotechnology in these areas.

8.1.2 Driver for Innovation in the Economy

Market pull is the most important driving force for all new developments in industry, thus also being for the future of bioeconomy. Over the last years, more and more customers ask for sustainable products that bioeconomy may deliver for them. Furthermore, even 40–60% of consumers (depending on the product) state that they are willing to pay a premium for green products, yet it must be noted that the consumers' intention is not always mirrored in their behavior. Actually, the majority of the consumers tends to buy biobased products only if they are comparable in cost and performance to conventional products. Only a minority is actually paying a green premium.[4] However, extensive market data of biobased products in Europe are not readily available. The Bioeconomy Observatory that was launched by the European Commission in February 2013 started to collect market data of biobased products and is supposed to give a comprehensive overview over the market and the development of the market for biobased products in Europe in the following years.

To identify measures that could support the market development of biobased products, Directorate General Enterprise and Industry of the European Commission initiated the Ad-hoc Advisory Group for Bio-based Products in frame of the Lead

Market Initiative. In 2009 and 2011 this advisory group published a bunch of recommendations on how market introduction of biobased products could be promoted.[5] Access to feedstock at competitive price was identified as prerequisite for competitive prices of biobased products. Balancing legislation and policies between bioenergy and biobased products could contribute to a favorable price development. Green public procurement programs that have already been implemented in the United States under the label BioPreferred® could be another instrument with which European governments can contribute to the advancement of the bioeconomy. Common European and international standards were identified as helpful to verify claims about biobased products and as basis for a credible certification. Transparent and harmonized certification and labeling schemes contribute to an honest dialogue between industry and society. Implementation of some of these priority recommendations already started. The European Commission mandated stakeholders to install a coherent framework for standardization of biobased products. General topics concerning the standardization of all biobased products like terminology, detection methods, sustainability, and certification are treated in the horizontal technical committee CEN/TC 411. Product-specific aspects, such as the threshold for biobased carbon content, are handled in working groups focused on individual groups of products: CEN/TC 19/WG 33 "Bio-Lubricants", CEN/TC 249/WG 17 "Bio-Polymers", CEN/TC 276/WG 3 "Bio-Surfactants", and CEN/TC 411/WG 2 "Bio-solvents". In 2013 Directorate General Enterprise and Industry mandated the EC expert group for biobased products to advice on the implementation of the remaining priority recommendations.

Importantly, market pull is not the only driver for bioeconomy. Technology push can also enhance the market uptake of biobased products. This aspect is also considered in the list of priority recommendations of the Ad-hoc Advisory Group for Bio-based Products. Public funding shall not only stimulate technological innovation but help to bridge the gap between innovation and commercialization by supporting, for example, demonstration and pilot plants. Beginning 2014 two public–private partnerships between the European Commission and industry were launched to address this challenge under Horizon 2020, the European framework program for research and innovation. Evonik is actively involved in the public–private partnership Sustainable Process Industry through Resource and Energy Efficiency (SPIRE). SPIRE provides a solid basis for academia, SMEs, and multinational companies to join forces in cross-sectoral initiatives and unusual value chains (Figure 8.4). The public–private partnership for Bio-based Industries (BBI) brings together research and industry partners along the whole value chain of biobased products and focuses on the concept of biorefineries.

The ambitions of SPIRE are:

- A reduction in fossil energy intensity of up to 30% by 2030.
- Up to 20% reduction in nonrenewable, primary raw material intensity by 2030.
- A significant contribution to a drastic efficiency improvement in CO_2-equivalent footprints of up to 40% by 2030.
- Potential improvements extend beyond "process industry."

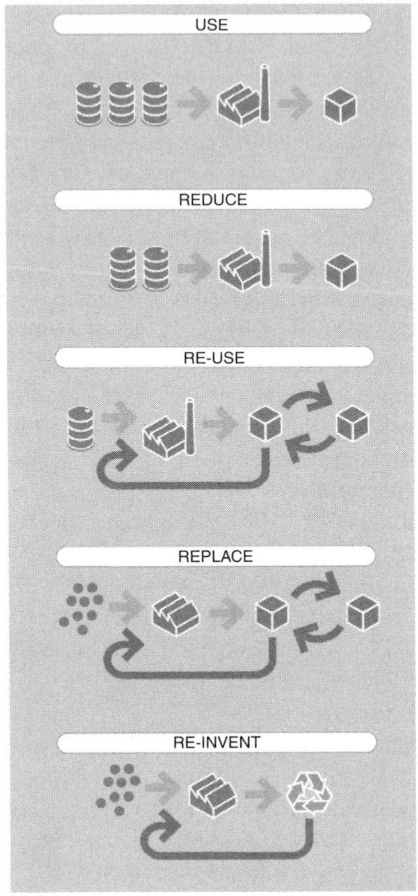

Figure 8.4 The public-private partnership SPIRE. *Source*: Reproduced with permission of SPIRE Roadmap (Adapted from the Rotterdam Port Vision 2030).

8.1.3 Innovation and Research in the Industrial Context

Innovations are a key driver of profitable growth for the specialty chemicals industry. They result in new products and solutions, open up attractive new business fields, and strengthen the leading market and technology positions of Evonik. To address ever shorter innovation cycles, more complex problems, and more demanding conditions, a culture of innovation is fostered, with courage to break new ground, in which employees are encouraged to take risks and which is based on trust, close cooperation, and openness. Thus, the specialty chemicals company's R&D pipeline includes over 500 projects. In 2013, €1.5 billion of the group's sales came from products and applications introduced in the past 5 years. Evonik has increased its spending on research and development by an annual average of 9% in the past 5 years, spending

€394 million in 2013. The operational units fund over 90%, while the group bears the remaining 10%. This is spent on strategic R&D projects with a mid- to long-term time horizon.

Creavis is the strategic innovation unit at Evonik. Its mission is to investigate topics new to Evonik that may possess high business potential but that could also carry high risks. Creavis gives high priority to sustainability as a driver of innovation in the future. Creavis' researchers also played an important role in setting up Evonik's biotechnology platform. Biotechnology is one of the technology platforms that are part of the growth strategy of Evonik. Biotechnological processes make biorenewables accessible for the production of industrial chemicals and additives, polymers, and consumer products. Thus, biotechnology is a key driver for the bioeconomy.

The biotechnology platform of Creavis comprises the biotechnological core competences and metabolic engineering as well fermentation technology and downstream process development. It cooperates closely with the corporate process technology unit as well as with the biotechnology competence centers of the business units. This decentralized innovation network enables to focus on regional market pull and technology push. At the same time a global coordination within this network ensures that the company benefits from cross-sectoral synergies.[6]

8.1.4 Scientific Excellence and Intellectual Property

Both scientific excellence and intellectual property are key for technology leadership. Evonik is a company with a strong focus on innovation. The group is very efficient in translating its research expenses into patents. That was confirmed by the Patent Asset Index (PAI), which rated Evonik as a leader in this area. In recent years, Evonik has advanced the value of its patent portfolio significantly. The specialty chemicals company has more than 26,000 patents and patent applications. The company filed patent applications for around 260 inventions in the year 2013.[7]

In addition to this, partnering is a core pillar of innovation at Evonik as well. This applies to joint research and development with academia, SMEs, and industry. For instance, Evonik cooperates with SMEs such as BioAmber, LanzaTech, OPXBIO, and Phytowelt (see chapter of Phytowelt as well).

8.2 BIOBASED AND BIOPROCESSED PRODUCTS (1)

8.2.1 Healthy and Efficient Nutrition for a Growing World Population by Amino Acid Fermentation (Commercial Stage)

According to United Nations estimations, the world's population is growing by about 80 million annually. By 2050 it will be necessary to feed 9 billion people, 70% of whom will live in cities. How will all these people be adequately fed in the future? With increasing affluence and urbanization, dietary habits will move into the direction of more meat, fish, dairy products, and eggs. Given these challenges, Evonik wants to help to decouple increasing meat consumption from resource consumption. Only in this way can a growing world population be assured of a healthy diet over the long term. In the "What if ..." brochure, Evonik shows how this can be done.[8] The brochure examines in detail various aspects of the nutrition question, including ecological footprint, overfishing, biodiversity, and land use. A study of the various possible scenarios clearly indicates that future meat requirements can be met sustainably – that is, in an environmentally, economically, and socially compatible way – only by the use of improved technologies in meat production. To that end, Evonik's solution is based on the composition of animal feeds. To be able to produce vital proteins, the organism requires amino acids in a certain ratio. If a particular component is missing, the animal cannot use the food optimally. For example, corn, wheat, and soy meal – often used as feeds for poultry and pigs – are short of the amino acids methionine, lysine, and threonine. If these are selectively added to the feed, factors such as emission of ammonia, nitrate, and greenhouse gases during meat production can be drastically reduced. Evonik has confirmed this by means of various life cycle assessment studies. Farm animals and fish can only use L-amino acids, except for DL-methionine. In this respect, at Evonik fermentation with bacteria as production organisms has for decades been the method of choice for enantiomerically pure production of Biolys® (L-lysine), ThreAMINO® (L-threonine), and TrypAMINO® (L-tryptophan). To increase production efficiency, deep biological understanding must be combined with process engineering expertise. Achieving a satisfactory balance between growth and product formation in the metabolism of the production organism is a fine art. And this must then be translated into an industrial process with both high productivity and high efficiency. Tackling these challenges, in the last 20 years, Evonik has succeeded in increasing the efficiency of Biolys production by more than 40% and by a few percentage points in 2013 alone. It must be noted that each increase in efficiency is also associated with a reduction in process-related CO_2 emissions.

Moreover, Evonik is working continuously on improvement of product properties and the development of specific product forms, such as for application in aquaculture. In fish and shrimp farming as well, it makes ecological and economic sense to replace costly and scarce raw materials such as fish meal by plant protein sources and amino acids. Evonik is already a leading global supplier of L-lysine for fish feeds. Current efforts in this area are directed toward developing customized and even more specific product forms – such as different particle sizes with varying solubility and absorption behavior – that meet the specific needs of fish and crustaceans. The first commercial product in this new range is L-lysine for aquaculture (AQUAVI®Lys) (Figure 8.5).[9]

Biotechnology particularly contributes to the megatrend health and nutrition. By 2020, the Evonik Group expects to achieve sales of €1 billion in the Health and Nutrition Business Unit alone from biotechnologically manufactured products. The largest fermentation-delivered product of Evonik is the feed additive Biolys (source of L-lysine).[10] The company completed the expansion of its existing capacities for Biolys to a capacity of 280,000 metric tons per year in Blair, United States in 2012.

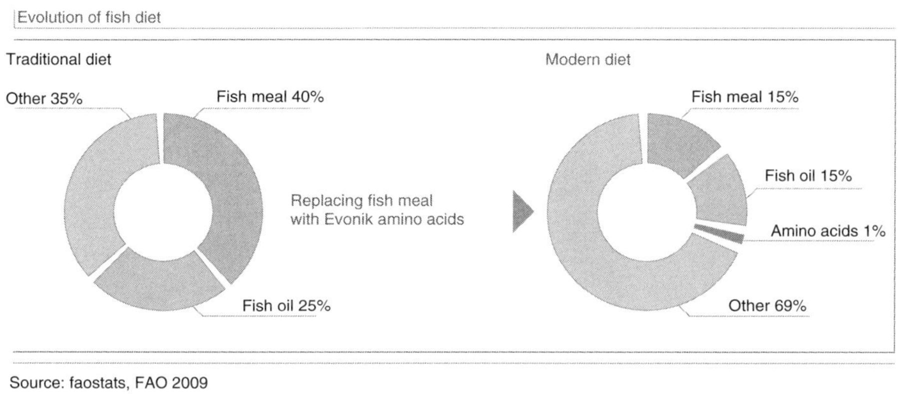

Figure 8.5 Traditional aquaculture requires up to 3 kg of wild-catch fish, which is fed in form of fish meal. Costly and scarce fish meal can be replaced by plant protein sources and amino acids. Supplementing amino acids effectively reduces fish meal requirements from 40% to only 15%. (*See color plate section for the color representation of this figure.*)

Figure 8.6 illustrates the Cargill biorefinery setup in Blair, Nebraska, United States, where Evonik operates fermentation for lysine production.

Cargill provides a starch bulk dextrose stream "over the fence" for amino acid production. Evonik also builds new facilities for the fermentative production of Biolys in Brazil and Russia, which will increase its annual global capacity to nearly 500,000 metric tons in 2015. Evonik is investing €350 million over the period 2011–2015 in building new production capacities for Biolys and expanding biotechnology research.[11] Whereas the new production site in Brazil will have the same setup as in the United States, the new plant in Russia will be based on wheat and Evonik will operate a wheat processing plant with its joint venture OOO DonBioTech (Figure 8.7).

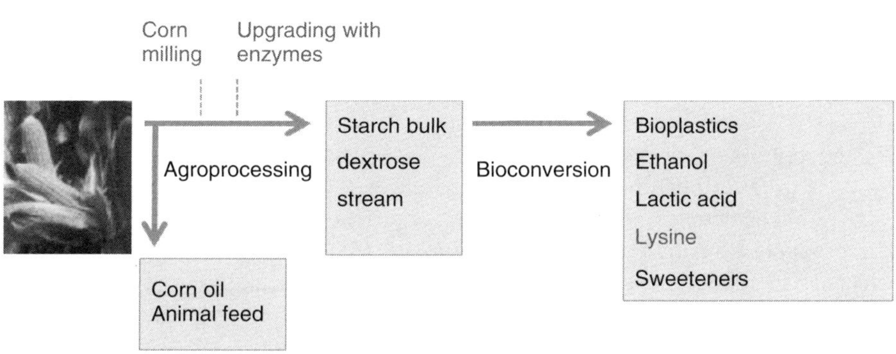

Figure 8.6 Cargill biorefinery concept at Blair, Nebraska, United States. Cargill will build a "Blair 2" in Castro, Brazil, startup 2015.

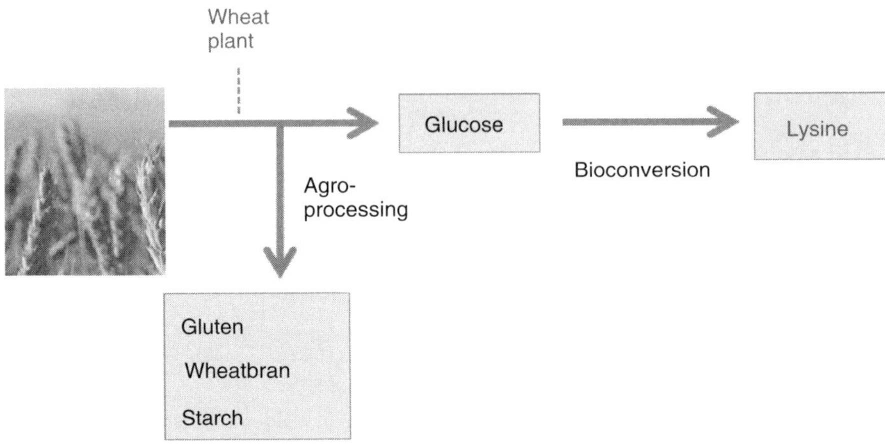

Figure 8.7 OOO DonBioTech biorefinery concept Volgodonsk, Russia.

8.2.2 Biocatalysis Improves the Sustainability of Emollient Ester Production (Commercial Stage)

The vegetable-based emollient Isoamyl Cocoate TEGOSOFT® AC represents the next generation of emollients. Monitoring the global cosmetic market, several trends can be observed. A most noticeable trend across all segments is the demand for natural-based products. In addition there is a development toward lighter textures in the skin care segment. Having identified these aspects, TEGOSOFT® AC is the emollient combining these trends. Sensory panel tests proved that TEGOSOFT® AC imparts a light and nonoily skin feel. In contrast, most natural oils have a quite heavy, oily, and tacky skin feel. Given that formulators working on certified natural cosmetics also need a certified emollient with a light skin feel, TEGOSOFT® AC provides the ideal solution for a more pleasant overall skin feel of the final cosmetic product. Therefore, TEGOSOFT® AC is of greatest interest for the modification of textures in certified formulas. Additionally to its light skin feel, TEGOSOFT® AC is a 100% vegetable-based emollient and supports with these benefits the market trend for modern and sustainable formulations. TEGOSOFT® AC is NATRUE as well as COSMOS certified. In addition to the vegetable-based sourcing, the production process for TEGOSOFT® AC is based on Evonik's unique, low-energy enzymatic technology. This ecooptimized production reduces energy consumption emissions compared to the standard esterification process.

Several cosmetic emollients by Evonik are produced via enzymatic esterification. The production of the emollient ester Myristyl Myristate TEGOSOFT® MM was selected as a model process to conduct the first environmental life cycle assessment of emollient ester production for cosmetic applications. It has been evaluated based on five standardized environmental categories:

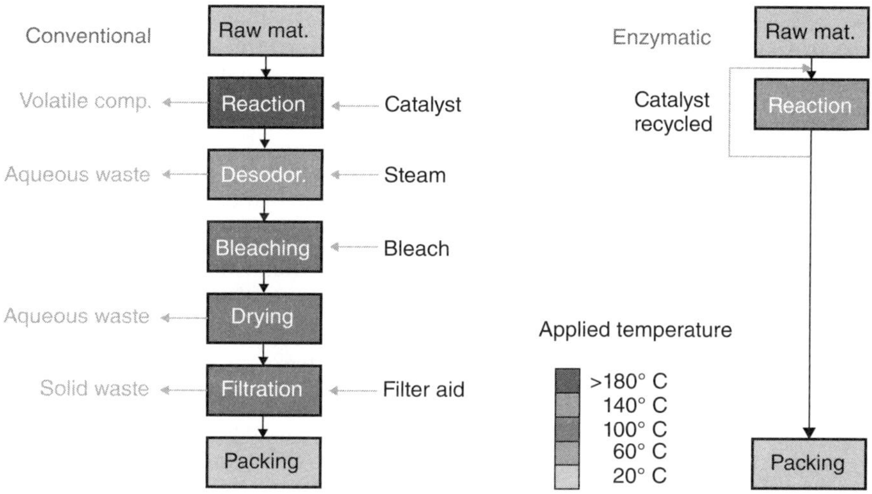

Figure 8.8 Biocatalysis improves the sustainability of emollient ester production.

- Energy consumption
- Influence on global warming (using greenhouse gas emissions, e.g., CO_2)
- Acidification (especially of rain)
- Effect on nutrient enrichment (eutrophication)
- Smog formation through volatile organic compounds (VOC)

The biocatalytic manufacturing process for the emollient ester can save more than 60% energy while reducing the formation of environmentally damaging impurities by as much as 88 percent. All these facts clearly support the sustainability of the biocatalytic process (Figure 8.8). Details on the corresponding life cycle assessment have been described elsewhere.[12]

8.2.3 High-Performance Polymers from Renewable Resources (Demonstration Stage)

Evonik is an important provider of high-performance polymers for a wide variety of applications. To be marketable and attractive, these polymers must perform special jobs and, in some cases, do so for very long periods of time. Examples of applications include the soles of athletic shoes made from polyamide 12 that have to withstand bending stress for extended periods, brake pressure lines that have to be flexible yet stable when exposed to environmental influences ranging all the way up to impact stress, and dishwasher racks that spend years in an aggressive environment of changing temperature and moisture levels. Furthermore, nowadays most plastics are manufactured on the basis of petroleum. Evonik's goal, however, is to move toward more flexible and sustainable sources of raw materials. Environmental awareness is likewise growing among the company's clientele. The challenge here is to develop biobased products that continue to meet our exceptionally high standards of quality yet remain competitive. Herein, Evonik is a leading manufacturer of high-performance polymer polyamide 12 (PA12). VESTAMID is Evonik's trade name for the construction material polyamide 12 and VESTOSINT® for coating powders from Evonik.

In this area, Evonik has managed to develop a biotechnological process for manufacturing PA12 utilizing palm kernel oil as a renewable raw material and significantly reducing the number of production steps. The new method offers a long-term option for replacing the current petroleum-based production of PA12. The conventional, multistage production process leads to the monomer lauryl lactam (LL); the new method, however, is based on palm kernel oil and generates an alternative PA12 precursor: ω-amino lauric acid (ALA), which can be subsequently polymerized to an identical PA12. The development of the fermentation process for the polyamide precursor stage ω-amino lauric acid (ALA) was subsidized by the German Federal Ministry of Education and Research (Promotion No. 0315205). The key conversion step is catalyzed by *Escherichia coli* bacteria in a fermenter (Figure 8.9).

The approach was obviously challenging, and developing the strain, the method, and the processing steps all began over 5 years ago. Developing this method meant bundling the extremely wide range of expertise that can only be found in correspondingly large companies. Thus, the bioprocess has been developed at Creavis in close collaboration with academia and SMEs (small and medium enterprises). Since

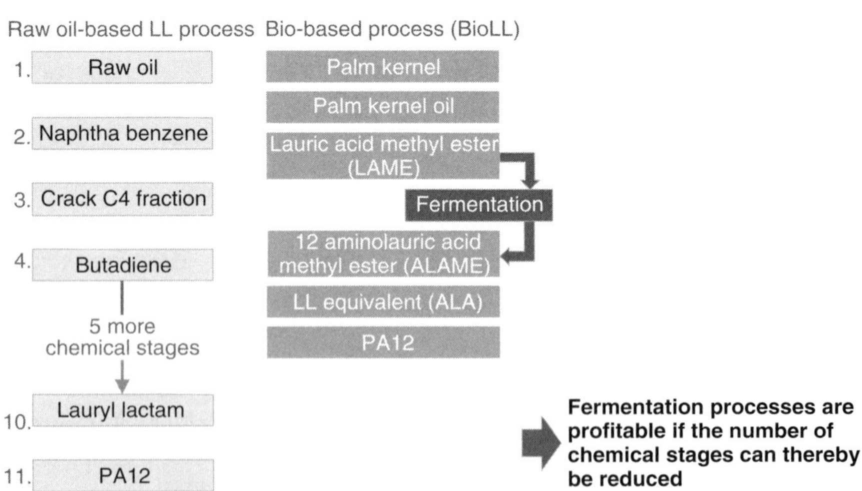

Figure 8.9 Comparison of the chemical and the new biotechnological way for the PA12 synthesis developed at Evonik.

2007 Creavis has been collaborating within an extensive scientific network in which, for example, the SME Phytowelt investigated key enzymes for transamination (see chapter of Phytowelt as well) and the University of Dortmund investigated terminal lipid oxidation.[13] For the purpose, Evonik started up a pilot plant for making ω-amino lauric acid in early 2013. The plant is located at Evonik's fermentation site in Slovenská Ľupča (Slovakia), which is home of highly specialized employees and an infrastructure geared specifically to fermentation.[14] Overall, this example illustrates how the biotechnology innovation network at Evonik contributes to value creation. The new technology is protected by roughly 20 patent families. Of major importance is one patent family covering aldehyde aminotransferases (AAT), for example, a bacterial transaminase that has been found and characterized by our cooperation partner Phytowelt along with a similar plant enzyme, which was discovered with their proprietary knowledge-based enzyme identification and characterization technology (phytomining). These newly discovered enzymes provided valuable guidance for identification of even better transaminase enzymes, which proved to be important for successful implementation of the biobased production process for 12-aminolauric acid methyl ester (ALAME).[15] The previously described project illustrates how the development of a new biobased process for a high-performance polymer works at Evonik.

Apart from the previously discussed example, in general oleochemicals play a central role for the future development of polyamides. A further project using fatty acids as raw material is the explorative innovation project SynRG with SME Phytowelt acting as coordinator as well as technology provider for plant enzymes with special properties, for example, hydroxylation of terminal methyl groups of fatty acids (ω-hydroxylation).[16] The main goal of the project is to develop a functionalization process for fatty acids preferably by ozonolysis. The unsaturated fatty acids, for example, oleic acid and erucic acid, are functionalized and the resulting monomers can be polymerized to polyamides and their compounds. An additional project in collaboration with the US-based startup OPXBIO targets at producing long-chain fatty acids by a sustainable and economical biotechnological process out of sugar. The joint-development agreement calls for OPXBIO to use its proprietary Efficiency Directed Genome Engineering (EDGE™) technology to develop the bioprocesses. This example illustrates how specialty chemicals corporations can benefit from platform technologies of SMEs.[17]

Taken these examples together, new intermediates that are produced by industrial biotechnology and traditional vegetable oil-based processes are key competencies for future polyamide development.

8.2.4 Specialty Chemicals from Waste Streams (Explorative Stage)

The aforementioned biotechnological processes described earlier are the so-called first-generation industrial biotechnology processes using plant oils, grains, and sugars as raw materials for fermentation. The first second-generation biotechnology plants are currently under construction, and these utilize residual plant materials from agriculture and forestry. Given the changing availability of resources, Evonik is committed to become more independent of fossil fuels and individual renewable resources. For this purpose, scientists at Creavis are exploring new biotechnological processes, which are capable of using various waste materials as a basis instead of sugar or plant oil. That makes one less dependent not only on fossil-based raw materials, but also on renewable resources that could potentially compete with the food supply.

These new bioprocesses use as raw material gasified biomass waste streams including gasified forestry and agricultural residues and gasified municipal solid waste or waste gases produced in industries such as steel production. These gasified waste streams contain synthesis gases or syngas. Syngases are gas mixtures consisting primarily of carbon monoxide and/or carbon dioxide and hydrogen. Syngas has been used for synthesizing chemicals for decades in petrochemical plants. In these new industrial biotechnology processes, microorganisms placed in fermenters are used to turn synthesis gas into chemical products.

In this area, Evonik has been screening several industrial biotechnology approaches that could enable the use of carbon dioxide or carbon monoxide as feedstock since 2006. With the support of Evonik, researchers at the Freiburg University (Germany) discovered a new highly efficient metabolic pathway in microorganisms to bind the greenhouse gas carbon dioxide. Not only plants can absorb the greenhouse gas carbon dioxide from the atmosphere and bind it via their metabolism, some bacteria are also capable of doing this. For such purpose, scientists have so far identified five different metabolic pathways in bacteria and plants. The working group headed by Prof. Dr. Georg Fuchs at the University of Freiburg (Germany) discovered a new metabolic pathway in *Archaea* in which this so-called biological carbon dioxide fixation is especially active. In plants, a key enzyme is responsible for the biological carbon dioxide fixation – namely, the ribulose bisphosphate carboxylase/oxygenase. However, in the metabolism this enzyme is also involved in other conversion processes, which reduces the efficiency of the carbon dioxide fixation. The new metabolic pathway that the Freiburg scientists discovered is more efficient due to reduced activity of the key enzyme in competing conversion processes. This kind of insights is essential to value new industrial biotechnology options and to choose the most promising approaches. The research work was supported by the state of North Rhine-Westphalia (Germany), the European Union, and Evonik as the industrial partner. The results have been published in the reputed scientific journal Science and have also drawn attention from outside scientific circles.[18] The work carried out by the Freiburg team could contribute to reducing the greenhouse gas carbon dioxide in the atmosphere. Along the same line, Evonik laboratories recently also demonstrated the feasibility of producing specialty chemicals from syngas using biotechnology. The laboratories of the strategic innovation unit Creavis succeeded in generating 2-hydroxy isobutyric acid (2-HIBA) from syngas with biotechnology methods. This research is partly funded by the German Federal Ministry of Food, Agriculture and Consumer Protection.[19] For the ability to convert carbon monoxide, carbon dioxide, and hydrogen into more valuable molecules, Evonik looked to bacteria from earth's earliest history – to times when oxygen was not yet present in earth's atmosphere. Certain microorganisms today still contain the genetic information for these processes. Evonik has used their enzymes to create a cell factory that generates specialty chemicals from syngas.

2-HIBA has been identified as a potential key intermediate for renewable methyl methacrylate (MMA) production at Evonik.[20] 2-HIBA can also be produced via chemical synthesis. Both the chemically produced and biotech-produced products can be converted to MMA. MMA is widely used in paints, varnishes, and antirust coatings, as well as in soft contact lenses and dental implants, to cite some industrial examples. Poly(methyl methacrylate) (PLEXIGLAS®) is used for creating sheets, profiles, roofs, soundproof walls, molded components for automotive engineering applications, and backlight units for illuminating flat-screen monitors and televisions. Evonik is one of the largest producers of MMA.

Basic research on 2-HIBA has been conducted in collaboration with Helmholtz Centre for Environmental Research and Technical University Munich (Germany). At Technical University Munich a recombinant *Cupriavidus necator* H 16 strain with the novel metabolic pathway using a cobalamin-dependent mutase was investigated. Different strains of *C. necator* H 16 deficient in the PHB synthase gene and genetically engineered to enable the production of 2-HIBA were evaluated in 48 parallel milliliter-scale stirred tank bioreactors with a volume of 11 mL each. The effects of media composition, limitations, pH, and feed rate were in-depth studied. 2-HIBA production was at a maximum at nitrogen limiting conditions and if the pH was controlled between 6.8 and 7.2 under fed-batch operating conditions. The final concentration of 2-HIBA was 7.4 g/L. Best reaction conditions identified on the milliliter scale were transferred to a laboratory-scale fed-batch process in a stirred tank bioreactor.[21]

Moreover, not only 2-HIBA but also 3-HIBA has been investigated. Scientists at Creavis succeeded in evolving a nonnatural reductase, which reduces methylmalonic acid to 3-HIBA.[22] The structural basis for a bispecific $NADP^+$ and CoA-binding site in an archaeal malonyl-coenzyme In this project, a reductase has been investigated in cooperation with the Max Planck Institute for Biophysics in Frankfurt (Germany).[23] Further characterization studies and research to implement the desired biosynthetic pathway in microbial hosts were carried out in close cooperation between Creavis and external partners such as Phytowelt. This research represents an example for scientific excellence, which is necessarily required to develop technology leadership in industrial biotechnology.

Furthermore, Evonik investigates an alternative microbial production route to carbon dioxide-based bioprocessed plastics via acetone fermentation.[24] With microbial acetone production from CO_2-containing waste gas streams, the project partners plan to provide a sustainable alternative to the petrochemical production route. This research is partly funded by the German Federal Ministry of Education and Research (BMBF-FKZ 01RC1105A).

In addition, Evonik is combining these existing biotechnology platforms with LanzaTech's synthetic biology and gas fermentation expertise for the development of a route to bioprocessed precursors for specialty plastics from waste-derived synthesis gas. Therefore, Evonik Industries and LanzaTech have signed a 3-year research cooperation agreement. LanzaTech, a leader in gas fermentation technology, uses proprietary microorganisms to capture waste gas streams, including synthesis gas for the production of low carbon fuels and chemical intermediates.[25]

In general, the choice of feedstock is of central importance to the sustainability of all three process generations, thus also for synthesis gas-based value chains. Evonik is continuously benchmarking sources such as carbon dioxide from power/cement generation in combination with renewable hydrogen against gasified municipal solid waste, gasified waste biomass, and natural gas. In addition, the gasification route has to be superior to the corresponding second-generation process using liquefaction/saccharification of biomass.

The work of Creavis is financially supported by the German Federal Ministry of Education and Research and the Federal Ministry of Food, Agriculture and Consumer Protection. Funding is also provided by the state of North Rhine-Westphalia (Germany) and is cofinanced by the European Union.

8.3 PRODUCTS PRODUCED FROM BIOBASED FEEDSTOCK BY CONVENTIONAL CATALYSIS (2)

8.3.1 Renewable Polymers: Terra Product Series

- VESTAMID® Terra – Under the brand name of VESTAMID® Terra, Evonik markets a group of new polyamides based on renewable raw materials: the monomers are obtained partly or entirely from castor oil, a raw material that is neither foodstuff nor animal feed and whose cultivation does not compete with that of food crops. The starting material castor bean (*Ricinus communis*) and its oil derivatives are synthesized into monomers that form the basis of the VESTAMID® Terra product range. Unlike other biobased polymers, biopolyamides are high-performance polymers. They are not disposable products, there are no restrictions on their service life, and they do not give rise to any technical weaknesses at all. On the contrary, biopolyamides are long-lived, durable, and can be used in demanding applications, for example, in the automotive industry.

VESTAMID® Terra products are typically applied in injection molding, fibers, powder, extrusion, and films. VESTAMID® Terra is offered in three versions with different property profiles, which fill the gaps in the previously accessible property spectrum of the polyamides.

VESTAMID® Terra HS is based on polyamide 610, which is the polycondensation product of 1,6-hexamethylene diamine (H) and 1,10-decanedoic diacid (sebacic acid – S). Sebacic acid is derived from castor oil, making Terra HS a 63% biocontent polymer. Due to its performance, VESTAMID® Terra HS fills a position between mass polyamides such as PA 6 or PA 66 and long-chain niche products. PA 610 is resistant to high temperatures and thus especially well suited for parts that must withstand hot conditions, such as the engine covers of cars.

VESTAMID® Terra DS is based on polyamide 1010, which is the polycondensation product of 1,10-decamethylene diamine (D) and 1,10-decanedoic diacid. Both decamethylene diamine and sebacic acid are derived from castor oil, making Terra DS a 100% biocontent polymer. VESTAMID® Terra DS fills a position similar to that of the long-chain high-performance polyamides PA 12 and PA 1212. It is extremely rigid and is therefore suitable, as a fiber-reinforced variant, for high-stress housings and similar applications.

The 45% biocontent polymer VESTAMID® Terra DD is based on polyamide 1012, which is the polycondensation product of 1,10-decamethylene diamine (D) and 1,12-dodecanedoic diacid (D). Due to its properties, it fills a position comparable to PA 12 and PA 1212. It is transparent and distinguished by high impact resistance. As an additive in the soles of high-quality sports shoes, it can improve the cushioning of the shoe and increase its resilience.

The VESTAMID® Terra family of polyamides from Evonik Industries has been certified as "biobased" by independent institutes. The biobased components have been certified per DIN ISO 10694; 1996-08 and by the USDA's BioPreferred program. These official laboratories used C14 trace carbon analytics to verify that the carbon in these products originates not from petroleum but biomass instead.[26] Figure 8.10 summarizes the syntheses of the different VESTAMID® Terra products.

- DYNACOLL® *Terra* – The DYNACOLL® product range began its success story back in the late 1980s. At that time, Evonik's development of copolyesters containing hydroxyl groups crucially influenced the engineering of one-component moisture-curable polyurethane hot melts (PUR-HM). The building block system of compatible amorphous, liquid, and crystalline types of DYNACOLL® enabled the precise formulation of PUR-HM for the most diverse applications in the adhesives and sealants industry. While this modular system of more than 20 medium molecular weight copolyesters has been constantly adapted and further developed to meet the latest market demands, the addition of a product range based on renewables introduced a revolutionary new chapter in the DYNACOLL® story. The aim was to develop a similar building block system of amorphous, liquid, and crystalline copolyesters based on a renewable feedstock, resulting in the new DYNACOLL® *Terra* product range. Still enabling customers to formulate their adhesive exactly to their requirements, the DYNACOLL® *Terra* product range partially even exceeds the performance of the standard grades. For instance, excellent adhesion properties have already been achieved on a variety of substrates. Evonik introduced nine products with a broad spectrum of properties. The individual products contain from 30% up to 100% of raw materials derived from renewable resources. The development is just a starting point. Scientists at Evonik are continuously screening for new biobased monomers from renewable resources in order to further expand and optimize the DYNACOLL® *Terra* product portfolio system.

One of the growth drivers for this development is the megatrend of resource efficiency. Therefore the adhesive industry is showing a growing interest in renewable raw materials, which are more and more requested by their customers. By developing DYNACOLL® *Terra,* Evonik emphasizes its commitment to sustainability and is driving the trend in the chemical industry to a greener future.[27]

- DYNAPOL®*Terra* – DYNAPOL *Terra* is another example for biobased binder systems at Evonik. Under the established DYNAPOL brand, scientists at Evonik developed a new product range of polyester resins for coil coating primers

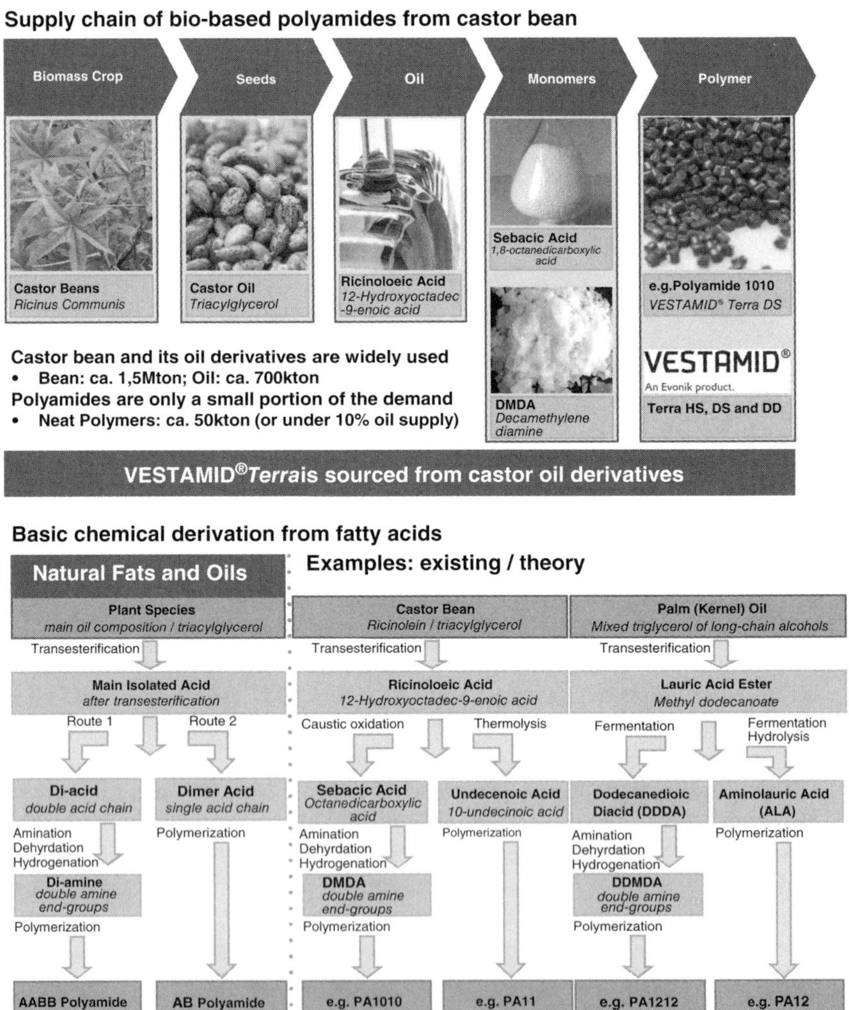

Figure 8.10 Synthesis of the different Evonik's VESTAMID® Terra products.

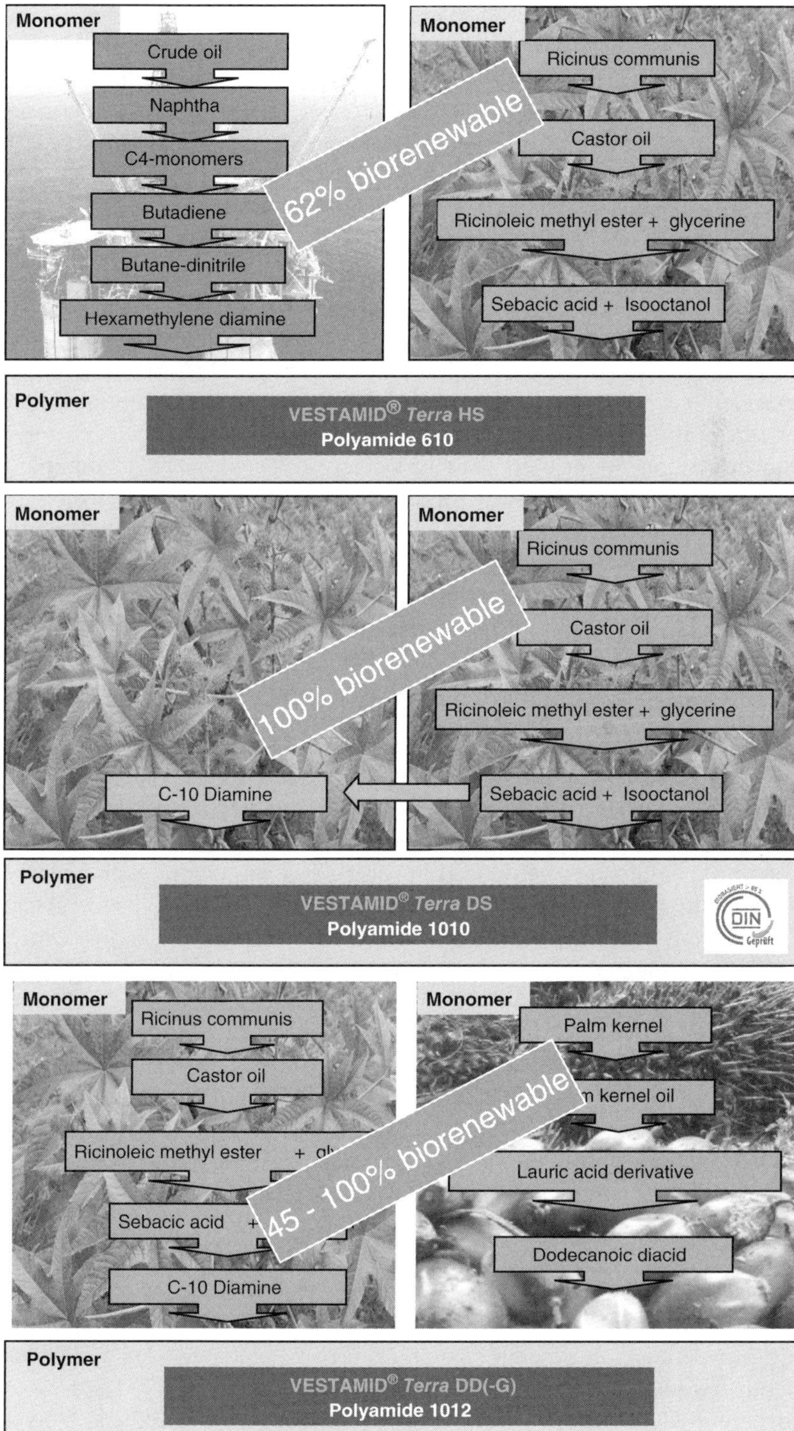

Figure 8.10 (*Continued*)

and topcoats to marketability. DYNAPOL *Terra* is fully or partially based on renewable raw materials, which sets it apart from other binders used for this type of coating. Most binder systems based on saturated, hydroxyl functional polyester polyols are produced mainly from petroleum-based raw materials. The coatings industry has been looking for suitable alternatives to these petrochemical building blocks for many years, since resource efficiency is an important growth generator. With this developed new range of products, Evonik has overcome two challenges at once. On the one hand, the coating experts identified suitable biobased polyester resin building blocks, and on the other, they ensured general applicability, fast implementation, and availability on an industrial scale. The first DYNAPOL *Terra* polyester resins have already been successfully introduced to customers. Paints based on DYNAPOL *Terra* exhibit excellent cross-linking properties and offer a well-balanced ratio of hardness and flexibility. The technical product properties are on a par with conventional binder systems. In the coil coating process, precoated steel and aluminum substrates are used on a large scale for roof and wall cladding, ceiling panels, household appliances, air-conditioning units, and for many other applications. Evonik supplies the paint, coatings, adhesives, and sealant industries with high-quality functional polymers and monomer specialties.[28]

8.3.2 Next-Generation Biofuel Component Methyl *tert*-Butyl Ether (mtbe)

Bio-MTBE is a biological version of a premium antiknock agent that is offered to oil companies as a new option for significantly increasing the biocontent of their fuels and reducing their carbon footprint. Bio-MTBE is the only commercially available, next-generation biofuel component for gasoline in Germany. It does not compete with food production and is produced in Marl (Germany) along with conventional MTBE. Evonik produces Bio-MTBE from isobutene and biomethanol. Because it is made from raw glycerine – which is itself a coproduct of the biodiesel manufacturing process – biomethanol is classified as a waste product according to the EU Renewable Energy Directive 2009/28/EU (RED); doubling its value for determining bioenergy content. That makes Bio-MTBE a promising option for fuel manufacturers looking to meet EU specifications for biofuel use and carbon dioxide reduction. MTBE has been a trusted antiknock agent for decades, and Bio-MTBE possesses the same technical advantages as its conventional counterpart: high energy density (86% of gasoline), low vapor pressure, low oxygen content, and very low solubility in water. That translates to excellent compatibility with other gasoline components and to its well-known positive effect on gasoline quality. It also means that Bio-MTBE can be handled safely in refineries and storage tanks and be conveyed by pipeline. While Evonik has primarily sold Bio-MTBE in Germany and the Netherlands, implementation of EU directives in other member states promises additional growth potential for this next-generation biofuel component. If needed and biobased methanol is available sufficiently, Evonik could shift the full capacity of its plant (550,000 metric tons per year) over to production of Bio-MTBE. Bio-MTBE has been certified as biobased by the independent certification body International Sustainability and Carbon Certification (ISCC) along the whole value chain.[29]

8.4 BIODEGRADABLE PRODUCTS (3)

8.4.1 Biodegradable Surfactants

REWOCARE® BDS 15 is a fully biodegradable silicone surfactant. Due to its excellent wetting and spreading properties – especially on nonpolar surfaces – REWOCARE® BDS 15 ensures a super-fast drying effect without leaving streaks or stripes on the treated surfaces. In addition REWOCARE® BDS 15 provides an extremely low surface tension, even when combined with other surfactants, which cannot be reached with any typical oleochemical surfactant.[30]

8.4.2 Biodegradable Polymers for Drug Delivery

RESOMER® polymers are fully biodegradable in the body. They are based on lactic acids that are biobased and bioprocessed and on glycolic acids. Biodegradable polymers are used for controlled release depot formulations (such as gel, microspheres, and rod implants) with a therapeutic effect for up to 6 months. The lactic acid polymers are also used in the manufacture of biodegradable medical devices such as surgical screws, nails, and plates, which disintegrate without trace after a defined life span in the body, making further surgery to extract them unnecessary.[31]

8.5 ENABLING CHEMICALS (4)

8.5.1 Catalysts for the Conversion of the Biobased Platform Chemical Succinic Acid

Biobased succinic acid is a platform chemical that can be used to make a number of products currently made from petrochemicals. Figure 8.11 illustrates the expected market development for biobased succinic acid. It is expected that on midterm a global capacity of about 300 kta will emerge.

Evonik is not involved in the production of the succinic acid molecule itself but in the catalytic conversion of succinic acid to various industrial chemicals. In 2012 Evonik's Catalysts Business Line agreed with US-based BioAmber Inc. on a long-term cooperation for the development and manufacturing of catalysts for making BDO (1,4-butanediol), THF (tetrahydrofuran), and GBL (γ-butyrolactone) from biobased succinic acid. BDO, THF, and GBL are large-volume industrial chemicals used in a wide range of applications including polymers, paints, adhesives, and solvents. The global market for these products currently made from petrochemicals is $4 billion. BioAmber is a leading manufacturer of biosuccinic acid from renewable feedstock. The company licensed BDO hydrogenation catalyst technology from DuPont in 2010, which has been developed further by Evonik and

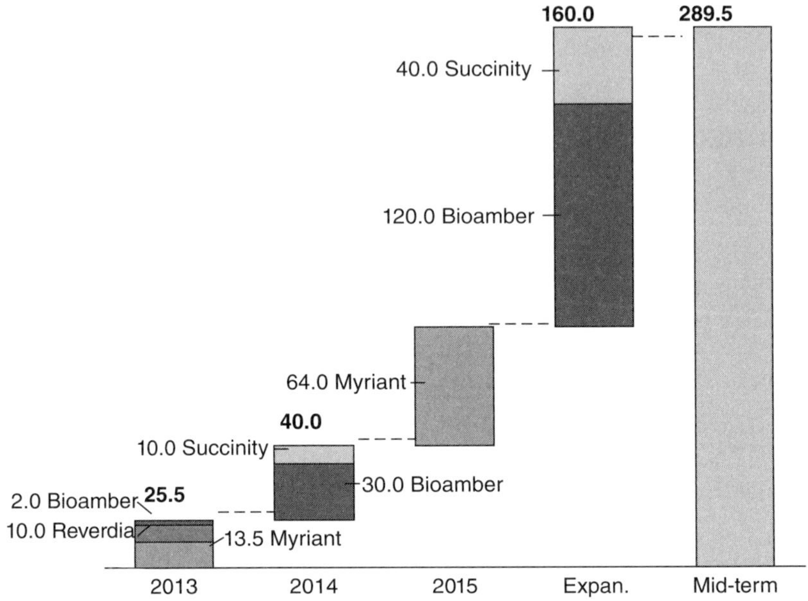

Figure 8.11 Global biobased succinic acid capacity. (*See color plate section for the color representation of this figure.*)

the Center for Applied Catalysis at Seton Hall University (South Orange, NJ, United States) to improve the technology. In parallel, BioAmber and Evonik have started to develop a new generation of BDO catalysts. BioAmber operates the world's first production plant dedicated exclusively to biosuccinic acid in Pomacle, France, with a 350,000 L commercial-scale fermenter. The company, together with Mitsui & Co., plans to expand capacity to 50,000 metric tons of succinic acid in Sarnia, Canada, in 2016. Furthermore, BioAmber announced to build a plant in North America to produce 100,000 metric tons per year BDO that will commission in 2017. A 100% offtake agreement with Vinmar International facilitates marketing of BDO produced by the new plant. Through its collaboration with Evonik, BioAmber has secured the expertise and capabilities it needs to rapidly bring competitive BDO and THF technology to market. Evonik will optimize and manufacture immediate catalyst needs and also develop a new generation of catalysts. The manufacturing of renewable chemicals very often entails biotechnological steps and catalytic conversion steps. It must be noted that catalysts are not off-the-rack products; they need to be specially tailored to every individual synthesis reaction and process condition.[32]

8.5.2 Membrane for Biogas Purification

Biogas, which consists mainly of the gases carbon dioxide and methane, is regarded as an environmentally friendly form of energy. Before biogas can be fed into the natural gas grid, it requires a considerable amount of processing and cleaning. The SEPURAN® Green membrane technology from Evonik makes this process much more efficient and environmentally friendly. The high-performance polymer membranes enable a level of purity of nearly 99%. A conditioning process for biogas, tailored especially to the membranes from Evonik, makes optimum use of their separation properties: in a three-stage process, the methane can be concentrated out of the crude gas with just one compressor and an especially high methane yield. In addition, the methane-rich gas does not have to be compressed further before it is fed into the natural gas grid. This membrane process is up to 20% more energy efficient than alternative methods. Besides, no auxiliary chemicals are required. No waste or wastewater is produced. Evonik initially trialed SEPURAN® Green in a test plant beside the Vöckla River in Neukirchen, Austria. Since then, several biogas processing plants using SEPURAN® Green technology have been put into operation. This application contributes to the overall sustainability of biogas production and opens the door to decentralized supply of biomethane for future low-emission mobility. Evonik is continuing to develop the SEPURAN® technology for new applications, such as separating hydrogen and recovering nitrogen from compressed air.[33]

REFERENCES

[1] European Commission, *Innovating for Sustainable Growth a Bioeconomy for Europe,* **2012**.

[2] (a) Bundesministerium für Ernährung, Landwirtschaft und Verbraucherschutz, *Politikstrategie Bioökonomie,* Berlin, **2013**; (b) German Federal Government, *Biorefineries Roadmap,* Berlin, **2012**; (c) German Federal Ministry of Education and Research, National Research Strategy BioEconomy 2030, Bonn/Berlin, 2011.

[3] Evonik Industries AG, *Corporate Responsibility Report 2012*, Essen, **2013**.

[4] GFK, GfK Green Gauge, New York, **2012**.

[5] (a) European Commission, Directorate-General for Enterprise and Industry, *Taking bio-based from promise to market*, Brussels, **2009**; (b) Ad-hoc Advisory Group for Bio-based Products, *Financing Paper*, **2011**; (c) Ad-hoc Advisory Group for Bio-based Products, *Recommendations on Communication*, **2011**.

[6] (a) A. Marx., V.F. Wendisch., R. Kelle, S. Buchholz, "Towards Integration of Biorefinery and Microbial Amino Acid Production, *in* Biorefineries – Industrial Processes and Products. Status Quo and Future Directions" (Eds. B. Kamm, P.R. Gruber, M. Kamm). Wiley-VCH, Weinheim, Germany, **2006**, 201–216; (b) A. Locher, *Folio.* **2014**, *4,* 12–17.

[7] Evonik Industries AG, Press Release, *Evonik starts innovation campaign*, can be found under http://corporate.evonik.com/en/media/search/Page/news-details.aspx?newsid=40329, **2013**.

[8] J. Krauter, D. Brennan, M. Binder, T. Kaufmann, M. Redshaw, *What if…*, Evonik Industries AG, **2011**.

[9] Evonik Industries AG, Press Release, *Healthy and efficient nutrition for a growing world population*, can be found under http://corporate.evonik.com/en/media/search/Pages/news-details.aspx?newsid=40566, **2013**.

[10] W. Pfefferle, B. Möckel, B. Bathe, A. Marx, *Adv. Biochem. Eng. Biotechnol.* **2003**, *79*, 59–112.

[11] Evonik Industries AG, Press Release, *Evonik to start up large-scale production facilities in 2014* can be found under http://corporate.evonik.com/en/media/search/Pages/news-details.aspx?newsid=40794, **2013**.

[12] O. Thum, *Elements* (Eds. Evonik Degussa GmbH), **2008**, *25*, 6–10.

[13] (a) M. Schrewe, A. O. Magnusson, C. Willrodt, B. Bühler, A. Schmid, *Adv. Synth. Catal.* **2011**, *353*, 3485–3495; (b) M. Schrewe, N. Ladkau, B. Bühler, A. Schmid, *Adv. Synth. Catal.* **2013**, *355*, 1693–1697; (c) M.K. Julsing, M. Schrewe, S. Cornelissen, I. Hermann, A. Schmid,. B Bühler, *Appl. Environ. Microbiol.* **2012**, *78*, 5724–5733.

[14] Evonik Industries AG, Press Release, *An alternative raw material for polyamide 12: Evonik is operating a pilot plant for bio-based ω-amino lauric acid* can be found under http://corporate.evonik.com/en/media/search/Pages/news-details.aspx?newsid=37328, **2013**.

[15] (a) M. Pötter et al. (Evonik Degussa GmbH) WO2011131420, **2011**; (b) S. Schaffer et al. (Evonik Industries AG), WO2013135650, **2013**; (c) S. Schaffer et al. (Evonik Industries AG), WO2013024114, **2013**; (d)M. Pötter et al. (Evonik Industries AG), US2013183725, **2013**.

[16] P. Welters, *BIOspektrum.* **2012**, *5*, 550–552.

[17] (a) Evonik Industries AG, OPXBIO, Press Release, *Evonik Industries Selects OPX Biotechnologies for Joint Development of Bio-Based Chemicals*, can be found under http://corporate.evonik.de/en/media/search/Page/news-details.aspx?newsid=35500, **2013**; (b) K. Aßmann, *Elements* (Eds.: Evonik Industries AG), **2013**, *45*, 48.

[18] I. A. Berg, D. Kockelkorn, W. Buckel, G. Fuchs, *Science* **2007**, *318*, pp. 1782–1786.

[19] Evonik Industries AG, Press Release, *Bacteria like the taste of syngas* can be found under http://corporate.evonik.com/en/media/search/Pages/news-details.aspx?newsid=40331, **2013**.

[20] L. Reinecke et al. (Evonik Röhm GmbH), US/2011/0171702, **2011**; M. Pötter et al. (Evonik Röhm GmbH), EP2291530, **2011**; R. H. Müller et al. (Evonik Röhm GmbH), EP1999264, **2010**; M. Buchhaupt et al. (Evonik Röhm GmbH), WO2011057871, **2012**; T. Haas et al. (Evonik Röhm GmbH), EP2294206, **2011**.

[21] (a) D. Przybylski, T. Rohwerder, H. Harms, R. H. Mueller. Sustainability and Society. *Energy,* **2012**, *2*, 11–20; (b) T. Hoefel, E. Wittmann, L. Reinecke, D. Weuster-Botz, *Appl. Microbiol. Biotechnol.* **2010**, *88(2)*, 477–84.

[22] (a) H. Gebhardt, T. Haas, A. Marx, S. Schaffer, A. Schraven, T. Tacke, *Elements* (Eds.: Evonik Degussa GmbH), **2008**, *25*, 26–32; (b) M. Pötter et al. (Evonik Röhm GmbH), EP 2152659, **2010**, 24.04.13; (c) A. Marx et al. (Evonik Röhm GmbH), WO2007141208, **2007**.

[23] U. Demmer, E. Warkentin, A. Srivastava, D. Kockelkorn, M. Pötter, A. Marx, G. Fuchs, U. Ermler, *J. Biol. Chem.* **2013**, *288(9),*6363–6370.

[24] *Elements* (Eds.: Evonik Industries AG), **2012**, *40*, 13.

[25] Evonik Industries AG, LanzaTech, Press Release, *Evonik and LanzaTech working on bio-processed precursors for specialty plastics*, can be found under https://bioeconomy.evonik.com/product/bioeconomy/en/pages/news-details.aspx?newsid=40447, **2013**.

[26] Evonik Industries AG, Press Release, *Bio-based high-performance polymer: A successful year for VESTAMID® Terra* can be found under http://corporate.evonik.com/en/media/search/Pages/news-details.aspx?newsid=31087, **2012**.

[27] Evonik Industries AG, Press Release, *Bio-based polyesters for Reactive Hot Melts – DYNACOLL® Terra* can be found under http://corporate.evonik.com/en/media/archive/pages/news-details.aspx?newsid=18468, **2011**.

[28] Evonik Industries AG, Press Release, *Resource efficiency through bio-based binder systems* can be found under http://corporate.evonik.com/en/media/press_releases/pages/news-details.aspx?newsid=34205, **2013**.

[29] Evonik Industries AG, Press Release, *Evonik launches next-generation biofuel component* can be found under http://corporate.evonik.com/en/media/archive/pages/news-details.aspx?newsid=30332, **2012**.

[30] Evonik Industries AG, Press Release, *Evonik brings locally manufactured products with global standard to PCHi 2014* can be found under http://corporate.evonik.com/en/media/search/Page/news-details.aspx?newsid=41754, **2014**.

[31] Evonik Industries AG, Press Release, *Evonik opens a new facility for its Health Care business in Darmstadt* can be found under http://corporate.evonik.com/en/media/press_releases/pages/news-details.aspx?newsid=41404, **2014**.

[32] Evonik Industries AG, BioAmber Inc.,Press Release, *Evonik and BioAmber Partner on Catalysts for Sustainable Chemicals Made from Bio-Based Succinic Acid* can be found under http://corporate.evonik.com/en/media/archive/pages/news-details.aspx?newsid=28779, **2012**.

[33] Evonik Industries AG, BioAmber Inc.,Press Release, *Prize awarded to new membrane technology from Evonik* can be found under http://corporate.evonik.com/en/media/press_releases/corporate/pages/news-details.aspx?newsid=40364, **2013**.

9

MARKET STRUCTURE AND GROWTH RATES OF INDUSTRIAL BIORENEWABLES

GUNTER FESTEL[1,2,3]

[1] *Festel Capital, Fueringen, Switzerland*
[2] *Technical University Berlin, Berlin, Germany*
[3] *University of Basel, Basel, Switzerland*

9.1 BACKGROUND FOR INDUSTRIAL BIORENEWABLES AND DATA SOURCES

Over the last years, the production of industrial biorenewables, such as chemicals and materials using renewable resources and using either emerging biotechnological conversion processes or more classical chemical ones – for one or more of the process steps – has increased significantly.[1,2] In these fields, and in contrast to conventional chemical production methods using energy-intensive processes with high

Industrial Biorenewables: A Practical Viewpoint, First Edition. Edited by Pablo Domínguez de María.

temperatures, biotechnology utilizes microorganisms or enzymes that enable conversion processes at lower temperatures with significant environmental advantages.[3] Using more and more renewable resources makes it possible to overcome the consequences of diminishing fossil resources and to master the transition from the current petro-based to a future bio-based economy.[4,5]

The market data disclosed in this chapter derive from a database built up since 2003 by Festel Capital, and it is based on interviews with experts from the industry, universities and research institutions, the investment sector, as well as representatives from governmental institutions. Additionally, the biorenewable markets are being analyzed periodically through desk research using public sources like databases, articles, and company disclosures (e.g., websites, annual reports, press releases). In this database, the sales figures for biorenewables are shown as part of chemical sales due to the fact that the most important market players are established chemical companies or companies in related industries, like the agro or pulp and paper industries. The database uses the segments and subsegments used by the Conseil Européen de l'Industrie Chimique (CEFIC) for global chemical sales between 2002 and 2009, for example, for their yearly publication "Facts & Figures" describing the chemical industry in Europe and its global position. The following segments – with the subsegments in brackets – are studied: base chemicals (anorganics, fertilizers and gases, organic chemicals, polymers and fibers), specialty chemicals (agrochemicals, adhesives and sealants, paints and coatings, food additives, other specialty chemicals), and consumer chemicals (detergents, cosmetics). The subsegment polymers and fibers, including building blocks and intermediates to produce these products, is also shown on a segment level due to its size and importance.

Within the database, the sales of biorenewables are estimated on a subsegment level (in some cases also on a more detailed product level) as rolling forecast. Subsequently, the subsegment or product-level data are aggregated to obtain the figures on a segment level. This is done separately for the regions Europe (EU-27 countries and Switzerland), North America (Canada, Mexico, and the United States), Asia Pacific, and rest of the world. Biomass-derived energy (including biofuels) and pharmaceuticals (pharmaceutical end products as well as active pharmaceutical ingredients and intermediates) are not included in the following sales figures.

An important aspect is the definition of which sales figures within chemical sales actually account for biorenewables. Naturally, the sales of all products made from renewable resources using biotechnological conversion processes are considered. Apart from this, if single chemical process steps are involved, these products are also considered, but renewable raw materials converted only through chemical processes and nonrenewable raw materials converted through biotechnological processes are not included in these market data. Furthermore, captive production and intercompany sales based on transfer prices are not considered for the assessment either. The sales figures in this database generally only show product sales between two independent companies based on market prices.

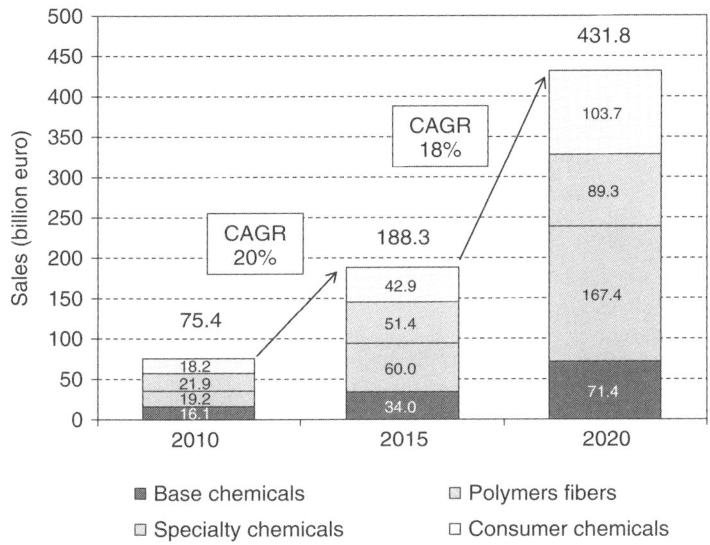

Figure 9.1 Market size of industrial renewables in 2010, 2015, and 2020.

9.2 MARKET OVERVIEW AND GROWTH RATES

Sales of industrial biorenewables in 2010 were 75.4 billion euros accounting for 5.3% of total chemical sales (Figure 9.1 and Table 9.1). The four categories – base chemicals, polymers and fibers, specialty chemicals, and consumer chemicals – contributed nearly equally with specialty chemicals being the largest segment with 21.9 billion euros, followed by polymers and fibers with 19.2 billion euros. In all chemical segments it is expected that the percentage of biorenewables will increase significantly during the next years. In 2015, sales of industrial biorenewables will be around 188.3 billion euros representing 10.5% of total chemical sales, resulting in a compound annual growth rate (CAGR) from 2010 to 2015 of 20.1%. The segment polymers

TABLE 9.1 Biorenewable Sales Including Growth Rates for the Different Segments in 2010, 2015, and 2020.

	Biorenewable Sales [billion Euro]			Compound Annual Growth Rates (CAGR) (%)	
	2010	2015	2020	2010–2015	2015–2020
Base chemicals	16.1	34.0	71.4	16.1	16.0
Polymers and fibers	19.2	60.0	167.4	25.6	22.8
Specialty chemicals	21.9	51.4	89.3	18.6	11.7
Consumer chemicals	18.2	42.9	103.7	18.7	19.3
Total/average	*75.4*	*188.3*	*431.8*	*20.1*	*18.1*

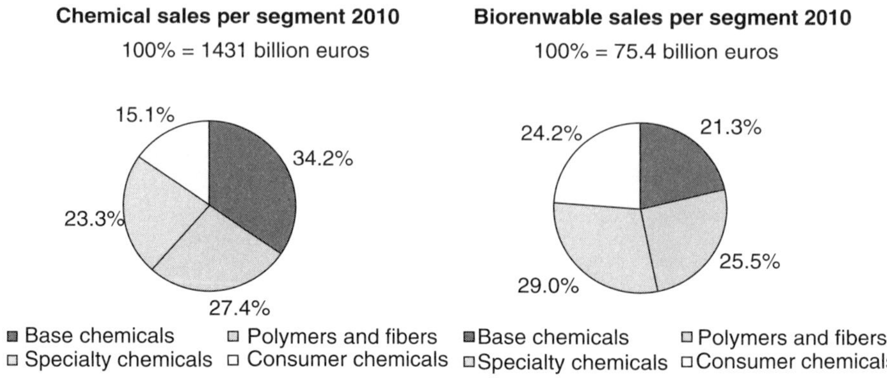

Figure 9.2 Chemical and biorenewable sales per segment in 2010.

and fibers, with 60 billion euros, will represent almost a third of total sales. Second will be specialty chemicals with 51.4 billion euros, and the smallest segment will be base chemicals with 34 billion euros. Five years later in 2020, 431.8 billion euros representing 19.5% of total chemical sales is expected to belong to industrial biorenewables. The CAGR from 2015 to 2020 reaches, with 18.1%, almost the same level as the CAGR from 2010 to 2015. The largest segment will be polymers and fibers with 167.4 billion euros and consumer chemicals, with 103.7 billion euros, the second largest segment.

A more differentiated picture regarding growth rates is possible by looking at the sales figures for each segment of biorenewables (Table 9.1). From 2010 to 2015, the segment polymers and fibers shows the strongest growth with 25.6% and base chemicals, with 16.1%, the smallest. The segment polymers and fibers also grows the strongest from 2015 to 2010 with 22.8%. Specialty chemicals have the smallest growth with 11.7%. It is important to note that all growth rates outperform by far the growth of traditional chemicals.

9.2.1 Detailed Market Analysis for 2010

Chemical sales in 2010 were in total 1431 billion euros with a 75.4 billion share for biorenewables (Figure 9.2 and Table 9.2). Although basic chemicals made up around 34.2% of total chemical sales in 2010 (=490 billion euros), only 3.3% of those (=16.1 billion euros) were biorenewables. Biorenewables had a share of 19.2 billion euros of sales equaling 4.9% of chemical sales in the segment polymers and fibers, which was in total 392 billion euros. Within specialty chemicals, which accounted for 333.2 billion euros, the share of biorenewables was 6.6% of chemical sales (=21.9 billion euros). The highest share of biorenewables within chemical sales with 215.6 billion euros had consumer chemicals with 8.5% (=18.2 billion euros). This is remarkable, as consumer chemical was the smallest chemical segment with only 15.1% of total chemical sales.

Biorenewable sales per region show that Asia Pacific is the strongest region with 37.9% (=28.6 billion euros), followed by Europe with 29.5% (=22.2 billion euros)

TABLE 9.2 Chemical and Biorenewable Sales per Segment in 2010.

	Chemical Sales 2010		Biorenewable Sales 2010		
	Sales [Billion Euro]	% of Total Sales	Sales [Billion Euro]	% of Total Sales	% of Chemical Sales
Base chemicals	490.0	34.2	16.1	21.3	3.3
Polymers and fibers	392.0	27.4	19.2	25.5	4.9
Specialty chemicals	333.2	23.3	21.9	29.0	6.6
Consumer chemicals	215.6	15.1	18.2	24.2	8.5
Total	*1430.8*	*100.0*	*75.4*	*100.0*	

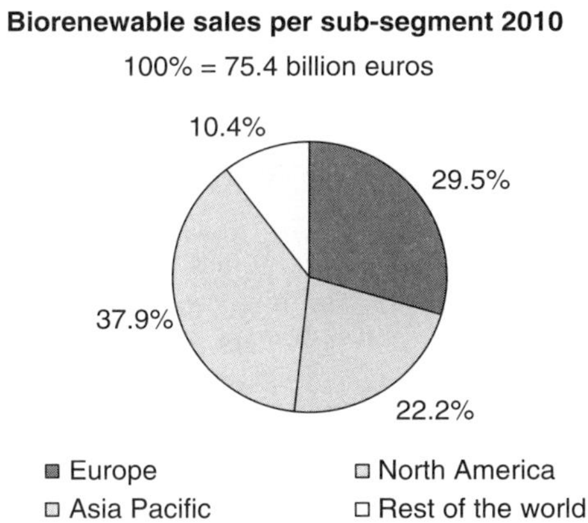

Figure 9.3 Biorenewable sales per region in 2010.

and North America with 22.2% (=16.8 billion euros) (Figure 9.3). The rest of the world accounts for 10.4% (=7.8 billion euros) and this figure is mainly driven by Brazil.

The most important subsegment in 2010 was bio-based polymers and fibers followed by the subsegments organic chemicals, cosmetics, and detergents (Figure 9.4). Bio-based polymers and fibers had almost 6 billion euros sales in Europe, around 4.5 billion euros sales in North America, and around 7.5 billion euros sales in Asia Pacific. Organic chemicals had more than 4 billion euros sales in Europe as well as Asia Pacific and around 3 billion euros sales in North America. Cosmetics and detergents accounted for around 3 billion euros sales in Europe and Asia Pacific. The sales figures in North America were around 2 billion euros. Important subsegments with around 2 billion euros sales in each major region were also agrochemicals and food additives.

9.2.2 Detailed Market Analysis for 2015

In 2015, it is predicted that chemical sales will reach 1785 billion euros (Figure 9.5 and Table 9.3). Sales for industrial biorenewables will be around 188.3 billion euros

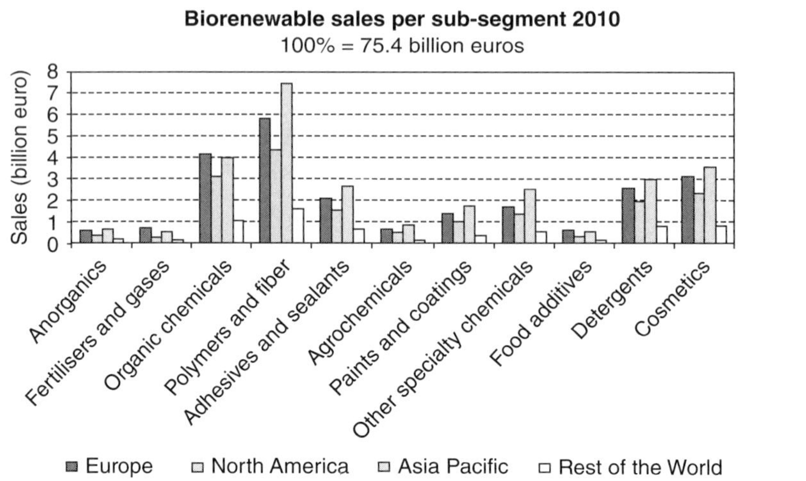

Figure 9.4 Biorenewable sales per subsegment in 2010.

representing 10.5% of total chemical sales. Whereas basic chemicals will contribute 30.5% of total chemical sales (=545.1 billion euros), only 6.2% (=34 billion euros) will be biorenewables. Biorenewable polymers and fibers will strongly increase to 60 billion euros, which is almost a third of total biorenewable sales and 12.1% of 495.9 billion euros as chemical sales in that segment. Specialty chemicals will show 446.3 billion euros including 51.4 billion euros for biorenewables, which is 27.3% of total biorenewable sales and 11.5% of chemical sales. Consumer chemicals with 297.5 billion euros will include biorenewable sales of 42.9 billion euros. With 14.4%, this is the largest share of biorenewable sales in a chemicals segment.

The most important subsegments in 2015 will remain the same as in 2010 (Figure 9.6). Polymers and fibers will account for 20 billion euros in both regions: Europe and Asia Pacific. In North America with 13 billion euros, this value is significantly lower. The reason is that producers in North America stick more to

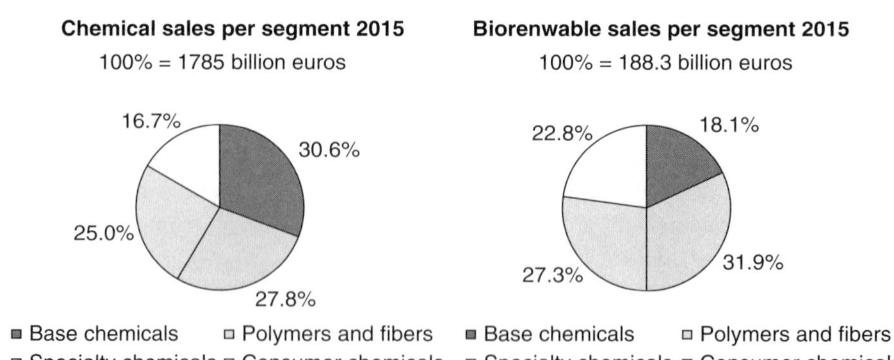

Figure 9.5 Chemical sales and biorenewable sales per segment in 2015.

TABLE 9.3 Chemical and Biorenewable Sales per Segment in 2015.

	Chemical Sales 2015		Biorenewable Sales 2015		
	Sales [Billion Euro]	% of Total	Sales [Billion Euro]	% of Total	% of Chemical Sales
Base chemicals	545.1	30.5	34.0	18.1	6.2
Polymers and fibers	495.9	27.8	60.0	31.9	12.1
Specialty chemicals	446.3	25.0	51.4	27.3	11.5
Consumer chemicals	297.5	16.7	42.9	22.8	14.4
Total	*1784.8*	*100.0*	*188.3*	*100.0*	

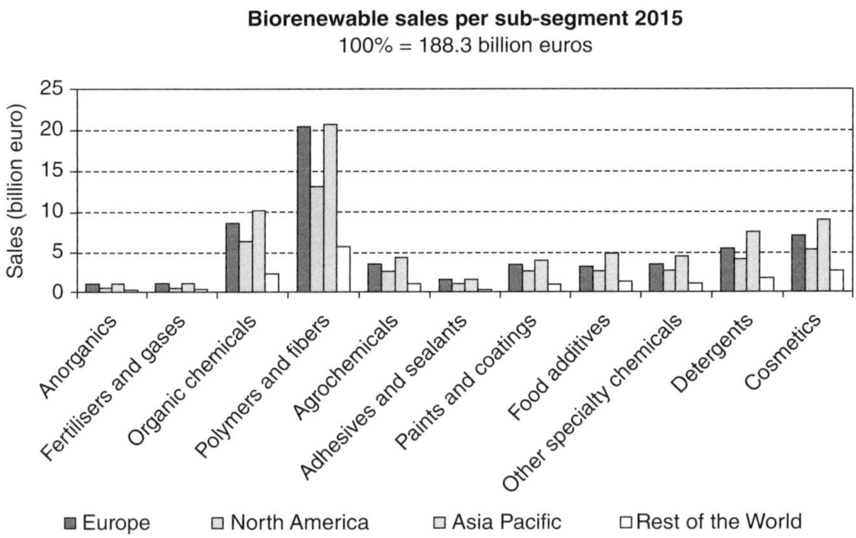

Figure 9.6 Biorenewable sales per subsegment in 2015.

traditional chemical routes, and a major part of the available biomass is used for biofuel production. Organic chemicals have between 7 and 10 billion euros in the three major regions and detergents as well as cosmetics between 4 and 9 billion euros. The subsegments anorganics and fertilizers and gases will see the lowest sales figures in the range of 1–2 billion euros.

9.2.3 Detailed Market Analysis for 2020

It is predicted that chemical sales will increase to 2225 billion euros in 2020, whereby 431.8 billion euros, representing 19.5% of total chemical sales, will belong to industrial biorenewables (Figure 9.7 and Table 9.4). The importance of base chemicals will further decrease with 26.8% of total chemical sales in 2020 compared with 34.2% in 2010. Predicted sales of 595.2 billion euros include 71.4 billion euros biorenewables, equaling a share of 12%. Polymers and fibers will achieve the highest biorenewable

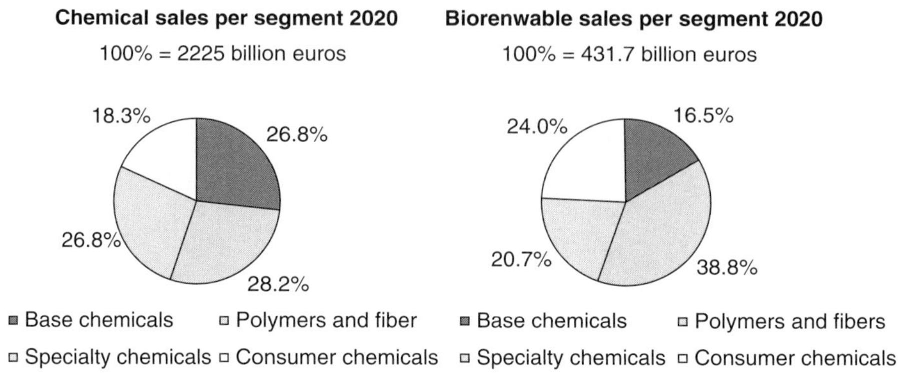

Figure 9.7 Chemical sales and biorenewable sales per segment in 2020.

TABLE 9.4 Chemical and Biorenewable Sales per Segment in 2020.

	Chemical Sales 2020		Biorenewable Sales 2020		
	Sales [Billion Euro]	% of Total	Sales [Billion Euro]	% of Total	% of Chemical Sales
Base chemicals	595.2	26.8	71.4	16.5	12.0
Polymers and fibers	626.8	28.2	167.4	38.8	26.7
Specialty chemicals	595.5	26.8	89.3	20.7	15.0
Consumer chemicals	407.4	18.3	103.7	24.0	25.5
Total	*2224.9*	*100.0*	*431.8*	*100.0*	

sales with 167.4 billion euros. This will be almost 40% of total biorenewable sales and 26.7% of chemical sales, which will be 626.8 billion euros. Specialty chemicals with 89.3 billion euros will account for 15% of 595.5 billion euros as total chemical sales in that segment. Consumer chemicals will be the second largest biorenewable segment with 103.7 billion euros. This is 24% of total biorenewable sales and 25.5% of 407.4 billion euros as total sales for consumer chemicals.

Looking at the subsegment level, polymers and fibers will be strongest with 50 billion euros in Europe, nearly 40 billion euros in North America and more than 60 billion euros sales in Asia Pacific (Figure 9.8). The second largest subsegment, organic chemicals, will account for around 20 billion euros in all major regions. Detergents and cosmetics will show between 10 and 22 billion euros. As in 2015, anorganics and fertilizers and gases will have the lowest sales figures in the range of 1–2 billion euros.

9.3 EXAMPLES FOR BIOTECHNOLOGY-BASED PRODUCTS RELATED TO BIORENEWABLES

The Chapters of this book deal with a broad number of emerging and consolidated biotechnology-based products at industrial scale. As an introductory section, to

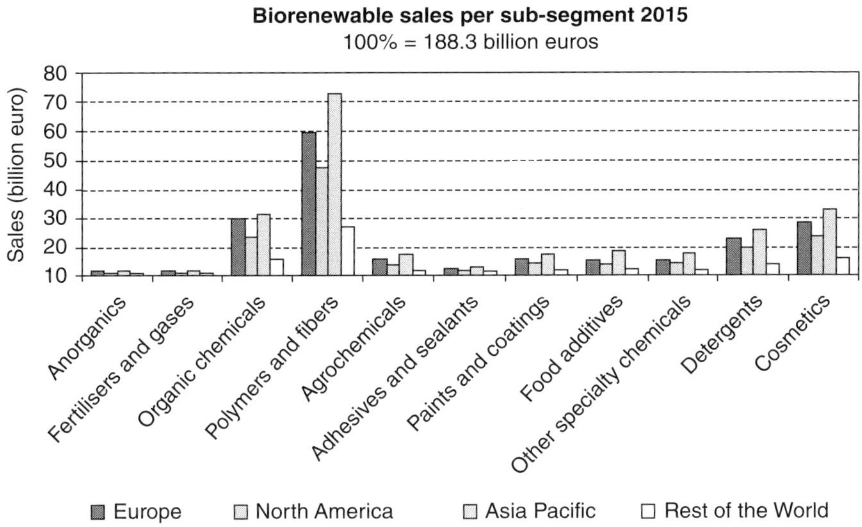

Figure 9.8 Biorenewable sales per subsegment in 2020.

start to illustrate the diversity of molecules made by biotechnology processes, some examples are described in this section (Figure 9.9).

1,3-Propanediol is a three-carbon diol mainly used as a building block in the production of polymers, such as polytrimethylene terephthalate. It can also be formulated into a variety of industrial products including adhesives, coatings and paints, as well as composites, laminates, and moldings. In the cosmetics industry it is used as a solvent, humectant, emollient, or hand-feel modifier. 1,3-Propanediol has a production volume of more than 100,000 tons per year. It can be chemically synthesized by the hydration of acrolein or by the hydroformylation of ethylene oxide and hydrogenation. A newer biotechnology process enables the conversion from corn syrup effected by a genetically modified strain of bacteria by DuPont Tate & Lyle Bio Products (for more information see chapter written by DuPont).

Succinic acid or butanedioic acid is a dicarboxylic acid, which is used as a precursor for polyesters and a component of alkyd resins. It is also applied in the food and beverage industry, primarily as an acidity regulator. Global production is estimated at 30,000 tons per year. This molecule is produced by several chemical methods, like the hydrogenation of maleic acid, oxidation of 1,4-butanediol, or carbonylation of ethylene glycol, but more and more succinic acid is being produced through the fermentation of glucose from renewable feedstock and purification of raw bio-based succinic acid. Companies like BioAmber, Reverdia, Myriant, BASF, and Purac are progressing from demonstration-scale production to viable commercialization.

Lysine is an essential amino acid for humans and an important additive to animal feed because it is a limiting amino acid when optimizing the growth of certain animals, such as pigs and chickens, for the production of meat. The production exceeds 600,000 tons per year and main producers are Archer Daniels Midland, BASF, and

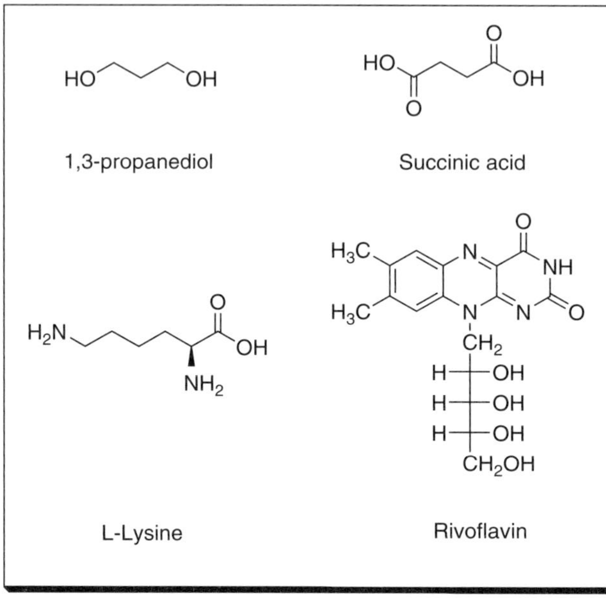

Figure 9.9 Structural formulas of molecules produced by biotechnological processes.

Evonik. Lysine is usually manufactured by a microbial fermentation process using bacteria from a base mainly of sugar. Genetic engineering research is actively pursuing bacterial strains to improve the efficiency of production and allow lysine to be made from other substrates.

The vitamin riboflavin, also known as vitamin B_2, is a micronutrient with a key role in maintaining health in humans and animals. Various biotechnological processes have been developed for industrial-scale riboflavin biosynthesis using different microorganisms, including filamentous fungi as well as bacteria, which are genetically modified to increase the bacteria's production of riboflavin. BASF, for example, produces riboflavin using filamentous fungi.

Overall, not only through biotechnology but also using chemical means, the number of examples of industrial products using or being based on biorenewables is continuously increasing. Chapters of this textbook deepen in different industry processes.

REFERENCES

[1] G. Festel, J. Knöll, H. Götz, H. Zinke, *Chem. Ing. Tech.* **2004**, *76*, 307–312.
[2] R. Coons, *Chem. Week* **2010**, *172*, 22–26.
[3] B.G. Hermann, K. Blok, M.K. Patel, *Environ. Sci. Technol.* **2007**, *41*, 7915–7921.
[4] G. Festel, *Ind. Biotech.* **2010**, *6*, 88–94.
[5] P. Nieuwenhuizen, D. Lyon, *J. Commer. Biotechnol.* **2010**, *17*, 159–164.

10

GÖTEBORG ENERGI: VEHICLE FUEL FROM ORGANIC WASTE

ERIC ZINN[1] AND HENRIK THUNMAN[2]

[1]*Göteborg Energi AB, Göteborg, Sweden*
[2]*Chalmers University of Technology, Göteborg, Sweden*

ABBREVIATION LIST

Barg: gauge pressure

bioDME: dimethyl ether from renewable resources

bio-FT-diesel: diesel fuel produced from renewable resources using the Fischer–Tropsch process

Industrial Biorenewables: A Practical Viewpoint, First Edition. Edited by Pablo Domínguez de María.

COS: carbonyl sulfide
CBG: compressed biogas
CNG: compressed natural gas
FGS: Fordonsgas Sverige, a Swedish retailer of methane gas for transportation
GoBiGas: Gothenburg Biomass Gasification Project
GWh: gigawatt-hours
GWhth: gigawatt-hours, thermal
H2S: hydrogen sulfide
kWh: kilowatt-hours
LBG: liquefied biogas
MW: megawatt
MWth: megawatt, thermal
MDE: methane-diesel engine
NOx: nitrogen oxides
SCR: selective catalytic reduction
THT: tetrahydrothiophene
TSO: transmission service operator
TWh: terawatt-hours
ZnO: zinc oxide

10.1 THE COMPANY

Göteborg Energi is Sweden's largest municipal energy company, owned entirely by the City of Gothenburg. Swedish legislation allows for it to compete equally on the open market. Where the company has a monopoly (e.g., distribution of electricity and gas), special regulations apply. The history of the company starts with the introduction of town gas for street lighting in 1846. Since then, it has come to incorporate several different products and services including electricity production and distribution, natural gas distribution, and district heating. The environmental benefits of the investments made in district heating in past decades are evident. Air quality has improved greatly in the city and by making use of waste heat from refineries and waste incinerators; energy efficiency has improved greatly as well. Since 2007, Göteborg Energi has grown to become Sweden's largest supplier of biogas. It is also the largest investor and developer of biogas in Sweden. In 2014, the company will increase its capacity to produce over 300 GWh of biogas per year and runs seven biogas plants ranging from 0.2 to 20 MWth.

10.2 SWEDEN'S RENEWABLE ENERGY TARGETS AND THE ROLE THAT BIOGAS WILL PLAY IN MEETING THESE

In 2010, biomass made up 131 TWh of the 411 TWh of energy used in Sweden. These were used in industries, in houses, and in transportation. Sweden's target is that at least 50% of the energy used in Sweden will come from renewable sources by 2020. The biogas industry in Sweden has estimated that 24 TWh of biogas could be produced annually by 2030, provided that the appropriate incentives have been put in place. In 2012, approximately 1 TWh of biogas from anaerobic digestion was used in transportation, with lesser amounts being used in other applications or just flared. Currently seven anaerobic digestion plants inject their biomethane into the Swedish natural gas grid. In all instances, the biomethane is injected into local distribution grids and not the overlying transmission grid. Gothenburg Biomass Gasification (GoBiGas) will be the first site where biomethane is injected into the higher-pressure transmission grid.

Due to the lower calorific value of biomethane, injection into local distribution grids requires an addition of propane to guarantee an even supply of energy. Adding propane is a costly business for biomethane producers. As the transmission grid carries much larger volumes of gas where dilution is greater, the transmission service operator (TSO) Swedegas has not required that propane is added. Swedegas, together with several other European gas grid operators, recently assumed the target of achieving 100% green gas grids by 2050.

10.3 BIOGAS IN TRANSPORTATION: CASE STUDIES WITHIN GÖTEBORG ENERGI

Sweden is a world leader in using biogas in transportation. Beginning in the late 1990s, biogas has been favored by several governments that have introduced support programs for biogas to be used in transportation. Several municipalities have spearheaded development by building anaerobic digestion plants for sewage sludge or municipal waste and used the gas for fueling local busses. The market has seen an annual growth rates of 20–30%, and there are now around 150 filling stations for CNG/compressed biogas (CBG). Currently, biogas makes up approximately 1% of the total energy used in road transport in Sweden.

10.3.1 Gasendal

Göteborg Energi's first biogas plant (called "Gasendal") is an upgrading plant installed in Gothenburg in 2007. The plant receives biogas produced at the regional wastewater plant (Gryaab) where sewage sludge has been anaerobically digested for decades, primarily to reduce sludge volumes. The gas is piped by two compressors 3 km between the wastewater plant and the upgrading plant. Biogas often contains high amounts of hydrogen sulfide (H_2S) for which an active carbon filter needs to be built. Chemical additives that the wastewater plant adds to its process result in very low concentrations of H_2S in the biogas received by Gasendal. The active carbon filter is regenerated every 2–3 years.

The biogas then moves on the removal of carbon dioxide. This is achieved though chemical absorption, where an amine is used to efficiently absorb the CO_2 present in the methane flow. The amine is in an aqueous solution (approximately 50%) and flows downward over column packings in order to maximize the surface uptake area.

The gas flow goes upward and the methane concentration increases to approximately 98%. The amine is regenerated and "stripped" of CO_2 by boiling it using natural gas. The amine then releases the CO_2, which is vented to the atmosphere. Each batch of amine solution can be used for 5 years. The residual heat is captured and used to keep the access road free of ice. After upgrading, the methane flow passes through two compressors, alternating gas dryer and particle filters. Propane is added to the methane to match the energy content of the natural gas in the grid. A smell additive (tetrahydrothiophene (THT)) is also added as a precaution in case of gas leaks.

The Gasendal upgrading plant has a very small methane slip (<0.1%). The plant produces on average 60 GWh of biomethane that is injected into the 4 bar natural gas grid of Gothenburg. The biomethane is sold mainly to consumers in the transport sector. Public transport (busses) is an important market sector. The investments at the Gasendal site and at the wastewater treatment plant Gryaab total approximately €4 million. The project has been supported by Swedish (KLIMP) funding as well as support from the European Union (through Biogasmax program) (Figure 10.1).

10.3.2 Lidköping Biogas

Lidköping biogas is one of the world's first plants for liquefied biogas (LBG). LBG enables heavy-duty trucks to choose a viable, renewable alternative, where there currently are few. The fuel production in Lidköping is based on local vegetable waste products from grain trade and food production.

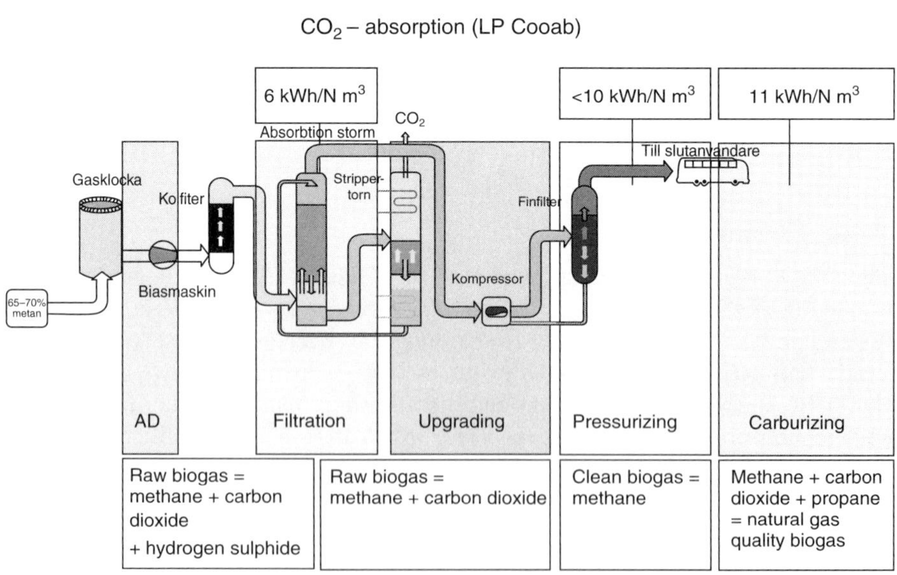

Figure 10.1 The Gasendal process. (*See color plate section for the color representation of this figure.*)

10.3.2.1 The Process The new biogas plant in Lidköping is a joint effort between Göteborg Energi AB, Swedish Biogas International Lidköping AB, and the Municipality of Lidköping. Göteborg Energi AB is Sweden's largest municipal energy company and one of the most active developers of the biogas fuel market, which has seen an annual growth of 20–30% over the past decade. The biogas production process is based on local vegetable waste products from grain trade and food production. The substrates are macerated, mixed, and heated to 38 °C before being pumped into two large digesters, each 4100 m^3. New substrate material is continually pumped into the process that produces biogas and biofertilizer. The biofertilizer is pumped to two 2500 m^3 covered storage pools for postdigestion. The entire process takes 60 days. Plant production is designed at 7.5 MWth with an annual target of 60 GWhth. The biogas is upgraded in a water scrubber to fulfill the Swedish standard for biogas as a vehicle fuel (SS 155428).

A majority of the biogas is liquefied in the condensation plant. The technology is supplied by Air Liquide. In order to liquefy the biogas, the majority of remaining small amounts of carbon dioxide are purged by pressure swing absorption (PSA) before the gas temperature is lowered using liquid nitrogen in a Brayton cycle. The developed technology allows for liquefaction in the span of −140 °C (at 4 barg) to −161 °C (at atmospheric pressure), depending on the developing requirements of the vehicle market. The design capacity is 12 ton LBG day^{-1} but will initially be run at around 70%. The energy requirements of liquefaction are in the vicinity of 1 kWh/Nm^3 upgraded biogas, which equals approximately 10% of the energy content of the biogas. The LBG is stored in a 115 m^3, 20 m tall insulated canister. The distributer, Fordonsgas Sverige AB (FGS), fills insulated 50 m^3 trailers every second day and transports the gas to filling stations in Göteborg with more to come. A smaller portion (around 30%) of the biogas produced and upgraded is delivered directly to FGS's two compressors, which fill mobile storage containers in one of six filling places. FGS distributes CBG to other filling stations in the vicinity using mobile storage containers. There is also a public filling station on-site where vehicles are able to refill CBG at 200 bar. As the demand for LBG increases in Sweden, CBG production at Lidköping will decrease.

This investment of €16 million in total has received €3 million in support from the national climate investment program (KLIMP). The biogas produced in Lidköping is enough to fuel 6000 cars that drive 17,000 km per year. This results in 16,000 tons of reduced CO_2 emissions annually. The distributor FGS is the world's only supplier of vehicle fuel approved by the Nordic Ecolabel (www.ecolabel.se), and production at Lidköping must fulfill these high standards.

LGB allows for a more efficient distribution and simpler storage of biogas. Biogas distribution becomes less reliant on being close to a gas grid. We can increase the availability of biogas to larger areas off the natural gas grid than what is possible with biogas in gaseous state. LBG also opens for new applications, such as for heavy-duty trucks. LBG has the energy density required for heavy vehicles to travel long distances (Figure 10.2).

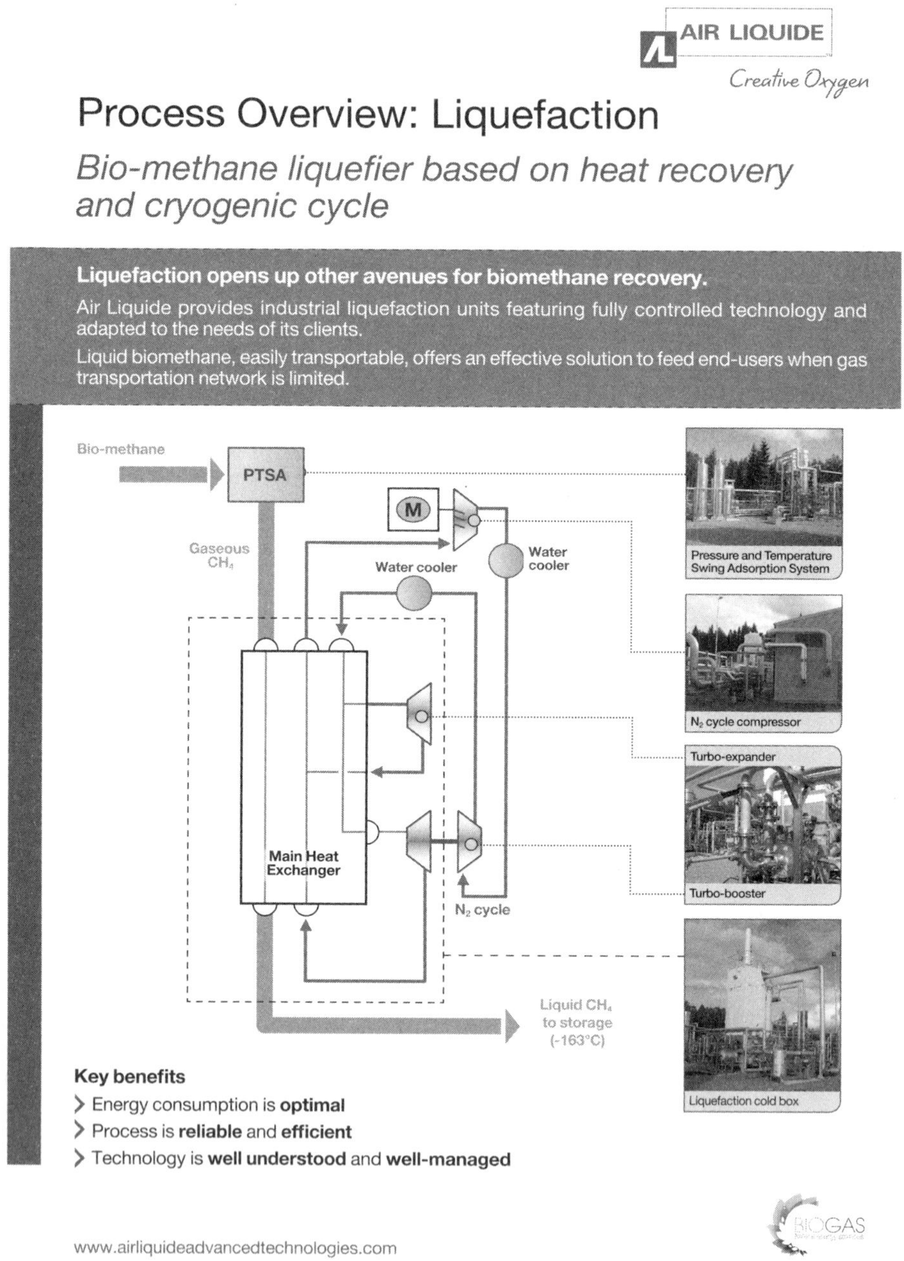

Figure 10.2 The Air Liquide technology. *Source*: Reproduced with permission of Air Liquide advanced Technologies.

10.3.2.2 The Market Potential for LBG Liquefied methane gas, either from biogas or natural gas, is a new vehicle fuel in Sweden. Volvo Trucks has begun the serial production of their FM line using methane-diesel engines (MDE). The target for the first year of production is 100 MDE trucks for sale in the initial markets of Sweden, Holland, and the United Kingdom. LBG production in Lidköping is sufficient for approximately 200 of these new trucks. At a national level, the biogas distributors FGS, E.ON, and AGA have in 2014 established five filling stations for LBG and liquefied natural biogas in Sweden, currently serving approximately 100 MDE trucks.

10.3.3 GoBiGas

10.3.3.1 Background The GoBiGas Project is the largest demonstration of utilizing forest residues for the production of a renewable fuel, in this case biomethane. The €135 million investment may unlock a substantial market and multiply the possible volumes of waste-based energy for transportation, industry, CHP, and heating. Biomethane is commonly produced today by upgrading biogas from anaerobic digestion plants in several European countries, where a variety of substrates (such as sewage sludge, food waste, manure, and energy crops) are used. Waste-based biomethane is an excellent energy source, which can be used to generate either heat or power, to fuel vehicle, or to replace natural gas in industrial processes, with unparalleled environmental benefits. The potential for biogas from waste through anaerobic digestion is nonetheless limited. In Sweden, this potential is estimated to be at 10% of the current use of fuel in road transportation. The potential in residues from the forestry industry is approximately six times larger in Sweden, but the high content of lignocellulose in forest residues makes it unsuitable for anaerobic digestion. Gasification as a production process is an alternative pursued by many renewable fuel options. One key aspect for each process route is the overall efficiency of the process, resulting in as much renewable energy as possible from the resource base. GoBiGas aims at proving an exceedingly high efficiency rate of transforming 65% of the energy contained in the forest residues into high-value biomethane. By utilizing the plant's waste heat in the district heating grid of Gothenburg, the total energy efficiency of the plant will reach 90% or more.

10.3.3.2 Biomethane from Forest Residues As stated earlier, the biomass to be used in the GoBiGas Project will be forest residues. The planned production of the demonstration plant is 160 GWh of biomethane per year (20 MW biomethane and 8000 operating hours). In order to produce this volume of gas, approximately 260 GWh of fuel is needed. Initially approximately 50,000 tons of wood pellets year^{-1} (less than 3% of the total Swedish pellet consumption) will be used within the GoBiGas plant. This is a fuel that already is used in the adjacent hot water plant, which allows GoBiGas to make use of the existing fuel feeding infrastructure. However, the use of wood pellets as fuel to the gasifier is only planned during the start-up and initial operation period. After this period the fuel will be changed to forest residues when a new fuel handling and dryer system has been built and is in

Figure 10.3 Facilities of the GoBiGas plant. *Source*: Pictures courtesy of Mr. Rob Vanstone.

operation. Forest residues are the main biomass for heat and power production in Sweden and consist mainly of branches and treetops. The fuel will probably come from the surrounding areas of Göteborg, the Lake Vänern, or the Baltic region. When running on wet forest residues, the amount that will be used in the GoBiGas plant will be approximately 110,000 tons (less than 0.2% of the Swedish consumption of forest residues) (Figure 10.3).

10.3.3.3 The Technology In gasification the chemically stored energy in the solid biomass is converted to a gas mainly consisting of hydrogen, carbon monoxide, carbon dioxide, and water depending on the gasification type and operating temperature with varying amounts of different hydrocarbons. Depending on what the gas is intended to be used various efforts of cleaning and further conversion of the hydrocarbons are needed. For instance, if the intention is to produce biofuels, all hydrocarbons – except methane if biomethane is the intended product – have to be converted to carbon monoxide and hydrogen before they are put together with the aid of a catalyst to the desired hydrocarbons, for example, biomethane, biomethanol, bioethanol, bioDME, and bio-FT-diesel. For this approach, the main technical challenge is to find efficient and economically acceptable solutions for the gasification and the downstream cleaning. Independent of future development, the complexity of the conversion process will be comparable to an oil refinery and, thereby, be favored by large-scale production units. The technology is available and the commercial rollout is mostly dependent on the long-term conditions for biofuels given through political decision. It should be remembered that for the foreseeable future biofuels will become more expensive than alternatives produced from raw oil natural gas or coal if they are not penalized for their negative environmental impact.

In the process, dry fuel is fed to the gasification reactor using screw conveyers. The gasifier is fluidized with superheated steam at a slight overpressure in the bottom of the reactor. In the gasifier the fuel is then converted into a so-called product gas and a small amount of unconverted char. The heat required for the gasification reactions is supplied by bed material circulating between the gasification reactor and a combustion reactor. In the combustion reactor char transported with the circulating bed material is combusted together with a small fraction of the produced gas. For this combustion, air is used as fluidizing gas and as oxidant. The heated bed material from the combustion reactor is recycled to the gasification reactor where it supplies heat to the gasification. The temperature in the gasification reactor is approximately 850 °C and in the combustion reactor about 930 °C.

After the gasifier, the product gas is cooled down to about 150 °C by generating hot water for use internally in the plant and for district heating. After the cooler the product gas is filtered in a baghouse filter with addition of limestone as a precoat. The dust-free gas from the filter is fed to a tar removal scrubber, where the gas is washed with biodiesel and cooled to about 40 °C. The removed tar together with spent biodiesel is fed to the combustion reactor as a fuel. After this scrubber the gas is fed to the methanation plant. The flue gas from the combustion reactor is cooled down in several stages by heating water, superheating steam, and preheating combustion air to the process. After the coolers, the flue gas is filtered in a baghouse filter, passing by through an selective catalytic reduction (SCR) process for NO_x reduction and released to the atmosphere.

Remarkably, some of the catalysts and processes used in the methanation plant have been commercially proven in other applications, but never before in this specific combination and with the purpose to produce a synthetic natural gas from a product gas produced by biomass gasification. The first step in Haldor Topsøe process is a tar removal step based on adsorption on activated carbon. After the tar removal the product gas is compressed to approximately 16 bar. The gas is fed to an olefin hydrogenation reactor where unsaturated hydrocarbons are saturated by reaction with hydrogen in a catalytic process. The product gas is then sent to a COS (carbonyl sulfide) hydrolysis reactor, where COS is converted to H_2S. The conversion is performed using a new and unique catalytic process. The gas is washed in an amine wash where the H_2S is removed together with a large part of the CO_2. The sulfur- and CO_2-saturated amine is regenerated in a stripper and returned to the scrubber, and the stripper off-gas is sent to the combustion reactor in the gasification process for destruction. The cleaned gas is sent to a guard bed with ZnO to remove the last traces of sulfur compounds that might still be in the gas. After this the gas is divided in two streams, where a small part is bypassing, while the major part of the gas is passing a shift reactor and a so-called premethanation reactor. The amount of the stream that is bypassing is controlled in such a way that the H_2/CO ratio is 3 when the streams meet again downstream the premethanation. The combined stream is then cooled down, water is condensed, and separated and the gas is fed to the CO_2 removal.

As stated earlier, the shift reactor and the shift catalyst as such are commercial technology used in processes for coal or oil gasification, but have not been previously used when processing biomass-derived gas. The premethanation reactor used is an entirely new construction developed by Haldor Topsøe in liaison with the GoBiGas Project. Like the H_2S removal, the CO_2 removal is also done in an amine wash where the remaining CO_2 is absorbed by the amine in a scrubber. The CO_2-saturated amine solution is flashed in a high-pressure and a low-pressure flash stage and is then completely regenerated in a stripper before it is sent back to the scrubber. The high-pressure flash gives a gas containing a considerable amount of methane, carbon monoxide, and hydrogen. This stream is routed to the suction side of the compressor. From the stripper and the low-pressure flash comes a clean CO_2 stream. A part of this stream is compressed, dried, and used as inert gas, while the excess is released to the atmosphere. The almost CO_2-free gas is then fed to four catalytic adiabatic methanation reactors connected in series. Between the reactors, the gas is cooled down by generating and superheating steam, which is consumed internally in the process. This system with four reactors is complex and critical for the quality of the end product, biomethane, and the efficiency of the entire plant. This system has never before been used for gas from biomass gasification. After the last methanation reactor the gas is dried to achieve a dew point low enough to allow feeding of the gas to the natural gas grid. The drying is done by adsorbing the water on molecular sieves. The produced biomethane is fed to a medium-pressure transmission grid for natural gas.

10.4 THE ROLE OF GASIFICATION TECHNOLOGY IN THE FUTURE AS THE DEMAND FOR BIOMASS-BASED ENERGY AND FUEL GROWS

A change from a fossil-based fuel model to biofuels produced from forest residues demands large-scale and highly efficient production processes, where as much as possible of the chemically biomass-stored energy goes into the final biofuel. In this context the gasification technology has an advantage over all other known conversion technologies, which is the possibility to convert all types of biomass or wastes to the desired end product and make use of the losses that arise in the conversion to drive the process. This advantage will be most evident as the demand of biofuels grows and markets of side products produced in, for example, various proposed biorefinery schemes reach market saturation.

10.4.1 Different Gasification Process Options

The choice of gasification technology is to a large extent dependent on which technology that first was put into operation in significant numbers. The two main technologies for large-scale biofuel production are both based on fluidized bed technology, where one is the indirect type chosen in the GoBiGas Project and described earlier (Section 10.3.3) and the second is the oxygen/steam-blown pressurized direct gasifier.

In the latter the heat is produced within the gasifier by burning some of the fuel with the added oxygen, and in theory full conversion can take place in one reactor, instead of in two reactors. However, the plant is complemented with an air separation unit for the oxygen production and a combustion unit to incinerate various unwanted side flows and to produce the electricity needed in the process. In relation to efficiency there are no theoretical differences between the two gasification technologies. The differences is more in relation to the complexity of the process, where fuel feeding, ash removal, and maintenance are simpler in an atmospheric indirect gasifier than for the pressurized direct gasifier, whereas the opposite is the case for the introduction of high-temperature filtration, secondary catalytic conversion of hydrocarbons, and pressurizing of the produced gas to the required pressure of the downstream catalytic process. Due to the smaller number of process units for an indirect system, the indirect gasifier has also an advantage at smaller scales (up to 100–150 MW fuel input). However, a commercial biofuel production unit will most likely always be larger than 100 MW, so in this context this advantage can be disregarded. So which of the two technologies that is best will depend on the experience and belief of the supplier and the buyer of the gasifier.

10.4.2 Investment and Milestones

The total cost of the production unit is approximately €135 million. The experiment also involves a period of seven years of operations during which add accumulated operating losses and accumulated calculated costs of capital. The total aggregated cost of the experiment is accordingly SEK 1.3 billion. The whole project, including the three-year building phase and the seven-year operations phase, is necessary to produce the required information regarding the feasibility and the viability of the technologies used in the GoBiGas Project. The development and building phase will produce know-how regarding technical solutions. The operations phase is necessary to prove the efficiency and sustainability of the production process.

RECOMMENDED ACADEMIC REFERENCES FOR FURTHER READING

[1] "Bio-SNG from Thermal Gasification – Process synthesis, Integration and Performance." Chalmers Technical University. Gothenburg, Sweden. **2013**. ISBN 978-91-7385-538-0.

[2] D. Deublein, A. Steinhauser. "Biogas from Waste and Renewable resources: An Introduction". 2nd Revised and Expanded edition. **2013**. ISBN: 978-3-527-32798-0.

11

GREASOLINE: BIOFUELS FROM NON-FOOD MATERIALS AND RESIDUES

GEORG DAHMEN[1], PETER HAUG[1], GUNTER FESTEL[1,2], AXEL KRAFT[3], VOLKER HEIL[3], ANDREAS MENNE[3], AND CHRISTOPH UNGER[3]

[1]*Greasoline GmbH, Oberhausen, Germany*
[2]*Festel Capital, Fueringen, Switzerland*
[3]*Fraunhofer UMSICHT, Oberhausen, Germany*

Industrial Biorenewables: A Practical Viewpoint, First Edition. Edited by Pablo Domínguez de María.

11.1 FUELS AND CHEMICALS: NECESSITY OF RENEWABLES

11.1.1 Carbon Dioxide and Other Emissions by Transportation: An Opportunity for Biofuels

11.1.1.1 ***Ground Transportation*** CO_2 originating from fossil fuel consumption in the transport sector is one of the main drivers of climate change. According to GIZ, a German government funding organization for international cooperation,[1] the transport sector is responsible for 27% of all energy-based CO_2 emissions across the world. Politics are therefore increasingly requesting industry to research and develop clean and sustainable solutions. In the case of automotive gasoline and diesel, the European Union, North and South America, as well as Asian countries have introduced mandatory blending of biofuels to fossil fuels. Next to climate change major drivers for biofuels are the generation of new jobs and income diversification in rural areas. For many countries, including the United States, a reduction of the dependence on fuel imports and the simultaneous improvement of their trade balance are important incentives as well.[2]

Currently available biofuels, in particular the first-generation biofuels biodiesel and ethanol, come with performance disadvantages. Both lower the energy density of the blended fuel and result in so-called blend walls due to unfavorable fuel characteristics. Those incompatibilities demand special blending facilities, dedicated fuel blending standards, and different materials of construction for feeding and storing biofuel blends in cars. Additionally, a separate refinery infrastructure for pipelines and tanks is required. Particularly first-generation biodiesel is challenging since a whole range of biobased and plant-dependent components, unlike in ethanol for gasoline, are acting sometimes in a not predictable long-term manner on the engine and its components like the carburetor and the control system. Moreover, processing of biodiesel in a refinery can be a risky endeavor when typical multipurpose pipelines are employed for diesel and jet fuel. A prominent example is the severe incompatibility of biodiesel blends with jet fuel. According to the jet fuel standards ASTM-D1655–12 and ASTM-D7566-14a already a contamination of 5 ppm of biodiesel (FAME) will render jet fuel out of spec.[2a,b]

11.1.1.2 ***Aviation Industry*** The aviation industry is the fastest growing sector in transportation growing globally with about 4.5% per year. EU27 alone consumes 53 million tons of jet fuel per year. Global consumption accrues to more than 200 million tons annually. The European Union has set the goal to replace 2 million tons per year of jet fuel by sustainable jet fuel until 2020. The voluntary self-commitment of global airlines aiming for climate-neutral growth provides for additional driving forces. Starting in 2020 the aviation industry wants to grow climate neutral, which means additional fuel demand has to be covered by renewables. Additionally, it is envisioned to curb carbon dioxide emissions by 50% in 2050 compared to 2005.[3]

Carbon dioxide certificates required for jet fuel in the European Union in the future will put additional pressure on the aviation industry. The recent decision to eliminate

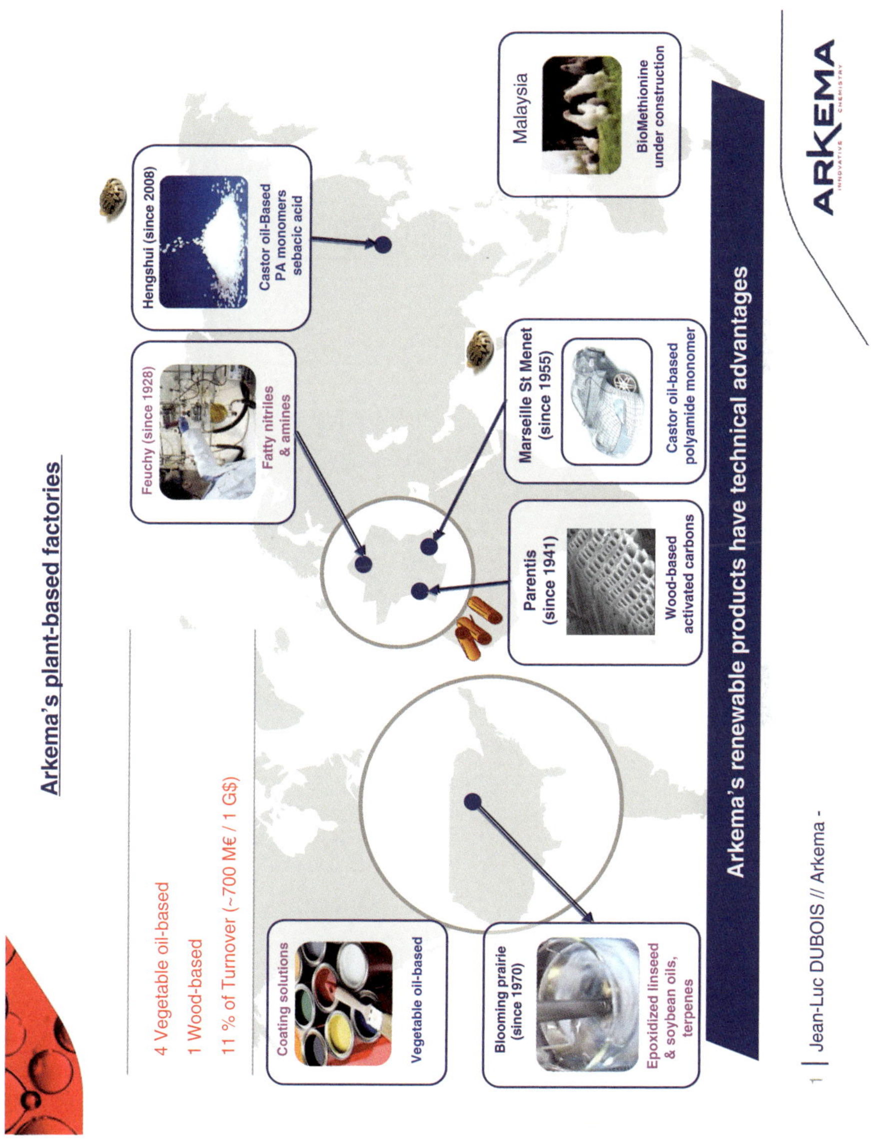

Figure 3.1 Overview of the Arkema biorefineries.

Industrial Biorenewables: A Practical Viewpoint, First Edition. Edited by Pablo Domínguez de María.

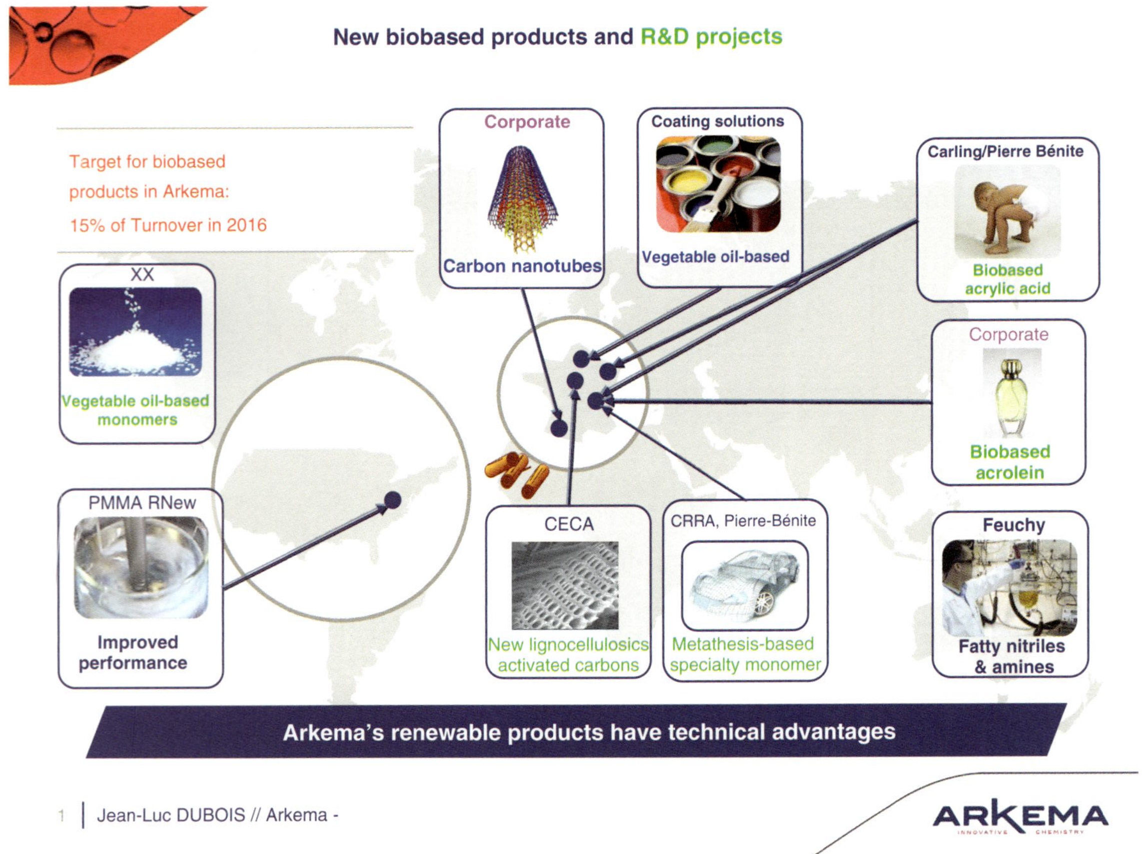

Figure 3.2 Biobased products at Arkema.

Figure 3.3 *Ricinus communis* plant. Source: Copyright by Dr. Jean-Luc Dubois.

Figure 6.1 International Petroleum Investment Company (IPIC).

Figure 6.10 Used frying oil and two hydrotreated products.

Figure 6.19 Coconut, camelina, and UFO.

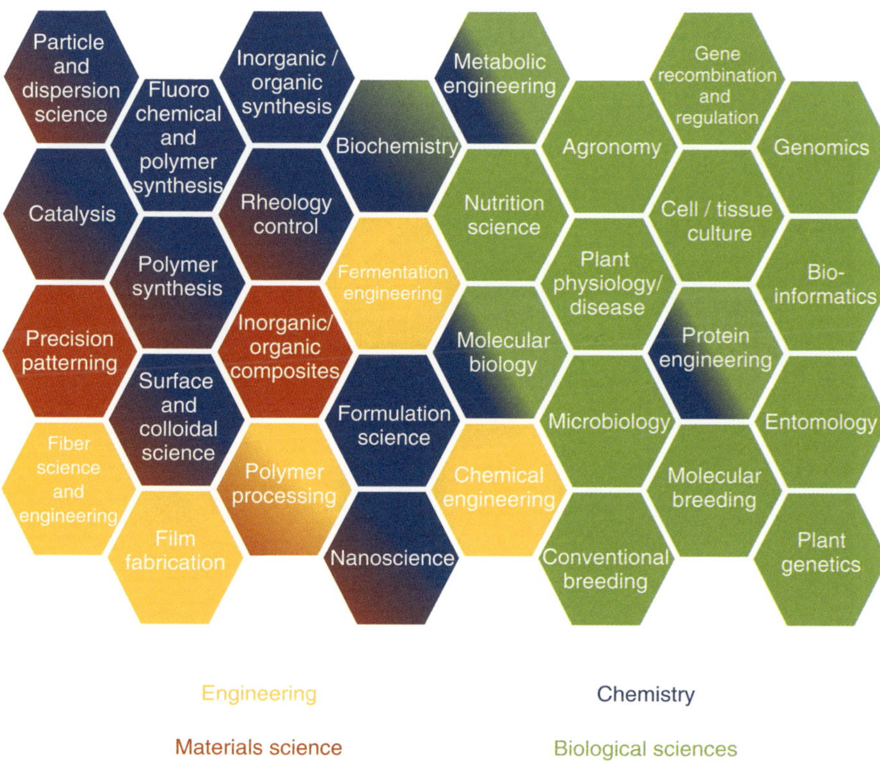

Figure 7.2 DuPont's scientific competencies. This breadth and depth in a variety of disciplines enables successful development of industrial biorenewables.

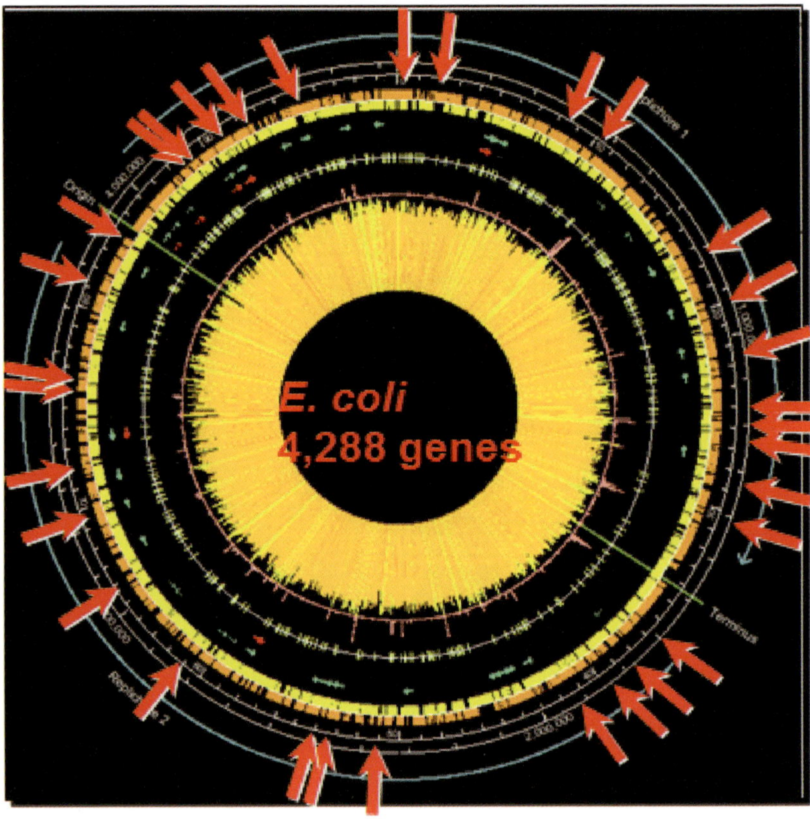

Figure 7.4 Alterations made in the *E. coli* genome in the course of making the PDO biocatalyst. Changes were made to glucose transport, glycerol metabolism, glycolysis (Entner–Doudoroff, pentose phosphate, and tricarboxylic acid cycles), respiration, amino acid biosynthesis pathways, anaplerotic reactions, and global regulators. Not shown – the plasmid containing the pathway genes.

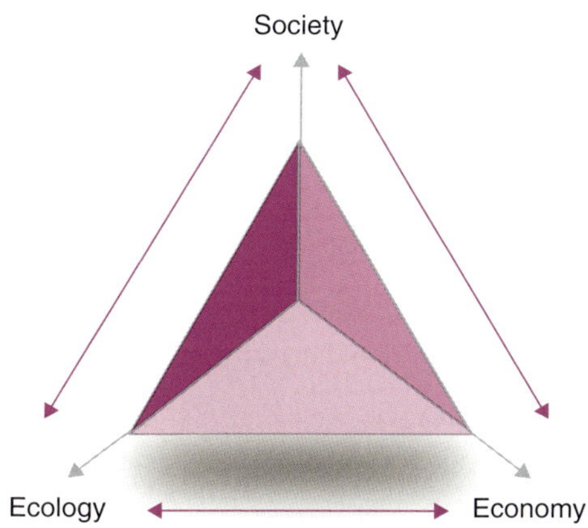

Figure 8.2 Each nation, institution, and company need to have a sustainability or CR strategy for keeping the balance of all three sustainability aspects in order to keep mankind's well-being on earth.

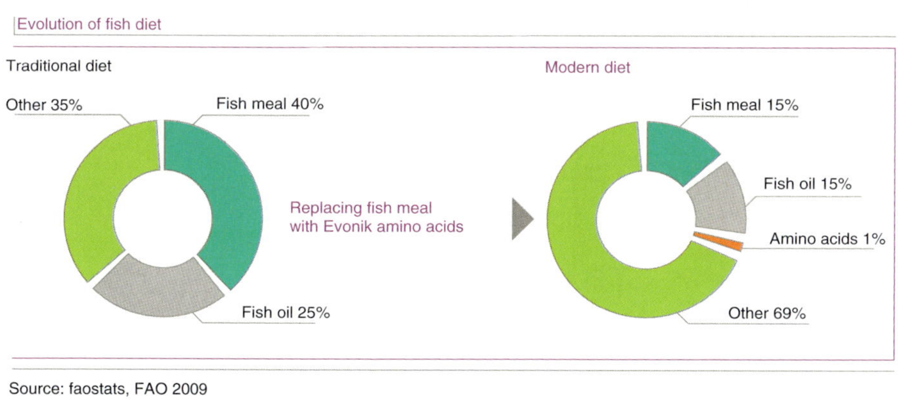

Figure 8.5 Traditional aquaculture requires up to 3 kg of wild-catch fish, which is fed in form of fish meal. Costly and scarce fish meal can be replaced by plant protein sources and amino acids. Supplementing amino acids effectively reduces fish meal requirements from 40% to only 15%.

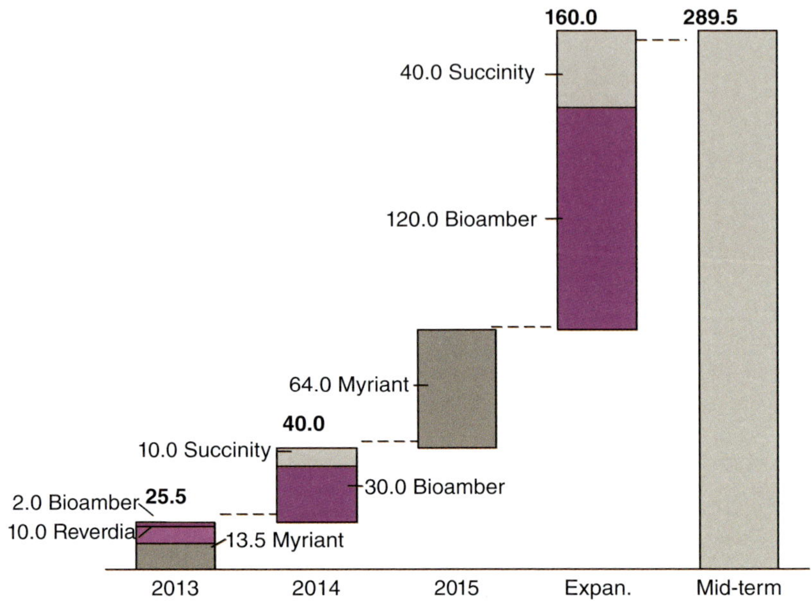

Figure 8.11 Global biobased succinic acid capacity.

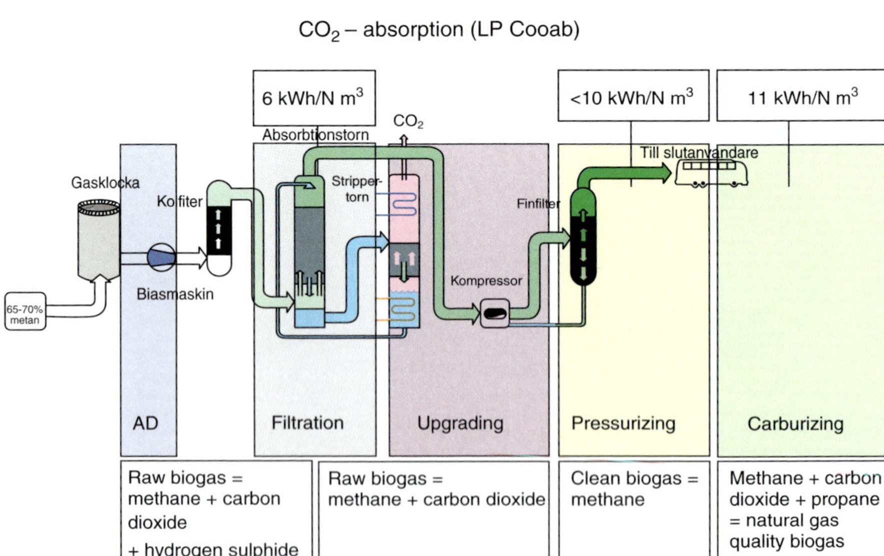

Figure 10.1 The Gasendal process.

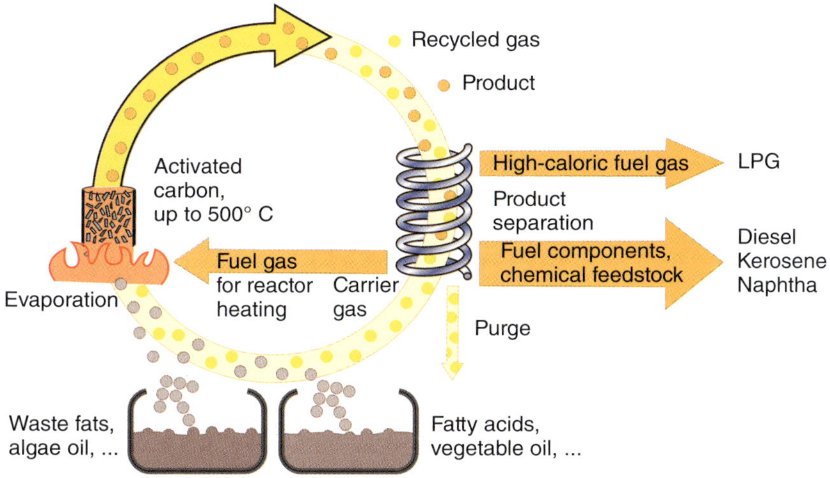

Figure 11.1 Simplified flow sheet of the Greasoline® technology.

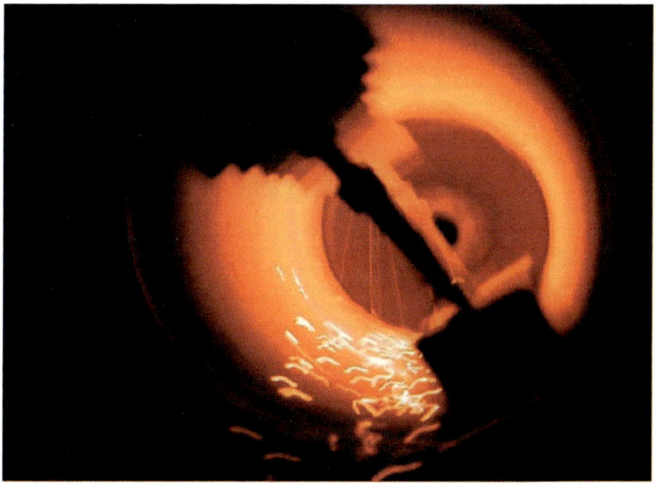

(b)

Figure 11.4 Pilot rotary kiln for catalyst reactivation. (Outside (a) and inside (b) view of kiln.)

Figure 12.1 Pretreatment of rice husk (a) to fine sieved powder (b) via milling prior to processing.

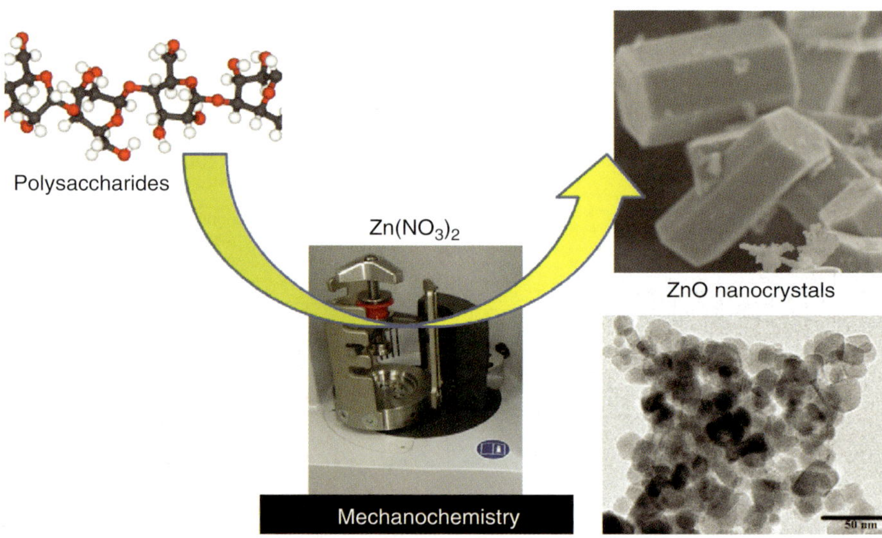

Figure 12.3 Mechanochemical production of designer nanomaterials from various polysaccharides. Source: from[31].

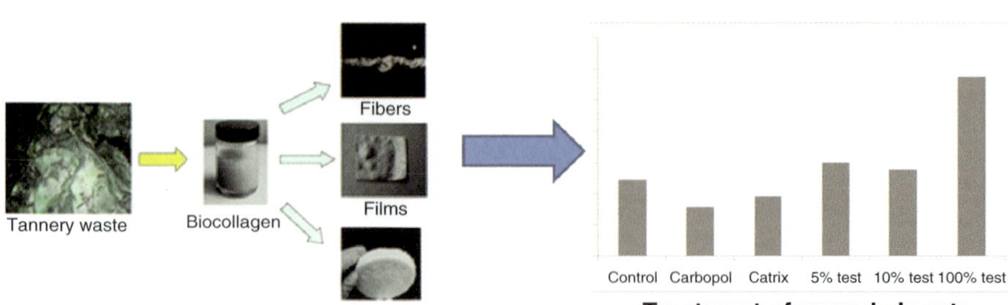

Figure 12.4 From slaughterhouse/tannery waste to biocollagenic polymers with applications in tissue regeneration.

Figure 14.7 Mixotrophic growth and biomass productivities of HS26 in a 100,000 L open reactor. Harvesting was made semicontinuously with 20–80% daily harvest throughout a 29 d of stable operation. The results reflect typical values obtained in production.[29] Source: Reproduced with permission of Heliae Development LLC.

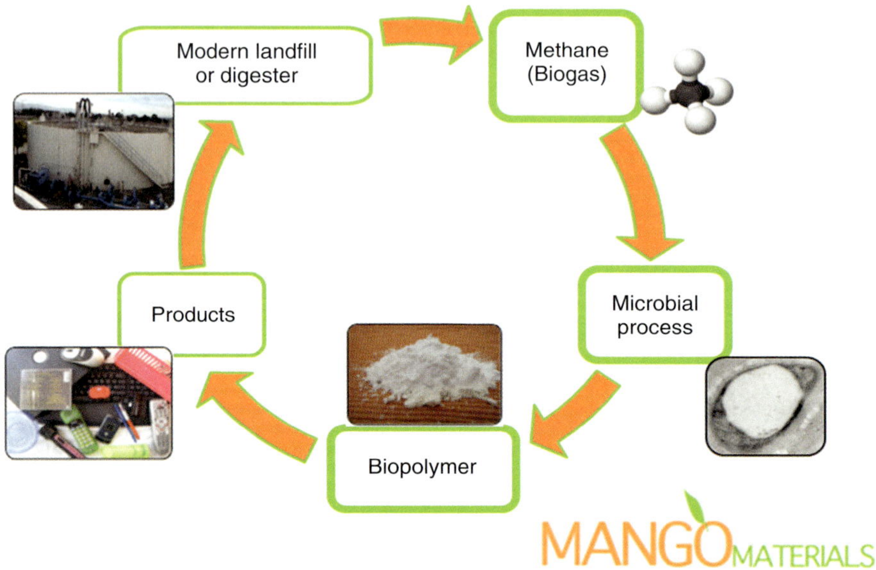

Figure 16.1 The closed-loop system of Mango Materials.

Figure 18.1 Novozymes overview.

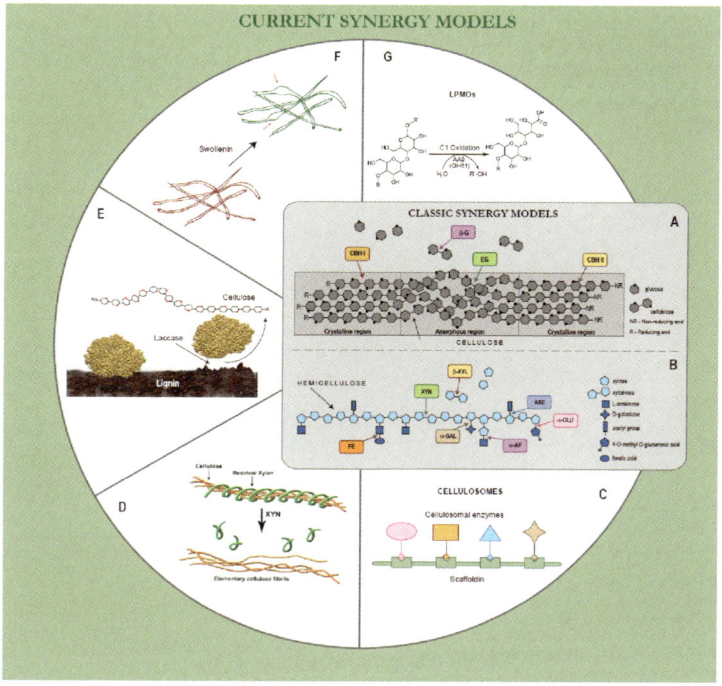

Figure 18.2 Enzymatic synergies in plant biomass deconstruction. Panel a – schematic representation of the enzymatic hydrolysis of cellulose by synergizing cellobiohydrolases (CBH I and CBH II), an endoglucanase, and a β-glucosidase; Panel b – schematic representation of the enzymatic hydrolysis of hemicellulose by synergizing xylanase, β-xylosidase, and xylan-debranching enzymes; Panel c – illustration of a typical bacterial cellulosome; Panel d – illustration of the increase of the accessible cellulose surface area in biomass as a result of hemicellulose by an endo-β-1,4-xylanase; Panel e – illustration of the release of a nonproductively bound to lignin cellulase by a laccase; Panel f – effect of a swollenin on cellulose microfibril bundles; Panel g – oxidation of the anomeric carbon by LPMOs. List of abbreviations used in the figure: NR – nonreducing cellulose end; R – reducing cellulose end; CBH I – an R-end-acting exo-β-1,4-glucanase (EC 3.2.1.176), the *Hypocrea jecorina* (formerly *Trichoderma reesei*) Cel7A (formerly cellobiohydrolase I or CBH I); CBH II – an NR-end-acting exo-β-1,4-glucanase (EC 3.2.1.91), the *H. jecorina* Cel6A (formerly cellobiohydrolase II or CBH II); EG – an endo-β-1,4-glucanase (EC 3.2.1.4); β-G – a β-glucosidase (EC 3.2.1.21); XYN – an endo-β-1,4-xylanase (EC 3.2.1.8), which cleaves the main xylan backbone; β-XYL – a β-D-xylosidase (EC 3.2.1.37), which hydrolyses the released xylooligosaccharides to xylose; α-AF – an α-L-arabinofuranosidase (EC 3.2.1.55), active on arabinans; α-GAL – an α-D-galactosidase (EC 3.2.1.22), active on nonreducing α-d-galactose residues; α-GLU – an α-D-glucuronidase (EC 3.2.1.131), active on α-D-1,2-(4-O-methyl) glucuronosyl links; FE – a feruloyl esterase (EC 3.1.1.73), which releases ferulic acid by cleaving ester bonds; AXE – an acetylxylan esterase (EC 3.1.1.72), which catalyzes the release of acetyl groups; LPMOs – lytic polysaccharide monooxygenases (formerly, GH61); AA9 – CAZy auxiliary activity family 9 comprising LPMOs; GH61 – CAZy glycoside hydrolase family 61; R′, R – glycosyl groups; C1 – anomeric carbon. (Illustration by Vera M. Gutman.)

Figure 21.1 Shading in a photobioreactor.

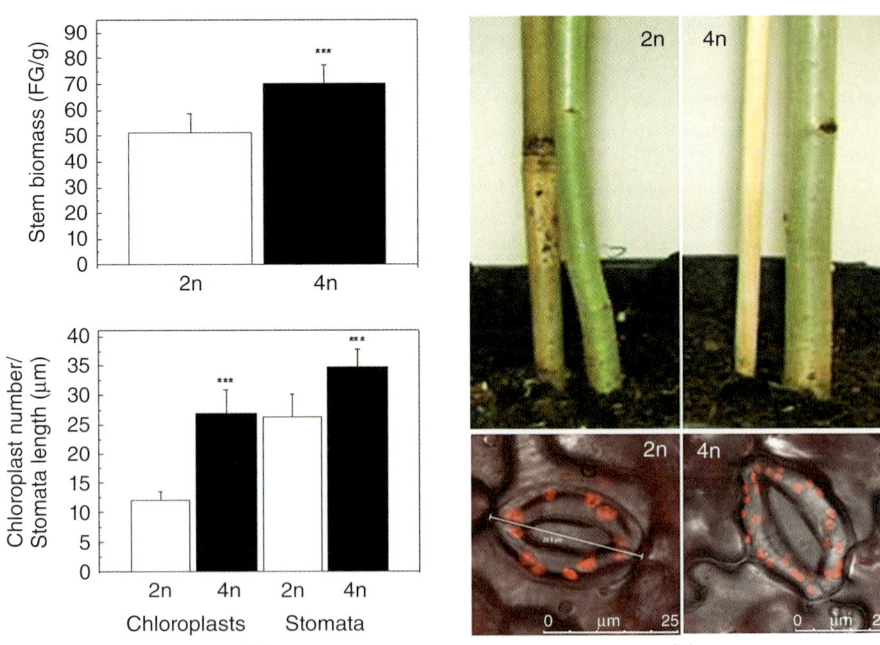

Figure 22.6 Enhanced biomass of tetraploid poplar plants derived from protoplast fusion. (a) New poplar line (tetraploid, 4*n*) shows significant increases in stem biomass, chloroplast number, and stomata length compared to diploid (2*n*) poplar. (b) Corresponding pictures of the plant materials. Source: Pictures and data provided by Dr. T. Teichmann, University of Goettingen

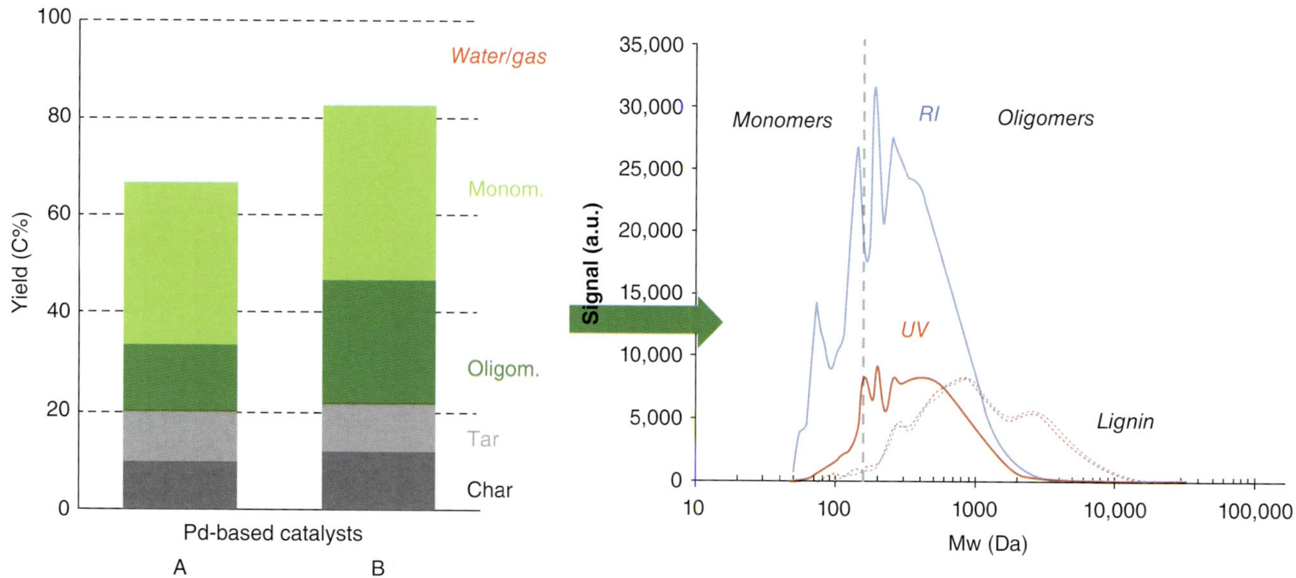

Figure 23.13 Acidic hydroliquefaction of lignocellulose (acetic acid as cosolvent, pH <2200–250°C, 40–100 bar H_2).

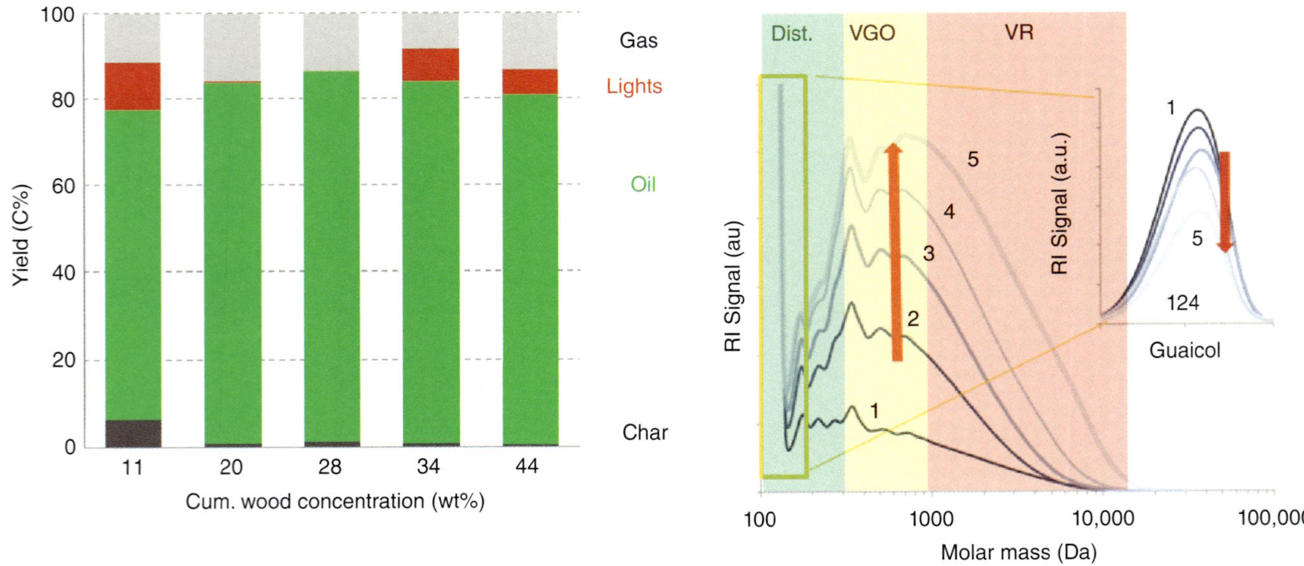

Figure 23.14 Liquefaction of lignocellulose in phenolic medium (30 min at 300 °C). *Source:* Source: Adapted from[97]. Reproduced with permission of Wiley.

a 4.7% import duty from January 2014 on will further increase the trend to jet fuel imports.

Those recent changes may provide an opportunity for biojet fuel technologies capable to source raw materials globally and make use of local production facilities and supply chains. More stringent quality requirements for sustainable jet fuel and the 5–10 times smaller demand of jet fuel compared to diesel will probably imply a higher product price compared to fossil product for new stand-alone production units churning out sustainable jet fuel.

11.1.1.3 Shipping Industry Environmental protection requirements for marine fuels – also called bunker fuels – are increasing constantly, especially with regards to sulfur- and nitrogen-based harmful substances, next to soot generating polyaromatic hydrocarbons (PAH). Those efforts are partially driven by the internationally agreed MARPOL regulations (Annex VI) as well as by national European and American legislation. New legislation taking effect in January 2015 in Europe will cause a shift in bunker fuels usage and sales. Bunker fuel market participants need to prepare for a cut in the fuel sulfur content from 1% to 0.1% by weight in *Sulfur Emission Control Areas* (SECAs) in Northern Europe and North America.[4]

Slightly different requirements of bunker fuel for cetane number, density, and viscosity compared to diesel for automotive applications according to EN590 or ASTM D975 open up more options for blending and product diversification for green diesel. Hence, biofuels capable of partially replacing clean fossil diesel or supplementing the market will turn out to be a viable alternative in the near future. Clean Marine Gas Oil (MGO) and the heavier and more viscous Marine Diesel Oil (MDO) will gain market share in the SECAs. Typically MGO and MDO sell for about US$250–300/t more than high-sulfur 380 and 180 cst fuel oil. Pricing of MGO and MDO in the European Union is very similar to heating oil.[3,5]

11.1.2 Chemical Industry

The shale gas boom in the United States and the focus to switch from expensive naphtha to cheap ethane to produce ethylene has recently also triggered unusual technology development to combat the lack of commodity alkenes other than ethylene. Several sources (TOTAL, IFP, University of Gent) report the hydroprocessing of fats and oils to long-chain alkanes that can replace naphtha for steam cracking in order to generate the commodity alkenes, ethylene, propylene, butenes, butadiene and benzene, toluene, and xylene.[6]

11.2 EVOLVING MARKETS FOR GREASOLINE® TECHNOLOGY

The Greasoline® biofuel technology has been developed to produce fossil-like drop-in fuels for automotive diesel. It stands for the *Gas-phase heterogeneous Catalytic Cracking* (GCC) of biobased fats, oils, waste oils, and oil residues into fossil-like drop-in diesel and naphtha/kerosene fractions. The technology offers significant advantages since a flexible range of raw materials, not being in competition with food, can be employed. The technology's main focus is on raw materials and residues, for example, from food processing, that are not suited for alternative fuel technologies due to either impurities, the degree of oxidation, or the content of free fatty acids. Additionally many of the aforementioned raw materials need to be processed on a relative small scale of several ten thousand tons per year due to their limited availability. Competing technologies, in particular hydrotreated vegetable oils (HVO), require several hundred thousand tons capacity per year to be profitable. The Greasoline technology is adaptable to future raw materials like algae oils and other toxic or low-quality feedstock like jatropha or macauba oil among others. Greasoline product can also serve as blendstock for blends of biodiesel and HVO with fossil fuels.

The upcoming demand for aviation fuels can be addressed by Greasoline technology, since mainly diesel boiling in the kerosene range is yielded. Depending somewhat on the type of oils and fats fed to the process, a certain degree of alkylated aromatics is produced. Those chemicals are particularly important for blending with most currently available biobased jet fuel like *hydroprocessed esters and fatty acids* (HEFA), which do not contain any of those components required for swelling of seals in the airplane fuel system. To limit the jet fuel price, the technology is ideally implemented into or near by a refinery and operated in a mode producing sustainable diesel and jet fuel.

Especially in the case of jet fuel, a whole industry is working on competitive drop-in technologies. Till now technologies producing biojet fuel are too expensive due to expensive raw materials or expensive conversion technology. Time till 2020 – when the self-commitment of the aviation industry enters into force – is rather short. Therefore the pressure on the whole value chain increases.

Furthermore, Greasoline can not only deliver a green and clean blending component for MDO and MGO in SECAs. Since the Greasoline technology allows simultaneous deoxygenation and cracking, the resulting mixture of naphtha and alkane-rich diesel can also serve as an additional supply partially replacing fossil naphtha.

11.3 TECHNOLOGY OVERVIEW GREASOLINE®

The Greasoline technology has been developed by the Fraunhofer Institute for Environmental, Safety, and Energy Technology UMSICHT in Oberhausen, Germany, and is exclusively commercialized via Greasoline GmbH, Oberhausen, Germany. The technology is currently operated in semiautomated pilot scale generating an output of 2–3 L crude liquid product per hour. The next step in scaling-up the technology will be to build a fully automated demonstration unit including catalyst recycling. Alternatively a direct implementation of the Greasoline process into existing fuel value chains of a refinery is discussed with value chain partners. Inexpensive activated carbon serves as catalyst, which is high performing, available in more than sufficient quantity and can be recycled after use or reactivated for other applications.

Compared to alternative technologies feeding on fats and oils, in particular HVO, the following points differentiate Greasoline from related technologies:

- It achieves high cetane numbers together with low CFPP, good lubricity, and a higher density compared to HVO.
- It offers a flexible selection of sustainable raw materials.
- It is already economical starting at a size of several 10,000 t annually.
- It operates at atmospheric pressure.
- It employs activated carbon as an inexpensive, recyclable catalyst (not depending on refining catalyst suppliers).
- It allows exhaustive deoxygenation to diesel without the addition of hydrogen.

The technology comprises several steps. Raw materials are evaporated and typically require additional superheating to reach reaction temperature. The degree of superheating and the evaporation temperature depends on the composition, in particular on the degree of free fatty acids, which start boiling in the range of 300–400 °C. In case of a low percentage of free fatty acids, partial decomposition of the feedstock takes place in the evaporator without leading to a significant degree of deoxygenation. Steam can be employed additionally to facilitate evaporation and in order to prevent the system from building up coke or oily layers. Reaction temperatures are between 400 and 500 °C. The deoxygenated product mixture leaving the activated carbon catalyst bed is quenched by cold water in the pilot plant.

Optionally the gaseous product stream can be incinerated to preheat the feed and a part of the gaseous product stream can be recycled. Designs for a retrofit into a refinery may differ. The catalyst recycle unit and purification steps are not shown in the following simplified sketch of the core technology in Figure 11.1.

Crude fuels mixture may require additionally posttreatment steps typical for cracked process streams in a refinery to adjust the oxidative stability and the sulfur content. This depends to a certain degree on the biobased raw material, residue, or waste stream. The liquid yield of Greasoline technology products boiling in the diesel and kerosene boiling range is about 55–65% on average plus about 5–15% boiling in the naphtha range. The remainder is comprised of the LPG fraction, water as well as carbon mono- and dioxide and coke. The suitability of new raw materials can be assessed by a simple test on small scale.

11.4 DESCRIPTION OF BUSINESS MODEL

International patents protect the Greasoline technology. For the basic technology a European patent has been granted and been nationalized in 12 EU member countries. Further patent applications have been filed and are granted or pending.[7]

Governmental funding, in-house efforts, and partner companies' efforts total several million €, which have been invested for developing and building a pilot plant system and for the generation of analytical and raw material-related process know-how. Figures 11.2–11.4 show different parts of the machinery. Revenues

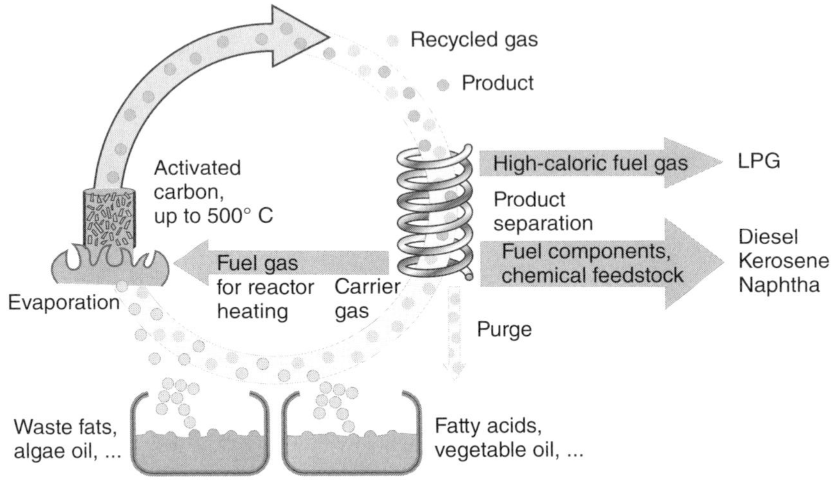

Figure 11.1 Simplified flow sheet of the Greasoline® technology. *(See color plate section for the color representation of this figure.)*

Figure 11.2 Lower part of pilot plant showing feed system, preheater, evaporator, and product quench.

for Greasoline GmbH are generated by licensing the technology and by providing services for development of an adapted technology integrating new raw materials or for retrofitting the technology to refineries together with the respective value chain partners and third party investors. Envisioned markets are diesel and jet fuel, with naphtha as outlet for any low boiling side stream.

Services provided by Greasoline GmbH are:

- Process development and generation of data for basic engineering of new or retrofitted plants with new or existing biobased raw materials together with engineering partners.

Figure 11.3 Pilot atmospheric distillation.

- Analytical service together with Fraunhofer UMSICHT and external partners.
- Provision of crude and refined product samples on 0-, 2-, or 10-L scale (per day).
- Catalyst screening, recycling, and related product optimization on lab and pilot scale.

(a)

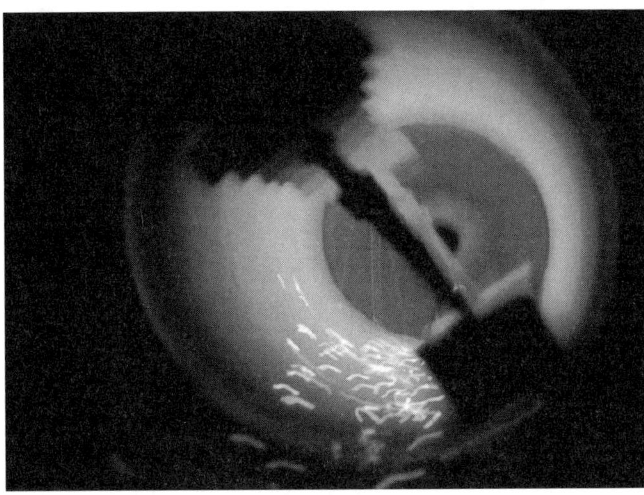

(b)

Figure 11.4 Pilot rotary kiln for catalyst reactivation. (Outside (a) and inside (b) view of kiln.) (*See color plate section for the color representation of this figure.*)

11.5 DIESEL FROM DIFFERENT RAW MATERIALS

The accessible feedstock supply for Greasoline production plants totals an estimated global volume of 50 million tons per year in the medium and long term. Suited raw materials and their origin are depicted in Figure 11.5.

Since residues and waste streams are chemically not well defined, the following section will only focus on a few suited Non-food feedstock with a more or less defined chemical composition.

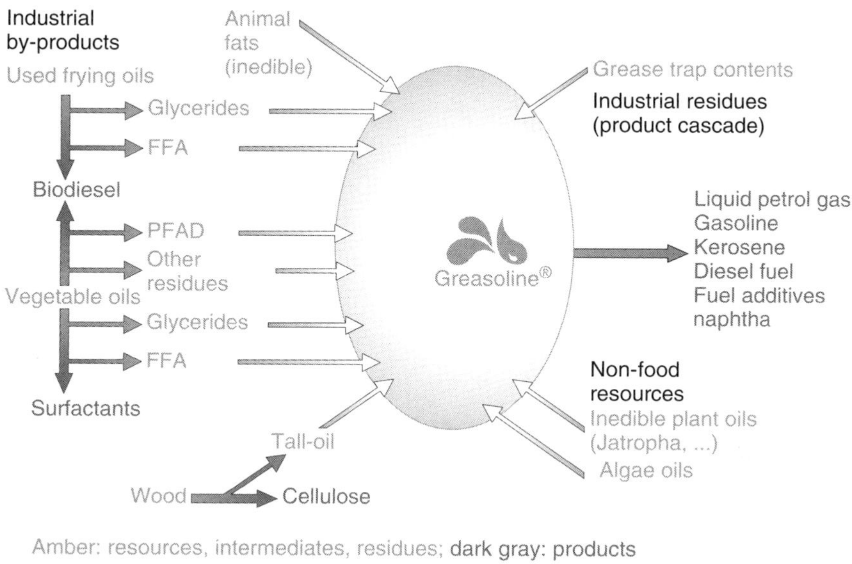

Figure 11.5 Greasoline: Suited raw materials and possible products.

11.5.1 Palm Fatty Acid Distillate: Existing Non-food Feedstock

The estimated availability of *palm fatty acid distillate* (PFAD) in Indonesia and Malaysia alone is around 2.5 million tons per year. The amount of available PFAD accrues to about 4-6 wt% of palm oil. The production of palm oil has been increasing during the last years by 5% annually (USDA, LEUPH) due to an increased demand for frying oils, surfactants, candles, and cosmetics and reached about 60 million tons in 2014.[8]

PFAD is a suited residue stream for Greasoline. However, since PFAD is also used in the manufacture of quality soaps, animal feed, as well as for HVO and biodiesel, it is risky to bank on it alone. The typical chemical composition of PFAD[6] is shown in Table 11.1.

11.5.2 Jatropha Oil: Partially Commercialized Non-food Feedstock

Jatropha curcas L. is an excellent oil feedstock for renewable energy that can be grown in Southeast Asia, Africa, and Latin America. The black jatropha kernels consist of about 40–50% oil, 20–25% protein, and 25–35% carbohydrates including fibers. Oil yields up to 1.5 t/ha are achieved, for example, in Indonesia. The annual yield of jatropha kernels varies globally depending on irrigation and cultivation management and type of jatropha. It is considered the next-generation Non-food energy crop because it can be grown on land of lesser agricultural value with low irrigation requirements. The oil is not edible since it contains toxic phorbol esters. Jatropha plants are much more tolerant to moderate climate conditions than palm

TABLE 11.1 Chemical Properties of PFAD (e.g., Malaysia).

Free fatty acid (%) (as C16:0)	72.3–89.4
Moisture content (%)	0.05–0.15
Unsaponifiable matter (%)	1.5–3.4
Fatty acid	(wt%) (mean values)
12:0	0.2
14:0	1.2
16:0	47.1
18:0	4.5
18:1	36.6
18:2	9.6
18:3	0.5

oil trees. Major sources of jatropha plants are found not only in countries in Southeast Asia like in Indonesia, India, China, Thailand; in Latin America like in Mexico, Brazil, and Ecuador; but also in African countries like Ethiopia and Ghana. Jatropha can typically not be grown in most parts of Europe and the United States; however first projects are launched to cultivate jatropha in the Mediterranean region. Jatropha attracts already for several years the interest of multinational companies coming from developed countries like Daimler, Bayer, and ADM, among others including many governmental organizations.[9]

Jatropha comes in many different phenotypes, which do have different traits and performance characteristics, including nontoxic varieties in Mexico, see Figure 11.6 for examples.[10]

Uses of jatropha oil include conversion to first-generation Biodiesel or hydroprocessing to HVO. Alternatively, the neat oil is used as lamp oil and as fuel for rural electricity generators.[11] The by-product utilization of the toxic press cake is still under investigation. One of the major issues for overall value generation to be tackled is the detoxification of the protein-rich press cake to make use of it as animal feed or as intermediate for value-added industrial products. Currently biogas seems to be the best option to fully utilize the press cake. Hulls and husk serve as fuel. According to FAO a very important factor for making food and fuels working together is to plant jatropha in so-called intercropping systems. This implies to plant jatropha together with plants like maize, beans, peas, cassava, or even bananas and vanilla. Since different crops have different needs in terms of inputs – water, soil, nutrients – and do react differently under the influence of other crops, potential synergies or competition will vary.[12]

Table 11.2 depicts the chemical composition of preprocessed jatropha oil used to make Greasoline diesel.

Table 11.3 shows the properties of a diesel fraction generated in the pilot plant in Oberhausen. The low boiling fraction has been removed by continuous atmospheric distillation until the boiling point curve and the flash point reached the required minimum values according the European standard for automotive Diesel fuels. The diesel fraction was clay treated to remove color and traces of polar substances.

Figure 11.6 Different phenotypes of jatropha.

Unlike for straight HVO, a low CFPP and a high cetane number as well as higher density and an in-spec viscosity are achieved. Not surprisingly a posttreatment is necessary to adjust the limited oxidation stability. Such a processing step is also necessary for catalytically cracked fossil feedstock in a refinery to adjust the amount of olefins. It should be noted here that the corresponding diesel standard to EN590:2013:12 in the United States, ASTM D975-11b, is less demanding in terms of requirements for cetane number, density, flash point, distillation curve, viscosity, and sulfur open up different regimes for blending.

The following Figures 11.7 and 11.8 are meant to illustrate the physical and chemical characteristics of the crude water quenched product mixture leaving the pilot plant reactor.

TABLE 11.2 Chemical Properties of Preprocessed Jatropha Oil for Greasoline. Physical and Chemical Properties of Jatropha Oil (Extracted by Hot Water Extraction with Aqua High® Technology).

Free fatty acid (%) (as C16:0)	0.5
Moisture content (%)	0.75
Fatty acid composition	(wt%)
16:0	14.4
16:1	1
18:0	6.4
18:1	40.5
18:2	37.1
18:3	0.2
20:0	0.2

Figure 11.7 reveals that about 75% of the quenched product belongs to the diesel fraction. The rest boils in the gasoline, respectively, naphtha boiling range.

It becomes evident from Figure 11.5 that *n*-alkanes, next to minor quantities of alkenes, show up as the main reaction products. The low boiling components display a typical spectrum of cracked naphtha components. Hence they can be employed as feed for a steam cracker to generate monomers like ethylene, propylene, butenes, butadiene and benzene, toluene, and xylene.

The GC-MS-derived results in Figure 11.5 also reveal that a breakthrough of oxygenated components, in particular ketones, can happen if the catalyst is close to the end of its lifetime. Ketones as such do not disturb diesel properties. However, extended catalyst use will lead to undesired fatty acids in the product mixture compromising the product quality. Different feedstock and process conditions allow boosting alkylated benzenes boiling in the diesel and kerosene range.

11.5.3 Macauba Oil: Not Yet Commercialized Non-food Feedstock

Macauba oil originates from a tree belonging to the same family as oil palm trees. The plant grows up to 15–20 m in height. Its fruits are yellowish green and contain a single, dark brown, nut-like seed. Harvesting is done similar to palm oil. The macauba oil contains a high degree of free fatty acids (over 35%). Macauba is widespread throughout Central and Latin America and often grows on poor soils. Its tolerance to drought makes it a suitable species for oil production in regions that are too dry for the African oil palm and coconut palm; the fruit yield can exceed 6 t/ha. Macauba can develop also in poor, salty, sandy, or rocky soils, although the best growth takes place in fertile soils. Wet–warm conditions with temperatures above 24 °C are required for optimal growth. The residue from the extraction, the "macaúba pie," can be used as organic fertilizer or as animal feed for cattle, goats, and sheep. Based on the chemical composition and the high degree of free fatty acid and partial glycerides found in the oil, it is a promising future feedstock for Greasoline.[13]

TABLE 11.3 Properties According to EN590 for Neat Jatropha Oil-Based Greasoline Diesel. Greasoline®-Derived Diesel Produced from Jatropha.

		Limits EN 590:2013–12				
Property	Method	Min.		Max.	Result	UNIT
Color	Sensoric				Clear, yellowish	–
Odor	Sensoric				Diesel-like	–
Cetane number	DIN EN 15195	51		–	68,5	–
Cetane index	DIN EN ISO 4264	46		–	65,9	–
Density at 15°C *	DIN EN ISO 12185	820 ●		845	800,1	kg/m3
Polycyclic aromatic hydrocarbons (PAH)	DIN EN 12916	–		8	0,5	% (m/m)
Sulfur Content	DIN EN ISO 20846	–		10	<5	mg/kg
Flashpoint	DIN EN ISO 2719	> 55		–	71	°C
Carbon Residue (10% D.)	DIN EN ISO 10370	–		0,3	0,05	% (m/m)
Oxid ash	DIN EN ISO 6245	–		0,01	<0,005	% (m/m)
Water content Karl-Fischer	DIN EN ISO 12937	–		200	42	mg/kg
Total contamination	DIN EN ISO 12662	–		24	–	mg/kg
Copper strip corrosion	DIN EN ISO 2160	Class 1		Class 1	Class 1	–
Oxidation stability (Rancimat)	DIN EN 15751	20 ●		–	0,9	h
Oxidation stability (Petrooxy) **	ASTM D 7545-09	70 ●			8,7	Min
HFRR (Lubricity) at 60°C ***	DIN EN ISO 12156-1	–		460	291	μm
CFPP (Cold Filter Plugging Point) ****	DIN EN 116	–			-16	°C
% (V/V) recovered at 250°C	DIN EN ISO 3405	–		< 65	43,2	% (V/V)
% (V/V) recovered at 350°C	"	85		-	>98	% (V/V)
95 % recovered at	"	–		360	317,4	°C
Kinematic viscosity (40 °C)	DIN EN ISO 3104	2		4,5	2,405	mm^2/s

* Fulfills arctic winter standard (>800); NesteOil: 770 – 790
** Not yet an accepted standard everywhere
*** NesteOil: 650 μm
**** Requirements CFPP depend on country and season: typically –20 to 0 °C; for nordic countries –32 to 0 °C

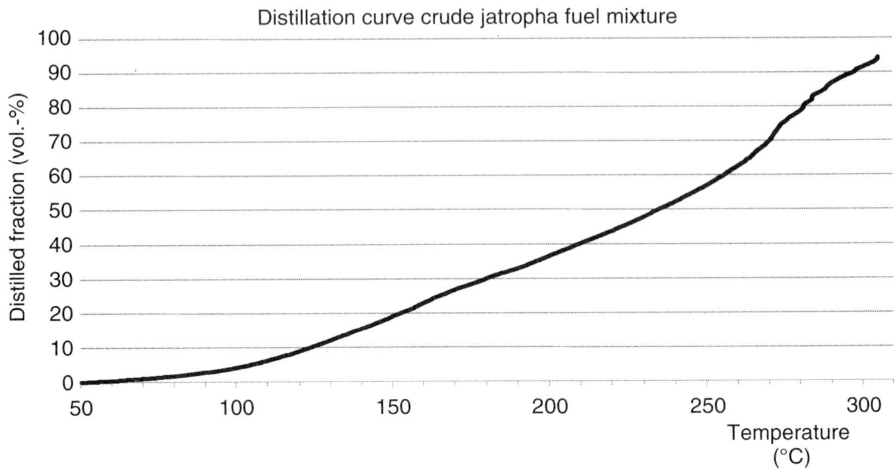

Figure 11.7 Distillation curve of water quenched crude product mixture in pilot plant.

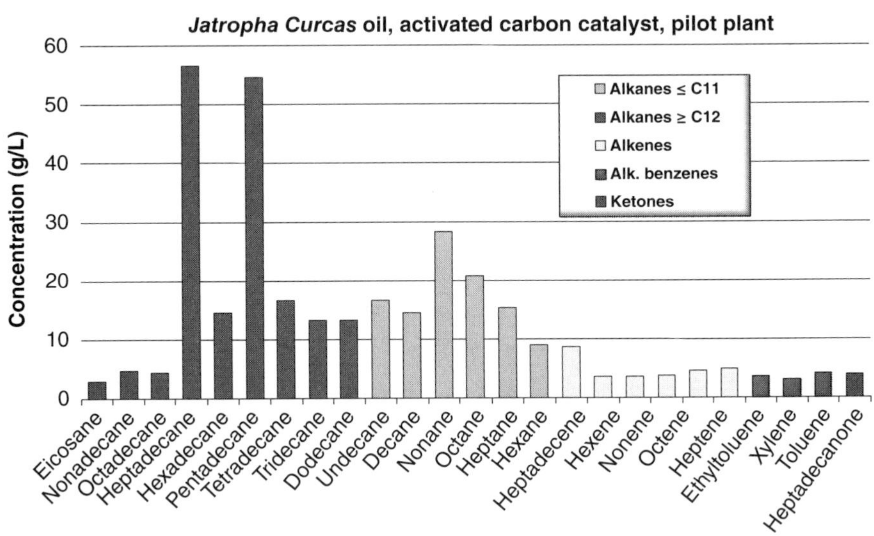

Figure 11.8 GC-MS: 25 biggest peaks of crude product mixture making up 41% by mass of total.

REFERENCES

[1] E. Hüttner, *Transport and Mobility,* http://bit.ly/1wgC0F4, accessed 25 Nov 2014.

[2] (a) ASTM D1655 – 12 Standard Specification for Aviation Turbine Fuels, http://bit.ly/1t62vCR, accessed 23 Dec 2014; (b) ASTM D7655 – 14a, *Standard Specification for Aviation Turbine Fuel Containing Synthesized Hydrocarbons*, http://bit.ly/1xGWei5,

accessed 25 Nov 2014; (c) EC, *European Advanced Biofuels Flight path Initiative*, http://bit.ly/1vJXrUV, accessed 25 Nov 2014; (d) IATA, *Fact Sheet: Climate Change*, http://www.iata.org/policy/environment/Pages/climate-change.aspx, accessed 10 Feb 2015.

[3] D. Elward, J. Fox, *EU to suspend jet fuel import duty from Jan 1, 2014*, http://bit.ly/15mGV28, accessed 25 Nov 2014.

[4] IMO, *Prevention of Air Pollution from Ships*, http://bit.ly/1wBRSVQ, accessed 25 Nov 2014.

[5] BWI, Bunker World Index, http://bit.ly/1GSqvfo, accessed 25 Nov 2014.

[6] (a) W. Vermeiren, N. Gyseghem, US 20140155665 A1, **2009**; (b) Q. Debuisschert, J. Cosyns, J. Chapus, D. Hudebine, US 20100292517 A1, **2010**; (c) W. Vermeiren, F. Bouvart, N. Dubut, US 20140114105 A1, **2009**; (d) T. Dijkmans, S.P. Pyl, M.-F. Reyniers, R. Abhari, K.M. van Geem, G.B. Marin, *Green Chem.* **2013**, *15*, 3064–3076.

[7] (a) C. Cinquemani, V. Heil, J. Jakob, A. Weber, EP1489157B1, **2004**; (b) V. Heil, E. Jelen, K. Keldenich, C.A. Unger, EP 1 724 325 A1, **2005**; (c) V. Heil, A. Kraft, W. Urban, C. Unger, K. Meller, U. Seifert, A. Menne, A. Fastabend, S. Rossow, WO 2010037756A3, **2010**; (d) C.A. Unger, A. Menne, A. Kraft, V. Heil, WO 2013053754A1, **2011**.

[8] (a) USDA, *Oilseeds: World Markets and Trade*, http://1.usa.gov/1AO202i, accessed: 25 Nov 2014; (b) A. Gapor, *Production and utilization of palm fatty acid distillate (PFAD)*, http://bit.ly/1vK1ZKN, accessed 25 Nov 2014.

[9] (a) V. Müller, *Archer Daniels Midland Company, Bayer CropScience and Daimler to cooperate in Jatropha biodiesel project*, http://bit.ly/1xSjCYs, accessed 25 Nov 2014; (b) SeedQuest, Archer Daniels Midland Company, Bayer CropScience and Daimler to cooperate in Jatropha biodiesel project, http://bit.ly/1xXz1YH, accessed 25 Nov 2014; (c) JatroMed, *Evaluation of the energy crop Jatropha curcas as a mean to promote renewable and sustainable energy for the Mediterranean region*, http://bit.ly/1viZn1G, accessed 25 Nov 2014.

[10] (a) K.T. One, P. Tanya, N. Muakrong, K. Laosatit P. Srinives, *Biomass Bioenerg.*, **2014**, *67*, 137–144; (b) V. Pecina-Quintero, J.L. Anaya-López, A. Zamarripa-Colmenero, C.A. Núñez-Colín, N. Montes-García, J.L. Solís-Bonilla, M.F. Jiménez-Becerril, *Biomass Bioenerg.*, **2014**, *60*, 147–155; (c) F.D. da Silva Araújo, I.C. Araújo, I.C.G. Costa, C.V. Rodarte de Moura, M. H. Chaves, E.C. Araújo, *Renew. Energy*, **2014**, *71*, 495–501.

[11] A. Grevé, L. Barbaranti, S. Fazio, *Handbook on Biofuels and Family Agriculture in Developing Countries*, 2011, http://bit.ly/1AUGs0Z, accessed 25 Nov 2014.

[12] FAO, Making Integrated Food-Energy Systems work for People and Climate, 2010, http://bit.ly/1wCyqKK, accessed 25 Nov 2014.

[13] (a) R.A. Ferrari, A.J. de Azevedo Filho, *J. Agric. Sci. Technol. B.* **2012**, *B2*, 1119–1126; (b) K. Averdunk, T. Zelt, P. Golka, M. Höpfner, C. Müller, I. Bettermann, Macauba – *Sustainable Palm Oil*, http://bit.ly/1trCaIA, accessed 25 Nov 2014.

12

GREEN APPLIED SOLUTIONS: CUSTOMIZED WASTE VALORIZATION SOLUTIONS FOR A SUSTAINABLE FUTURE

CHUNPING XU[1] AND RAFAEL LUQUE[2]

[1]*School of Food and Biological Engineering, Zhengzhou University of Light Industry, Zhengzhou, Henan, P.R. China*
[2]*Green Applied Solutions S.L.U., Cordoba, Spain*

12.1 INTRODUCTION

Human societies are facing a daunting future in terms of water, food, and resource scarcity. This has been clearly evidenced by recent studies showing a preoccupying

Industrial Biorenewables: A Practical Viewpoint, First Edition. Edited by Pablo Domínguez de María.

diminishing in fossil fuel resources and the increasing generation of waste as well as the expected increase in population in the future years. Facing these challenges is not an easy task. Multidisciplinary team efforts from many disciplines are needed to come up with suitable alternatives for a future more sustainable society, which is able to deal with these important issues.

Waste production is currently one of these alarming problems that the planet is currently experiencing and will be facing further in the future years. Waste generation in Spain as well as internationally has dramatically increased in the recent years due to important factors including a larger population and the exponential resource consumption that society has experienced. More and more products are available in our markets, with a largely increased offer and demand of enhanced and sophisticated of high-performance products – better presentation, packaging, and societal perspective – which also significantly contribute to such recent increase in waste generation. There are several staggering figures only in Europe in this respect, although values are even more unbelievable in Asia.

In Europe, waste generation was reported to be over 2.25 billion tons in Western European countries between 1998 and 2001, as well as 550 million in Eastern European ones.[1] Main sources of such waste were construction (31%) and mining and quarrying (15%) as well as agricultural and forestry waste, which accounted roughly for 30% of the total generated waste. The remaining 25–40% correspond to municipal solid waste (MSW) including household waste, plastic films, and unrecyclable packaging, accounting for a staggering 1.3 billion tons of food wasted worldwide yearly.[2] Such figure implies that one third of the globally produced food for human consumption is wasted every year. Even worse, in many cases organic residues are derived from supply chains (e.g., outdated products, products that did not pass the strict quality controls in industry, and even badly labeled products), which are difficult to manage and generally end up in authorized landfilled sites. These terrifying facts are telling us that something needs to be changed in our society to tackle an obvious problem that needs solution in a future expected scenario of resource scarcity and ever increasing population.

In this regard, sustainable industrial development can significantly contribute to such issues, relying on the efficient exploitation of agricultural and forestry residues, aquatic biomass, and various waste streams by means of integrated biorefining, industrial symbiosis, cascade processing, on-site processing of seasonal waste streams, extraction of specialty products, and conversion through green chemical or biotechnological processes.[3,4]

The successful transition to the bioeconomy era will be achieved by restructuring significantly current manufacturing practices. For instance, in the food industry, the traditional meaning of "processing" is associated with transformation of the initial raw material into a food product with optimum structure, properties, quality, safety, and nutritional requirements. However, in the dawn of a biobased society, food processing should also provide viable alternative schemes that combine food production with minimization of energy consumption and water usage as well as valorization of waste and by-product streams. Packaging should also be included as an essential processing aspect considering both the production of a safe, fresh, and high-quality food product and its recyclability or reusability in a cascade system for different value-added applications.

Industrial symbiosis should be accomplished in order to enable utilization of coproducts and processing streams by different industrial sectors. Furthermore, integration of chemical production in a traditional industry is a new concept that will create significant added value and investment opportunities. This argument could be supported by the fact that, in 2007, the industrial sectors of food and beverages and chemicals including pharmaceuticals were among the largest three industrial sectors in EU-27 regarding added value and investments.[5] Waste valorization via integrated biorefinery schemes should target the production of both commodity and specialty products. As an example, producing high-value products for cosmetic, pharmaceutical, and nutraceutical applications could provide the needed profitability margin required for the generation of low-value higher-volume products including chemicals and biodegradable polymers. The main problem for ensuring the profitability of the whole integrated process is to identify diversified market applications to avoid saturation of a single market outlet.

Furthermore, integration of chemical production in existing industries using waste-derived streams and feedstocks can also provide a novel low-cost source of carbon. For instance, the development of large-scale fermentative production of commodity chemicals is mainly hindered by the high cost of raw materials – mainly purified sources of carbon such as glucose or sucrose – used in traditional fermentation processes. Furthermore, integrated production of chemicals in existing industries could reduce fixed capital investment, which is also significant in traditional fermentation processes and hinders industrial implementation.

Taking into account the previously discussed premises, Green Applied Solutions S.L.U. (from now on denoted as GAS) emerged as a spin-off company specifically targeting the important issue of waste valorization to useful materials, chemicals, and fuels utilizing a wide range of green chemical conversion strategies, as well as biotechnological applications. This contribution has been aimed to showcase some of the key aspects of the company as well as ongoing projects for future development of integrated biorefineries for the valorization of different types of waste to valuable products, from the beginning of the company in 2012 to advances and progress in these 2 years that will probably lead to the first set of commercial products in early 2015.

12.2 THE COMPANY

GAS was originally founded back in 2012 in Cordoba (south of Spain) as an ambitious joint venture of four scientists with various expertise and backgrounds related to waste and biomass valorization, sustainable agricultural practices, and bioenergy.[6] The company emerged in an attempt to provide customized solutions and possibilities in the field of biomass/waste valorization and bioenergy based on the significant experience of the cofounders Dr. Tomás David Conesa (Ph.D. in chemistry), Dr. Alberto Perea (Ph.D. in agricultural engineering), Mr. Jesús Gandía (BSc in chemistry), and

Prof. Rafael Luque (Ph.D. in chemistry). GAS started as a spin-off company aiming to design new biobased products with a 100% green policy to change current perception of waste from our society as a problem rather than as an extraordinary source of valuable products.

After about 1 year (early 2013), the current CEO of the company Prof. Rafael Luque – with affiliation to Universidad de Cordoba (Spain) – made an effort to rearrange and refocus the company by acquiring all participation and shares, aiming to restructure the original goals and objectives of GAS into more realistic and stepwise growing approach. These include three different and well-structured business lines:

1. *Consultancy.* GAS was created to provide consultancy solutions to industry and academia in topics related to biofuels and bioenergy, biomass and waste valorization, and agriculture engineering based on the expertise of all cofounders. The company has been also involved in partnerships for proposal/project writing and consultancy in projects mostly at international level, particularly in South America and Asia.[6]
2. *Formation and training.* Prof. Luque is a senior experienced scientist with an extensive experience on waste and biomass valorization, nanomaterials, and heterogeneous catalysis. Currently affiliated at Universidad de Cordoba, Prof. Luque delivers worldwide tailor-made courses for academia and industry on several important topics including bioproduct engineering from biomass and waste, scientific manuscript/report/patent writing skills, heterogeneous catalysis, and materials engineering. These courses have been imparted in different countries worldwide, including China, Philippines, Hong Kong, and Spain, to cite some.[6]
3. *Product development and activities related to R&D.* 2014 marked the beginning of a new business line involving product development and R&D. GAS will be able to provide customer-tailored efficient solutions focused on biomass/waste valorization practices to high added value products. GAS can offer know-how and R&D for the extraction, fractionation, and separation as well as isolation and production of chemicals, materials, and fuels from many types of feedstocks using clean and low environmental impact technologies (e.g., ball milling, microwaves, ultrasounds, etc.). The company is also able to provide detailed analysis of feedstocks in terms of composition and valorization potential as consultancy service to industry and research institutions.[6]

GAS currently counts with two full-time individuals as part of the company, as well as a number of visitors/contract scientists depending on various ongoing projects with different institutions. GAS also has a close collaboration with the university and scientifically breeds from the group of Prof. Luque at UCO in Spain, contributing to have scientific inputs into GAS through contracts and licensing.

As it can be observed, over the past 2 years, GAS has evolved from a purely scientific consultancy company – from 2012 to early 2013 – toward a spin-off that is able to contribute to research/scientific formation (through dedicated courses, talks, events, etc.) and more recently in 2014 started the first R&D projects on various topics including rice husk valorization to valuable products, valorization of tobacco waste via fermentation for the production of polysaccharides (in collaboration with the School of Food and Biological Engineering at Zhengzhou University of Light Industry), and subsequent utilization of engineered polysaccharides for the preparation of nanomaterials and nanocomposites with remarkable photocatalytic activities for wastewater purification as well as advanced polymers with antimicrobial activities. These areas will be further discussed later in this chapter.

12.3 PROJECTS AND FUTURE

This section aims to describe some recent projects that are currently under development in the company, as well as some others currently under negotiation, which will serve as the basis for future products of the company. As stated earlier, even after a very short time after its foundation, GAS is prepared and qualified to emerge as one of the key international waste valorization companies in the coming years.

12.3.1 Silicates and Carbonates from Rice Husk: Food Waste to Materials and Chemicals

Rice husk and rice straw are two of the most important by-products of rice cultivation and processing. The global rice production increases at an average of 16 million tons per year.[7] Rice by-products have essentially different compositions but similar in terms of the different present fractions, namely, cellulose (30–40%), hemicelluloses (20–30%), lignin (10–15%), and ash (8–20%).[8] The metal content in rice husks is of particular commercial interest, as it is very much enriched in silica (SiO_2), which can be potentially extracted from rice husk using alkali solutions. The proposed approach has been previously attempted utilizing waste ashes from a commercial biomass combustion facility in the United Kingdom,[9] as well as related ash-type waste feedstocks.[10–12] In general terms, the methodology involved a simple hydrothermal extraction with KOH to generate potassium silicate solutions under reproducible 60–70% extraction yields, which were subsequently analyzed and validated using a straightforward infrared method for simple quantification.[9] Upon extraction, the alkali silicate solutions were used as silica source for the preparation of high surface area mesoporous silica materials of typical MCM-41 structure by the addition of a structure-directing agent (viz., cetyltrimethylammonium bromide (CTAB)) and similar synthetic conditions to those generally employed for the preparation of such mesoporous silicate materials.[9]

Based on the aforementioned previous work, the company has been interested in developing an analogous system for the production of different high added value materials and chemicals from rice husk. Clearly, being a commercial approach to generate revenues, the sophistication level typically present in research scientific projects and publications has not been the main aim of this project. Furthermore, it must be noted that the composition of rice husk is remarkably more complex as compared to that of waste ashes (>90% metal oxides with over 80% silica content). Hence

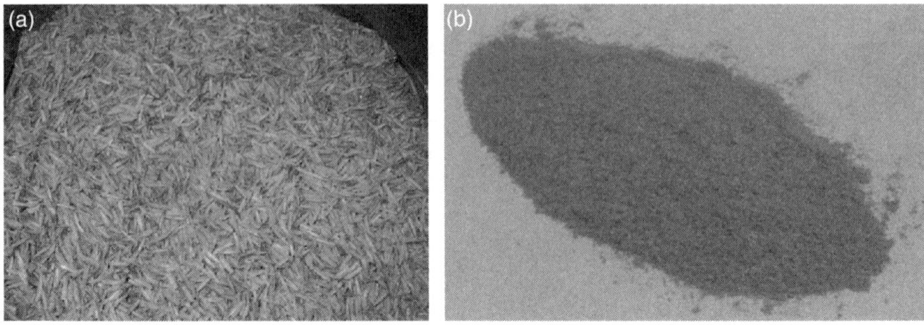

Figure 12.1 Pretreatment of rice husk (a) to fine sieved powder (b) via milling prior to processing. (*See color plate section for the color representation of this figure.*)

preliminary experiments have been aimed to find out more or less optimum conditions to produce valuable compounds from rice husk through the previously described methodology. For such purpose, the starting material is generally milled in a planetary ball mill to a fine powder, trying to minimize the large volume of rice husk for a simplified and more efficient processing (Figure 12.1).

Gratifyingly, under optimized conditions – including extraction, workout, and treatment – the first developed protocol can yield a highly pure and crystalline sodium carbonate material (Na_2CO_3) in about 20% yield (by dry weight). Further experiments and investigations are currently ongoing in this direction to optimize both production yields and product quality.[13] Remarkably, sodium carbonate is a commonly utilized solid base in chemical processes and glass manufacture (it acts by decreasing the melting point of the mixture with silica to generate the so-called soda lime glass), as well as the commonly known application as water softener.

The proposed technology – from a commercial viewpoint of being simple, efficient, and clean – has the potential to provide a promising commercialization alternative to the most extended Solvay process for sodium carbonate production (soda ash) or other technologies including the Leblanc process (from brine), new Asahi, or Akzo (dry lime) process among others.[14] To compare the current commercial production process with our newly reported system, a pictorial representation of the developed methodology has been depicted in Figure 12.2, showing all the way from unprocessed rice husks – actually a biobased waste – to deliver sodium carbonate.

In comparison, for the particular example of alkali silicate solutions from rice husk, the foreseen application relates to their further use in binders and construction-related applications rather than any silicate-type materials. Research studies have previously reported the potential of wheat straw combustion ashes as source of silicate solutions to replace traditional formaldehyde binders in construction boards.[15]

12.3.2 Advanced Nanocomposites and Nanomaterials from Tobacco Waste

Tobacco waste is also an important issue in both developed and developing countries. For example, in China, tobacco sales leave millions of tons of tobacco wastes yearly,

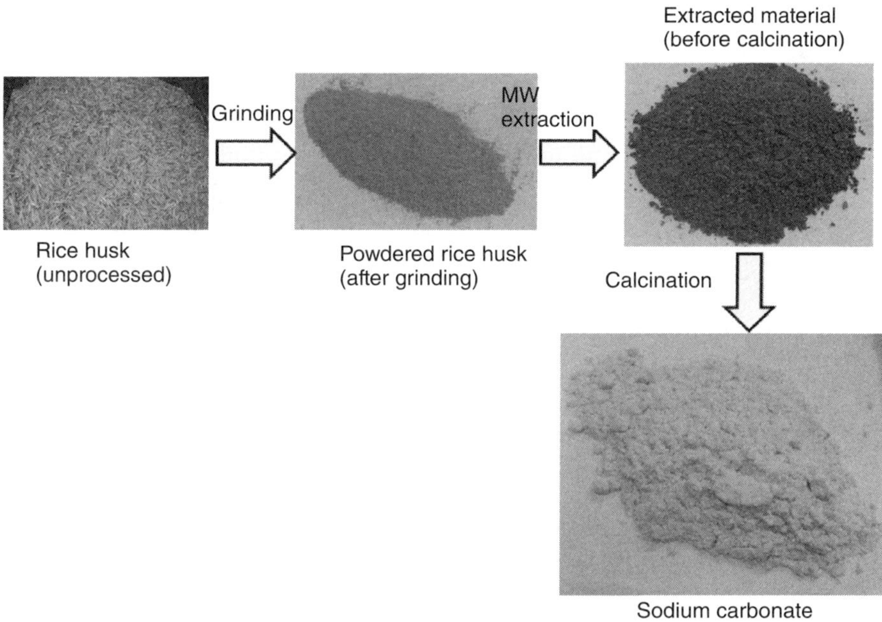

Figure 12.2 Conversion of rice husks into sodium carbonate. *Source*: From[13].

which find little uses that are different from disposal/landfilling and burning. Having identified this opportunity, and in collaboration with the School of Food and Biological Engineering at Zhengzhou University of Light Industry (Prof. Chunping Xu), GAS has found an interesting number of applications of bioengineered polysaccharides from a simple but highly effective aqueous extraction from tobacco,[16–18] a technology that more recently also extended to the utilization of tobacco waste. (C. Xu et al. unpublished work.) In their proposed approach, different sources of tobacco were utilized including burley (cured) and flue cured. The most extended practice for the extraction and separation of polysaccharides from tobacco leaves was a simple aqueous extraction process assisted by ultrasounds.[16–18] Upon extraction, polysaccharides were subsequently separated and isolated using Sepharose-type columns and eventually characterized and quantified. Several fractions of different polysaccharide extracts could be isolated from tobacco leaves, with interesting biological activities.[16–18] These extracts were mainly tested for antioxidant activities, with results pointing to excellent radical scavenging properties (for both hydroxyl and DPPH) in a dose-dependent manner.[16–18] In some other cases, analogous polysaccharides extracted from tobacco or other biomass-derived sources have been reported to have promising antimicrobial and neuroprotective activities as well.[19,20]

In a similar approach, exopolysaccharides (EPS) can also be produced via fermentation of certain organisms.[21–29] The group of Xu et al. have extensively worked with EPS produced by submerged cultures of fungi including *Boletus aereus*,[21]

Phoma sp.[22] (isolated from the leaves of *Taraxacum ongolicum* – a Chinese medicinal plant), *Stropharia rugosoannulata*,[23,24] *Pleurotus geesteranus*,[25] *Incutis tamaricis* (Pat.),[26] and *Trametes gibbosa*.[27] Importantly, EPS generated upon fermentation exhibited remarkable biological properties both *in vitro* and *in vivo*, ranging from antioxidant (e.g., radical scavenging),[21,24,26] antimicrobial (due to the presence of 2-hydroxy-6-methylbenzoic acid),[22] and antitumoral[24,26] to most interesting hypoglycemic and hypolipidemic effects.[24, 25, 26, 28] EPS hypoglycemic and hypolipidemic effects were investigated in streptozotocin-induced diabetic rats, including a decrease in plasma concentrations of glucose (17–37%), total cholesterol (14–26%), and triacylglycerol (12–24%) as well as aspartate aminotransferase activity (up to 20%).[23,25,27]

EPS were produced from batch lab fermentations (1–15 L),[22–30]] although successfully scaling up for the production of mycelial extracellular products was proved in large-scale industrial fermentors (2000 L) from *Ophiocordyceps sinensis* strain Cs-HK1.[28,29] Apart from the aforementioned antimicrobial, antitumoral, antioxidant, and hypoglycemia and hypolipidemia activities of the reported EPS, the production of important chemical compounds (e.g., betulin, 2-hydroxy-6-methylbenzoic acid, etc.) has also been reported.[22,29] Importantly, large-scale fermentations could lead to relevant extracts with remarkable antifatigue effects on forced animal activities (e.g., swimming).[28] These included increasing endurance of mice and glycogen levels under stress conditions as well as significantly reducing lactic acid and blood urea nitrogen levels.[28] These can offer additional possibilities for further developments in important market products for antioxidant, antimicrobial, and ultimate sportive formulations for athletes as well as antitumoral and cytotoxicity protection.

In view of the relevant properties and biological activities of these extracts and EPS, GAS recently teamed up with the groups of Prof. Luque at UCO in Spain and Heihua Biotechnotic Ltd. from Zhengzhou (China) to develop advanced applications for extracted polysaccharides and related compounds from tobacco as well as from other sources (e.g., macroalgae).[30]

The first initiative recently under development at GAS relates to the utilization of such polysaccharides as sacrificial templates for the production of a wide range of nanomaterials including nanocrystals of metal oxides. Preliminary research results from the group indicated that the utilization of pure polysaccharides including starch and alginic acid as templates in a dry ball milling methodology could lead to advanced nanocrystals of metal oxides (ZnO) in high purities and with a highly crystalline nature (Figure 12.3). Importantly, the synthesized porous ZnO nanostructures were found to have excelling photocatalytic properties for the degradation of pollutants (e.g., phenols) in water[32] as well as very promising antimicrobial activities when utilized as formulations in polymer nanocomposites.[33]

These results were initial proof of concept from pure polysaccharides, but in some cases algae-extracted polysaccharides show a remarkable potential as replacement of pure compounds in the proposed technologies.[32] Hence, the concept currently under development in GAS deals with the fabrication of not only ZnO nanostructures but also related nanomaterials and porous metal oxides (e.g., Ag_2O, TiO_2, Fe_3O_4) for biomedical research. GAS and the group of Prof. Luque are also currently working in

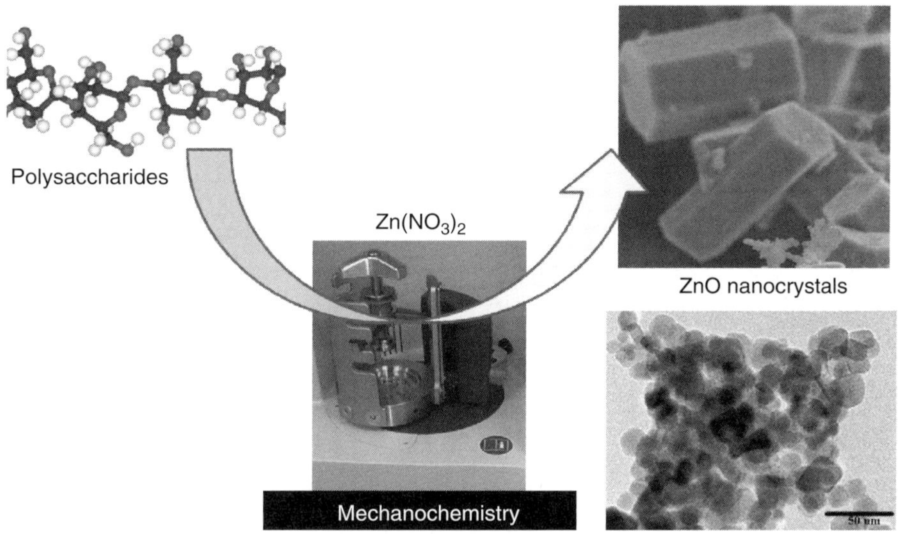

Figure 12.3 Mechanochemical production of designer nanomaterials from various polysaccharides. (*See color plate section for the color representation of this figure.*) *Source*: from[31].

close collaboration with TecnATox Centre from Universitat Rovira i Virgili (Spain) to test the toxicological properties of selected nanomaterials in living tissues (e.g., lung carcinoma cells) as well as related possibilities with highly promising results so far.[34]

12.3.3 Other Projects

More recently, GAS has been interested in expanding its portfolio of applications and projects to other types of feedstocks including collagenic-like biopolymers. The Prof. Luque's group has an extensive experience working with the extraction, isolation, and applications of biocollagen from slaughterhouse residues as well as from leather waste.[35] The utilized protocol involved a simple hydrolytic extraction process using acetic acid in which under optimized conditions biocollagenic polymers of different molecular weight distributions could be obtained depending on the conditions. In general, these biocollagenic polymers could be formed into various morphologies including fibers, films, and sponges, with a relevant number of applications in tissue regeneration, biomedicine, and even cosmetics (Figure 12.4).[36]

In this regard, we have been in contact with companies that are able to produce waste products with a high protein content (>80%). These have been not only related to the meat and leather industries but also to the production of collagenic products both in a gel-like phase (protein content 5–10%) and in a solid form (with protein content over 20%). These may have interesting potential uses for various applications (e.g., amino acid solubilization) as dietary supplements in cattle and even for health and care products.

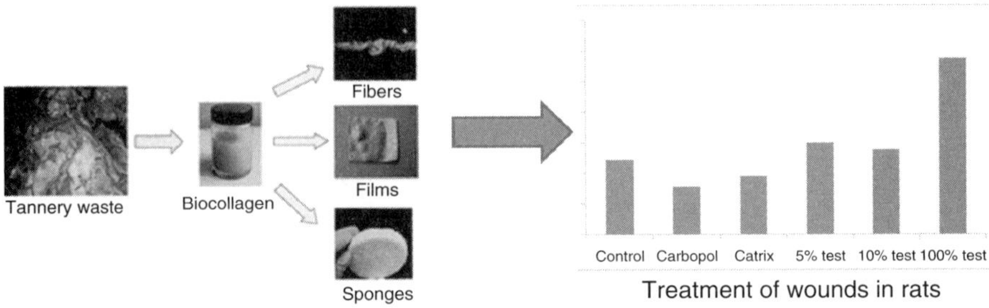

Figure 12.4 From slaughterhouse/tannery waste to biocollagenic polymers with applications in tissue regeneration. (*See color plate section for the color representation of this figure.*)

12.4 CONCLUSIONS AND PROSPECTS

This chapter has been aimed to provide an overview activities of the recently created GAS in the production of new families of biobased products as product development and R&D, together with important consultancy service and formation and training to academia and industrial partners in the fascinating field of biomass and waste valorization to valuable end products. GAS was only very recently created but has emerged with a clear strategy and business plan based on *efficiency* (solutions are quickly implemented and developed, within months), *commitment* (to produce top-quality products and services that have an impact in the community), *innovation* (key to success), *openness to changes* (to fit better into customer needs), *multidisciplinarity*, and *passion* to round up a cocktail that guarantees future success.

Various current research lines – mostly related to materials engineering from biomass and waste – have been discussed in this chapter, which essentially illustrate the present and near future of the company, surely to be expanded in the coming years to other feedstocks and products depending on customer needs and market demands. GAS is constantly seeking opportunities at both national and international levels, finding partnerships with different entities and institutions worldwide. In this regard, the CEO of the company has recently started another set of activities as part of another Spanish company (POSIDONIA OCEANICA[37]). The aim of this company is to raise awareness on the potential and possibilities of this marine sea grass species, which is causing a significant issue in beaches and landscape in several places in Europe and Spain (Mallorca) after being washed ashore in several forms (dead leaves, "furry" lignocellulosic balls, etc.). Ongoing efforts and joint projects between GAS and POSIDONIA OCEANICA will be signed in the near future to provide also various possibilities for the valorization of algal species that grow together with *Posidonia* and may also have further opportunities to produce high added value compounds.

ACKNOWLEDGMENTS

Rafael Luque gratefully acknowledges efforts from past cofounders and owners of the company in the first stage of creation. Big thanks are extended to his friends, colleagues, and initial business partners Tomás David Conesa, Alberto Perea, and Jesús Gandia who in fact pushed him to "*start our sustainable dream together*," something that the author of this contribution hopes to take to another future success. David Cuenca (current company manager) and Rick A.D. Arancon are also gratefully acknowledged for the support and hard work in these past months as well as the future manager Alina M. Balu who will join the team of GAS in the coming years. The author also wants to thank Heihua Biotechnotic Ltd. based in Zhengzhou (China) for the scientific and economic support to the company and the coauthor of this contribution Prof. Xu for his kind consideration and highly appreciated partnership from 2013.

REFERENCES

[1] http://epp.eurostat.ec.europa.eu/cache/ITY_OFFPUB/KS-55-03-471/EN/KS-55-03-471-EN.PDF. **2003**, accessed Jul 2014.

[2] FAO (Food and Agriculture Organization of the United Nations). *Global food losses and food waste - Extent, causes and prevention*. **2011**, http://www.fao.org/docrep/014/mb060e/mb060e00.pdf, accessed 09 Dec 2013.

[3] C.S.K. Lin, L. Pfaltzgraff, L. Herrero-Davila, E.B. Mubofu, S. Abderrahim, J.H. Clark, A. Koutinas, N. Kopsahelis, K. Stamatelou, F. Dickson, S. Thankappan, Z. Mohamed, R. Brocklesby, R. Luque, *Energy Environ. Sci.* **2013**, *6*, 426–464.

[4] J.C. Serrano, R. Luque, A. Sepulveda-Escribano, *Chem. Soc. Rev.* **2011**, *40*, 5266-5281.

[5] The European chemical industry in a worldwide perspective. European Chemical Industry Council, **2011**.

[6] www.greenappliedsolutions.com.es.

[7] http://www.fao.org/economic/est/publications/rice-publications/rice-market-monitor-rmm/en/. **2014**, accessed Jul 2014.

[8] P. Kaparaju, M. Serrano, A.B. Thomsen, P. Kongjan, I. Angelidaki, *Bioresour. Technol.*, **2009**, *100*, 2562–2568.

[9] J.R. Dodson, E.C. Cooper, A.J. Hunt, A. Matharu, J. Cole, A. Minihan, J.H. Clark, D.J. Macquarrie, *Green Chem.* **2013**, *15*, 1203–1210.

[10] S. Chiarakorn, T. Areerob, N. Grisdanurak, *Sci. Technol. Adv. Mater.* **2007**, *8*, 110–115.

[11] K. S. Hui, C. Y. H. Chao, *J. Hazard. Mater.* **2006**, *137*, 1135–1148.

[12] L. Tzong-Horng, *Chem. Eng. J.* **2011**, *171*, 1458–1468.

[13] A. Franco, A.A. Romero, R. Luque, Pat. Appl. pending.

[14] http://www.inclusive-science-engineering.com/sodium-carbonate-manufacturing-synthetic-processes-chlor-alkali-industry/. **2013**, accessed Jul 2014.

[15] V.L. Budarin, J.R. Dodson, K.J. Milkowski, J.H. Clark, WO/2009/087360. **2009**.

[16] C. Xu, Y. Xiao, D.B. Mao, *Biosci. Biotechnol. Biochem.* **2013**, *77*, 2100–2104.

[17] C. Xu, Y. Xiao, D.B. Mao, *Croat. J. Food Sci. Technol.* **2013**, *5*, 46–52.

[18] C. Xu, C. Yang, D.B. Mao, *PhCog. Mag.* **2014**, *10*, 66–69.

[19] H.Y. Wang, M.M. Zhao, B. Yang, Y.M. Jiang, G.H. Rao, *Food Chem.* **2008**, *107*, 1399–1406.

[20] K. Castagnoli, J.B. Petzer, S.J. Steyn, C.J. Van der Schyf, C.J. Castagnoli, *Inflammopharmacology* **2003**, *11*, 183–188.

[21] J.-Q. Zheng, J.-Z. Wang, C.-W. Shi, D.-B. Mao, P.-X. He, C.-P. Xu, *Process Biochem.* **2014**, *49*, 1047–1053.

[22] H. Zhang, Y. Xiong, H. Zhao, Y. Yi, C. Zhang, C. Yu, C.-P. Xu, *J. Taiwan Inst. Chem. Eng.* **2013**, *44*, 177–181.

[23] X. Zhai, A. Zhao, L. Geng, C.-P. Xu, *Ann. Microbiol.* **2013**, *63*, 1013–1020.

[24] P. He, L. Geng, J. Wang, C.-P. Xu, *Int. J. Med. Mushrooms* **2012**, *14*, 365–376.

[25] D.-B. Mao, Y.-P. Ma, L. Geng, A. Zhao, J.-Q. Zheng, C.-P. Xu, *An. Acad. Bras. Cienc.* **2013**, *85*, 1473–1481.

[26] J.-Q. Zheng, X.-J. Mao, L.-J. Geng, G.-M. Yang, C.-P. Xu, *J. Taiwan Inst. Chem. Eng.* **2014**, *45*, 725–733.

[27] Y. Ma, D. Mao, L. Geng, Z. Wang, C. Xu, *Carbohydr. Polym.* **2013**, *96*, 460–465.

[28] J.-Y. Wu, H.-P. Leung, W.-Q. Wang, C.-P. Xu, *Int. J. Med. Mushrooms* **2014**, 16, 121–130.

[29] Y.-H. Bai, Y.-Q. Feng, D.-B. Mao, C.-P. Xu, *J. Taiwan Inst. Chem. Eng.* **2012**, *43*, 663–669.

[30] M. Francavilla, A. Pineda, C.S.K. Lin, M. Franchi, P. Trotta, A.A. Romero, R. Luque, *Carbohydrate Polym.* **2013**, *92*, 1555–1560.

[31] http://commons.wikimedia.org/wiki/File:Cellulose-3D-balls.png#mediaviewer/Archivo:Cellulose-3D-balls.png, access Jul 2014.

[32] M. Francavilla, A. Pineda, A.A. Romero, J.C. Colmenares, C. Vargas, M. Monteleone, R. Luque, *Green Chem.* **2014**, *16*, 2876–2885.

[33] A.M. Diez-Pascual, C. Xu, R. Luque, *J. Mater. Chem. B* **2014**, *2*, 3065–3078.

[34] C. Xu, R.A.D. Arancon, M. Gomez, J.L. Domingo, A.A. Romero, R. Luque, in preparation.

[35] M. Catalina, J. Cot, A.M. Balu, J.C. Serrano-Ruiz, R. Luque, *Green Chem.* **2012**, *14*, 308–312.

[36] M. Catalina, J. Cot, M. Borras, J. De Lapuente, J. Gonzalez, A.M. Balu, R. Luque, *Materials* **2013**, *6*, 1599–1607.

[37] www.posidoniaoceanica.com.

13

GROVE ADVANCED CHEMICALS: FLOX® COAGULANTS – ENVIRONMENTALLY FRIENDLY WATER AND WASTEWATER TREATMENT USING BIODEGRADABLE POLYMERS FROM RENEWABLE FORESTS

BÁRBARA VAN ASCH, PAULO MARTINS, FILIPE SANTOS, ELISABETE SEPÚLVEDA, PEDRO CARVALHO, RICHARD SOLAL, CARLOS ABREU, RUI SANTOS, JORGE VASCONCELOS, PHILIPPE GEYR, AND HENRIQUE VILLAS-BOAS

Grove Advanced Chemicals SA, Maia, Portugal

Industrial Biorenewables: A Practical Viewpoint, First Edition. Edited by Pablo Domínguez de María.

13.1 INTRODUCTION

The historic period commonly known as the Industrial Revolution started in the second half of the 18th century with the invention of new machine-based manufacturing processes and the harnessing of powerful energy sources. This prolific era allowed for unprecedented economic, technical, and scientific progress and profound demographic and social changes globally. A burst of innovation, which was never to be stopped, rapidly reached all sectors of human activity and literally changed the face of the planet with the advent of large cities, industries, transportation, railroads, mines, dams, and bridges – along with water, air, and soil pollution – and the disappearance of large areas of natural ecosystems.

One of the most remarkable differences between the preindustrial era and the present one is the systematic large-scale exploitation of natural resources to meet the increasing needs and aspirations of a growing human population. However, another striking difference between preindustrial and our modern times is the awareness that the natural resources necessary to maintain human growth and development are not unlimited and should be used wisely. This perception led to the construction of the philosophical and analytic framework of sustainability, a concept that can be defined as "meeting the needs of the present without compromising the ability of future generations to answer their own needs."[1] In recent years, the view that sustainable development requires the reconciliation of the demands for economic growth, environmental protection, and social equity has found extraordinary support in many society domains, albeit the political adhesion has been less than ideal.

Adequate management of renewable resources is one of the keys to address the challenges of balancing economic growth and sustainable development. It has become clear that only the sensible exploitation of renewable resources such as water, energy, soil, and biological patrimony can prevent catastrophic environmental, economic, and social erosion. It is estimated that 10 billion people will be living on this planet by 2100, many of which will work their way toward better life standards and achieve prosperity, health, and education. These people will be our children's children and most of us will not live to know them. Still, they will inhabit a world they will legitimately inherit. Our legacy of sustainability has to be built now.

13.2 COMPANY OVERVIEW

Grove® Advanced Chemicals™ was founded in 2009 with the objective of offering a range of innovative environmentally friendly and totally biodegradable coagulants of vegetal origin (the Flox® product range) to the water and wastewater treatment industry. Since then, the company has steadily conquered a growing share of the market for coagulants in Europe, Asia, and the Middle East and presently supplies chemical products and services to a vast range of clients operating in the manufacturing and transforming sector such as the food, ceramics, wood and wood-derived products, pulp and paper, petrochemical, mining, metallurgy, glass, and textile industries.

Grove Advanced Chemicals is proud to have become a reliable long-term partner by cultivating a business philosophy strongly focused on meeting the needs of the client and developing customized solutions for each particular case. In order to achieve excellent results both in the improvement of the efficiency of the water treatment and in the overall reduction of operational costs, the implementation of our innovative products is supported by preliminary laboratory tests, industrial-scale trials, and periodic follow-up visits and technical assistance. We seek to fully integrate water treatment by also offering technical solutions to optimize processes upstream and downstream of the physicochemical step. Our innovative products are also distributed through business partnerships with representatives that share our values and have the constant technical support and commercial assistance from Grove Advanced Chemicals.

13.2.1 Vision, Purpose, and Values

We exist to make water clean while protecting the environment, natural resources, and people. Our vision is to become the number one worldwide provider of eco-friendly chemicals and technology for the water and wastewater treatment by selling and aiding industries to implement the use of innovative noncontaminant, nondangerous, biodegradable, and renewable products. We are committed to creating value for our clients, employees, and stakeholders. We stimulate teamwork, innovation and communication, and delivery of high-performance products and services and maintain and expect the highest level of ethical behavior.

13.2.2 Scientific Cooperation and Technical Partnerships

Grove Advanced Chemicals actively seeks technical and scientific partnerships in cooperation with public and private institutions. Our present technical and commercial partnership with Oulu Water Alliance (Finland) allows us to develop research projects on water and wastewater treatment at the European level with the objective of exploring the full technical potential of natural coagulants.

13.3 COAGULATION AND FLOCCULATION IN WATER TREATMENT

Water and wastewater treatment typically aims at removing variable amounts of organic matter, solids, nutrients, microbial pathogens, and other contaminants, depending on the final use or destination of the treated water.

Most water and wastewater treatments include an essential coagulation and flocculation step promoted with the aid of specific chemical composts. This physicochemical process focuses primarily on the separation of very fine particles (<1 μm) of a colloidal nature from the solution by altering their physical state, thus allowing for aggregates to form. This strategy is surprisingly ancient; in potable water treatment, clarification has been performed since antiquity using a variety of coagulating agents. As remarkable examples, almonds smeared around a vessel were used to clarify river water in Egypt as early as 2000 B.C., and alum was mentioned as a coagulant by the Romans around 77 A.D.[2] Aluminum sulfate was reportedly used in England for coagulation in drinking water treatment (DWT) in 1757 and in municipal treatment of public water supplies in 1881.[3] In modern times, the most commonly used coagulants are composts based on iron and aluminum that have the capacity to form multicharged complexes with enhanced adsorption characteristics. The most commonly used coagulants are ferric chloride and ferric sulfate, aluminum sulfate and aluminum polychloride, and sodium aluminate mixtures of organic and inorganic compounds. Among the inorganic coagulants, polyaluminum sulfate and ferric chloride have been to this day the most widespread coagulants for the treatment of water and wastewater. Organic coagulants have been gaining widespread use since their introduction in the early 1950s in the United States.[4]

Currently, the most widely used organic coagulants are polydiallyldimethylammonium chloride (polyDADMAC) and epichlorohydrin dimethylamine (epi-DMA) polymers, generically referred to as polyamines.[5] The coagulation process is generally complemented with a synthetic polyelectrolyte flocculation agent to assist in the separation of the suspended solid particles from the solution. A physicochemical process based on coagulation, flocculation, and solid/liquid separation may constitute the sole process in water and wastewater treatment, or it may be part of a sequence of processes.

13.4 FLOX® COAGULANTS

13.4.1 Vegetal Origin

Flox® coagulants consist of totally biodegradable organic cationic polymers with low molecular weight produced from tannins extracted from the bark of the black wattle plant (*Acacia mearnsii* De Wild.). The tannin extraction and subsequent chemical process leading to the product coagulants is an intellectual property under protection by international laws. Black wattle is a fast-growing leguminous tree of great socioeconomic relevance with 60% of the plantations being explored by small farmers and representing an important source of income to these families. In some world regions, it is considered a useful plant for the remediation of degraded sites.[6] Black wattle has been extensively grown as a renewable industrial source of the most widely used high-quality tannin for the tanning of many types of skins and hides and as a source of cellulose. In some world regions, black wattle wood is also extensively used as fuel. The plantations can be explored as multifunctional systems: they aid in the recuperation of low-fertility soils by the fixation of atmospheric nitrogen through the action of nitrobacteria associated with the roots of the plant and allow for consortium cultivation with food crops, and the type of flower and late flowering season make the black wattle suitable for domestic bee forage. The black wattle plantations at the origin of Flox® coagulants are certified by the Forest Stewardship Council® (FSC) International, an independent organization that promotes responsible management of forests worldwide to meet the sustainability pillars of environmental appropriateness, social worth, and economic viability.

13.4.2 Environmentally Friendly Use

Flox® coagulants are specifically manufactured for their use in the treatment of wastewater and industrial process water and water for human consumption. They are mainly used in the primary physicochemical treatment process, specifically in the direct substitution of conventional synthetic organic and inorganic coagulants (e.g., polyamines, polyDADMAC, aluminum sulfate, iron chloride, and polyaluminum chloride (PAC)). These coagulants typically contain nondegradable aluminum and iron salts, polyamines, and other molecules of petrochemical origin that subsist in the water under treatment. Likewise, the sludge produced using conventional coagulants incorporate those nondegradable substances and may be subject to restrictions and difficulties in the final discharge, deposition in landfill site, or recycling in agricultural soil. By contrast, Flox® coagulants are biodegradable and can be quickly and totally digested in biological systems.

This environmentally friendly nature is an extremely important argument that would justify the application of Flox® coagulants in a wide range of water and wastewater treatment contexts per se. Remarkably, the excellent news is that Flox® coagulants also have superior technical performance and allow for significant economic savings in the overall operational cost of water and wastewater treatment. In conclusion, the eco-friendly nature of Flox® coagulants associated with the economic advantages and technical performance constitutes a real response to the increasingly growing demand for such products by industrial and governmental entities.

13.4.3 Advantages of Flox® Coagulants

Flox® coagulants are natural eco-products derived from plants cultivated under renewable forestry practices, contributing to the safety and sustainability of water and wastewater treatment. The application of Flox® coagulants in water and wastewater treatment plants evidences comparative advantages that directly lead to environmental, economic, and social improvements.

13.4.4 Decreased Sludge Volume

Water treatment produces vast amounts of sludge that more than often represent a significant part of the total operational cost, generally including the sludge dewatering process and the transport to its final destination. Flox® coagulants significantly decrease the volume of the sludge produced in the coagulation and flocculation process by comparison with coagulants based on metallic salts. These are originally anionic and form metallic hydroxides (e.g., aluminum hydroxide, iron hydroxide) in contact with water, becoming cationic in a chemical reaction with the alkalinity of the solution. In the hydroxide form, metallic salts acquire the colloidal physical state (gel) that contains a significant amount of water at the intramolecular level (e.g., $AlOH{\cdot}XH_2O$) responsible for hampering sludge dewatering. On the contrary, Flox® coagulants are constituted by organic molecules produced and ready for application in the cationic form, thus originating gels that also contain water at the intramolecular level but in much lesser amounts. This is the reason why Flox® coagulants generate lower sludge volumes compared to conventional coagulants, and sludge dewatering is facilitated.

13.4.5 pH and Conductivity

The absence of chemical reaction between Flox® coagulants and the alkalinity of the water under treatment are also the reasons why pH remains unaltered when Flox® are used, in contrast with metallic salts. In addition to this, Flox® coagulants do not increase the conductivity of the water under treatment also because they do not contain metallic salts. This latter aspect is particularly relevant when physicochemical processes are followed by osmosis, as well as in cases of closed recirculating water circuits where conductivity and total dissolved solids (TDS) originating from the inorganic coagulants increase at each cycle.

13.4.6 Human, Equipment, and Environmental Safety

Flox® coagulants are nontoxic and not classified in the United Nations register of hazardous products, thus contributing to decrease the risk for damage to human health and the environment. Flox® coagulants are noncorrosive and contribute to a longer useful life of equipment. Flox® coagulants are efficiently used in a broad range of pH (4–9) and do not alter the pH of the water under treatment. This characteristic allows for significant reduction or elimination of acidic and alkali agents commonly used for pH correction, resulting in increased safety of the operators and economic savings. Flox® coagulants contribute to the efficiency of downstream biological processes due to the absence of inhibitor agent (e.g., organic salts) while contributing to the microbial growth with organic material.

13.4.7 Flox® Technical Information

In a broad sense, Flox® coagulants act upon neutralization and destabilization of electric charges, thus triggering the formation of compact fuzzes with lower water content. The product range portfolio is composed of Flox®-QT, Flox®-QTH, and Flox®-QTP, all compatible with most of the compounds commonly used in water and wastewater treatment. The main difference between the three versions of Flox® is the net positive charge, which is more accentuated in Flox®-QTH and Flox®-QTP. The coagulation efficiency of the different Flox® versions depends on the many factors affecting the composition of the wastewaters. Therefore, Grove Advanced Chemicals recommends that previous testing is performed in specific wastewaters in order to find the most suitable Flox® among the three available options. Overall, Flox® coagulants are an environmentally friendly alternative to conventional products such as inorganic coagulants and organic synthetic compounds. Flox® coagulants are supplied to the final client in the form of a ready-to-apply aqueous solution at a 25% (weight/weight) concentration and stored in 1000 L refillable plastic containers (IBCs) or high-volume cisterns (Figure 13.1).

Moreover, the coagulants can also be supplied in the solid form (powder), in which case the client must implement an adequate dilution system that may enable the production of standard Flox® aqueous solutions. Importantly, Flox® products are not reusable as coagulants, but nevertheless are recyclable in the sense that they are just constituted by decomposable organic matter. Therefore, the ultimate disposal of unused Flox® coagulants should be performed according to local legislation regarding nondangerous chemicals.

The most relevant physicochemical characteristics of the commercial solution of Flox®-QT are presented in Table 13.1 as an example. The general chemical structure of Flox® coagulants can be found in Figure 13.2.

Figure 13.1 Samples of the commercial solution of coagulants Flox®-QT, Flox®-QTH, and Flox®-QTP (25%, w/w).

TABLE 13.1 Most Relevant Physicochemical Characteristics of the Commercial Solution of the Flox®-QT Coagulant (25%, w/w).

Chemical Composition	Ammonium Quaternary Tannate
Ionic charge	Cationic
Origin	Vegetal tannin
Presentation	Dark brown liquid
Solids	25%
Density (at 20 °C)	1.065–1.100 g/cm^3
pH	1.3–2.3

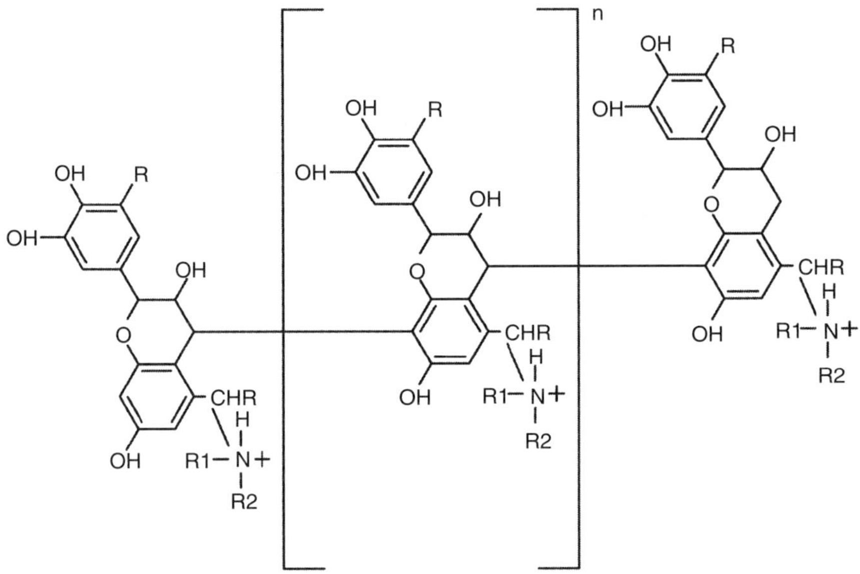

Figure 13.2 Generic representation of the chemical structure of Flox® coagulants.

13.5 COMPANY AND PRODUCT CERTIFICATIONS

13.5.1 Quality, Environment, and Human Resources

Grove Advanced Chemicals S.A. is certified by the Quality Standard NP EP ISO 9001, Environment Standard NP EN ISO 14001, and Human Resources Standard NP 4427.

13.5.2 National Sanitation Foundation Certification

Flox®-QT coagulant is certified for the treatment of water for human consumption by the National Sanitation Foundation (NSF) International, an accredited independent third-party certification body that tests and certifies products that meet public health and safety standards. The certification process allowing for the use of Flox®-QT in DWT within the European Union is currently in progress.

13.5.3 FSC Certification

All Flox® coagulants are produced under sustainable management of the forest resources certified by the FSC (https://ic.fsc.org/), an independent, nongovernmental, nonprofit organization established to promote the responsible management of the world's forests.

13.5.4 ISO 9001 and ISO 12001

All Flox® coagulants are manufactured by a producer certified according to the ISO 9001 and ISO 12001 norms.

13.5.5 Halal Certification

All Flox® coagulants are manufactured Halal according to the Islamic laws and meet the requirements of Muslim communities.

13.5.6 Kosher Certification

Flox®-QT and Flox®-QTH coagulants are Kosher/Pareve certified excluding Passover under the regulations that serve the requirements of Jewish communities.

13.6 CASE STUDIES

In this section, we present an overview of the already implemented industrial application of Flox® coagulants in several different water treatment contexts in the form of case studies. The cases here presented are real-life reports that were selected to illustrate a range of industries, types of water effluents, and the treatment improvements achieved through the use of Flox® coagulants. It should be noted that wastewaters may vary greatly from plant to plant, even within the same industrial area. Therefore, the results here presented should be regarded as indicative for each industry and should not be directly extrapolated to different processes, equipment, and conditions. A case-by-case assessment is thus always needed.

13.6.1 Ceramics Industry

According to the European Best Available Techniques Reference Document (BREF) in the Ceramic Manufacturing Industry,[7] water is an important raw material and a key environmental issue as well. Water quality is essential to the preparation of clays and glaze slips, clay bodies for extrusion, "muds" for molding, preparation of spray dried powders, wet grinding/milling, and washing and cleaning operations. Most of the produced wastewater in the ceramics industry is generated during the manufacturing processes – for example, cutting and polishing – and the subsequent cleaning of equipments. This wastewater generally carries inorganic suspended solids (clays, frits, and insoluble silicates), dissolved anions, suspended and dissolved heavy metals, small amounts of boron, and traces of organic matter. The minimization of freshwater consumption is an important environmental objective attainable through water recycling, and a zero effluent discharge is technically viable in specific closed circuits for process wastewater.

A particular ceramic tiles industry generates approximately 37,500 m^3 per month of wastewater in the rectification process composed of nine independent production lines. The wastewater mainly originates from the cutting of ceramic tiles. The main objective of the treatment was to maximize the reuse of the treated water in the production process. Physicochemical parameters such as conductivity, pH, chlorides, hardness, COD, and TSS were monitored to ensure the quality of the treatment, which was performed as follows:

- *Step 1: Homogenization.* The wastewater is homogenized in a tank prior to the physicochemical treatment.
- *Step 2: Coagulation/flocculation.* The coagulation and flocculation stage takes place in a canal where the inorganic coagulant PAC 18% and an anionic flocculant are added separately to promote the coagulation and flocculation of solid particles. The wastewater then proceeds to a coagulation/flocculation reaction tank and from there to a clarifier tank. The needed adjustment of pH was performed using caustic soda.
- *Step 3: Sedimentation unit/clarifier.* Flocculated particles enter a small turbulence zone in the clarifier where the separation of suspended solids is achieved by gravity. The clarified water is directed to a storage tank prior to be reused in the manufacturing process.
- *Step 4: Sludge dewatering.* The sludge generated in the clarifier is extracted and directly forwarded to a filter press. The dewatered sludge is reincorporated in the tile manufacturing process.

Preliminary tests were performed in the internal laboratory of Grove® Advanced Chemicals™ in order to assess the potential application of Flox® coagulants for the treatment of the wastewater produced in the rectification process and to estimate the optimal dosages needed to replace the inorganic coagulant typically used in the plant. Based on the laboratory results, in this case a full-scale treatment was proposed using Flox®-QTH as coagulant and an anionic flocculant as adjuvant. The wastewater was directed from the homogenization tank to an open canal where the commercial form of Flox®-QTH (25%, w/w) was added at an average dosage of 50 ppm. The flocculant solution (0.1%, w/w) was added at 2 ppm concentration immediately upstream of the coagulation and flocculation tank. The wastewater was then forwarded to a sedimentation unit for the separation of solids, without pH adjustment. The treated

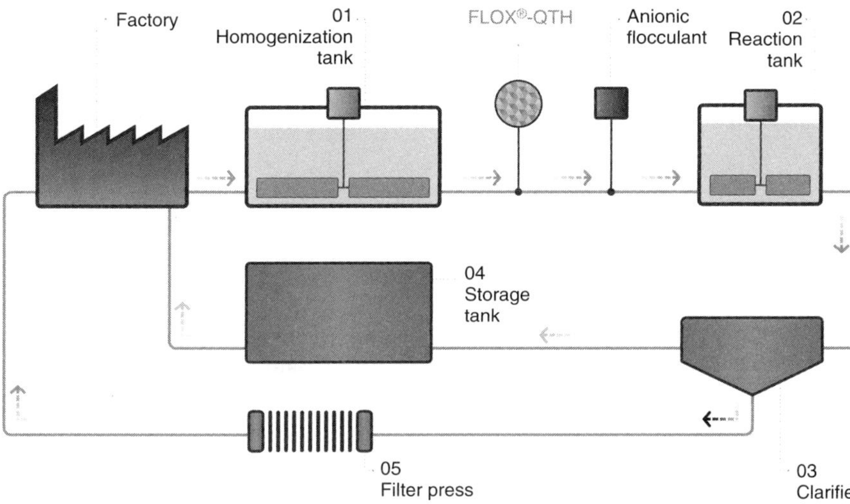

Figure 13.3 Schematic representation of the wastewater treatment process in a particular ceramics industry after the implementation of Flox®-QTH coagulant.

water was directed to a storage tank and reused in the industrial process. The resulting sludge was reincorporated in the manufacturing process after dewatering in a filter press. A schematic representation of the wastewater treatment process after the implementation of Flox®-QTH coagulant is presented in Figure 13.3.

Gratifyingly, Flox®-QTH replaced PAC 18% as a coagulant in the wastewater treatment of the rectification line with success. Compared to the previous treatment, it was possible to observe a 36% decrease in the conductivity of the treated effluent (3.900–2.500 mS/cm, average values), thus allowing for a 100% reuse of the treated wastewater. Moreover, the fact that Flox®-QTH can be applied in a wide range of pH (4–9) without significant alteration of the pH of the water under treatment allowed for the maintenance of a modal pH value of 7.7 without the need for adjustment. Consequently, the alkalinization agent (NaOH) previously used with PAC 18% was eliminated from the treatment process. It was also possible to observe a 59% and a 4% average decrease in chlorides and hardness levels, respectively. The overall quantities of chemicals used in the wastewater treatment plant decreased significantly (50% in coagulant and 100% in sodium hydroxide). The anionic flocculant dosage increased from 1.3 to 2 ppm. The use of Flox®-QTH slightly increased the content of suspended solids (from 26 to 38 mg/L) and COD (from 5 to 23 mgO_2/L) in the treated wastewater (Tables 13.2 and 13.3). It must be noted that this increment did not affect the quality requirements for the reuse of the treated wastewater in the manufacturing process.

Notably, the new treatment resulted in an improved quality of the treated wastewater, thus allowing for further maximization of wastewater reuse. Therefore, the demand for freshwater input to dilute and replenish the closed water recirculating circuit process decreased significantly. This aspect led to remarkably important

TABLE 13.2 Comparison between the Wastewater Treatment Dosages of Coagulants PAC 18% and Flox®-QTH and Alkalinization Agent NaOH in a Ceramics Industry.

	PAC 18%	Flox®-QTH
Coagulant (ppm)	100	50
Anionic polymer (ppm)	1.3	2
NaOH (ppm)	26	Eliminated

TABLE 13.3 Comparison of Quality Parameters (Average Values) of Treated Wastewater Using Coagulants PAC 18% and Flox®-QTH in a Ceramics Industry.

	PAC 18%	Flox®-QTH
Conductivity (mS/cm)	3.900	2.500
pH	7.7	7.7
Chlorides (mg/L)	700	288
Hardness (mg/L $CaCO_3$)	730	700
COD (mgO_2/L)	5	23
TSS (mg/L)	26	38

economic savings upon decreased electromechanical equipment wear and maintenance costs due to the low contribution of Flox®-QTH to conductivity in the treated wastewater. Furthermore the use of hazardous chemicals (sodium hydroxide and PAC 18%) was totally eliminated from the wastewater treatment process, and the overall quantity of other chemicals generally decreased, thus potentially improving the health and operational safety of the workers.

13.6.2 Fish Canning

The fish canning industry consumes a large amount of clean water in operations such as washing, cooking, cooling, and general plant cleaning operations. According to European BREF in the Food, Drink and Milk Industries,[8] this consumption reaches 25 m^3 of water for each ton of processed fish, of which about 70% is spent during manual skinning and cutting. Moreover, fish canning typically generates large quantities of wastewater particularly difficult to treat due to the high load of organic matter, salts, and oils and greases. Wastewater is generated in all processes starting with the reception of fish, usually stored in salt water. Fish preparation and cleaning by removal of scales, head, and viscera also requires a high amount of clean water, thus wastewater rich in blood, oil, salt, and other fish tissues. Likewise, cooking and cooling wastewaters also have high loadings of organic matter and fish oils. General plant clean-in-place (CIP) operations also contribute to the wastewater flow and typically contain caustic and acid detergents, surfactants, chelating agents, and sanitizers.

In this context, the Flox® series was assessed. A particular fish canning plant produced 62,000 m^3 per month of wastewater with an average COD of 7000 mgO_2/L and pH between 6.0 and 8.5. The treatment employed PAC 18% as coagulant and an anionic emulsion polymer as a flocculation adjuvant. The efficiency of the treatment was negatively impacted by pH fluctuations in the wastewater caused by CIP procedures. The wastewater treatment was performed as follows:

- *Step 1: Homogenization.* The wastewater generated in the canning processes – rich in organic matter – and plant cleanup (loaded with caustic and acid/alkaline detergents and disinfectants) is homogenized in a tank upstream of the physicochemical treatment.
- *Step 2: Physicochemical treatment.* PAC 18% and an anionic polymer were added sequentially in line upstream of the dissolved air flotation (DAF) tank. These reagents are used to promote the coagulation of particles in the DAF tank where flocculated particles are combined with a pressurized stream of water and extremely fine air bubbles that easily incorporate the suspended solids. The air–solid particles float to the surface of the flotation cell and are separated by a skimmer and scraper mechanism into the sludge chamber.
- *Step 3: Sludge dewatering.* The sludge originated in the DAF is temporarily stored in a storage tank and then sent to a decanter to be dewatered with the aid of a cationic polymer. The dewatered sludge had a final solid content of approximately 20–30%.

Flox®-QT and Flox®-QTH were tested in the internal laboratory of Grove Advanced Chemical using standard jar-test procedures and COD quantification. The main objectives were to decrease the COD load of the treated wastewater and to decrease the volume and increase the efficiency of sludge dewatering. Good results were obtained, leading to successful industrial-scale implementation of Flox®-QT and approval by the client for permanent use. During the weekend periods, when Friday-to-Sunday CIP procedures caused major changes in the effluent's pH, Flox®-QT was replaced by Flox®-QTH with excellent results. The previous consumption of anionic polymer in the form of commercial emulsion was extremely high; thus equipment for the dilution of powder polymer was installed allowing for approximately 70% savings in the cost of this product.

The replacement of PAC 18% with alternate application of Flox®-QT and Flox®-QTH allowed for the elimination of the alkalinizing agent (caustic soda) used for pH adjustment of the raw wastewater, thus decreasing the consumption and handling of dangerous chemicals. The impact of pH fluctuations in the wastewater caused by CIP procedures that caused difficulties with PAC 18% was minimized by the use of Flox® coagulants, which have a broader pH application range. Importantly, the volume of sludge generated in the wastewater plant decreased by 30% because Flox® coagulants generate physically compact particles with lower water content in the coagulation and flocculation process. The installation of equipment for the dilution of powder polymer also allowed for significant savings (Table 13.4). A

TABLE 13.4 Comparison of Industrial Wastewater Treatments Using Coagulants PAC 18% and Flox®-QT/Flox®-QTH in a Fish Canning Plant (Average Values).

	PAC 18%	Flox®-QT/Flox®-QTH
Coagulant (ppm)	350	150
Anionic polymer (ppm)	20	6
pH adjustment	NaOH	Eliminated
COD of treated water (mgO_2/L)	1.900	1.900
Sludge (ton per month)	300	210

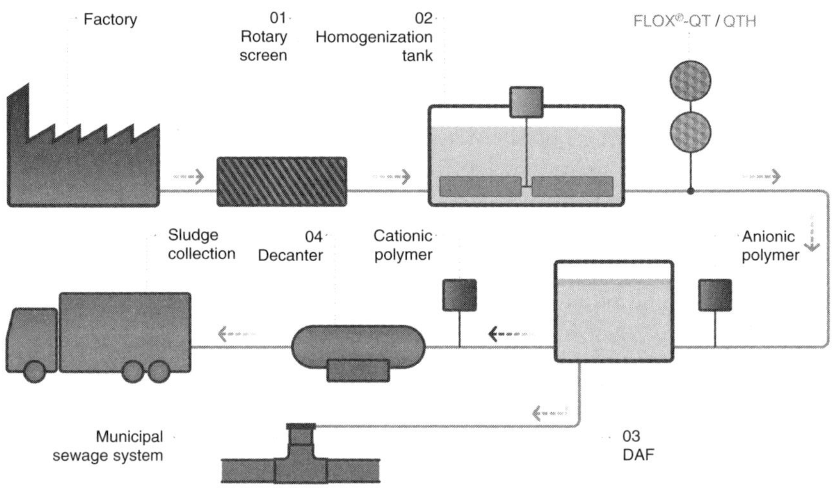

Figure 13.4 Schematic representation of the wastewater treatment process in a particular fish canning industry after the implementation Flox®-QT/ Flox®-QTH coagulants.

schematic representation of the wastewater treatment process after the implementation of Flox®-QTH coagulant is presented in Figure 13.4. The total wastewater treatment operation cost decreased by 24% (€ m^{-3} of treated effluent + sludge).

13.6.3 Poultry Abattoir

According to the European BREF in Slaughterhouses and Animal By-Products Industries, poultry slaughtering and processing industries can use 5–67 m^3 of water per ton of poultry carcass. The main source of water consumption is related to the cleaning and chilling processes. For example, it has been reported that in a particular Danish poultry slaughterhouse that processed 9000 birds per hour, the minimum water demand for external and internal wash and for chilling was approximately 56% of the total. Cleaning and washing of carcasses can represent more than 80% of the total amount of the water consumed and the wastewater generated.[9]

Poultry slaughtering and processing plants frequently produce 4000–6000 m^3 of wastewater daily. That wastewater is generally overloaded with oil, protein, and carbohydrates from blood, fat, meat, skin, feathers, and fecal material, and the major contributions to the total waste load result from the evisceration and cleaning steps. Minor contributions to the total waste load are produced in the receiving areas and vary greatly depending on the quantity of dirt, fecal material, and feather deposits generated during the holding time. The wastewater also contains inorganic matter and particulates such as grit. General plant cleanup operations can generate up to 50% of the biodegradable organic matter, or BOD. In order to comply with local and national environmental regulations, poultry processing plants have to remove significant amounts of all soluble and particulate organic materials in their wastewater.

A particular poultry abattoir industry generates approximately 9600 m^3 per month of wastewater from the slaughtering, preparation, and general plant cleanup processes (average 8 m^3/t of chicken carcass). The raw effluent had a COD of 3850 mgO_2/L and pH between 6.3 and 6.5 (average values). The effluent treatment was performed using an inorganic coagulant (PAC 18%), a cationic flocculant, and sodium hydroxide (NaOH) for pH adjustment of the treated water prior to discharge in the municipal sewer system. The treated water had a COD of 710 mgO_2/L and a pH of 5.3 (average values). The treatment was performed as follows:

- *Step 1: Removal of solids.* Solids from the raw wastewater are removed using a rotary screen.
- *Step 2: Homogenization.* The wastewaters generated in the slaughtering process (rich in organic matter) and plant cleanup (loaded with detergents and disinfectants) are homogenized in a tank upstream of the physicochemical treatment step.
- *Step 3: Coagulation and flocculation.* The coagulation and flocculation stage takes place in a tubular reactor, where PAC 18% and a cationic flocculant were used to promote the coagulation and flocculation of solid particles. The wastewater is then directed to the DAF chamber.
- *Step 4; DAF.* In the DAF chamber, flocculated particles are combined with a pressurized stream of water and extremely fine air bubbles that incorporate the suspended solids. These air–solid particles float to the surface of the flotation cell where they are separated by a skimmer and scraper mechanism and allowed to accumulate in the sludge chamber.
- *Step 5: Sludge dewatering.* The accumulated DAF sludge provides a full batch for dewatering using a sludge centrifuge and a cationic polymer to aid the thickening process. The resulting sludge cake had approximately 20–35% solids with a high content of oil and grease.

Flox®-QT potential application as a coagulant in this particular wastewater was assessed in the internal laboratory of Grove Advanced Chemicals using jar-test

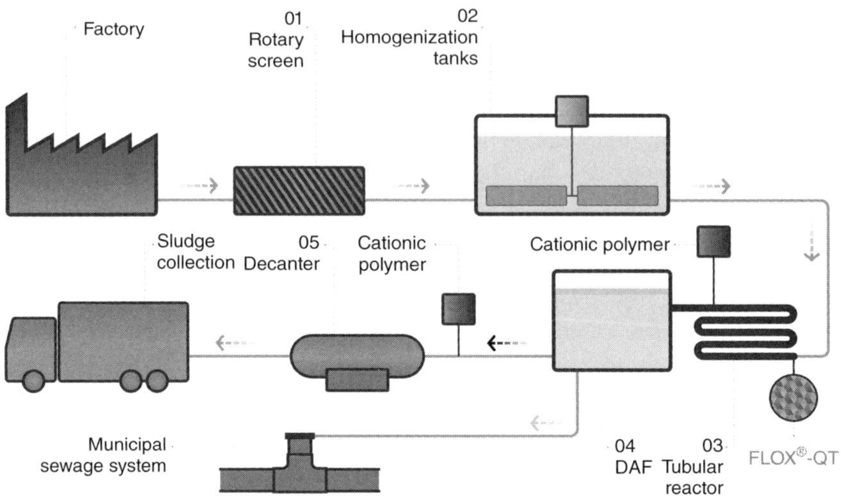

Figure 13.5 Schematic representation of the wastewater treatment process in a particular poultry abattoir after the implementation of Flox®-QT coagulant.

procedures, followed by standard COD quantification, oil and grease content assessment, and pH measurements. The preliminary results showed increased COD removal relative to the PAC 18% treatment and allowed for the elimination of the pH adjustment step of the treated water prior to discharge in the municipal sewer system. The percentage of removal of oils and greases using Flox®-QT also increased relatively to the inorganic coagulant. The subsequent full-scale tests at the wastewater treatment plant confirmed a significant decrease in sludge volume and water content when using Flox®-QT. A schematic representation of the wastewater treatment process after the implementation of Flox®-QTH coagulant is presented in Figure 13.5.

The application of Flox®-QT impacted positively the poultry abattoir wastewater treatment. The percentage of removal of COD and oils and greases increased significantly by comparison with the previous PAC 18% treatment. The use of sodium hydroxide as alkalizing agent for pH adjustment could be totally eliminated because Flox® coagulants do not lower the pH of the water under treatment (Table 13.5). The elimination of a dangerous chemical (NaOH) and the use of Flox®-QT represented potential improved health and increased operational safety for the wastewater plant workers. The total volume of the sludge was decreased by 45%, thus allowing for important economical savings in the collection, transport, and disposal of this residue. Decreasing the volume and achieving a higher dewatering of the sludge are important characteristics of the wastewater treatments that employ Flox® coagulants. This is possible because Flox® coagulants generate particle aggregates that are physically more compact and contain a lower amount of intramolecular water. The total cost of the wastewater treatment operation was reduced by 30% (€ m^{-3} treated effluent and sludge).

TABLE 13.5 Comparison of Industrial Wastewater Treatments in a Poultry Abattoir Using Coagulants PAC 18% and Flox®-QT.

	PAC 18%	Flox®-QT
Coagulant (ppm)	1110	320
Cationic polymer (ppm)	12	6
Average COD of treated water (mgO_2/L)	710	660
Average pH of treated water	5.3	6.4
pH adjustment	NaOH	Eliminated
COD remotion	81%	84%
Oil and grease remotion	60%	85%

13.6.4 Medium-Density Fiberboard

Medium-density fiberboard (MDF) is an engineered wood-based material primarily composed of softwood fibers bonded with a synthetic resin. The large-scale production of MDF panels started in North America and Europe in the 1980s. At present, MDF is heavily used in the furniture and construction industries. The magnitude of the environmental impact of this industry is significant as the production of MDF in a modern plant represents total water consumption, on site level, up to 1.52 m^3 of water for each m^3 of MDF boards produced. MDF industry wastewater is characterized by a high organic load with a normal COD of 30,000–40,000 mgO_2/L. The pH of the wastewater is typically low (around 5) because soluble organic acids are released from the wood in the fiber production stage. Chemicals added during the manufacture of MDF such as synthetic resins are also present. The nitrogen load is normally too low to support an effective biological treatment downstream to the physicochemical process.

A particular manufacturer produces 200,000 m^3 of MDF boards per year. In the manufacturing process, the wood is washed, debarked, and subsequently sent to a digester where it is processed into fibers. The water used for the washing and cooking steps in the processing of wood chips is sent to the wastewater treatment plant. The physicochemical treatment was conducted with PAC 18% as coagulant, an anionic polymer as flocculation adjuvant, and hydrated lime for pH adjustment. The raw effluent has a flow of 11,000 m^3 per month with a COD of 35,000 mgO_2/L and pH between 4.5 and 5.3 (average values). The tertiary phase of the treatment (described below) was problematic because the use of PAC 18% and bentonite clay caused frequent costly obstructions of the sand filter and the microfiltration system upstream of the reverse osmosis process:

- *Step 1: Remotion of solids.* Coarse solids are removed from the raw wastewater using a filtration grille.
- *Step 2: Homogenization 1.* The wastewaters are retained in the first homogenization tank for 24 h with the purpose of equalization of the effluent by reducing the physicochemical variability of the different factory effluents and flows. The equalized wastewater is then directed to the physicochemical treatment.

- *Step 3: Physicochemical treatment.* PAC 18% coagulant and an anionic flocculant were used to promote the coagulation of suspended particles and the formation of flocs in two reaction basins (coagulation and flocculation) upstream of the clarifier. The separation of the suspended solid particles by precipitation was enhanced by the aforementioned reagents. Soda lime was used for pH adjustment at this stage.
- *Step 4: Homogenization 2.* The clarified water is conducted to a second homogenization basin that allows the stepwise feeding of the two sequential batch reactors (SBR) and the necessary cycle adjustments to optimize the subsequent biological treatment.
- *Step 5: Biological treatment by SBR.* The SBR system is fed from homogenization tank 2 and promotes the biological removal of the dissolved and suspended organic matter. Essential nutrients required by the biological process are generally lacking in this type of effluent and are added to the water under treatment at this stage.
- *Step 6: Tertiary treatment.* After the biological treatment step, the wastewater under treatment is forwarded to a tertiary treatment consisting of a lagoon and a precipitation system that promotes the physicochemical separation of suspended solids using sand and activated carbon filters, microfiltration, and reverse osmosis. PAC 18% and bentonite were used in the tertiary treatment to promote the preliminary removal of organic and solid materials that remained in the effluent downstream of the biological treatment.
- *Step 7: Sludge dewatering.* The sludge generated by the physicochemical clarifier and the biological process (waste activated sludge) is temporarily stored in a storage tank and sent to a filter press for dewatering with the aid of a cationic polymer.

Flox® coagulants were tested in the internal laboratory of Grove® Advanced Chemicals™ using jar-test procedures and standard COD quantification and pH measurements for the assessment of optimal treatment dosages. Excellent results were also achieved in subsequent industrial-scale tests in the wastewater treatment plant, leading to the definitive adoption of Flox®-QT in the physicochemical process (Table 13.6). At a later stage, after the stabilization of the biological process, PAC 18% and bentonite were replaced by the conjoint application of Flox®-QTH and an anionic flocculant in the tertiary treatment. The tertiary treatment also showed significant improvements, leading to the definitive adoption of Flox®-QTH also in this step of the wastewater treatment (Table 13.7).

The use of Flox® coagulants impacted positively the MDF wastewater treatment plant. The biological treatment achieved greater stability through the use of Flox®-QT and allowed for an average decrease of 20% in the organic load (COD) of the treated water. Improvements were particularly significant in the tertiary phase because of increased efficiency in the removal of suspended solids. The removal of COD in the osmosis step was comparable to Flox®-QTH and PAC 18% treatments (90–120 mgO_2/L). However, Flox®-QTH does not increase the conductivity of the

TABLE 13.6 Comparison of Wastewater Physicochemical Treatment in a Medium-Density Fiberboard (MDF) Plant Using PAC 18% and Flox®-QT Coagulants.

	PAC 18%	Flox®-QT
Coagulant (ppm)	6.700	2.120
Anionic polymer (ppm)	8	2
Average COD (mgO_2/L)	6500	6500
Hydrated lime (kg per day)	600	200

TABLE 13.7 Comparison of Industrial Wastewater Tertiary Treatment in a Medium-Density Fiberboard (MDF) Plant Using PAC 18% and Flox®-QTH Coagulants.

	PAC 18%	Flox®-QTH
Coagulant (ppm)	600	800
Bentonite	1800	Eliminated
Anionic polymer (ppm)	None	2
Tertiary treatment (m^3 per month)	300	450

water under treatment, thus allowing for increased efficacy and durability of costly equipment such as filters and membranes. This important characteristic of Flox® coagulants allowed for a 50% increment in the flow of the wastewater subjected to osmosis. A schematic representation of the wastewater treatment process after the implementation of Flox®-QTH coagulant is presented in Figure 13.6. Rationalization of manpower and improved health and operational safety of the workers through the use of a nontoxic product and decreased use of hydrated lime were also important improvements.

The use of Flox® coagulants allowed for an overall 27% (€ m^{-3} treated effluent) savings in the wastewater treatment. Notably, the impact on the operation cost of the tertiary system was more than 50% of the total savings, mostly due to the decreased use of wastewater treatment chemicals and a longer useful life of the filters and membranes. Importantly, part of the wastewater treated in the tertiary phase was reused in the MDF plant.

13.6.5 Dairy Industry

Milk is composed of approximately 87% of water, with the remainder being protein, fat, lactose, calcium, phosphorus, iron, and vitamins. Cow's milk primarily is consumed and transformed into dairy products such as cheese, butter, and cream, but goat, sheep, and buffalo's milk are also consumed in significant quantities worldwide. Accordingly to European BREF in the Food, Drink and Milk Industries, water consumption in the dairy industry may vary due to different portfolios of products, batch

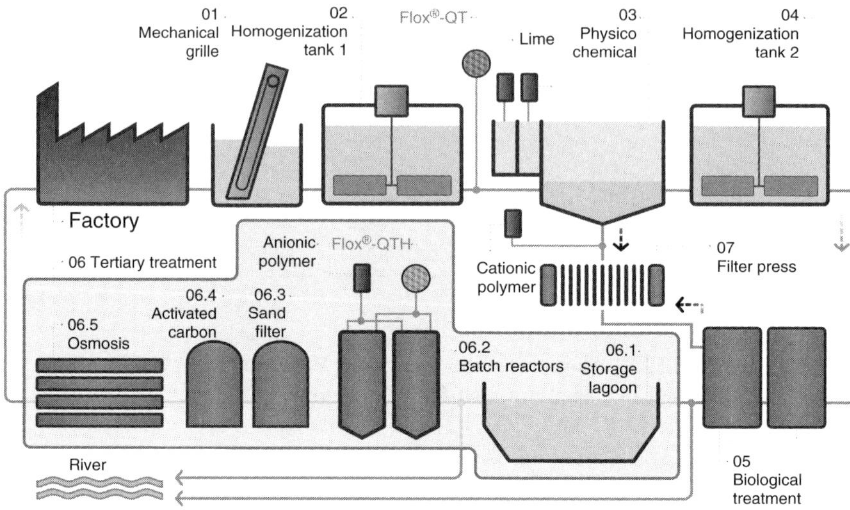

Figure 13.6 Schematic representation of the wastewater treatment process in a particular medium-density fiberboard (MDF) industry after the implementation of Flox®-QT coagulant.

sizes, and cleaning processes.[8] For example, the water consumption in the production of market milk from 1 liter of received milk is 0.6–1.8 L/L. The production of milk powder from 1 L of received milk using advanced equipment and good operation methodologies may consume 0.8–1.7 L/L of water. Water consumption in standard procedures may vary between 0.8 and 60 L/kg of milk processed, including water for cooling, depending on the final products. Wastewater emissions can vary between 0.4 and 60 L/kg of processed milk, also depending on the final products. The main contributor to wastewater production in the dairy industries is the cleaning of transport cisterns and productive areas. The implementation of CIP procedures can result in up to 65% savings in clean water consumption. Another important contribution to the wastewater is the contaminated condensate from milk and whey drying towers.

For the assessment of Flox® series, a particular dairy industry generates approximately 1200 m^3 per day of wastewater with an average COD of 2500 mgO_2/L. The wastewater treatment used PAC 18% as coagulant and an anionic polymer as a flocculation adjuvant. Due to a normally alkaline pH, the process required the constant use of CO_2 to decrease pH ranges and ensure the efficient performance of coagulation and flocculation chemicals. The treatment generated large sludge volumes that led to the need of using two decanters simultaneously for dewatering. This continuous process largely exceeded the installed dewatering capacity and led to the need of liquid sludge storage equipment in the event of mechanical failure of the decantation system.

The treatment was performed as follows:

- *Step 1: Homogenization.* The wastewaters generated in the manufacturing processes and general plant cleaning operations are homogenized in a tank upstream of the physicochemical treatment.

- *Step 2: Coagulation and flocculation.* PAC 18% and an anionic flocculant were used to promote the coagulation and flocculation of particles in a coagulation and flocculation reactor. Adjustment of pH prior to the addition of PAC 18% was performed with CO_2. The wastewater is then directed to the DAF chamber.
- *Step 3: DAF.* The flocculated particles are combined with a pressurized stream of water and extremely fine air bubbles that incorporate in the suspended solids. The air–solid particles float to the surface of the flotation cell and are separated from the water by a skimmer and scraper mechanism and directed into the sludge chamber.
- *Step 4: Sludge dewatering.* The thickened DAF sludge accumulated provides a full batch for dewatering using two sludge decanters and a cationic polymer to aid the thickening process. The disposal of the sludge must comply with local environmental regulations.

The potential use and treatment dosage estimates of Flox®-QT were tested by Grove Advanced Chemicals at the laboratory scale using jar-test procedures and standard COD and pH quantification. The objective consisted of eliminating a corrosive chemical from the process, nevertheless guaranteeing the appropriate values for the parameters legally defined for wastewater discharge at a lower cost. The good results obtained in the laboratory led to successful industrial-scale implementation of Flox®-QT and its approval for permanent use in the wastewater treatment plant.

The use of Flox®-QT allowed for a decrease in the organic load (COD) of the treated water without alteration of the original pH and conductivity. The replacement of PAC 18% by Flox®-QT also allowed for savings in the consumption anionic flocculant of approximately 68%. The volume of sludge generated by the wastewater treatment with Flox®-QT decreased by 43%. A decreased volume of sludge resulted in lower consumption (50% less) of cationic polymer for sludge dewatering and also enabled a more efficient operation of the dewatering system (Table 13.8). It became possible to alternate the use of the two decanters, thus allowing for maintenance operations and energy-saving stop periods of several hours a day. A schematic representation of the wastewater treatment process after the implementation of Flox®-QTH coagulant is presented in Figure 13.7.

The treatment did not eliminate CO_2 for pH adjustment of the normally alkaline wastewater because Flox®-QT does not cause pH alterations. In fact, CO_2 consumption increased with the use of Flox®-QT because PAC 18% has the property of lowering the pH of water under treatment. Several months after the successful implementation of Flox®-QT in the wastewater treatment plant, an adjuvant was added to improve the coagulation efficiency in higher pH ranges, thus lowering significantly CO_2 consumption while achieving good results even above pH 8.0. Within a normal pH range (6–7.5) the adjuvant dosage was negligible or unnecessary; at higher pH values (>7.5), a small dosage of adjuvant allowed for a stable and efficient treatment with savings in CO_2.

The overall cost of the wastewater treatment and sludge dewatering significantly decreased even though an adjuvant was added to the process and CO_2 consumption increased. The total monthly operation cost decreased by 10% (not including the

TABLE 13.8 Comparison of Wastewater Treatments in a Dairy Plant Using PAC 18% and Flox®-QT Coagulants.

	PAC 18%	Flox®-QT
Coagulant (ppm)	778	268
Anionic polymer (ppm)	11	8
Adjuvant (ppm)	None	72
COD of treated water (mgO_2/L)	1000	800
pH of treated water	6.2–6.3	6.2–7.5
CO_2 (ton per month)	3	7
COD removal (%)	60	68
Cationic polymer (kg per month)	1000	500
Sludge (ton per month)	150	85

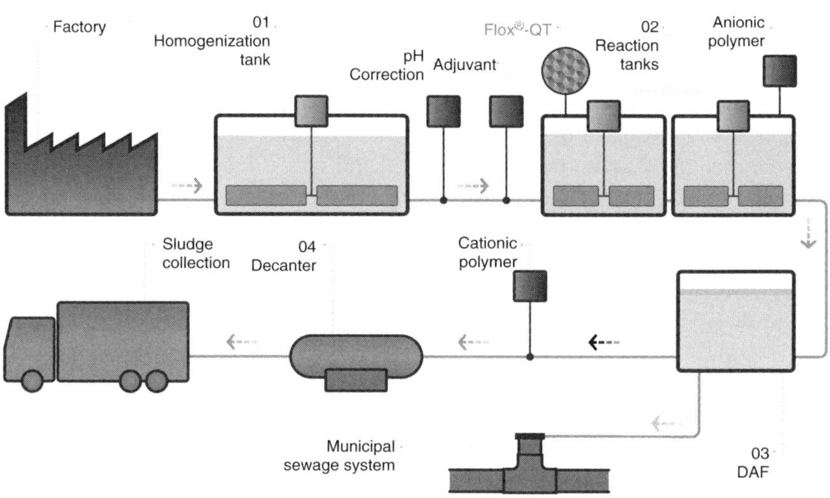

Figure 13.7 Schematic representation of the wastewater treatment process in a particular dairy industry after the implementation of Flox®-QT coagulant.

energy savings due to reduced operation time of the decanter), distributed mainly between the savings in anionic and cationic polymers, and the decreased volume of the sludge generated in the treatment.

13.6.6 Treating Drinking Water with Flox®-QT

The quality of drinking water is one of the most important public health issues worldwide. In many cases, drinking water is produced by treating freshwater obtained from natural resources such as groundwater and surface water bodies. The coagulation process applied on the clarification of freshwater to produce drinking water presently relies on inorganic chemicals, most often aluminum sulfate. The intake of aluminum has growingly been accepted as representing a serious risk factor for the development or acceleration of the onset of Alzheimer's disease in humans.[10]

A possible health-based guideline for the maximum aluminum concentration in drinking water is 0.9 mg/L, taking into account that 20% of the aluminum provisional tolerable weekly intake (PTWI) originates from residual aluminum added to drinking water during the coagulation process.[11] However, practicable levels based on the optimization of the coagulation process in drinking water plants using aluminum-based coagulants are less than 0.1 mg/L in large water treatment facilities and less than 0.2 mg/L in small facilities. In order to minimize the deposition of potentially soluble aluminum in the distribution systems, these values of residual aluminum should not be exceeded.[11] The European Directive 98/83/EC on the quality of water intended for human consumption set the indicator parameter for aluminum at 0.2 mg/L. Nevertheless, there has been presently a general effort to limit the intake of aluminum and thus the quantity of aluminum salts used in DWT not only through regulatory limits but also by the search for alternatives to the conventional aluminum-based coagulants such as aluminum sulfate.[12] Accordingly to the NSF/ANSI Standard 60 (DWT Chemicals – Health Effects), Flox®-QT (trade name) is safe for DWT.

The results of the laboratory trials performed by Grove Advanced Chemicals to study the potential application of Flox®-QT in the coagulation process of a DWT plant in the north of Portugal, in comparison with an aluminum-based coagulant therein implemented, are here reported. The coagulation and flocculation assays with monitoring of pH and turbidity were performed on the raw water samples pumped from an adjacent river over a period of six months. An anionic flocculant was used to aid the process of precipitation of solids, mainly in the tests performed with the aluminum-based coagulant. The results showed that Flox®-QT achieved higher turbidity reduction than aluminum sulfate and did not decrease the pH of the water after the treatment. Furthermore, the use of Flox®-QT allowed for the elimination of the anionic flocculant in the coagulation process (Table 13.9). An analogous study was also performed to compare the performance of Flox®-QT relatively to aluminum sulfate ($Al_2(SO_4)_3$, 8.5% commercial solution) in the treatment of river water and lake water to produce water for human consumption. Optimal coagulant dosage, turbidity,

TABLE 13.9 Turbidity and pH of River Raw Water and Treated Water Comparative Results Obtained in Laboratory-Scale Tests Using Flox®-QT and Aluminum Sulfate ($Al_2(SO_4)_3$) in the Coagulation and Flocculation Process of Drinking Water Treatment (Average Values).

	Raw Water	Flox®-QT	$Al_2(SO_4)_3$
Dosage (ppm)	–	16.6	47.0
Turbidity (NTU)	2.5	0.68	0.91
Turbidity decrease (%)	–	72.8	63.7
pH	7.39	7.34	6.98
pH variation (%)	–	1.35	5.07

TABLE 13.10 Turbidity and pH of Raw River Water Comparative Results in the Treated Water Obtained in Laboratory-Scale Tests Using Flox®-QT and Aluminum Sulfate in the Coagulation Process of Drinking Water Treatment.

	Raw River Water	Flox®-QT	$Al_2(SO_4)_3$
Optimal dosage	–	20	40
Turbidity	7.03	0.43	1.51
UV absorbance (m^{-1})	5.2	5	4.25
UV absorbance reduction (%)	–	1	18.27
pH	8.23	8.28	7.98
Al^{3+} (mg/L)	–	0	1.76

TABLE 13.11 Turbidity and pH of Raw Lake Water Comparative Results in the Treated Water Obtained in Laboratory-Scale Tests Using Flox®-QT and Aluminum Sulfate in the Coagulation Process of Drinking Water Treatment.

	Flox®-QT	$Al_2(SO_4)_3$	Raw Lake Water
Optimal dosage	120	140	
Turbidity	1.95	0.98	14.2
UV absorbance (m^{-1})	21.03	9.95	19.82–26
UV absorbance reduction (%)	20.13	49.8	
pH	7.65	6.68	7.65
Algae (cells/L)	5500	12,460	1,031,200
Algae reduction (clarified water/raw water) (Log)	2.27	1.92	
Al^{3+} (mg/L)	0	6.87	

metal content (Al^{3+}), pH, and algae removal were monitored. Organic matter content of treated water was measured through UV absorbance at 254 nm. The best results for turbidity reduction obtained for each coagulant tested are presented in Tables 13.10 and 13.11.

Results from river water show a good turbidity reduction with Flox®-QT and a pH almost unaltered, with lower dosages, when compared to the inorganic coagulants. The metal concentration analysis showed increased and Al^{3+} content on each respective trial tests as already expected. Organic matter content was removed at higher rates with inorganic coagulants.

The trials performed with lake water and river water showed similar results in terms of increased metal (Al^{3+}) concentration and pH changes for the metal-based coagulant. Flox®-QT performed better than aluminum sulfate in algae removal, but its performance in terms of turbidity and organic content was inferior. Both in river water and lake water, the comparatively lower performance of Flox®-QT in the decrease of organic matter content may be explained by the persistence of unreacted dissolved coagulant in the clarified treated water after the precipitation of solids. In fact, Flox®-QT molecules are tannin based and known to strongly absorb

UV light.[13] Further tests should be performed using alternative methods for the quantification of organic matter.

In conclusion, Flox®-QT stands as a promising alternative to inorganic metal-based coagulants, namely, aluminum sulfate, regarding the treatment of drinking water for human consumption. Flox®-QT does not contribute with metals (such as aluminum), is efficient for decreasing turbidity, and allows for the elimination of an alkalizing agent because it does not alter the pH of the treated water.

Due to its bridging action between colloidal particles, Flox®-QT also eliminates the need for a flocculation adjuvant (e.g., acrylamide-based polymer). Further trials should be performed to study the organic matter elimination capabilities of these coagulants (Flox®-QT). A different method for organic matter quantification should be used in order to eliminate possible quantification bias due to the inherent error of UV absorbance at 245 nm. The synergistic use of Flox®-QT with lower dosages of an inorganic coagulant constitutes a promising application that deserves further studies.

13.6.7 Recycling Sludge from Flox® Coagulants: Beneficial Uses and Valorization

Water and wastewater treatment plants produce extremely large quantities of sludge every year. Although there are many alternatives for the final disposal of sludge, landfilling is regrettably still a technically and economically viable solution for these biodegradable wastes in many countries.[14] Remarkably, depending on their origin, water and wastewater treatment sludge (WTS) can be rich in organic matter and essential nutrients and micronutrients, rendering it extremely valuable for land disposal and soil enrichment and amendment. Despite the economic, technical, and environmental advantages, this alternative is not a simple issue because these biosolids (especially sewage sludge) are potentially contaminated with many unknown substances and microorganisms, including unstudied contaminants such as the contaminants of emerging concern (CECs) that are presently raising many questions regarding their impact on public health and environment.[15] Additionally to the original water and wastewater contamination, the chemicals used in the treatment itself can also be a source of contamination. This treatment-derived contamination is a well-known reality in WTS, resulting from the use of aluminum and iron-based coagulants. For example, it has been shown that the sludge generated in DWT with ferric chloride and aluminum sulfate exceeded the soil and groundwater thresholds for these compounds, as well as trace elements of arsenic and vanadium. The same study suggested a major source of the arsenic and vanadium in the WTS could be the chemical treatment itself.[16]

Studies on soil application of sludge originated from water treatment using aluminum sulfate showed that this strategy may, in determined quantities, hinder plant growth because alum chemically adsorbs soil phosphorus, rendering it unavailable for plants, creating a deficiency in an essential nutrient.[17] Flox® coagulants do not share this feature with inorganic metal-based products so that binding with phosphorus, thus creating phosphorus compounds not available for uptake by plants uptake, is not an issue in soil disposal of sludge-containing residues of Flox®. We have recently performed characterization studies of WTS from a dairy industry and fish canning industry where Flox® coagulants were used in physicochemical phase of the wastewater treatment (data not shown). The results showed that these WTS were safe for disposal in agricultural soil according with the Portuguese legislation DL n° 118/2006 reflecting the European Council Directive 86/278/EEC on the protection of the environment, and in particular of the soil, when sewage sludge is used in agriculture. Additionally, these WTS have shown a very high energetic potential, that is, high lower calorific value, and could be valorized for energy recovery. In conclusion, WTS generated with the use of Flox® coagulants has an interesting potential for valorization through different applications that deserves further assessment.

13.7 FUTURE PERSPECTIVES

Water is a vital substance that was viewed as an infinite resource in the past. At present, misuse and growing demand have had such a strong impact on reserves of fresh clean water that conflicts over access to water may reach gigantic dimensions and seriously disrupt the peace in some parts of the world. It widely agreed that water crisis will increase worldwide, calling for our best technical and management innovations in the next decades, of which appropriate treatment for reutilization or low-risk disposal in the environment will certainly play a central role in the discussion of a sustainable future. Pure water is rare in nature because of its ability to solubilize a wide range of composts (e.g., colloidal particles, organic matter, microorganisms, and other substances) that are detrimental to human health and the environment and have to be removed before consumption or disposal. Any compost introduced in the treatment of water should contemplate the possibility that residues may be carried to the consumer or the environment. Natural polymers such as Flox® coagulants represent real alternatives for the elimination of products potentially dangerous to public health with the advantage of being readily obtained dispensing the use of complex chemical synthesis.

Although natural products for water treatment have been known since ancient times, their large-scale application remains at its infancy. Flox® coagulants have played a pioneer role in water treatment and are still to this day one of the very few products that are vegetal-derived, sourced from renewable plantations, and employed at the industrial scale with excellent performance and economic advantages.

A review of the scientific literature shows that plant-based coagulant sources, processes, and effectiveness and relevant coagulating mechanisms for the treatment of water and wastewater are not being ignored by researches in many parts of the world. However, these coagulants are generally being employed only as point-of-use technology in less developed communities where they are relatively cost-effective compared to chemical coagulants.[18] Frequently studied plant-based coagulants include *Strychnos potatorum,*[19] *Moringa oleifera,*[20] tannins,[21] *Schinopsis balansae,*[22] and *Opuntia ficus-indica.*[23] In general, these products are reported to be comparably or superiorly efficient in the removal of turbidity of water to metal coagulants and at lower dosage and were shown to be independent of temperature, pH correction, and alkalinity of the water under treatment.

We believe vegetal-derived polymers are certainly part of the avenues to pursue in the quest for green water and wastewater treatment. The success of Flox® coagulants is, hopefully, one of many future examples in the paradigm change that we like to express in a simple way: "Why treat contaminated water with a contaminant when you have a natural alternative?"

REFERENCES

[1] United Nations, *Report of the World Commission on Environment and Development, Our Common Future*, (Ed.: United Nations), World Commission on Environment and Development, **1987**.

[2] J. Bratby, *Coagulation and Flocculation in Water and Wastewater Treatment*, 2nd Edition, IWA Publishing, London, **2006**.

[3] S.D. Faust, O.M. Aly, *Chemistry of Water Treatment*, 2nd Edition, CRC Press, Boca Raton, **1998**.

[4] J.C. Crittenden, R.R. Trussell, D.W. Hand, K.J. Howe, G. Tchobanoglous, *MWH's Water Treatment: Principles and Design*, 3rd Edition, John Wiley and Sons Inc, New Jersey, **2012**.

[5] D.W. Hendricks, *Water Treatment Unit Processes: Physical and Chemical*, 1st Edition, CRC Press, Boca Raton, **2006**.

[6] J.C. Doran, J.W. Turnbull, P.N. Martensz, L.A.J. Thomson, N. Hall, In "*Australian Trees and Shrubs: Species for Land Rehabilitation and Farm Planting in the Tropics*", Vol. ACIAR Monograph n°24 (Eds. J.C. Doran, J.W. Turnbull). Australian Centre for International Agricultural Research, Canberra, **1997**.

[7] European Commission, *Integrated Pollution Prevention and Control (IPPC) Reference Document on Best Available Techniques in the Ceramic Manufacturing Industry,* (http://eippcb.jrc.ec.europa.eu/reference/BREF/cer_bref_0807.pdf), **2007**.

[8] European Commission, *Integrated Pollution Prevention and Control (IPCC) Reference Document on Best Available Techniques in the Food, Drink and Milk Industries* (http://eippcb.jrc.ec.europa.eu/reference/BREF/fdm_bref_0806.pdf), **2006**.

[9] European Commission, *Integrated Pollution Prevention and Control (IPCC) Reference Document on Best Available Techniques in the Slaughterhouses and Animal By-products Industries,* (http://eippcb.jrc.ec.europa.eu/reference/BREF/sa_bref_0505.pdf), **2005**.

[10] L. Tomljenovic, *J. Alzheimers Dis.* **2011**, *23*, 567–598.

[11] WHO, *Guidelines for Drinking-water Quality*, 4th Edition, World Health Organization, **2011**.

[12] European Commission, *Council Directive 98/83/EC of 3 November 1998 on the quality of water intended for human consumption,* Official Journal of the European Communities, **1998**, 330–332.

[13] A.D. Eaton, M.A.H. Franson, In "*Standard Methods for the Examination of Water and Wastewater*", 21st Edition (Joint Editorial Board: American Public Health Association, American Water Works Association and Water Pollution Control Federation), American Public Health Association, Washington, DC, **2005**.

[14] R.J. LeBlanc, P. Matthews, R.P. Richard, *Global Atlas of Excreta, Wastewater Sludge, and Biosolids Management: Moving Forward the Sustainable and Welcome Uses of a Global Resource,* United Nations Human Settlements Programme (UN-HABITAT), Kenya, **2008**.

[15] (a) A. Luczkiewicz, *Pol. J. Environ. Stud.* **2006**, *15*, 869–876; (b) J.M. Mathney, *New Solut.* **2011**, *21*, 43–56.

[16] P. Jain, Y. Jang, T. Tolaymat, M. Witwer, T.G. Townsend, *J. Residuals Sci. Tech.* **2005**, *2*, 13–23.

[17] H. Rigby, D. Pritchard, D. Collins, K. Walton, N. Penney, *Environ. Technol.* **2013**, *34*, 1359–1368.

[18] C.-Y. Yin, *Process Biochem.* **2010**, *45*, 1437–1444.

[19] R. Babu, M. Chaudhuri, *J. Water Health* **2005**, *3*, 27–30.

[20] J. Sánchez-Martin, K. Ghebremichael, J. Beltrán-Heredia, *Bioresour. Technol.* **2010**, *101*, 6259–6261.

[21] J. Sánchez-Martín, J. Beltrán-Heredia, J. Gragera-Carvajal, *Ind. Crop Prod.* **2011**, *34*, 1238–1240.

[22] J. Beltrán-Heredia, J. Sánchez-Martín, G. Frutos-Blanco, *Sep. Purif. Technol.* **2009**, *67*, 295–303.

[23] (a) D.I. Fox, T. Pichler, D.H. Yeh, N.A. Alcantar, *Environ. Sci. Technol.* **2012**, *46*, 4553–4559; (b) A.L. Buttice, J.M. Stroot, D.V. Lim, P.G. Stroot, N.A. Alcantar, *Environ. Sci. Technol.* **2010**, *44*, 3514–3519.

14

HELIAE DEVELOPMENT, LLC: AN INDUSTRIAL APPROACH TO MIXOTROPHY IN MICROALGAE

ENEKO GANUZA[1], ANNA LEE TONKOVICH[1] AND BÁRBARA VAN ASCH[2]

[1]*Heliae Development, LLC, Gilbert, AZ, USA*
[2]*Grove Advanced Chemicals SA, Maia, Portugal*

14.1 PREAMBLE

When I moved out of the university environment into my first industry job, I had the chance to participate in the acquisition of a 300 L fermentation unit. Our culturing process was well defined for bench reactors, and we were ready to jump to the next level. The budget was small and we ended up acquiring a second-hand fermentation unit for a very reasonable price. On the downside, the unit was missing pieces and required some work before the unit would be operational. When I first met with the technical services personnel who were responsible for commissioning the unit, I was asked a question that I did not expect: "How clean do you want the unit?" That question really surprised me because my experience with bench-scale fermentation

Industrial Biorenewables: A Practical Viewpoint, First Edition. Edited by Pablo Domínguez de María.

work in the academic setting led me to consider only sterile or nonsterile conditions. I never thought about having to consider the risk of contamination (sterility assurance level) for a reactor. As they explained, there are choices related to the stirrer seal, the sampling setup, and other unit configurations that would significantly affect the end budget of the commissioning process. Of course I was scared that I would fail on my first assignment due to embarrassing contaminations, and we could not exceed our approved budget. With only minor modifications during the setup, we were able to get the unit running and within the approved budget, and later modifications to the process improved robustness to a satisfactory level. In other words, to achieve the production and economic targets, we implemented restrictive growth conditions that favored our organisms of interest. This early learning helped me to better understand the challenges that I will be facing later in my career. The production of microalgae with restriction on the cost per kilogram of biomass required the development of open commercial system where the microalga is in constant competition for resources with contaminating organisms.

14.2 INTRODUCTION TO HELIAE DEVELOPMENT LLC

In 2008, Heliae was founded on the premise that microalgae could provide significant value to society if commercialized with the appropriate technology. With financial support from Mars family members and the Salim group – both strong leaders in the food, agriculture, and industrial sectors – Heliae had an opportunity to drive technology success into microalga product development and commercialization.

Microalgae are a very large and diverse group of prokaryotic and eukaryotic microorganisms. Most of these microorganisms are capable of photosynthetic growth using an inorganic mineral salt medium, and this presents some interesting biotechnological opportunities. With industrial microbiology being dominated by heterotrophic fermenter-based processes, Heliae began growing microorganisms using sunlight as a free and sustainable energy source. The lack of an organic substrate in phototrophic cultures not only reduces costs of production medium but also more significantly imposes growth restrictive conditions for potential contamination by heterotrophic microorganisms. As a result, sterility requirements for a photoautotrophic system are not as crucial as with systems using organic substrates in the medium as an energy source. Microalgae can be cultivated commercially outdoors in photoautotrophic systems comprising an open pond, thus circumventing the costs associated with axenic operation, including the financial restraints of high sterility conditions mentioned above in "Preamble," which are typically achieved through steam-sterilized stainless steel vessels.

Having showed our investors that microalgae have the capacity to create valuable products, the primary approach of Heliae has been to develop breakthrough cultivation technologies that will radically decrease production costs by breaking the slow autotrophic growth paradigm. Heliae believes that cost reduction is key to increasing a microalgal presence in our society in the broader food, agriculture, and industrial sectors. A strategy for reducing the production costs will likely have multiple components. As one component, optimization of the microalgal culture medium has been important in an effort to reduce raw material costs. In addition, Heliae has developed inexpensive cultivation and harvesting systems to further reduce the microalga production costs, closer to those used in agriculture, rather than the systems used in industrial microbiology. Despite advances made by Heliae in those components, microalgal economics is still determined by the low titer or biomass yields attained in photoautotrophic systems.[1] The density of photoautotrophic cultures is typically below 1 g cell dry weight L^{-1} for open ponds and less than 10 g cell dry weight L^{-1} for photobioreactors, which is one or two orders of magnitude below the density found in other heterotrophic fermenter-based cultures.[2,3] The reason behind this disparity is the inherent limitation of the energy source being supplied to a photoautotrophic culture, as 90% of the incident light in a cell culture containing microalgae at a density of 1 g/L cell dry weight would only penetrate the first millimeter depth through the culture.[4] Most of that light would be dissipated in heat and would not be available for the cells deeper into the culture; thus photoautotrophic cultures are typically dilute. An improvement in titer or specific production rate will impact every economic aspect of the bioprocess, from the upstream to the downstream requirements. In this context, Heliae has focused on the development of technological breakthroughs to boost microalgal operating densities while simultaneously maintaining process simplicity to achieve the targeted microalga biomass production costs.

14.3 MIXOTROPHY

Microalgae comprise a group of microorganisms with remarkable metabolic versatility. Microalgae have been traditionally described as photoautotrophic organisms and were classified taxonomically using photosynthetic pigment types and morphological features. Nevertheless, many microalgal species uptake and oxidize organic substrates for energy or for a carbon source.[5] Furthermore, some microalgae are incapable of photosynthesis. For example, *Prototheca*, *Nitzschia leucosigma*, and *Astasia* are strains that were isolated using chemoheterotrophic conditions in the absence of light. Detailed taxonomic analysis of these microalgae revealed that even though they lacked photosynthetic pigments (apochloric), they were close relatives of known photoautotrophic strains, *Chlorella protothecoides*, *Nitzschia* sp., and *Euglena*, respectively. Each of these microalgae belongs to very different taxonomic phyla (Chlorophyta, Heterokontophyta, Euglenozoa, respectively), which suggests that heterotrophic metabolisms could drive growth of a wide taxonomic spectrum of microalgae. In fact, the line defining something as essential as the energetic metabolisms of the cell seems to be narrow among microalgae. The evolutionary origin of the chloroplast, the light-harvesting organelle, is theorized as a symbiotic episode between a heterotrophic eukaryote that captured and enslaved a phototrophic endosymbiont.[6] In this context, a photoautotroph – such as the eukaryotic diatom *Phaeodactylum tricornutum* – has an ancestral history that includes a heterotrophic eukaryote. Oligotrophic conditions in the ocean may explain why heterotrophy was no longer advantageous during the evolutionary history leading to *P. tricornutum*. However, the biochemical pathways for heterotrophy were not completely lost despite millions of years of evolution. For example, Zaslavskaia et al.[7] demonstrated through genetic engineering that a glucose transporter was the only missing link that would allow this particular photosynthetic organism to "recover" its heterotrophic capacity and utilize glucose as an energy source for growth in the absence of light. Microalgae comprise a group of mostly photoautotrophic organisms, to which the separation from organisms utilizing heterotrophic metabolism is exceedingly narrow as illustrated by mixotrophic, heterotrophic microalga species or obligate autotrophic species such as *P. tricornutum*. Consequently for biotechnological advances, the combination between two trophic types (i.e., heterotrophic and phototrophic) can be exploited as a property genuine to many microalgae.

The heterotrophic metabolism of microalgae has major implications in the commercialization of biotechnology. As an example, *Crypthecodinium cohnii* is a dinoflagellate that is usually found in decaying seaweeds, feeding on organic material for an energy source.[8] *Schizochytrium* is a marine heterokont microalgae that is frequently isolated from mangrove areas, often associated with dead leaves.[9] The heterotrophic capacity of these species has been a key to supporting the development of bioprocesses based on the industrial fermentation of these strains, using available fermentation technology that is well established for bacteria, fungi, or yeast. Martek Biosciences Corporation and Omega Tech, which were ultimately acquired by Royal DSM, successfully developed a commercial bioprocess using the heterotrophic capacity of those microalga species to produce omega-3 docosahexaenoic acid (DHA), a key ingredient in infant formulae and dietary supplements. Heterotrophic growth in fermenters resulted in larger titre than in photoautotrophic systems, with the heterotrophic cultures typically attaining two orders of magnitude higher yields for both biomass and lipids than photoautotrophic cultures.[3] While other photoautotrophic microalgae are capable of accumulating DHA,[10] the high titrate obtained in heterotrophy has been important to sustain the economics of this commercial process. In addition, DHA oil is considered a specialty lipid since it cannot be produced from traditional agricultural crop sources.[11] Other microbial fermentation processes have failed in the past to compete with agriculture sources, such as in the production of cocoa butter equivalent to compete with cocoa or in the production of an oil rich in gamma-linoleic acid to compete with borage. Thus, DHA oil is considered a high-value product due to limited supply and is being successfully commercialized in a niche market with high demand.[12] This allows for the high production costs associated with axenic operation in stainless steel fermenters and utilization of organic carbon substrate as an energy source in the production of DHA to be offset by the high product value. While heterotrophic technology can be economically justified for some high value niche product markets in biotechnology, the high costs may not be justifiable for broader food, agriculture, and industrial sectors of biotechnology.

As discussed previously with heterotrophy, the metabolic energetics has major implications in the design and economics of a commercial bioprocess. Photoautotrophic microalgae can be grown in nonaxenic low-cost open ponds in an inorganic medium, while heterotrophic microalga are produced in more expensive closed fermenters that are steam sterilized. When comparing these two systems, *a priori* is uncertain which type will be best for scaling up mixotrophic microalgae. Mixotrophy has been the subject of much academic interest[13] but has so far received little attention from industry. Mixotrophic microalgae metabolize inorganic and organic carbon sources simultaneously, with the energy source driving growth being supplied by both light and an organic carbon substrate (see Table 14.1 for more details on the different mixotrophic modes and definitions).

Potentially, microalgae utilizing mixotrophic metabolism provide the versatility to adapt to different scale-up technologies. For instance, *Chlorella* sp. is grown in either a heterotrophic mode using fermenters (e.g., Solazyme Inc.) or in photoautotrophic mode using most frequently open pond systems. Nevertheless, mixotrophic cultures combining heterotrophic and phototrophic growth have rarely been scaled up, and mixotrophy may provide more technological options for achieving the economic target for commercial production.

Laboratory studies have shown that the addition of an organic carbon substrate to an illuminated culture increases the amount of building blocks and energy (ATP) available to the mixotrophic algal cell, thus improving the culture's productivity over that obtained by photoautotrophic conditions. Heterotrophic metabolism minimizes the microalga's dependency on sunlight or climatic conditions and allows mixotrophic cultures to achieve cell densities well above those limited by the photosynthetic compensation point. Therefore, mixotrophy could potentially increase the titre, which if implemented correctly improves microalgal process economics and expands the range of microalgae for commercial development. An independent open pond sensitivity analysis concluded that microalga production costs were primary affected by the titre, as opposed to other engineering aspects that have a much lower impact on the production costs.

Mixotrophy has also benefits over heterotrophy where organic carbon substrates are one of the key components in the economics of the microalga production process. When light is added to a heterotrophic culture, creating mixotrophic conditions, the overall microalga growth and the organic carbon substrate yields are typically better than those observed for a conventional heterotrophic mode (i.e., in the absence of light). The mixotrophic growth model is often experimentally described as the superposition of the photoautotrophic and heterotrophic growth. For example, the growth of *Chlorella* with light and organics was the same as the sum of the photoautotrophic and heterotrophic cultures, and therefore more biomass was produced per unit of substrate in the mixotrophic mode. Under mixotrophy, the organic substrate is used as an energy source for heterotrophic growth; importantly, the inorganic carbon released during cellular respiration is used directly as CO_2 to support photoautotrophic growth as well.[14] Complementarily, oxygen produced through photosynthesis is available in turn for cell respiration. This metabolic versatility supports an elegant method for CO_2 fertilization, circumventing the difficulties of dissolving CO_2 gas into a culture broth through conventional technologies such as bubbling. CO_2 enrichment has been shown to improve photoautotrophic growth, but such enrichment will not improve mixotrophic growth and is a redundant addition to the process (see Figure 14.1).

TABLE 14.1 Microalgae Classified According to Their Trophic Metabolism and Other Terms Used in the Chapter.

	Growth Driving Energy Source		
Concept	Photosynthesis	Uptake of Organic Carbon	Details
Obligate photoautotrophs	Yes	None	Note that an obligate photoautotroph does not exist except with respect to specified conditions
Obligate heterotrophs	None	Yes	–
Facultative photoautotrophs	Yes	Yes	Separately, consecutively
Mixotroph[a]	Yes	Yes	Simultaneously, implies the concept of facultative heterotroph
Photolithotrophic heterotrophy	Yes, primary energy source	Yes, secondary energy source	Implies obligate mixotrophy
Photoheterotrophy	Yes, secondary energy source	Yes, primary energy source essential	Implies obligate mixotrophy. Also known as photoorganotrophy, photoassimilation, photometabolism
Amphitrophs			
Apochloric microalgae	No	Yes	Implies obligate heterotrophy or facultative photoautotrophy due to the lack of photosynthetic pigments
Bacteriovory	Maybe	Yes, bacteria	Implies phagotrophy
Phagotrophy	Maybe	Yes, particulate organic carbon	Includes bacterivory
Auxotrophy	Maybe	Maybe	Requires the uptake of organic but does not specifically refer to the energetic metabolism

[a]*Mixotrophic microalga* or mixotrophy is used in this chapter according to the broader definition in which heterotrophic and photoautotrophic system are simultaneously driving growth of the algae and directly contributing to the energetic metabolisms. That is to say, a mixotrophic culture assimilates organic compounds while using inorganic compounds as electron donors.

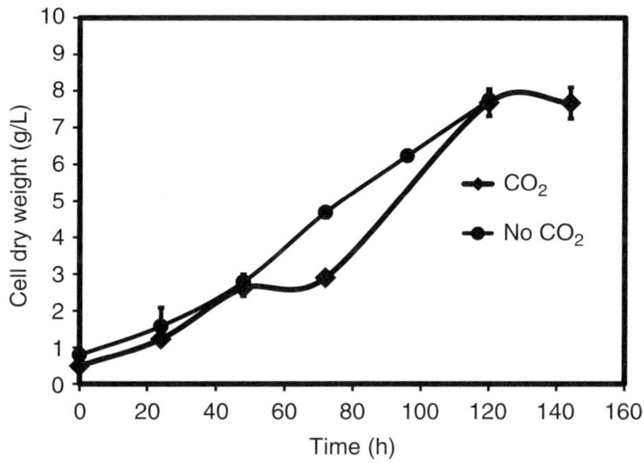

Figure 14.1 The supplementation with CO_2 does not improve the mixotrophic growth of HS26 in 20 L carboys. Inorganic carbon enrichment is only necessary when HS26 was grown phototrophically. *Source*: Reproduced with permission of Heliae Development LLC.

Furthermore, mixotrophy may benefit specific cellular processes and metabolite production that is associated with light, but this discussion should be described individually for each specific synthesis process. Finally potential benefits of the mixotrophic biotechnology are summarized in Table 14.2, and benefits of mixotrophy are compared to photoautotrophic and heterotrophic microalgal production models. Nevertheless, capitalizing on the benefits and economic opportunities of a mixotrophic process first requires an applicable scale-up technology that will allow this opportunity to materialize.

Scaling up mixotrophic cultures is a major challenge because the organic substrate dramatically increases the chances of undesired contamination from heterotrophic microorganisms (e.g., bacteria, fungi, zooflagellates, ciliates, rotifers). The traditional industrial approach to mitigating undesired contamination is to operate under sterile conditions typically involving steam-sterilized fermenters; however, a stainless steel fermenter is not designed to deliver light in significant amounts to sustain meaningful mixotrophic operation. In turn, to sterilize and axenically operate large illuminated areas would be impractical, if not economically prohibitive. Therefore, industrial mixotrophy must address the contamination risk caused by using an organic substrate in a nonaxenic environment. As challenging as this sounds, industrial microbiology was born in nonsterile conditions,[15] and today many microbial processes can operate with simple sanitation while exposed to spores present in the culture medium. Some of those processes are summarized in Table 14.3 and include algae, yeast, fungi, or bacteria that are cultivated in conjunction with environmental heterotrophic contaminants. For example, ethanol production is a transformation process catalyzed by yeast even though contaminant microbes such as lactic acid bacteria are present in the culture broth.

TABLE 14.2 Potential Benefits of Mixotrophic Microalga Growth When Compared to Heterotrophic and Photoautotrophic Growth Models.

	Photoautotrophy	Mixotrophy	Heterotrophy
Average process productivity	Low	**Medium**	**High**
Base operation	Monoalgal	Monoalgal	Axenic
Bioreactor	**Open pond**	**Open pond**	Fermenter
Capital costs	**Low**	**Low**	High
Contamination risk	Medium Grazers, algae	High heterotrophs	Low heterotrophs
Footprint/light path (cm depth)	<30	**>30**	–
Growth-limiting factor	Light	**Oxygen–light**	**Oxygen**
Night growth	No	**Yes**	**Yes**
Nutrient costs	Low	**Medium**	**High**
Outdoor seasonal affects	High	**Medium**	**Low**
Sustainable solar energy	Yes	Yes	No
Typical cell density (g/L)	0.5–1.5	**5–15**	**5–150**

The qualities that mixotrophy shares with either heterotrophy or photoautotrophy are represented in bold and plain font respectively.

Additionally, steam sterilization used in fermentation is not compulsory when the stability of the process can be preserved with cheaper disinfection and chemical treatment techniques.[16] For instance, in Brazil the yeast inoculum is typically harvested from the fermentation batch, diluted with water (1:1), and treated in order to reduce bacterial contamination before reusing for the next fermentation cycle. The robustness of the process significantly impacts operational costs and provides opportunities for scaling up mixotrophic technology. For example, a stainless steel closed fermenter can be replaced with a polymeric-liner-based system when steam sterilization is not mandatory. Steam sterilization, mechanical seals, magnetic drives, sampling ports, steam traps, and other fermentation protocols can be supplanted by simpler chemical disinfection procedures. As an example, the production of mushroom mycelia is an industrial fermentation process that does not even require a stainless steel vessel and is more often associated with agriculture rather than fermentation. Other food production processes are included among the microbial processes that operate under nonaxenic conditions (Table 14.3). Interestingly, those processes to which sterility is not compulsory include products that are commercially oriented toward the commodity markets and products consumed on a daily basis. Only a small number of processes are robust enough to deal with contamination at an industrial level, but the methods and strategies inspire the scaling up of mixotrophic microalgae. The following section addresses a case study of industrial mixotrophy at Heliae and describes the strategies adopted to address the challenges of scaling up mixotrophic microalgae.

TABLE 14.3 Brief Descriptions of Industrial Microbiological Processes That Due To Their Robustness Are Not Necessarily Operated under Axenic Conditions.

Industrial Process	Culture		Product		Fermentation		Species Complexity	
	Aerobic	Anaerobic	Microbial product	Transformation process	Liquid	Solid state	Monoculture	Poly-culture
Acetone		X		X	X		X	
Biological filters	X			X		X		X
Butanol		X		X	X		X	
Cellulosic ethanol		X		X	X		X	
Cheese ripening	X			X		X		X
Dairy derivatives		X		X	X	X		X
Ethanol acetification	X			X	X			X
Fermented foods		X		X	x	X		X
Mushroom (*A. bisporus*)	X		X			X	X	X
Phototrophic microalgae	**X**		**X**		**X**		**X**	
Silage production		X		X		X		X
Wastewater treatment	X	X		X	X			X

14.4 IMPLEMENTATION OF INDUSTRIAL MIXOTROPHY: A CASE STUDY

Bioprospecting. Heliae is located in the Sonoran Desert, southwest USA, in Gilbert, Arizona. Early mixotrophic efforts started locally with the bioprospecting for soil microalgae. Microalgae are typically found in freshwater and marine systems, but arid areas are also an important habitat for a wide variety of microalgal species.[17] Heliae hypothesized that inland soil is a good source of mixotrophic and heterotrophic microalgae. Ramsay and Ball[18] found that most soil microalgal cells isolated from depths below the photic layer were viable, which might indicate cohabitation with other soil heterotrophs. Without any light, those microalgae would have to compete for organic nutrients with fungi and bacteria. Parker[19] demonstrated that soil-based microalgae belonging to *Bracteacoccus* could grow in the dark in xenic conditions and compete with bacteria. Furthermore, Safonova and Reisser[20] studied the bacterial inhibition by allelochemical compounds released by soil microalgae, which further inspired our bioprospecting strategy in the search for robust mixotrophs.

Isolation. Soil samples from various bioprospecting trips were transferred to inorganic liquid medium and incubated photoautotrophically in order to selectively increase the microalgal populations. Microalga strains were then axenically isolated using traditional techniques as described by Andersen and Kawachi.[21] The isolates were tested for axenity using an enriched liquid medium containing yeast extract proteose peptone and glucose. Some of the algal isolates were not cultivable after isolation and were discharged. The remaining axenic isolates were transferred to a 50 mL culture using a minimal inorganic medium before evaluating their mixotrophic potential.

Selection of the substrate. The selection of the organic carbon substrate is a key choice that determines the viability of the process, and consequently this selection was conducted early in the design. The decision was based on both economic and biological criteria. Glucose is the most frequently used organic carbon source for heterotrophic microalgae, but it is also the ubiquitous fuel for living organisms, and it is readily used by a broad range of heterotrophic species, including potential microalga contaminants. Other organic carbon substrates preferred by microalgae include acetate and glycerol.[22] Unlike glucose, which is an agriculturally derived commodity, glycerol and acetate are inexpensive substrates that do not directly compete with food-based feedstock. Glycerol is the main by-product of biodiesel production, and current trends in the sector are driving the prices of glycerol down.[23] Likewise, acetic acid is cheaply produced from petrochemical feedstock through methanol carbonylation.[24] Preliminary economic evaluations suggested that utilizing glycerol or acetic acid was inexpensive compared with gassing photoautotrophic cultures using liquefied CO_2.

The organic carbon substrates were also screened for its capacity to resistance to heterotrophic contamination. Tests were conducted inoculating a mineral-rich medium with nonsterilized soil, using either glycerol or sodium acetate as a carbon substrate. Bacterial growth in the acetate-containing flasks was repeatedly lower than in the glycerol-containing flasks, even though treatments were isoenergetic. Tests were later repeated using nonsterile microalga cultures, and these test confirmed that bacterial contamination was lower in acetate-based mixotrophic cultures. The bacterial inhibitory effect of acetate in microalgal cultures had been earlier observed. The growth inhibitory effect of acetate to certain bacteria has been associated with the deregulation of homoserine trans-succinylase, an enzyme responsible for the synthesis of methionine, an essential amino acid.[25] Even though many bacteria such as *E. coli* do uptake and grow on acetate, the authors suggested that those probably would be more susceptible to oxidative damage when fed with acetate than when fed with glucose or glycerol. Finally, acetate is a nonfermentative substrate and does not support anaerobic growth of bacteria.

Acetate is transported into the cell through a monocarboxylic/proton transporter. Once inside the cell, the acetate is converted into acetyl CoA, which would be processed through the Krebs cycle or its alternative glyoxylate route. Consequently, acetate assimilation differs from the assimilation of any other carbohydrate by the fact that glycolysis is circumvented. Glycolysis intermediates such as pyruvate play an important role in the response against oxidative stress in *E. coli*[26] and eukaryotic cells.[27] In acetate grown media, however, such intermediates would be in short supply, leaving bacterial cells potentially less protected against oxidative damage. Mixotrophic microalgae in turn obtain their energy from the oxidation of acetate[13] and from the oxidation of carbohydrates produced through photosynthesis. In addition, microalgae cope with oxidative damage through many different molecules including polyunsaturated fatty acids, pigments, and enzymes.[28]

Stage zero economic assessment. The photoautotrophic production of microalgae typically relies on the addition of CO_2 as a source of inorganic carbon and as a way to prevent the pH from rising above optimal levels for growth. Typically CO_2 is bubbled into photoautotrophic ponds in response to rising pH levels. Open ponds typically operate with depths no greater than 30 cm, which results in poor CO_2 delivery efficiencies, because most of the CO_2 gasses off into the atmosphere causing a CO_2 yield of 0.05 kg of biomass per kg of CO_2. Our preliminary economic assessment was based on the hypothesis that we replace the CO_2 budget for pH control in photoautotrophic production with an acetic acid-based pH control system in a mixotrophic mode (see details in Table 14.3). Cost analysis was based on pilot scale data, and CO_2 and acetic acid yields were empirically validated and replicated multiple times (Figure 14.2). Analysis of our mixotrophic system suggests that spending the CO_2 budget on acetic acid results in a threefold increase in annual productivity while keeping other production costs stable. Unlike CO_2 gas, acetic acid readily dissolves in water, regulates pH, serves as an energy source, and supports higher yields.

Screening for mixotrophic capacity. Microalga isolates were often first tested using well plates to determine the response to acetate supplementation. Isolates selected from the well plate screening tests were then cultivated in Erlenmeyer flasks to evaluate mixotrophic productivity and capacity to grow and survive under heterotrophic conditions. The fastest growing strains were chosen, and experiments were conducted to optimize the culture medium and the growth conditions. During the optimization phase, the most robust strain was selected and named HS26 following a strain database code. The strain was cultivated in a minimal defined

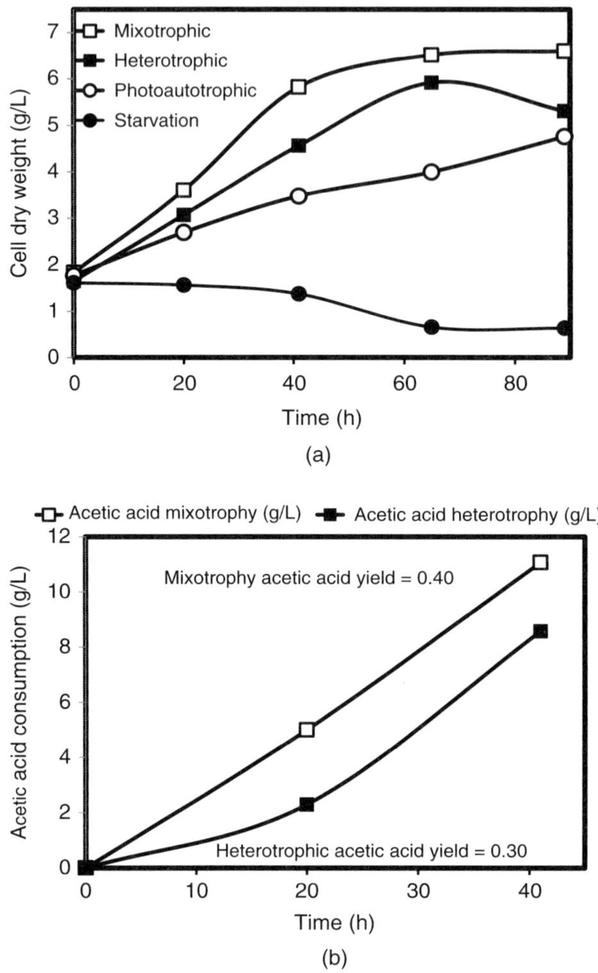

Figure 14.2 The mixotrophic growth of HS26 in shake flasks is above the photoautotrophic as well as the heterotrophic treatment (a). Light in mixotrophic cultures improves acetic acid yields (0.4) above those obtained in heterotrophic cultures (0.3 g CDW per g of acetic acid) (b). All treatments were analyzed in duplicates. Data and figure retrieved from literature.[29] *Source*: Reproduced with permission of Heliae Development LLC.

medium in order to narrow the spectrum of potential heterotrophic contaminants. The growth response to the trophic configuration of strain HS26 was characterized in the laboratory. Its mixotrophic growth on acetate was greater than the heterotrophic growth (without light) or for photoautotrophic growth (without acetate) individually (Figure 14.2). Furthermore growth with light improved the acetate yields from 0.3 up to 0.4 g biomass per gram of acetate added to the medium.

Strain identification. The total genomic DNA was extracted from the strain HS26, and the nuclear-encoded rRNA operon was amplified by PCR and sequenced for 18S, ITS1, 5.8S, ITS2, and 28S. A phylogeny tree was built by Dr. Barbara Melkonian using Randomized Axelerated Maximum Likelihood (RAxML; 1000 bootstraps; 1762 aligned characters used). The phylogenetic analysis compared strain HS26 against the sequences from 285 taxa. Based upon these results, we conclude that strain HS26 represents a new or undescribed species. Furthermore, strain HS26 and eight other strains represent a new genus (74% bootstrap).

Scale-up technology. Typically acetate is transported into the cell along with a proton (see monocarboxylic/proton transporter). Acetic acid is preferred as a raw substrate because, unlike acetate, the undissociated form contributes with the required proton to the culture medium. Conversely, the utilization of acetate salts results in the sequestration of a proton for every acetate molecule transported into the cell. As protons are taken up, hydroxyl anions increase the pH of the culture, and potentially the pH may rise to growth inhibitory levels. When the acetate uptake increases the pH, then the medium should be titrated with an acid. The titration problem is solved by using acetic acid initially because it acts as both the titration agent and the organic carbon substrate. In practice, the acetic acid reservoir is connected to the culture vessel through a feeding line that is linked to a dosing pump and a pH probe controller. This arrangement will feed acetic acid to the mixotrophic culture "on demand." The microbes consuming acetic acid will activate the pH-controlled pump and replenish the acetic acid consumed from the medium. This system will help to maintain a relatively constant concentration of acetic acid that would allow microalgae to grow at the fastest possible rate for the given conditions. The alkalinity of the medium would determine the base acetate concentration of the fed batch and could be easily modified to match the requirements of the organisms. The system was first named "acetic acid/pH auxostat" by Martin and Hempfling in 1976[30] and later successfully implemented for the heterotrophic microalgae *Crypthecodinium cohnii*. We adapted the acetic acid/pH auxostat fed batch system for mixotrophic growth of strain HS26 (Figure 14.3), and results showed growth and productivity improvements over cultures grown mixotrophically with acetate salts (Figure 14.4).

Acetic acid stock or feeding solutions are self-sterilizing and reduce the risk of source contamination. The auxostat system helps to minimize organic carbon substrate waste resulting from a batch failure, from unspent medium, or simply from evaporation. When disposed, it possesses a low biochemical oxygen demand (BOD) in the disposed residual medium. Finally, the system prevents organic carbon substrate limitation and allows monitoring growth indirectly through the acetic acid consumption. To our knowledge, this is the first description of mixotrophic pH auxostat system, and it is likely the first industrial mixotrophic production system (see Figures 14.5, 14.6, and Table 14.4).

Product development. Algae as fertilizers have long been used to augment plant productivity of agricultural crops. The nutrient release rates of algae organic fertilizers are typically lower than in an inorganic fertilizer, which helps to prevent premature leaching of the nutrients. In addition to the "manurial value" or micronutrient composition, algae fertilizers increase the abundance of microbial community.[31] The mechanism and interactions of microalgae with the soil community are complex to evaluate, and they might change depending on the type of soil and crop in which they

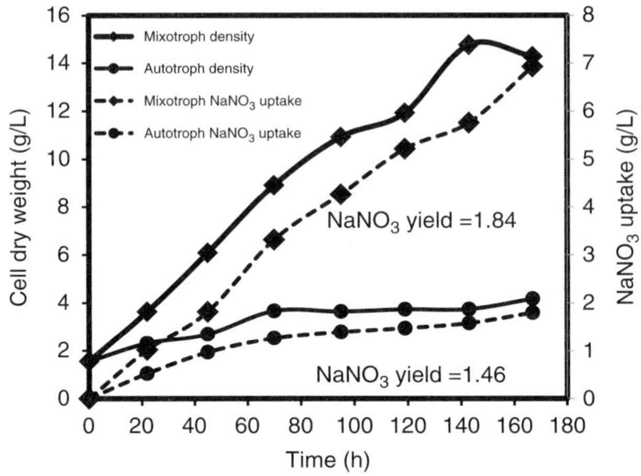

Figure 14.3 Mixotrophic growth of HS26 in 20 L flat panel photobioreactor with an acetic acid auxostat system (squares) is threefold more productive than the traditional phototrophic system (cycles). All treatments were operated in duplicates. Retrieved from literature.[29] *Source*: Reproduced with permission of Heliae Development LLC.

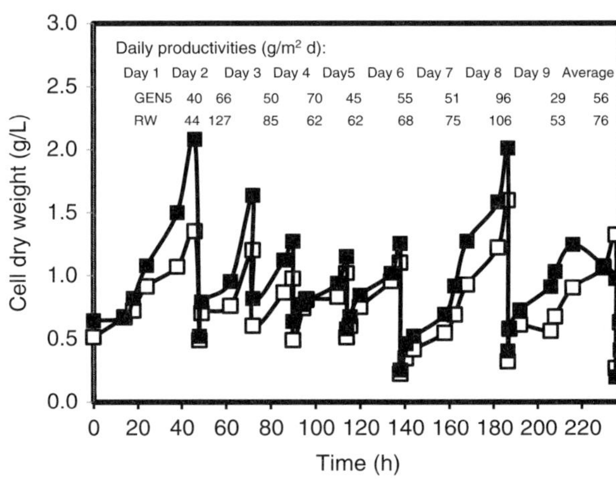

Figure 14.4 Mixotrophic growth and biomass productivities of HS26 in two replicate (1000 L) open reactors. Harvesting was made semicontinuously with 50–80% daily harvest throughout a 10 d period, whereas bacterial levels were maintained below 10^6 CFU/mL. The results reflect typical values obtained in production.[29] *Source*: Reproduced with permission of Heliae Development LLC.

are tested. In particular, HS26 is a soil-based microalgae and therefore would be a good candidate to increase the abundance of microbial community in the agricultural fields. Most common algae fertilizer is based on seaweed extracts such as *Ascophyllum nodosum*, which are naturally harvested from the littoral. In contrast, HS26 can be industrially cultivated and its production is not constrained by regulatory affairs

Figure 14.5 Large-scale bioreactors for the mixotrophic cultivation of microalgae.

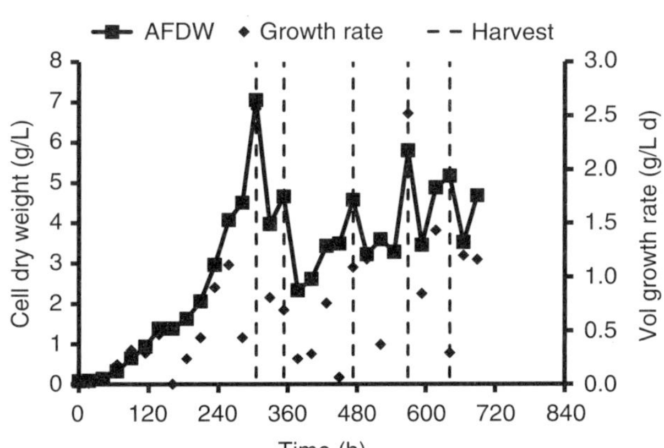

Figure 14.6 Mixotrophic growth and biomass productivities of HS26 in a 100,000 L open reactor. Harvesting was made semicontinuously with 20–80% daily harvest throughout a 29 d of stable operation. The results reflect typical values obtained in production.[29] *Source*: Reproduced with permission of Heliae Development LLC.

TABLE 14.4 Preliminary Economic Assessment of Mixotrophic Process Viability. Using the CO_2 Budget Observed in Photoautotrophic Open Pond Cultures with Acetic Acid (Mixotrophic Scenario) Results in Almost a Threefold Increase in Productivity per unit of Investment/Production.

	Titrate Used for pH Control	
	CO_2 Photoautotrophic	Acetic acid Mixotrophic
Bulk substrate price (USD $ per kg)	0.12	0.4
Substrate yield (kg cell dry weight per kg substrate)	0.05	0.4
Validated annual productivity (metric tons per ha per year)	15	–
CAPEX feeding line (USD $ per ha; 10-year depreciation)	13.500	1.300
Substrate costs per product (USD $ per kg cell dry weight)	2.4	1
Substrate cost per production unit (USD $ per ha per year)	36.000	36.000
Calculated annual productivity (metric tons per ha per year)	–	36

associated with the exploitation of natural seaweed resources. In order to evaluate HS26 potential as a fertilizer, several studies were conducted to refine our product and demonstrate its beneficial effects in crop productivity. Independent trials were conducted in the field by fertilizing zinfandel grape plants, cherry trees, and Thai basil crops. We prepared an aqueous product (10–30% solids) using strain HS26 so that we could directly apply the product onto crop plants. The whole cell biomass was applied directly to the leaves and fruits by spraying an aqueous product. The treatment of zinfandel grape vines increased the grape yield from 10.9 up to 18.7 ton per acre. Preliminary results for strain HS26 applied directly on cherry tree leaves resulted in larger and darker green-colored leaves when compared to untreated leaves. The Thai basil that we tested was hydroponically cultivated; a lysed suspension of strain HS26 was applied directly into the hydroponic solution, and it was compared to a commercial product of reference. The application of HS26 at 5 mL/L resulted in a 32% increase of the total weight of the Thai basil plant.[32]

Future prospects. The trophic metabolism of microalgae is versatile. Both light and organic substrates could be used as an energy source by numerous microalga species associated with diverse taxonomic classifications. Mixotrophic species are found in almost all major taxonomic groups, which open the door for more mixotrophic processes and products. The strain HS26 is a first example of many more to come. Strain HS26 is currently cultivated in 1000, 20000, and 100000 L mixotrophic reactors, where it achieves productivities that are at least 50 times greater to those achieved in photoautotrophic reactors. The robustness of this process allows up to 30 days of semicontinuous operations with 80–20% partial harvest (see Figure 14.7).

Figure 14.7 Overview of Heliae algae cultivation facilities in Phoenix, AZ. (*See color plate section for the color representation of this figure.*)

The case studies herein described help illustrate the mixotrophic approach that Heliae is capitalizing on through a broader technological platform named **Volaris**^{TM}. Mixotrophic metabolism is still poorly understood, and additional basic research will be necessary to better understand the interactions between heterotrophic and photoautotrophic metabolisms. This would help to optimize and better design of our processes. Current efforts are based on optimizing the process and expanding the mixotrophic platform toward new strains and other valuable products.

ACKNOWLEDGMENTS

We thank Robert A. Andersen, Albert G. J. Tacon, and Justin Kniep for their critical and thoughtful comments on the earlier version of the manuscript.

REFERENCES

[1] R. Davis, A. Aden, P.T. Pienkos, *Appl. Energ.* **2001**, *88*, 3524–3531.

[2] E. Ganuza, A.J. Anderson, C. Ratledge, *Biotechnol. Lett.*, **2008**, *30*, 1559–1564.

[3] F. Bumbak, S. Cook, V. Zachleder, S. Hauser, K. Kovar, *Appl. Microbiol. Biotechnol.* **2011**, *91*, 31–46.

[4] C.G. Lee, *Biotech. Bioprocess Eng.* **1999**, *4*, 78–81.

[5] W.F. Danforth, "Substrate assimilation and heterotrophy", In "*Physiology and Biochemistry of Algae*" (Ed. R.A. Lewin) Academic Press, New York-London, **1961**, 99–123.

[6] P.J. Keeling, *Am. J. Bot.* **2004**, *91*, 1481–1493.

[7] L. Zaslavskaia, J.C. Lippmeier, C. Shih, D. Ehrhardt, R. Grossman, K.E. Apt, *Science* **2001**, *292*, 2073–2075.

[8] C.A. Beam, M. Himes, *J. Protozool.* **1982**, *29*, 8–15.

[9] G.B. Bremer, *Hydrobiologia* **1995**, *295*, 89–95.

[10] Z. Cohen, I. Khozin-Goldberg, "Searching for PUFA-rich microalgae", In "*Single Cell Oils*" (Eds. C. Ratlege, Z. Cohen) AOCS Press, Champaign, Illinois, **2005**, 53–72.

[11] T. Sanders, DHA status of vegetarians. *Prostaglandins Leukot. Essent. Fatty Acids* **2009**, *81*, 137–141.

[12] W. Barclay, C. Weaver, J. Metz, "Development of a docosahexaenoic acid production technology using Schizochytrium*: a historical perspective*", In "*Single Cell Oils*" (Eds. C. Ratlege, Z. Cohen) AOCS Press, Champaign, Illinois, **2005**, 36–52.

[13] A. Neilson, R. Lewin, *Phycologia*, **1974**, *13*, 227–264.

[14] L.O. Ingram, J.A. Calder, C. Van Baalen, F.E. Plucker, P.L. Parker, *J. Bacteriol.* **1973**, *114*, 695–700.

[15] S. Knight, W. Frazier, *J. Bacteriol.* **1945**, *50*, 505–516.

[16] L. Basso, T. Basso, S. Rocha, "Ethanol production in Brazil: the industrial process and its impact on yeast fermentation", In "*Biofuel Production-Recent Developments and Prospects*" (Ed. M.A.S. Bernardes) 1st Edition, **2011**, Vol. 1530, http://www.intechopen.com/books/biofuel-production-recent-developments-and-prospects.

[17] M.L. Nagy, A. Pérez, F. García-Pichel, *FEMS Microbiol. Ecol.* **2005**, *54*, 233–245.

[18] A.J. Ramsay, K.T. Ball, *New Zeal. J. Sci.* **1983**, *26*, 493–501.

[19] B.C. Parker. *Ecology*, **1961**, *42*, 381–386.

[20] E. Safonova, W. Reisser, *Phycol. Res.* **2005**, *53*, 189–193.

[21] R.A. Andersen, M. Kawachi, "Traditional microalgae isolation techniques", In "*Algal Culturing Techniques*" (Ed. R.A. Andersen) Elsevier Academic Press, San Diego, California, **2005**, 83–101.

[22] O. Perez-García, F.M.E. Escalante, L.E. De-Bashan, Y. Bashan, *Water Res.* **2011**, *45*, 11–36.

[23] C.G. Quispe, C.J.R. Coronado, J.A. Carvalho Jr., *Renew. Sust. Energy Rev.* **2013**, *27*, 475–493.

[24] J. Jones, *Platin. Met. Rev.* **2000**, *44*, 94–105.

[25] M. Price-Carter, T.G. Fazzio E. Ibáñez Vallbona, J.R. Roth, *J. Bacteriol.* **2005**, *187*, 3088–3099.

[26] M. Valdivia-González, J.M. Pérez-Donoso, C.C. Vásquez, *Biometals* **2012**, *25*, 451–458.

[27] S. Desagher, J. Glowinski, J. Prémont, *J. Neurosci.* **1997**, *17*, 9060–9067.

[28] J.T. Cirulis, J.A. Scott, G.M. Ross, *Can. J. Physiol. Pharmacol.*, **2013**, *91*, 15–21.

[29] E. Ganuza, J.D. Licamelle, A.L. Tonkovich, US 2013/069046, **2013**.

[30] A.A. Martin, W.P. Hempfling, *Arch. Microbiol.* **1976**, *107*, 41–47.

[31] S.F.I. Haslam, D.W. Hopkins, *Appl. Soil Ecol.* **1996**, *3*, 261.

[32] U.S. Provisional Application No. 61/980723 Entitled Methods of Growing Vascular Plants with Mixotrophic Microalgae as Filed April 17, **2014**.

15

InFiQuS: MAKING THE BEST OF LEFTOVERS

INMACULADA ARANAZ, NIURIS ACOSTA, MARÍA N MENGÍBAR, LAURA CALDERÓN, RUTH HARRIS AND ÁNGELES HERAS

Sustainable Physical-Chemical Innovation S.L. (InFiQuS), Madrid, Spain

Industrial Biorenewables: A Practical Viewpoint, First Edition. Edited by Pablo Domínguez de María.

15.1 BRIEF DESCRIPTION OF InFiQuS

Sustainable Physical-Chemical Innovation S.L. (InFiQuS) is a start-up company created within the Complutense University of Madrid (UCM) by Dr. Angeles Heras in 2010. InFiQuS team is composed of professionals in the field of chemistry, biology, pharmacy, food science, and technology. The firm works in the isolation and development of functional ingredients to be used for food, pharmaceutical, and cosmetic applications. Our company is allocated at the facilities of Madrid Science Park (PCM) and Complutense University and we are members of the European Technology Platform for Sustainable Chemistry (SusChem).[1]

Dr. Heras is group leader of the "Chitin and Chitosan Research Group" at UCM, actually one of the pioneer groups in Spain in the field of chitin and chitosan, with more than 20 years of experience in the polymer extraction from fisheries waste, as well as in its characterization and application in different fields such as biotechnology, biomedicine, food industry, agriculture, and cosmetics.[2] From the very beginning, we have been working in very close collaboration with several companies belonging to very different sectors as biotechnological, food industry, or agro-industry with the aim of delivering the highest added value of the very interesting properties showed by chitin and chitosan (as it will be described along this chapter). Over the last 5 years, our research group has spread out its expertise to other biomass sources such as yeast, oil cake, olive leaves, and cardoon. Moreover, our core research fields have also been extended to the cosmetics industry.

In the 21st century the development of products that may improve our quality of life while being free from by-products is rather a challenge. This is what we tackle, from our core product – the chitosan family – to find out synergies with other natural by-products to design functional compounds for fields of large socioeconomic future projections. Thus, InFiQuS works as an open business model in the development of compounds with high added value from chitosan as principal passenger, along with other by-products – calling them copassengers – with the aim of improving their individual properties.

InFiQuS was created with the aim of developing functional compounds from by-products with applications in food, pharmacy, and cosmetics. The raw materials usually employed are industrial waste products from shellfish farms, breweries, oil mills, and agro-industries. A summary of InFiQuS activity is shown in Table 15.1. Within these by-products, chitosan is the main product used by InFiQuS not only for its exceptional biological properties, as it will be described in the following section, but also for its technological properties and good compatibility with other components. The source and method of preparation determine chitosan composition and size and therefore its functional properties. The preparation of specific chitosan samples leading to standardized and normalized functionalities is a very expensive process. Therefore, the number of companies able to commercialize these tailor-made products is small, and there is a lack of these products in the market since they are highly demanded.

In a first stage, the aim of InFiQuS was to launch new functional compounds – the cosmetic and health markets being its target. Later on, due to the broad interesting

TABLE 15.1 InFiQuS in a Nutshell.

Size	Start-up: (5 people)
Founded	2010
1st Year of Benefits	2012
Patents	1+2[a]
Services	• Characterization of specific chitin and chitosan[b] • Functional properties of specific chitin and chitosan[c]
R+D+i	• Tailor-made chitin, chitosan, and oligosaccharides • Development of specific controlled release systems for pharmacy and cosmetics • Development of functional compounds for food
Collaborations with Industries	• Beverage industry • Pharmaceutical industry • Cosmetic industry
Collaboration with Research Groups	5
Awards	• Acces it new ideas and new EBTs. UCM 2009 • Finalist RedEmprendia 2012

[a]One patent under revision and two in final stages of development.
[b]Molecular weight (gel permeation chromatography and viscometry), deacetylation degree (first derivative UV-spectrophotometric method and NMR spectroscopy), crystallinity (X-ray diffraction), humidity, ash content, and protein content. Thermal stability studies (DSC). Technological properties: emulsifying and gelling capacity.
[c]Fat binding and bile salt binding capacity, antioxidant activity (Folin–Ciocalteu, ORAC, FRAP, DPPH, and TEAC), and antimicrobial activity. (Other analysis on demand.)

properties of chitosan, the portfolio of InFiQuS has increased with the development of new applications of tailor-made chitosans in food, pharmacy, agriculture, bioremediation, or wastewater treatment.

As it can be seen in Table 15.1, what InFiQuS can offer in the chitin and chitosan market are products with the following characteristics:

- Tailor-made chitosan with specific functional properties and high purity.
- High-quality product with guarantees regarding origin and productive process and therefore also regarding its functional properties.
- Chitosan and derivatives with functional properties adapted to the market requirements.

The main advantage of InFiQuS is the *know-how* acquired after more than 20 years of research experience in the field of chitin and chitosan isolation, production, and characterization and in their application in different fields (Figure 15.1).

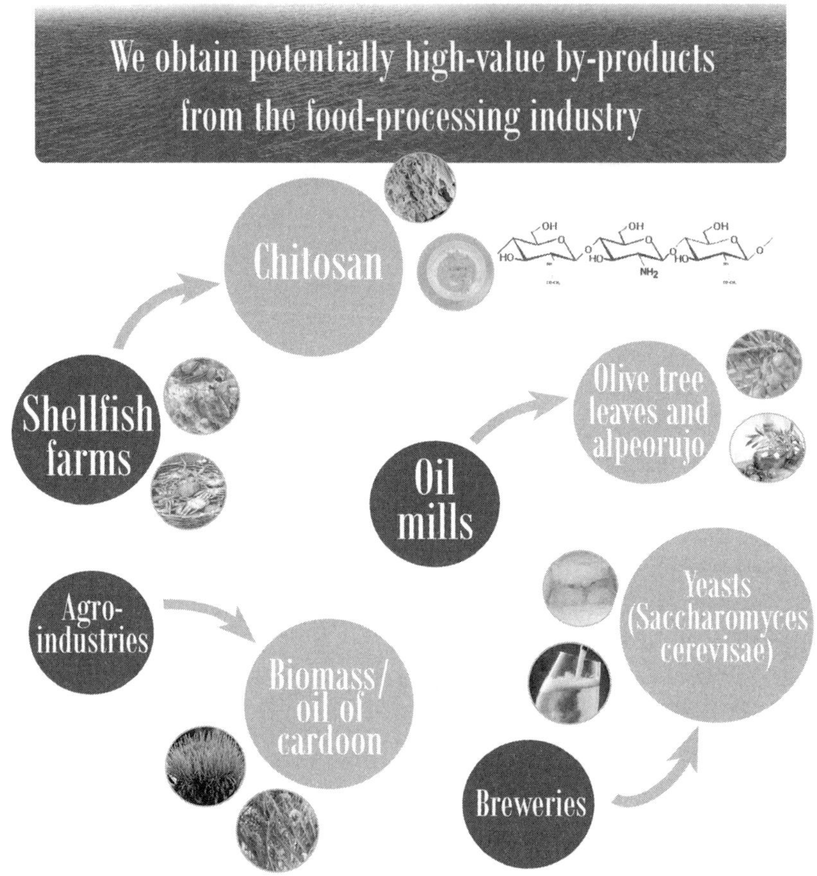

Figure 15.1 By-products used by our company from different industries.

Regarding the food industry, we are working in the design and development of functional ingredients with chitosan properties such as fat binder, dietary fiber, or antioxidant, as well as an antibacterial agent alone or as coadjuvant of other ingredients, which adds further functional properties. Other chitosan technological properties such as emulsifying, gelling, and stabilizing capacity allow InFiQuS the development of other ingredients for food applications. Thus, our potential clients are companies from the food industry, which will use our products as nutraceuticals or dietetic supplements or due to their technological properties as substitutes of current additives.

For the pharmaceutical and cosmetics industry, we are working in the design and development of specific controlled drug delivery systems for topical and oral administration on demand for pharmaceutical companies and in the development of natural and antiaging or topical skin cosmetic formulations.

15.2 VALUABLE BY-PRODUCTS UNDER RESEARCH BY InFiQuS

In this section the different by-products used by InFiQuS are described. Data regarding their production, legal information, market details, and examples of their application in different fields are discussed.

15.2.1 Valuable By-Products from Fisheries and Aquaculture

The aim of this section is to give a general overview of the current status of the crustaceans' waste management in fisheries and aquaculture. A brief description of the process used to isolate valuable by-products from this waste is given. Moreover, data about regulatory aspects and market situation are included. At this stage, InFiQuS research focuses on the use of chitosan and chitosan oligosaccharides (COS) in food, pharmacy, and cosmetics industries, and therefore examples of the application of these polymers in these industries are discussed.

15.2.1.1 Introduction The manufacture of crustaceans from the fish industry and aquaculture generates tons of solid residues mainly composed of crustacean shells, the waste of these crustaceans being a major source of surface pollution in coastal areas. The amount of captured crustaceans in 2011 is shown in Table 15.2.[3] Note that these data do not include aquaculture production, which also produces solid residues. The management of this waste is a serious issue due to the high volume of material; as an example, just in Europe, the Food and Agriculture Organization FAOSTAT estimates that more than 750,000 t of crustacean shell waste is produced every year.[4]

The legislation regarding crustacean shell disposal depends on each country, but, in general, the tendency is to set up more restrictive legislations that may increase the cost of management. In this area, several strategies can be used for managing crustacean shells (Figure 15.2).

Traditionally, crustacean shells are disposed in the ocean or landfills, which obviously are not disposal methods bringing an added value. Moreover, disposal of crustacean shells in landfills may result in risks to human health as well as to the environment due to bacterial contamination and strong odor during decomposition. Crustacean shells can also be burned, but due to their low burning capacity no heating is economically produced in this way, yet generating greenhouse gas emissions.

TABLE 15.2 Worldwide Crustaceans Harvested in 2011.

Species Group	Capture Production, T
Freshwater crustaceans	424,613
Crabs, sea spiders	1,472,961
Lobsters, spiny/rock lobster	284,957
King crabs, squat lobsters	43,628
Shrimp, prawns	3,288,467
Krill, planktonic crustaceans	180,986

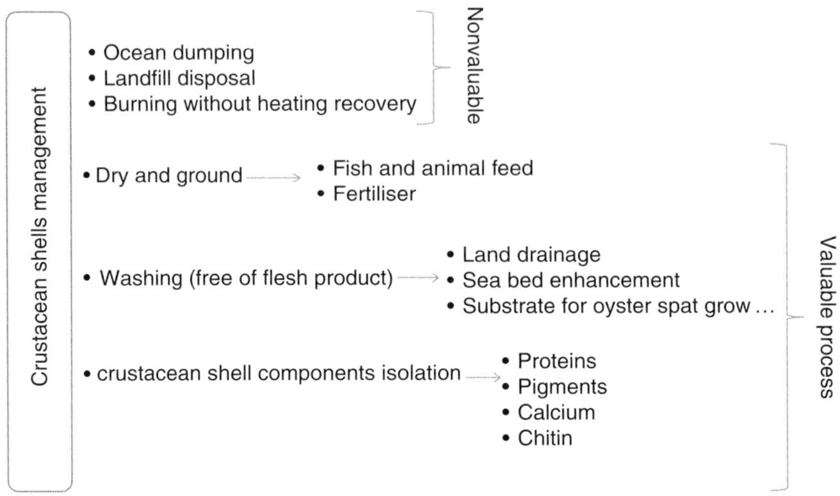

Figure 15.2 Overview of crustacean shell management.

Low-valuable uses of crustacean shells include their use as fertilizer after composting and their use as flour to feed animals in small amounts due to their high composition in inorganic material. Conversely, the isolation of crustacean shell components to be used in high-valuable applications represents another way to reduce the amount of waste generated with a positive economic impact.

15.2.1.2 Crustacean Shell By-Product Isolation Crustacean shells are composed of proteins, chitin, inorganic salts (mainly $CaCO_3$ and phosphates), pigments (i.e., astaxanthin and carotenoids), and lipids in a variable proportion. As it can be seen in Table 15.3, the proportion depends on the type of crustacean, molt stage, the season

TABLE 15.3 Crustacean Shell Composition.

Source	Chemical Composition				
	Protein, %	Chitin, %	Ash, %[a]	Lipids, %	Astaxanthin, mg/Kg
Blue shrimp (head)	42	35.5	20.5	2.4	88.2
Blue shrimp (shell)	58	16.4	24.2	1.4	n.d.
Lobster	23.1	20.2	33.7	2.2	983
Krill	41	24	23	11.6	n.d
Blue crab	13.1	13.8	72.4	0.8	36
Blue crab (A1)	–	24.4	–	–	–
Blue crab (D1)	–	14.3	–	–	–
Blue crab (E)	–	8.5	–	–	–

[a]Ash content is an indirect way to measure inorganic material. The percentage is calculated on dry matter.
A1, postmolt stage; D1, premolt stage; E, intermolt stage.

of harvesting, the location and the environmental conditions, and the body part under consideration (head, thorax, claws, etc.).[5] The most interesting by-product found in crustacean shells is chitin from which its deacetylated derivative chitosan can be obtained. Chitin is also a source for *N*-acetylglucosamine, a molecule with applications in different fields such as biomedicine (treatment of inflammatory bowel disease and joint damage) or cosmetics (improves skin quality, reduces the appearance of facial hyperpigmentation, and has a moisturizing effect).[6]

Chitin is the second most abundant natural polymer in nature after cellulose, and it is found not only in the structure of crustacean shells but also in a wide number of invertebrates (insects' cuticles and krill) and in the cell walls of fungi, among others. On the contrary, the occurrence of its deacetylated derivative chitosan is apparently restricted to fungi (*Absidia glauca*, *Absidia coerulea*, *Aspergillus niger*, *Mucor rouxii*, *Gongronella butleri*, *Phycomyces blakesleeanus*, *Absidia blakesleeana*, *Rhizopus oryzae*, *Trichoderma reesei*, and *Lentinus edodes*). Among all investigated fungal species, the most commonly researched is *M. rouxii*. The amount of chitin and chitosan in fungi can reach 35% of the cell wall dry weight. [7]

Although it was considered that Antarctic krill and cultured fungi would become major sources of chitin and chitosan, at this moment, from an industrial point of view, chitin is normally extracted from crustacean's shells, mainly shrimp, prawn, and crab shells.[8] On the other hand, chitosan is chemically prepared by deacetylation of chitin. To our knowledge, only one Belgium company, KitoZyme, used fungi to produce chitosan and chitin–glucan.[9]

Several chemical and/or enzymatic steps are involved in the isolation of crustacean shell by-products (Figure 15.3). The first step involves material preprocessing, and it is of great importance to keep the quality of the isolated by-products, since it

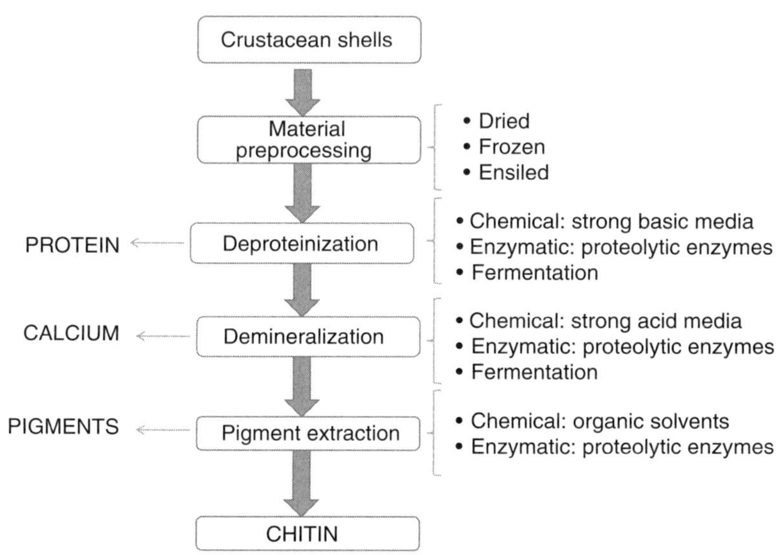

Figure 15.3 Isolation of valuable by-products from crustacean shells.

prevents product deterioration during the transport and/or storage of the crustacean shells. Previous to isolation, the shells are washed and the particle size is reduced to improve the subsequent treatments. The following steps are deproteinization, demineralization, and decoloration, and the order and type of process used (chemical, enzymatic, or mixed) depend on the type of products that one expects to isolate: either only chitin or chitin plus other by-products as high-quality proteins, pigments, or calcium. Although harsh chemicals cause environmental pollution and reduce the quality and amount of other by-products isolated along with chitin at this moment, they are the most used in an industrial scale due to the high cost of the nonchemical processes.

When high-quality proteins are required, deproteinization is carried out before demineralization. In this step, proteins are removed by chemical treatment in basic media or by enzymatic treatments with proteolytic enzymes or microorganisms. The demineralization can be carried out by chemical treatment in acid media or enzymatically. In this step, calcium and pigments can be recovered. Finally, decoloration of the isolated chitin can be carried out chemically using H_2O_2 and NaOCl; however this treatment seems to degrade chitin.

Likewise, chitin can be further processed into chitosan by chemical or enzymatic deacetylation (Figure 15.4). From both chitin and chitosan, oligomers can be produced by chemical, physical, or enzymatic methods. Chitin can also be used as source to *N*-acetylglucosamine by strong acid treatments or enzymatic processes.[10]

15.2.1.3 Overview of Chitin and Chitosan and Their Applications Chemically, chitin and its deacetylated derivative chitosan are natural polymers composed of randomly distributed β-(1–4)-linked D-glucosamine (deacetylated unit) and *N*-acetyl-D-glucosamine (acetylated unit) (Figure 15.5). Chitin is insoluble in aqueous media, while chitosan is soluble in acidic conditions due to the free protonable amino groups present in the D-glucosamine units. Being soluble in acid

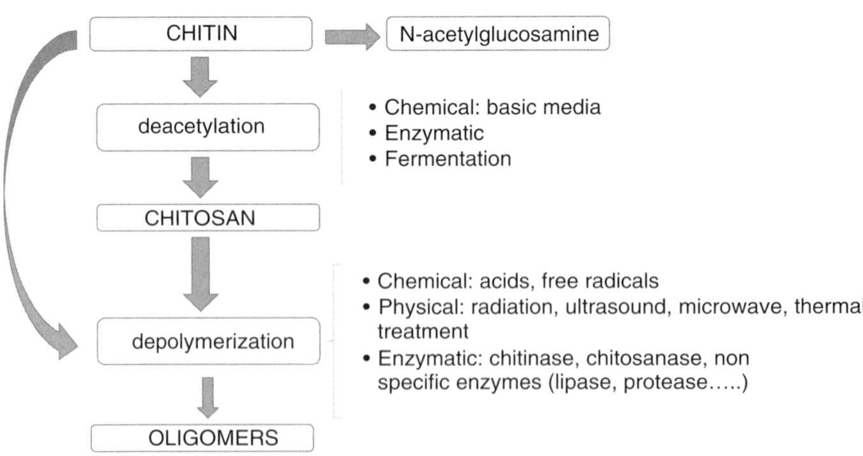

Figure 15.4 Preparation of chitin derivatives.

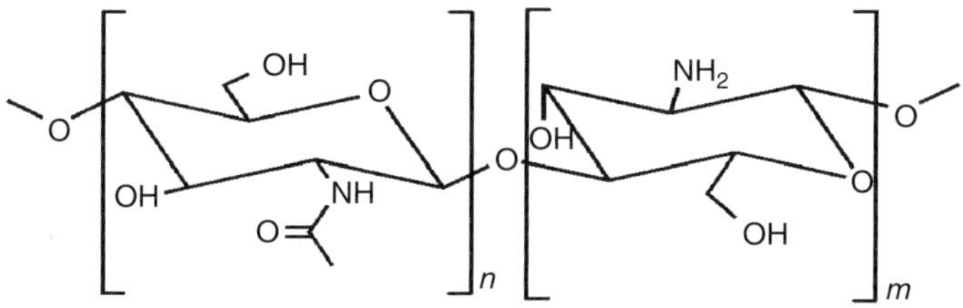

Figure 15.5 Chemical structure of chitin and chitosan (chitosan $m > 0.6$).

solutions, chitosan is easily to manufacture in different forms of technological interest depending on the application such as films, microspheres, nanospheres, sponges, threads, hydrogels, scaffolds, etc.[11] The positive charge also allows chitosan to interact with polyanions such as alginate, pectin, or polyacrylic acid, producing polyelectrolyte complexes.[12]

Due to their natural origin, both chitin and chitosan cannot be defined as a unique chemical structure, but as a family of polymers presenting a high variability in their chemical and physical properties. This variability is related not only to the origin of the samples but also to their method of preparation. Several properties of interest such as mucus adhesion, biodegradability, biocompatibility, and anticholesterolemic, antimicrobial, antioxidant, hemostatic, or adsorption enhancer activities have been described – all these properties being related to the characteristics of the polymers, that is, molecular weight (MW) and deacetylation degree (DD).[13] Likewise, more than 200 potential and usual applications of chitin and chitosan and their derivatives have been described in virtually every sector.[14] A summary of some of these applications is given in Table 15.4.[14,15]

Applications of Chitosan in the Food Industry In the food industry, chitosan offers a broad spectrum of unique applications. Most of them are related to its antibacterial activity, for example, in its use as packaging preservative or edible coatings or in its use to preserve emulsions.[16] Due to that preservative effect, chitosan is also studied in pathogen control in meat product storage and in the maintenance of good quality organoleptic properties, which is improved with the chitosan antioxidant capacity.[17]

Likewise, the ability of chitosan to bind fat allows its use as hypocholesterolemic agent in dietary products.[18] Moreover, the technological properties of chitosan as emulsion stabilizer, gelling agent, or emulsifier have been studied. For example, chitosan has been studied as stabilizer in peanut butter, and the mix of chitosan with olive oil, pectin, and glycerin produces a low fat mass with consistency, flavor, and butter texture.[19] As emulsifier, chitosan is able to form emulsions, and in the presence of whey proteins the stability of the emulsion is improved.[20] Chitosan flocculating and coagulating properties are used in the wastewater treatment of effluents from food industry or to clarify beverages (juices, wines, etc.).[13]

In addition, chitosan can be used to produce glycosylated proteins through Maillard reaction in order to improve the technological properties of the proteins used in the food industry. This reaction occurs spontaneously without using any kind of

TABLE 15.4 Examples of Chitin and Chitosan Applications.

Application	Use
Wastewater treatment	Recovery of metal ions, pesticides, phenols, or dyes, among other water pollutants (as adsorbent). Recovery of solid materials (as coagulant/flocculant)
Paper technology	Fixing agent in color photography; strengthen recycled paper
Biomedicine	Wound dressing, surgical sutures, antitumor activity, antiviral activity, hypocholesterolemic activity, scaffold for tissue engineering, burn treatment, matrix for controlled drug delivery, ophthalmology (contact lenses)
Biotechnology	Support to immobilized biocatalysts, bioremediation, affinity chromatography protein separation, cell sheet production, solid-state batteries, potential component in LEDs
Agriculture	Seed and fruit protection (film) fertilizer, plant protection (general elicitor), growth enhancer, bacteriostatic or fungistatic agent
Food industry	Food preservative and stabilizer, color stabilizer, animal feed additive, flavor retention, antioxidant, emulsifier, dietary supplement, weight control (fat trapper)
Cosmetic and toiletries	Moisturizing agent, hair care (increment of hear softness, smoothness, and mechanical strength); prevent the formation of plaque and tooth decay

chemicals, and therefore it has been proposed as an alternative and safe method for protein glycosylation. Maillard reaction between chitosan and different proteins such as lysozyme, casein, soya proteins, or β-lactoglobulin improves their emulsifier and antibacterial properties.[21] The use of COS are of great interest in the food industry and biomedical field due to their antibacterial, antioxidant, antitumor, and immunological effects.[2j,22] Due to their small size, the viscosity of COS solutions is low, and they are more soluble than chitosan, therefore being more efficient as antimicrobial agents against a wide number of microorganisms.[2k,23]

Despite these useful properties, the use of chitosan has to face some limitations. On one hand, it is necessary to develop more research regarding *in vivo* applications of chitosan, and on the other hand, legal limitations that will be discussed in Section 15.2.1.4 do not allow the broad use of chitosan at the international level.

Cosmetic Applications of Chitosan Chitosan finds application as cosmetic ingredient in different fields as hair, teeth, and skin care or as deodorant.[15,24] It has been included in shampoos, rinses, and permanent hair coloring agents due to its solubility in organic acids and its fungistatic activity. A clear solution of chitosan forms smooth and elastic materials on hair and thus can be used in hair gels. Chitosan acts as a deodorizing agent since it possesses antibacterial properties and prevents the excessive generation of odor-forming substances.

In addition, chitosan and its derivatives have found application in skin care because they are highly compatible with the skin and act as a physical barrier against UV radiation and dehydration. Chitosan can function as a moisturizer agent not only preventing dehydration but also supplying water that improves the quality of their hydrating effect. Creams based on chitosan take advantage of its filmogenic properties and reduce expressive wrinkles. Chitosan has been proposed as carrier for antiaging molecules. The development of delivered systems that allows the selective delivery of active molecules in low doses is of great interest since the reduction of the amount of active principle reduces the cost and also possible side effects. For systems below the millimeter, the delivery systems are able to be inserted between the corneal cells, this characteristic being of great interest in the cosmetic field.[25]

The use of nanoparticles in the cosmetic field is due to their appropriate size to enter the skin pores to deliver the active ingredient (i.e., antioxidants) in a controlled and selective manner protecting the skin against the damage due to free radicals that attack the lipid membrane, proteins, and DNA producing skin disruptions.[2m,26]

Pharmaceutical Applications of Chitosan Over the past few years, the interest on controlled release systems has grown in the pharmaceutical field. Controlled-release dosage forms enhance the safety, efficacy, and reliability of drug therapy. They regulate the drug release rate and reduce the frequency of drug administration. In this area, the use of nontoxic and biodegradable polymers is the basis for the design of innovative systems for controlled release of active principles. Chitosan has been considered for pharmaceutical formulation and drug delivery applications where the attention has been focused on its absorption-enhancing, controlled release, and bioadhesive properties.[2f–h,13]

Chitosan can be used to obtain nanoparticles[2m] or microparticles[27] as a vehicle for the encapsulation of active ingredients. Chitosan-based release devices may prolong the duration of active agent activity, improving its efficiency and reducing side effects under desired conditions. Chitosan has a variety of promising pharmaceutical uses, and it is presently considered as a novel carrier material in drug delivery systems for parenteral, oral, or even topical administration. These properties, together with the very safe toxicity profile, make chitosan an exciting and promising excipient for the pharmaceutical industry for present and future applications.

15.2.1.4 Regulatory Aspects The use of chitosan as food additive was approved by Japan and Korea in 1983 and 1995, respectively.[28] The inclusion of chitin and chitosan was considered in 2003 by the Codex Alimentarius Commission, but it is not currently listed in the General Standard for Food Additives since neither FDA nor EFSA has approved its use as a food additive. Although several studies have shown that this compound is not toxic, long-term studies of human safety have not been reported. Apart from its use as a food additive in some Asian countries, one ultrapure form of Norwegian shrimp-derived chitosan (ChitoClear®, Primex ASA, Norway) was self-affirmed generally recognized as safe (GRAS) in 2001.[29] In 2011, following a request from the European Commission, the EFSA was asked to provide a scientific opinion on a list of health claims in relation to chitosan and reduction in body weight, maintenance of normal blood LDL cholesterol concentrations, reduction of intestinal transit time, and reduction of inflammation. The Panel considered that only the claim related to maintenance of normal blood LDL cholesterol concentrations showed a cause–effect relationship between the consumption of chitosan and maintenance of normal blood low-density lipoprotein (LDL) cholesterol concentrations. The Panel considered that in order to obtain the claimed effect, 3 g of chitosan should be daily consumed by the adult target population. This was approved by European Commission Regulation (EC N° 432/2012).[30]

The use of glucosamine hydrochloride from *A. niger* as food additive in beverages and a commercial product ("KiOnutrime-CG™") based on a chitin–glucan complex as fiber has been successfully evaluated by the EFSA.[31] The use of chitosan in the biomedical and pharmaceutical industries requires the fulfillment of a variety of regulations depending on the final intended use (excipient, medical device, or active pharmaceutical ingredient). In 2002, chitosan hydrochloride was included as excipient in the European Pharmacopoeia.[32] More recently, in 2011, chitosan has been added in the list of excipients in the US Pharmacopoeia (USP).[33] Therefore, in the last years, the use of chitosan and its derivatives as excipients is being well established. Conversely, chitosan has not been approved yet by the FDA or similar agencies in the biomedical field. However, the American Society for Testing and Materials (ASTM F04 Division IV) is making a concerted effort to establish standard guidelines for tissue-engineered medical products based on chitosan or its derivatives. The F2103 guide covers the evaluation of chitosan salts suitable for use in medical applications considering aspects such as control of protein content and, hence, potential for hypersensitivity, endotoxin content, and total bioburden. The F2260 guide covers the determination of DD, while the F2602 guide covers the determination of MW of chitosan and chitosan salts.[34]

15.2.2 Valuable By-Products from Oil Mills

In this section an overview of the current status of olive oil production and oil mill waste management will be provided. A brief description of the process to produce olive oil and the generation of waste and the by-products with commercial interest is included. Finally, data about regulatory aspects are discussed.

15.2.2.1 Introduction The olive oil extraction is an economic activity of great importance in the Mediterranean area. According to the International Olive Oil Council, more than 95% of olive oil production in 2013 occurred in this particular area, that is, European Union (EU) countries (Spain, Italy, Greece, and Portugal), Turkey, Tunisia, Morocco, Syria, and Algeria (Table 15.5).

15.2.2.2 Olive Oil Production and Its By-Products Traditionally, olive oil processing was mainly carried out with discontinuous press-type mills producing high-purity extra virgin olive oils; but due to the high demand of this product, new processing methods based on higher-capacity centrifugal process have been developed (two- and three-phase decanter centrifuge methods). The three-phase

TABLE 15.5 World's Olive Oil Production in the Period 2007–2013 (International Olive Oil Council (November 2013)).[35]

Country	Percentage of Oil Production, %
EU	71.9
Tunisia	5.8
Syria	5.6
Turkey	5.2
Morocco	3.8
Algeria	1.7
Other Mediterranean countries	3.0
Argentina	0.8
Chile	0.5
Australia	0.5
China	0.2
United States	0.2

decanter centrifuge method is a continuous process that requires the addition of warm water to improve extraction, generating a high amount of wastewater. As observed in Table 15.6, the production of olive oil by this method generates three phases and two wastes: oily phase (20%), solid phase (30%), and aqueous phase (50%). The solid phase, so-called olive oil cake or orujo, is a combination of olive pulp and stones. The aqueous phase, so-called alpechin or olive mill wastewater, comes from the vegetation water and the soft tissues of the olive fruits, with water added during processing. The presence of large amounts of organic substances – oil, polyphenols, protein, polysaccharides, etc. – responsible for the high COD values (up to 220 g/L) and mineral salts represents a significant problem for the treatment of such wastewater.[36]

In order to attenuate this problem a two-phase decanter centrifuge method, in which water is not added, was developed. In this method, an oily phase and a solid waste called "alperujo" or two-phase olive mill waste (i.e., a combination of liquid and solid waste), is generated. Alperujo contains 80% of the olive fruit, including skin, seed, pulp, and pieces of stones, which is later separated and usually used as solid fuel.[37] In Spain, over 90% of olive oil mills operate with this system, which means that annual production of alperujo is approximately 2.5–6 million of tons depending on the season.[38] On the contrary, in other olive oil-producing countries, such as Italy or Greece, the implementation of this system is almost negligible (<5%). Presently, there is non-EU legislation regulating olive mill waste management, and standards are left to be set by individual countries.[39]

In the olive fruits, there is a large amount of bioactive compounds and substances of high interest. Many of them are known by owing health beneficial properties that contribute to the protective effect of the virgin olive oil. During olive oil processing, most of them remain in the olive oil wastes (Table 15.7). Up to now

TABLE 15.6 Three-Phase Systems versus Two-Phase Systems.

System	Phases	Phases Composition
Three-phase system	Oleous	Mainly olive oil with small amounts of water and fine solid particles
	Solid (so-called orujo)	Mixture of olive press cake (47%) with water (50%) and small amount of oil (3%)
	Aqueous (so-called alpechin) around 48% in weight of initial mixture of olives and water	Mainly water (90%) with small amount of oil (1%) and solid particles (9%)
Two-phase system	Oleous	Mainly olive oil with small amounts of water and fine solid particles
	Solid (so-called alperujo)	Water (54.5%) with olive press cake (42%) and small amount of oil (3%)
	Aqueous (around 1.5% in weight of initial mixture of olives and water)	Mainly water (86.7%) with solid (13.3%) and no traces of oil

TABLE 15.7 High-Valuable Components Found in Olive Oil Waste.[40]

By-Product	Location	Use
Phenolic components:	Alpechin	Antioxidant properties
Hydroxytyrosol	Alperujo	Antioxidant, antibacterial, antifungal, antiviral, and anti-inflammatory properties
Oleuropein	Leaves	
Squalene	Alpechin	Antioxidant properties
Tocopherols		
Triterpenes:	Solid waste	Anti-inflammatory and antioxidant properties
Erythrodiol		
Oleanolic acid		Vasodilation activity
Maslinic acid		
Pectins	Alperujo	Gelling properties
Oligosaccharides	Alperujo	Potential source of prebiotics Enhance growth of bifidobacteria
Mannitol	Alpechin Olive twig leaves Alperujo	Excipient in pharmacy Anticaking and free flow agent Low-calorie sweetener Bread products for diabetics
Polymer	Alpechin	Use in agriculture as bioamendment and metal integrator

emphasis has been put on detoxifying these wastes prior to disposal, feeding, and fertilization/composting – as they are not easily degraded by natural processes or even used in combustion as fuel.[37] However, the recovery of high-value compounds or the utilization of these wastes as raw matter for new products is a particularly attractive way to reuse it, if the recovery process is of economic and practical interest. This, added to the alternative proposals to diminish the environmental impact, will allow the placement of the olive market in high competitive position, and these wastes should be considered as by-products.[41] By using an adequate technology, the olive mill waste in general, and alperujo in the particular case of the Spanish industry, can be converted into value-added products. It is important to evaluate the economic conditions to obtain reproducible and safety products (essential that the pesticides and other agrochemicals are excluded) that can be developed on a large scale.[42]

Apart from the waste generated during olive oil production, around 15 million metric tons of vegetable waste (leaves and twigs) is produced from pruning olive trees each year.[43] As summarized in Table 15.7, this waste is also a source of products with commercial interest.

15.2.2.3 Regulatory Aspects Although there are several products in the market of several countries such as the United States, Australia, or Italy based on olive leave extracts claiming that they can act against free radicals and help to prevent cardiovascular diseases, these products have not been evaluated by the FDA. Following a request from the European Commission, the EFSA Panel on Dietetic Products, Nutrition and Allergies was asked to provide a scientific opinion on a list of health claims pursuant to Article 13 of Regulation (EC) No 1924/2006. The food constituent, which was the subject of the health claims, was polyphenols (hydroxytyrosol and oleuropein complex) in olive (olive fruit, olive mill wastewaters or olive oil, *Olea europaea* L. extract and leaf). The Panel considered that polyphenols in olive were sufficiently characterized in relation to the claimed effects. They concluded that a cause–effect relationship has been established between the consumption of olive oil polyphenols and protection of LDL particles from oxidative damage (5 mg of hydroxytyrosol and its derivatives in olive oil should be consumed daily). Regarding maintenance of normal blood pressure and the maintenance of normal blood HDL-cholesterol concentrations, the Panel concluded that the evidence provided is insufficient to establish a cause–effect relationship between the consumption of olive oil polyphenols and the intended claims. Regarding the anti-inflammatory effect of polyphenols, the Panel considered that the reduction of inflammation in the context of diseases such as osteoarthritis or rheumatoid arthritis is a therapeutic target for the treatment of the disease and does not comply with the criteria laid down in Regulation (EC) No 1924/2006. Regarding the contribution to the upper respiratory tract health to maintain a normal function of gastrointestinal tract and the contribution to body defenses' against external agents, the Panel considered that the claimed effects are general and nonspecific and does not refer to any specific health claim as required by Regulation (EC) No 1924/2006.[44]

15.2.3 Valuable By-Products from Brewery Industry

The aim of this section is to give a general overview of the current status of the beer production and the management of the by-products produced during the process. A brief description of beer production, the generation of yeast, and its use as valuable by-product is included. Finally, some data about regulatory aspects are discussed.

15.2.3.1 Introduction Beer is an alcoholic drink made from malt and flavored with hops. The two main varieties of yeast used for beer brewing are top-fermenting yeast (*Saccharomyces cerevisiae*) and bottom-fermenting yeast (*Saccharomyces uvarum*). The names of both are descriptive of where the fermentation takes place. A scheme of beer production is shown in Figure 15.6.

In 2012, global beer production reached 190.70 million kL with an increase of about 1.77 million kL (0.9%) from the previous year. Spain with a production of 3.300.000 kL is the 10th producer in the world ranking.[45] Once the fermentation is finished, it is important to remove as much yeast as possible before cleaning the tanks to reduce its presence in the residual water and to increase the yield of the process, since around 1.5–2.5% of the total amount of the beer production is imbibed in the yeast. The recovered yeast can be reincorporated in the beer production to start a new fermentation (usually 4–6 times). Yeast biomass is the second major by-product from brewing industry, after brewer spent grain. Albeit its disposal represents and environmental problem – increasing the COD of residual water and generating solid waste –, its use is very limited. In general, brewer's yeast is sold as inexpensive animal feed product after inactivation by heat. Dried yeasts are an excellent source of protein for swine and ruminant – this being an inexpensive nitrogen source with good nutritional characteristics and a very bitter taste. It has also been proposed as a substitute for live food in the production of certain fish. However, its use in human food is limited due to its high content in nucleic acid. Other uses of yeast biomass included bioremediation,

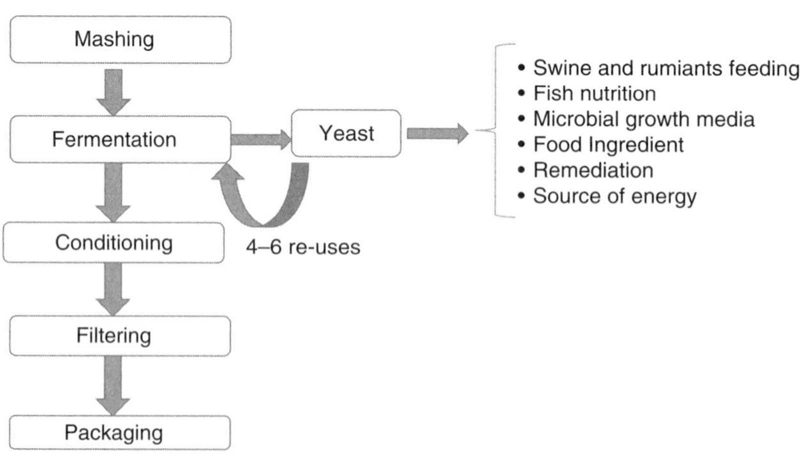

Figure 15.6 Scheme of beer production and main uses of yeast.

energy production, and nutrient source for the growth of fastidious microorganisms as microalgae, among other uses.[46]

15.2.3.2 By-Product Isolation and Their Applications The cell wall of yeast is composed of polysaccharides (~85%) and proteins (~15%). From the cell wall of yeast, mannoproteins (35–40%), glycogen (1–23%), β-D-glucans (50–60%), *N*-acetylglucosamine, chitin (1–3%), and chitosan can been extracted.[47] From these by-products, we are going to focus on β-glucans since they are the by-products used by our company. β-Glucans are polysaccharides of D-glucose monomers linked by β-glycosidic bonds that can be isolated from cereals or yeast, among other sources. Yeast beta-glucans differ from their cereal counterparts in that they comprise a mixture of beta-1,3- and beta-1,6-glucans, compared to the cereal derivatives, which are a mixture of beta-1,3- and beta-1,4-glucans. They are a diverse group of molecules that can vary with respect to molecular mass, solubility, viscosity, and three-dimensional configuration. β-D-Glucans have been included in several products from the food industry such as salad toppings, frozen desserts, sauces, and dairy products. Moreover, its technological properties such as high apparent viscosity, water retention, gelling properties, emulsion stabilizing capacities, and oil binding characteristics make them very useful in the meat industry and mayonnaise production, among other applications.[47b] β-D-Glucans from cereal enrichment of juices have been studied, showing the improvement of bowel transit time and the increase of satiety compared to biscuits enriched with β-D-glucans from cereals.[48] Their use in the biomedical field has also been proposed since possible activation of the immune system and radioprotective, anti-inflammatory, and antitumor effects have been claimed.[49]

15.2.3.3 Regulatory Aspects β-D-Glucans isolated from different sources such as food-grade baker's yeast, mushroom, and oat bran have been determined to be GRAS under specific conditions of use as food ingredients.[50] Furthermore, it has been reported that β-D-glucans isolated from cereals reduce LDL *cholesterol* levels and glycemic index that are considered major risk factors for cardiovascular disease. In 2009, EFSA Panel concluded that a cause–effect relationship has been established between the consumption of beta-glucans from cereals and the reduction of blood cholesterol concentrations. Foods should provide at least 3 g/d of β-glucans to be effective. In the same document, EFSA concluded that a cause–effect relationship has not been established between the consumption of β-glucans and the maintenance or achievement of a normal body weight.[51] In a similar way, in 2011 EFSA Panel concluded that a cause-effect relationship has been established between the consumption of barley β-glucans (at least 3 g/d) and the lowering of blood LDL cholesterol concentrations.[52] In 2011, EFSA Panel evaluated the use of baker's yeast, both soluble and insoluble, β-glucans as novel food ingredient in food supplements, in foods for particular nutritional uses (PARNUTs, as specified by Directive 2009/39/EC4), and in conventional foods.[53] The Panel concluded that yeast beta-glucans are safe at the proposed conditions of use.

15.2.4 Agro-Industry: Cardoon

The aim of this section is to give a general overview of the current status of cardoon production and the management of the by-products produced during its manufacturing including details about their potential applications.

15.2.4.1 Introduction Cardoon (*Cynara carduncculus*) is an herbaceous perennial plant, capable of producing 15–25 t of biomass per hectare and year that has been traditionally cultivated for centuries as vegetable for human food in Mediterranean countries. Flowers of cardoon are rich in proteases, namely, cardosins A and B, and therefore, aqueous extracts of its flowers have been used for centuries in the Iberian Peninsula for manufacturing of ovine and/or caprine milk cheeses. Cardoon is known since Roman times for its therapeutic application as liver protector and diuretic. Despite its use in folk medicine, the therapeutic potential of cardoon is poorly studied, and its putative application as source of bioactive compounds is completely unexplored. In the last years, several alternative uses have been proposed for cardoon. In particular, the use of the seeds for oil and biodiesel production has been widely investigated. As biodiesel the quality of the final product shows high-quality characteristics, similar to those of diesel no. 2, according to DIN EN 14214 standard specifications (European Standard requirements for biodiesel). Cardoon has also been proposed as solid biofuel, in paper pulp production, and in remediation (Figure 15.7).[54]

15.2.4.2 By-Product Isolation and Its Applications The use of cardoon as food for human and animals produces vegetable waste composed mainly of external leaves. Analytical chemical studies carried out on cardoon extracts and infusions have reported high phenolic contents. Most of the identified compounds – namely, 3-, 4-, and 5-caffeoylquinic acids, 1,3-dicaffeoylquinic acid

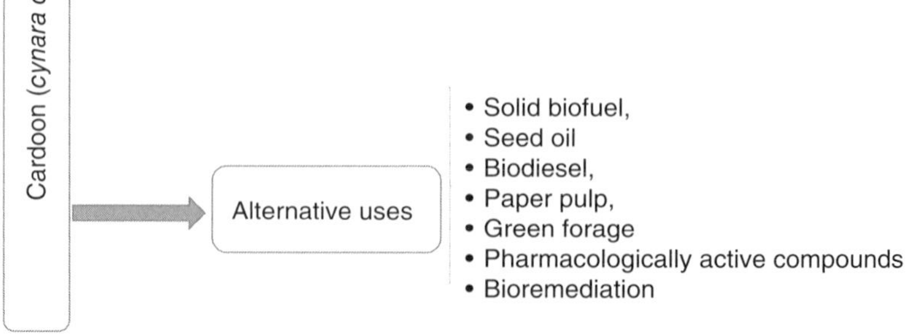

Figure 15.7 Uses of cardoon.

(cynarin), 1,5-dicaffeoylquinic acid, luteolin, luteolin 7-glucoside, luteolin 7-rutinoside, luteolin 7-malonylglucoside, luteolin aglycone, ferulic acid, and coumaric acid – are known to carry important biological roles in terms of liver protection, antioxidative, anti-inflammatory, antimicrobial, antimutagenic, and antiproliferative activity.[55] Inulin has also been isolated in the roots of cardoon.[56] Inulin is considered as functional food ingredient since it affects the physiological and biochemical processes in rats and human beings, resulting in better health and reduction in the risk of many diseases. Experimental studies have shown its use as bifidogenic agent, stimulating the immune system of the body, decreasing the pathogenic bacteria in the intestine, relieving constipation, decreasing the risk of osteoporosis by increasing mineral absorption, reducing the risk of atherosclerosis by lowering the synthesis of triglycerides and fatty acids in the liver, and decreasing their level in serum. It modulates the hormonal level of insulin and glucagon, thereby regulating carbohydrate and lipid metabolism by lowering the blood glucose levels. Inulin also seems to reduce the incidence of colon cancer. It has a wide application in various types of foods like confectionery, fruit preparations, milk desserts, yogurt and fresh cheese, baked goods, chocolate, ice cream, and sauces. It can also be used for the preparation of fructose syrups. Apart from its interesting properties in health and nutrition, it has also interest in the chemistry industry where its microbial fermentation is used to produce alcohols, including ethanol. It has also been used as precursor in the production of a number of chemicals such as glycerol, as a component in detergents and related products.[57]

Apart from the vegetable waste produce during its manufacturing in the food industry, the application of cardoon in the production of oil and biodiesel also generates a solid waste so-called solid cardoon cake, which includes several components that can be of commercial interest (Figure 15.8).

15.2.4.3 Regulatory Aspects From the aforementioned applications, the use of cardoon as source of nutritional and pharmacological bioactive compounds is very

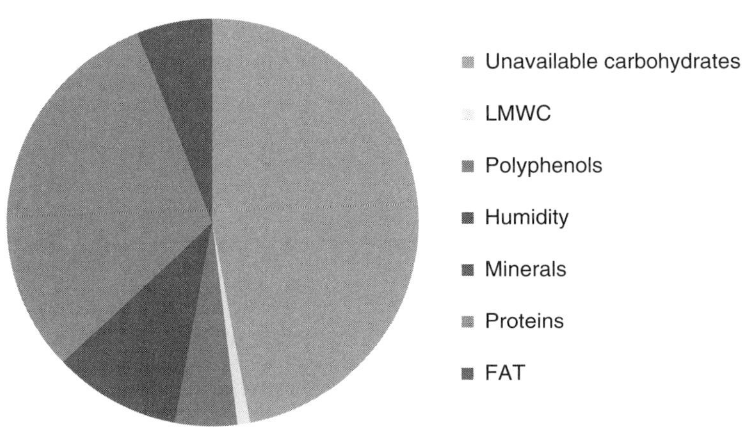

Figure 15.8 Solid cake cardoon composition.

interesting from an economical point of view. Moreover, it would add even more value to its cultivation. However, since research in this product is quite recent, no data regarding phenolic compounds or inulin isolated from cardoon is found at EFSA or FDA databases.

15.3 EXAMPLES OF PRODUCTS CO-DEVELOPED BY InFiQuS

In this section, some examples of products that are being developed by InFiQuS are briefly discussed. Since InFiQuS is a young company established in 2010, at this moment our products have not yet reached the commercialization phase.

15.3.1 Tailor-Made Chitosan and Chitosan Derivatives

Our company has broad experience in the isolation of chitin from crustacean shells and in the production of chitosan and chitosan oligomers with specific characteristics (controlled MW, DD, N-acetylation pattern (PA), low ash content, low metal content). As previously mentioned, the control of these characteristics is of great importance when highly valuable applications are intended, for instance, biomedical or food applications. InFiQuS can produce high-quality chitin and chitosan from lab scale (g) to pilot plant scale (kg) with controlled physicochemical properties to be used for research groups and companies. InFiQuS also provides the functional characterization of these high-quality samples with the aim of customer to acquire the adequate chitin or chitosan for specific applications since not all chitin and chitosan products deliver equally effective results. Recently, our company is focusing on the isolation of chitin from new sources and on the characterization of the physicochemical properties of these samples. In particular, we are also focusing on chitin isolation from insects as their use as a source of proteins for humans and animal feed is expected in the future.[58]

15.3.2 Development of Functional and Technological Ingredients for Food Applications

Our company is well balanced in the innovation in the area of food and considers the very importance of combining technological innovation – product formulation and process engineering – with social and cultural innovation (consumer needs and food qualities). The adequate innovation must be based on the complete food systems including production, harvesting, primary and secondary processing, manufacturing, and distribution and is important also taking into account environmental changes and sustainability. With this aim, our company is working in the development of various products based on chitosan and other copassengers also isolated from waste. At this stage, we have identified the most suitable chitosan and bioactive compounds for each specific application.

For example, taking advance of chitosan technological properties, we are drawing up dessert formulations in which chitosans are included as stabilizer and emulsifier agent. Chitosans with different properties (molecular weight, deacetylation degree, and N-acetylation pattern) have been successfully added to the mixture of ingredients in these desserts to improve their texture and body. Apart from the improved technological properties, functional properties are expected, considering that the specific chitosan added in this type of products has the capacity to bind fat and bile acids as well as antioxidant capacity.

We are also developing functional beverages employing by-products from the brewery industry and vegetal extracts with antioxidant activity from by-products isolated from olive and cardoon waste. The aim of this research line is to provide a product with hypocholesterolemic effect, besides to reduce total cholesterol and to increase antioxidant capacity.

Finally, the development of natural food additive for meat product preservation is being studied with the aim of replacing chemical preservatives included in different types of meat. To prevent or at least slow down lipid and protein oxidation, products with antioxidant properties composed of polyphenols found in vegetal extracts along with chitosan and its oligosaccharides are being tested. Moreover, due to their antimicrobial activity, an improvement of quality during the storage is expected.

InFiQuS also adds to its innovation other research areas on bioactive compounds: micro- and nanoencapsulation to produce different types of nutraceuticals. The large trajectory in micro- and nanoencapsulation of "Chitin and Chitosan Research Group" supports the company and allows dealing with the problems of solubility, low stability, and nondesirable taste and odors of different extracts used in food to produce functional effects. New formulation strategies and advances in production of micro- and nanostructured materials are possible to obtain stable particles with potential applications in the food technology.

15.3.3 Development of Specific Systems for Cosmetics and Pharmacy Applications

Our company is developing chitosan-based microspheres and nanoparticles for the encapsulation of active principles with cosmetic and pharmacological interest. Its use as an encapsulation vehicle and the interesting biomedical properties that chitosan exhibit by itself have provided a perfect combination for the application of these systems in these areas. Different chitosans and natural biopolymers are being used to obtain hydrophilic systems, and different types of active principles with antioxidant properties are being encapsulated to protect the active principle from biodegradation, to control its release through the matrix in which it is encapsulated, and to facilitate its arrival at the site of action.

Our chitosan micro- and nanoparticles are suitable for the encapsulation of natural antioxidants, such as olive leaf and cardoon extracts, and display good encapsulation efficiency. Moreover, the value of zeta potential of these particles is positive, providing a better absorption of the active principle on the skin surface. The company is currently studying the release of these active principles from the chitosan matrix and their antioxidant activity once released and the stability of the microspheres and nanoparticles once included in the creams.

15.4 MARKET SITUATION

From all the by-products that can be isolated from the sources previously mentioned, we focus on the market of chitin and chitosan as well as polyphenols since these are the products our company is currently working with. More details about β-glucan, pectin, mannitol, tocopherol, inulin, and squalene market can be found in the references.[59]

15.4.1 Chitin and Chitosan Market

In 2003, the Global Industry Analysts Inc. (GIA) estimated a 16% per year growth rate for the chitin and chitosan market with a production of 6000 metric tons.[60] By the end of 2010, the production of chitosan doubled this estimation with a production of 13,000 metric tons. In its last report (2010), the GIA is forecasting that the global market for chitin derivatives will reach $63 billion, while the global chitosan market will reach more than $21 billion by 2015.[61] The prices obtained for chitin and chitosan products can vary substantially in market price. The low-end chitin market typically supplies water purification and the food supplements industries and obtains $10/lb. High-end biomedical uses can draw $2000/lb (Figure 15.9). Differences in quality of chitin and chitosan set the prices between these extremes.

In the same report, it is said that water treatment represented the largest end-use segment for chitosan. However, the agrochemical end-use category was projected to witness the fastest growth from 2007 to 2015. Major issues impacting the chitosan market, according to the last GIA report, include:

- Production shortages.
- A lack of quality chitosan available in the market.
- High production costs.
- Heavy pollution during the production process.

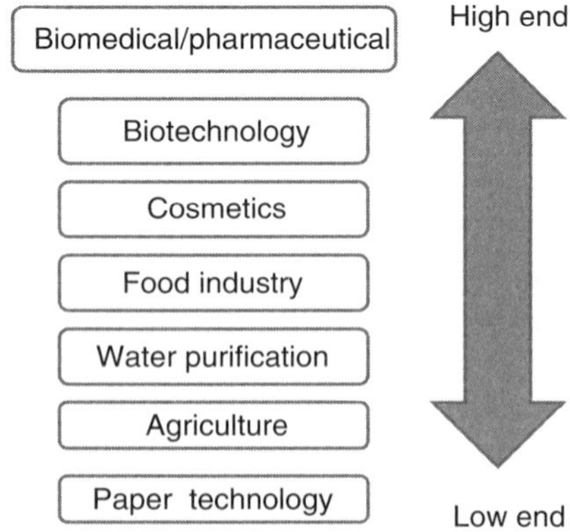

Figure 15.9 Chitin and chitosan market.

The supply of crustacean shells is highly dependent on seasonal and environmental factors, leading to unpredictable limitations on production capacity. Two alternatives are possible solutions to the seasonal changes in waste availability: plants built to accommodate and process the waste during the short season that would remain idle for the rest of the year or facilities designed to dry and storage the waste to be process along the year. The use of chitin and chitosan produced by mycelium waste during fermentation is also considered a very interesting alternative to crustaceans as a stable, nonseasonal source of raw material that would be more consistent in character, but this solution has not been applied industrially.

Depending on the end application of chitin and chitosan, they can be classified mainly in three categories according to the requirement of purity: technical-grade samples are used for agriculture and water treatment, pure-grade samples are used for the food and cosmetics industries, and ultrapure-grade samples are used for biopharmaceutical and biomedical applications.

The production of high-quality samples implies the strict control of the process with a raw material of high quality (i.e., low heavy metal and other pollutants' content in the exoskeleton due to water contamination), controlled isolation process, and detailed characterization of the samples. From an industrial point of view, controlling these variables is very expensive and most of the products in the market are samples of low quality with an inadequate characterization. However, a small number of companies offer controlled products of high quality. The design of more economical and fewer pollutants methods to produce controlled chitin and chitosan is one of the fields that could improve the chitin and chitosan market.

15.4.2 Polyphenol Market

According to Transparency Market Research, the global polyphenol market was valued at USD 580 million in 2011 and is expected to reach USD 873.7 million by 2018, growing at a compound annual growth rate (CAGR) of 6.1% from 2012 to 2018.[62] In terms of volume, global consumption was 12,214.4 t in 2011 and is expected to reach 21,032.7 t by 2018, growing at a CAGR of 8.2% from 2012 to 2018.

The global demand for polyphenols is being primarily driven by the increasing aging population in Japan, Europe, and the United States. Along with this, growing consumer awareness of polyphenol consumption as an avenue to maintain good health has also been a major factor in the growth of the polyphenol industry in the recent years. The antioxidant property of polyphenols and their possible application in the prevention of cancer, diabetes, and cardiovascular diseases have further contributed to the growth of the polyphenol market. However, the limited supply of raw materials due to other overlapping commercial applications such as wine manufacturing and preparation of juices, jams, and jellies is expected to inhibit the market growth over the next 6 years.

At this moment, polyphenols are mainly isolated from grape seed (accounting for over 50% of the total polyphenol demand in 2011), followed by polyphenols from green tea and apple (together accounted for over 32% of the total market in 2011). Green tea polyphenol, due to its high polyphenolic content and high consumption in Asia Pacific, is expected to be the fastest-growing product segment of polyphenols.

Functional beverages were the largest application segment of polyphenols and accounted for over 44% of the total polyphenol demand in 2011 followed by functional foods, which accounted for over 33% of the total consumption in 2011. Polyphenol demand for dietary supplements was 2056 tons in 2011 and is expected to grow at a CAGR of 7.8% from 2012 to 2018. In terms of volume, functional beverages are expected to be the fastest-growing application segment at an estimated CAGR of 8.5% from 2012 and 2018.

Asia Pacific accounted for 40.3% of the global polyphenol demand in 2011, followed by Europe with 27.8%. North America polyphenol demand is expected to reach 5294.3 tons by 2018. Asia Pacific is also expected to be the fastest-growing polyphenol consumer at an estimated CAGR of 8.8% from 2012 to 2018.

15.5 NEEDS OF RESEARCH: SYNERGIES BETWEEN INDUSTRY AND ACADEMIA

Over the last century, human development has improved our daily life. However, during the period in which many modern processes were developed and implemented, environmental questions were not the major issue that they have become nowadays. In this century, humankind has to face the following challenges:

- Natural resources are finite.
- Waste production contaminates soils, water, and air—this being a serious risk for the environment and our health.

In the past, waste was often disposed in relatively cheaply landfills or landspreading. However, subsequent to the Kyoto agreement, the issue of waste in our modern society has become more prominent since it contributes to many of the problems of global environmental sustainability.

Landfill taxes and increasing cost of disposal are driving the industry to new ways of waste management: the most obvious is new processes that reduce both natural resource consumption and waste production. However, large amounts of waste are still produced. Once generated, this waste can be a source of serious problems or a source of incomes. A simple way to use agro-industry waste is the use of biomass in low-valuable applications such as compost or animal feed or as source of energy. Going further, waste can be use as source for the isolation of highly valuable compounds (such as proteins, carbohydrates, monomers, polyphenols, etc.). In this way, not only the amount of waste ended in landfills is reduced, but also the amount of natural resources required is reduced since these compounds can be used for further industrial production.

Although the management of waste involves an increase in the production costs that companies have to afford, not all companies are in the situation of carrying out research to reduce the amount of natural resources and/or energy or to increase the value of the waste generated. For small–medium companies the insufficient R&D capability and the lack of markets for new products are a handicap. Moreover, these companies may have the perception of an insufficient return of investment. In the case of large companies, with sufficient R&D capability and ability to open or introduce themselves in new markets, the diversification is a risk, and the scheduling of research is too long in some cases.

In both scenarios, academia is an essential piece in the complex jigsaw of human development. Firstly, it has the possibility of carrying both applied and basic research, the latter being not profitable from a company point of view but being essential as the foundations for further development. Secondly, academia's scheduling is longer than the one in companies, and therefore they can stay with problems longer, which allow them to face more complicate challenges and carried out long-term research. Thirdly, they can also research in fields of low market return but high human interest as in the case of drugs for rare diseases. And last but not least, academia can supply the lack of facilities and qualified personnel in the small–medium companies.

Yet, in the past both academia and companies have mainly work in a separate manner. It is clear, specially taking into account the serious reduction of R + D + i budget in companies due to the crisis, that companies have to consider the important contribution that academia can offer them. The role of start-up companies born from academia – as InFiQuS – is to highlight products developed by academia working as a bridge between big companies and applied research. In the field of research in which InFiQuS is involved, the main needs that have been identified with the natural compounds under research are the following:

- Safety.
- The biological activity of these compounds needs to be demonstrated by *in vivo* studies and clinical assays.
- Control the content of the bioactive compounds in the final products.
- Improvements in chitin and chitosan production.
- Evaluation of the potential market and price.

It is of great importance to establish the risk/benefit balance of using these compounds for health-related purposes. Although some of these products have been widely used by folk medicine, both their safety and activity need to be established by *in vivo* studies and clinical assays. This work is fundamental to drive all the questions related to regulatory aspects and can be carried out in a complementary manner by both academia and companies. It is very remarkable that in the case of chitin and chitosan, their biological properties, biodegradability, etc. depend on the physicochemical characteristics of the samples. In the case of academia it is still very usual to find articles in which a poor or nonexistent characterization does not allow to reproduce the claims regarding specific activities. In other cases, such as products isolated from cardoon, their properties are not sufficiently studied. This confusion regarding the real activities of biological compounds plus the lack of control in the commercialized products in which the specific bioactive compounds are only present as traces, and therefore are inefficient for any purpose, is a serious danger for the future of the market since it introduce doubts in the clients regarding the benefits of these products. Therefore, a more complete and detailed study of the possible claims associated with these products will be highly beneficial, taking into account that their behavior is associated with their physicochemical characteristics (MW, chemical composition, etc.).

In the particular field of chitin and chitosan, as described by the GIA report, the main drawbacks in chitin and chitosan development are the high cost of production and the large amount of waste produce during this process. Moreover, there is a lack of first matter and its quality is not always the most appropriate. Therefore, the design of more economical and fewer pollutants methods to produce controlled chitin and chitosan samples is one of the hot topics in this field. The use of new sources to isolate by-products such as cardoon in the case of polyphenols or insects in the case of chitin and chitosan will help to reduce the problems associated with the lack of raw materials for the isolation of these products.

REFERENCES

[1] SUSCHEM web Page: http://www.suschem.org/, accessed Mar 2014.

[2] (a) N. Acosta, C. Jiménez, V. Borau, A. Heras, *Biomass Bioenerg.* **1993**, *5*, 145–153; (b) G. Galed, B. Miralles, I. Paños, A. Santiago, Á. Heras, *Carbohydr. Polym.* **2005**, *62*, 316–320; (c) G. Galed, E. Diaz, F. M. Goycoolea, A. Heras, *Nat. Prod. Commun.* **2008**, *3*, 543–550; (d) I. Aranaz, V. Ramos, S. De La Escalera, A. Heras, *Biocatal. Biotrans.* **2003**, *21*, 349–356; (e) I. Aranaz, N. Acosta, A. Heras, *J. Mol. Catal. B: Enzym.* **2009**, *58*, 54–64; (f) R. Harris, I. Paños, N. Acosta, A. Heras, *J. Control. Release* **2008**, *132*, e76-e77; (g) I. Paños, N. Acosta, A. Heras, *Curr. Drug Discov. Technol.* **2008**, *5*, 333–341; (h) R. Harris, E. Lecumberri, A. Heras, *Mar. Drugs* **2010**, *8*, 1750–1762; (i) L. Calderón, R. Harris, M. Cordoba-Diaz, M. Elorza, B. Elorza, J. Lenoir, E. Adriaens, J. P. Remon, A. Heras, D. Cordoba-Diaz, *Eur. J. Pharm. Sci.* **2013**, *48*, 216–222; (j) M. Mengíbar, I. Mateos-Aparicio, B. Miralles, A. Heras, *Carbohydr. Polym.* **2013**, *97*, 776–782; (k) M. Mengíbar, M. Ganan, B. Miralles, A. V. Carrascosa, A.J. Martínez-Rodriguez, M.G. Peter, A. Heras, *Carboh. Polym.* **2011**, *84*, 844–848; (l) G. Galed, M. E. Fernández-Valle, A. Martínez, A. Heras, *Magn. Reson. Imaging* **2004**, 22, 127–137; (m) R. Harris, E. Lecumberri, I. Mateos-Aparicio, M. Mengíbar, A. Heras, *Carbohydydr. Polym.* **2011**, *84*, 803–806.

[3] FAO, *Fishery and Aquaculture Statistics. Year Book,* **2011**.

[4] http://faostat.fao.org/, accessed Mar 2014.

[5] (a) C.J. Brine, P.R. Austin, *Comp.Biochem. Physiol.* **1981**, *69*, 283–286; (b) C.J. Brine, First International Conference on Chitin Chitosan (Eds. R.A.A. Muzzarelli, E.R. Pariser), **1978**, 509–516.

[6] J.K. Chen, C.R. Shen, C.L. Liu, *Mar. Drugs* **2010**, *8*, 2493–2516.

[7] (a) S.A. White, P.R. Farina, I. Fulton, *Appl. Environ. Microb.* **1979**, *38*, 323–328; (b) W. Teng, E. Khor, T.K. Tan, L.Y. Lim, S.L. Tan, *Carbohydr. Res.* **2001**, *332*, 305–316; (c) K.H. No, Y.N. Park, H.S. Lee, P.S. Meyers, *Int. J. Food Microbiol..* **2002**, *74*, 65–72; (d) Y.J. Joen, P.J. Park, S.K. Kim, *Carbohydr. Polym.* **2001**, *44*, 71–76; (e) N.R. Sudharshan, D.G. Hoover, D. Knorr, *Food Biotechnol.* **1992**, *6*, 257–272; (f) N. Shibuya, E. Mimami, *Physiol. Mol. Plant.* **2001**, *59*, 223–233; (g) E.A. EL-Glaouth, J. Arul, C. Wilson, N. Benhamor, *Postharvest. Biol. Tec.* **1997**, *12*, 183–194.

[8] G. G. Allan, J. R. Fox, N. Kong, In "*Proceedings of the First International Conference on Chitin and Chitosan*" (Eds. R.A.A. Muzzarelli, E.R. Pariser), **1978**, 64–78.

[9] http://kitozyme.com/eng/about-us/what-we-do, accessed Mar 2014.

[10] (a) A. Percot, C. Viton, A. Domard, *Biomacromolecules* **2003**, *4*, 12–18; (b) G.H. Jo, R.D. Park, W.J. Jung, In "*Chitin, Chitosan, Oligosaccharides and their Derivatives*" (Ed. S.K. Kim). CRC Press, Boca Raton, **2011**, 37–45; (c) N. Pacheco, M. Garnica-Gonzalez, M. Gimeno, E. Bárzana, S. Trombotto, L. David, K. Shirai, *Biomacromolecules* **2011**, *12*, 3285–3290; (d) M. S. Rao, W. F. Stevens, *J. Chem. Technol. Biotechnol.* **2005**, *80*, 1080–1087.

[11] (a) M.Z. Elsabee, E.S. Abdou, *Mater. Sci Eng: C* **2013**, *33*,1819-1841; (b) A. Rampino, M. Borgona, P. Blasi, B. Bellich, A. Cesaro, *Int. J. Pharm.* **2013**, *455*, 219–228; (c) N. Bhattarai, J. Gunn, M. Zhang. *Adv. Drug Deliv. Rev.* **2010**, *62*, 83–99; (d) C.K.S. Pillai, W. Paul, C.P. Sharma, *Progr. Polym. Sci.* **2009**, *34*, 641–678; (e) I. Aranaz, M. C. Gutiérrez, L. Yuste, F. Rojo, M.L. Ferrer, F. del Monte. *J. Mater. Chem.* **2009**, *19*, 1576–1582.

[12] (a) H. Vogt Sæther, H.K. Holme, G. Maurstad, O. Smidsrød, B.T. Stokke, *Carbohydr. Polym.* **2008**, *74*, 813–821; (b) F. Bigucci, B. Luppi, T. Cerchiara, M. Sorrenti, G. Bettinetti, L. Rodriguez, V. Zecchi, *Eur. J. Pharm. Sci.* **2008**, *35*, 435–441; (c) H.C.L. De Oliveira, J.L.C. Fonseca, M.R. Pereira, *J. Biomater. Sci.* **2008**, *19*, 143–160.

[13] I. Aranaz, M. Mengíbar, R. Harris, I. Paños, B. Miralles, N. Acosta, G. Galed, Á. Heras, *Curr. Chem. Biol.* **2009**, *3*, 203–230.

[14] M.N.V.R. Kumar, *React. Funct. Polym.* **2000**, *46*, 1–27.

[15] (a) P. K. Dutta, J. Dutta, V.S. Tripathi, *J. Scientific Ind. Res.* **2004**, *63*, 20–31; (b) M. Rinaudo, *Prog. Polym. Sci.* **2006**, *31*, 603–632.

[16] (a) P.K. Dutta, S. Tripathi, G.K. Mehrotra, J. Dutta, *Food Chem.* **2009**, *114*, 1173–1182; (b) P. Fernández-Saiz, G. Sánchez, C. Soler, J.M. Lagaron, M.J. Ocio, *Food Contr.* **2013**, *34*, 61–68.

[17] (a) V.C. Valenzuela, J.I. Arias, *Avances en Ciencias Veterinarias.* **2012**, *27*,33-47; (b) S.R. Kanatt, M.S. Rao, S.P. Chawla, A. Sharma, *LWT-Food Sci. Technol.* **2013**, *53*, 321–326; (c) Y.Y. Qin , J.Y. Yang, H.B. Lu, S.S. Wang, J. Yang, X.C. Yang, M. Chai, L. Li, J.X. Cao, *Int. J. Biol. Macromol..* **2013**, *61*, 312–316.

[18] M. Bangoura, X. Wenshui, Z. Jiali, *Am. J. Food Technol.* **2009**, *4*, 126–135.

[19] (a) M.S. Rodríguez, L.A. Albertengo, E. Agulló, *Carbohydr. Polym.* **2002**, *48*, 271–276; b(b) E.W. Schumacher, US Patent 6,153,250. **2000**; (c) N.A. Qinna, F.T. Akayleh, M.M. Al Remawi, B.S. Kamona, H. Taha, A.A. Badwan, *J. Funct. Food.* **2013**, *5*, 1125–1134.

[20] V. Speiciene, F. Guilmineau, U. Kulozik, D. Leskauskaite, *Food Chem.* **2007** *102*, 1048–1054.

[21] (a) B. Miralles, A. Martínez-Rodríguez, A. Santiago, J. van de Lagemaat, A. Heras, *Food Chem.* **2007**, *100*, 1071–1075; (b) Y. Song, E. E. Babiker, M. Usui, A. Saito, A. Kato, *Food Res. Int.* **2002**, *35*, 459–466; (c) A. Kato, K. Minaki, K. Kobayashi, *J. Agric. Food Chem.* **1993**, *41*, 540–543; (d) M. Usui, H. Tamura, K. Nakamura, T. Ogawa, M. Muroshita, H. Azakami, S. Kanuma, A. Kato, *Nahrung – Food* **2004**, *48*, 69–72.

[22] (a) Y.J. Jeon, S.K Kim, *J. Microbiol. Biotechnol.* **2002**; *12*, 503–7; (b) S.K. Kim, N. Rajapakse, *Carbohydr. Polym.* **2005**, *62*, 357–368; (c) Q. Caiquin, W. Wei, P. huei, H. Rong, L. Wei, *Carbohydr. Polym.* **2008**, *72*, 701–706.

[23] S. B. Lin, S. H. Chen, K. C. Peng, *J. Sci. Food Agric.* **2009**, *89*, 238–244.

[24] (a) V. Dodane, V. D. Vilivalam, *Pharm. Sci. Technol. Today* **1998**, 1, 246–253; (b) G. Lang, T. Clausen, In "*Chitin and Chitosan: Sources, Chemistry, Biochemistry, Physical Properties and Applications*" (Eds. G. Skjak-Braek, T. Anthonsen). Elsevier Sciences Publishers, London, **1989**, 139–147; (c) E.M. Costa, S. Silva, A. Cardelle-Cobas, A.R. Madureira, M.O. Ferreira, F.K. Tavaria, M.M. Pintado, *J. Chiin Chitosan Sci.* **2013**, *1*, 251–259.

[25] A. Vogt, B. Combadiere, S. Hadam, K.M. Stieler, J. Lademann, H. Schaefer, B. Autran, W. Sterry, U. Blume-Peytavi, *J. Invest. Dermatol.* **2006**, *126*, 1316–1322.

[26] F. Shahidi, J. Kamil, Y.J. Jeon, S.K. Kim, *J. Food Lipids* **2002**, *9*, 57–64.

[27] B. N. Estevinho, F. Rocha, L. Santos, A. Alves, *Trends Food Sci. Technol.* **2013**, *31*, 138–155.

[28] (a) M. L. Weiner, In "*Advances in Chitin and Chitosan*" (Eds. C.J. Brine, P. Standford, J. Zakakis). Elsevier, **1992**, 663–670; (b) *Additive Code. Korea Food and Drug Administration* **1995**.

[29] G. Raafat, H. G. Sahl, *Microb. Biotechnol.* **2009**, *2*, 186–201.

[30] *EFSA J.* **2011**, *9*, 2214–2235.

[31] (a)*EFSA J.* **2009**, *1099*, 1–19; (b)*EFSA J.* **2010**, *8*, 1687–1704.

[32] *European Pharmacopeia* **2002**, 1774–1775.

[33] *Supplement n°2, "Chitosan", United States Pharmacopeia Convention* **2011**, 5361–5365.

[34] (a)ASTM, *F2602-13 Standard Test Method for Determining the Molar Mass of Chitosan and Chitosan Salts by Size Exclusion chromatography with Multi-angle Light Scattering Detection (SEC-MALS),* **2013**; (b)ASTM, *F2260-12. Standard Test Method for Determining Degree of Deacetylation in Chitosan Salts by Proton Nuclear Magnetic Resonance (1H NMR) Spectroscopy,* **2012**; (c)ASTM, *F2103-11. Standard guide for characterization and testing of chitosan salts as starting materials intended for use in biomedical and tissue-engineered medical product applications,* **2001**.

[35] http://www.internationaloliveoil.org/estaticos/view/131-world-olive-oil-figures, accessed Mar 2014.

[36] R. Borja, J. Alba, C. J. Banks, *Process Biochem.* **1997**, *32*, 121–133.

[37] A. G. Vlyssides, M. Loizides, P. K. Karlis, *J. Clean. Prod.* **2004**, *12*, 603–611.

[38] J.M. Aragon, M. C. Palancar, *Improlive 2000. Present and future of Alperujo*, Editorial Complutense, S.A., Madrid, **2001**.

[39] *Integrated Strategy of actions, measures and means suitable for Mediterranean Countries Analysis of national and European legislative frameworks for Oil Olive Waste and Soil Protection.* http://www.prosodol.gr/sites/prosodol.gr/files/ACT15_1.pdf.

[40] J. Fernández-Bolaños, R. Rodríguez, R. Guillén, A. Jiménez, *Grasas y aceites* **2006**, 57, 95–106.

[41] M. Niaounakis, P. Halvadakis, "*Olive-mill waste management: literature review and patent survey*", Typothito-George Dardanos Publications, Athens, **2004**.

[42] (a) F. A. Tomas-Barberan, R. Lorach, J. C. Espin, F. Ferreres, In "Proceeding Total Food", Norwich (UK), **2004**; (b) K. Waldron, In "*Proceeding Total Food*", Norwich (UK), **2004**.

[43] H. G. Helal, M. M. Hassan, in *Animal Feed Technical articles.* **2013**. http://en.engormix.com/MA-feed-machinery/formulation/articles/effect-sprouted-corn-grains-t2568/800-p0.htm, accessed Mar 2014.

[44] *EFSA J.* **2011**, *9*, 2033–2058.

[45] http://www.kirinholdings.co.jp/english/news/2013/0822_01.html, accessed Mar 2014.

[46] (a) I.M.P.L.V.O. Ferreira, O. Pinho, E. Vieira, J.G. Tavarela, *Trends Food Sci. Technol.* **2010**, *21*, 77–84; (b) A.F. Campos, O.G. Pereira, K.G. Ribeiro, S.A. Santos, S.D.C. Valadares Filho, *Animal Feed Sci. Technol.* **2014**, *190*, 38–46; (c) S. Phugare, P. Patil, S. Govindwar, J. Jadhav, *Int. Biodeterior. Biodegradation* **2010**, *64*, 716–726; (d) N. Aimaretti, C. Ybalo, *Biomass Bioener.* **2012**, *42*, 18–23; (e) R. Han, H. Li, Y. Li, J. Zhang, H. Xiao, J. Shi, *J. Hazard. Mater.* **2006**, *137*, 1569–1576.

[47] (a) G. Lesage, H. Bussey, *Microbiol. Mol. Biol. Rev.* **2006**; *70*, 317–343; (b) T.M. Tam, N.Q. Duy, N.P. Minh, D.T.A. Dao, *Am. J. Res. Commun.* **2013**; *1*, 149–158.

[48] S. Pentikäinen, L. Karhunen, L. Flander,. K. Katina, A. Meynier, P. Aymard, S. Vinoy, K. Poutanen, *Appetite* **2014**;*75*, 150–156.

[49] (a) A.L. Jenkins , D.J. Jenkins, U. Zdravkovic, P. Würsch, V. Vuksan, *Eur. J. Clin. Nutr.* **2002**, *56*,622-628; (b) D.L. Williams, *Mediators Inflamm.*, **1997**, *6*, 247–250; (c) M. Hofer, M. Pospisil, *Int. J. Immunopharmacol.* **1997**, *19*, 607–609; (d) J. Soltys, M. Quinn, *Infect. Immun.* **1999**, *67*, 244–252; (e) C. Li, H. Tuanzhu, J. Kelley, X. Gao, Y. Qiu, R.L. Kao, W. Browder, D.L. Williams, *Cardiovasc. Res.* **2004**, *61*, 538–547; (f) J.L. Turnbull, M.L. Patchen; D.T. Scadden, *Acta Haematol.* **1999**, *102*, 66–71; (g) F. Hong, J. Yan, T. Baran, D.J. Allendorf, R.D. Hansen, G.R. Ostroff, P.X. Xing, N.K. Cheung, G.D. Ross, *J. Immunol.* **2004**, *173*, 797–806.

[50] *Gras Notices. FDA. GRN n° 437, GRN n° 413, GRN n° 239. Available at* http://www.fda.gov/Food/IngredientsPackagingLabeling/GRAS/NoticeInventory/default.htm.

[51] *EFSA J.* **2009**, *7*, 1254–1272.

[52] *EFSA J.* **2011**, *9*, 2471–2484.

[53] *EFSA J.* **2011**, *9*, 2137–2159.

[54] (a) J. Fernandez, M.D. Curt, P.L. Aguado, *Ind. Crops Prod.* **2006**, *24*, 222–229; (b) E. Maccarone, B. Fallico, F. Fanella, G. Mauromicale, S.A. Raccuia, S. Foti, *Ind. Crops Prod.* **1999**,*10*, 229–237; (c) M.D. Curt, G. Sànchez, J. Fernández, *Biomass Bioener.* **2002**, *23*, 33–46; (d) S.A. Raccuia, M.G. Melilli, *Field Crops Res.* **2010**, *116*,147-153; (e) S.A. Raccuia, I. Piscioneri, N. Sharma M.G. Melilli. *Biomass Bioener.* **2011**,*35*, 3167–3173; (f) A. Ierna, G. Mauromicale, *Biomass Bioener.* **2010**, *34*, 754–760; (g) M. Llugany, R. Miralles, I. Corrales, J. Barceló, C. Poschenrieder, *J. Geochemical Exploration* **2012**, *123*, 122–127; (h) J. Gominho, J. Fernandez, H. Pereira, *Ind. Crops Products* **2001**,*13*,1–10.

[55] (a) F. Fratianni, M. Tucci, M. de Palma, R. Pepe, F. Nazzaro, *Food Chem.* **2007**, *104*, 1282–1286; (b) P. Valentão, E. Fernandes, F. Carvalho, P.B. Andrade, R.M. Seabra, M.L. Bastos, *J. Agric. Food Chem.* **2002**, *50*, 4989–4993; (c) S.P. Trajtemberg, N.M. Apóstolo, G. Fernández, *In Vitro Cell Dev. Biol. Plant* **2006**, *42*, 534–537; (d) P. Pinelli, F. Agostini, C. Comino, S. Lanteri, E. Portis, A. Romani, *Food Chem.* **2007**, *105*, 1695–1701; (e) L. Bezáková, D. Grancai, M. Oblozinsky, M. Vanko, I. Holková, I. Paulikovà, V. Garaj, M. Gáplovský, *Acta Fac. Pharm. Univ. Comen.* **2007**, *Tomus LIV*, 48–52; (f) P. Sevcikova, Z. Glatz, J. Slanina, *Electrophoresis* **2002**, *23*, 249–252; (g) X. Zhu, H. Zhang, R. Lo, J. Agric. *Food Chem.* **2004**, *52*,7272–7278; (h) L. Velez, M.A. Campinho, A.R. Guerra, L. García, P. Ramos, O. Guerreiro, L. Felício, F. Schmitt, M. Duarte, *Agriculture* **2012**, *2*, 472–492.

[56] (a) S.A. Raccuia, M.G. Melilli, *Aus. J. Agric. Res.* **2004**, *55*, 693–698.

[57] (a) T. Barclay, M. Ginic-Markovic, P. Cooper, N. Petrovsky, *J. Excip. Food Chem.* **2010**, *1*, 27–50, and references in; (b) N. Kaur, A.K. Gupta, *J. Biosci.* **2002**, *27*, 703–714.

[58] A. van Huis, J. Van Itterbeeck, H. Klunder, E. Mertens, A. Halloran, G. Muir, P. Vantomme, FAO Technical Papers, **2013**.

[59] (a) *The world beta-glucan ingredient market. Ubic Consulting* **2014**; (b) *Global Market Report of Pectin. Research and Markets* **2014**; (c) *Mannitol (CAS 87-78-5) Market Research Report. Research and Markets* **2013**; (d) *Mixed Tocopherols Market. Global Industry Analysis Market Size Share Trends Analysis Growth and Forecast 2013–2019. Transparency Market Research,* **2013**; (e) *Global Market for Inulin to 2017 - Market Size, Growth and Forecasts in Over 50 Countries. Research and Markets,* **2013**; (f) *Squalene Market – Global Industry Analysis, Size, Share, Growth, Trends and Forecast, 2012–2018. Transparency Market Research,* **2012**.

[60] *Chitin and chitosan a global strategic business report. Global Industry Analysis,* **2003**.

[61] *Chitin and chitosan a global strategic business report. Global Industry Analysis,* **2010**.

[62] Polyphenols Market by Product (Grape seed, Green tea, Apple and Others), by Application (Functional beverages, Functional food, Dietary supplements and Others) – Global Industry Analysis, Size, Share, Growth, Trends and Forecast, 2012 – 2018. Transparency Market Research, **2013**.

16

BIORENEWABLES AT MANGO MATERIALS

Allison Pieja, Anne Schauer-Gimenez, Ann Oakenfull, and Molly Morse

Mango Materials, Oakland, CA, USA

Industrial Biorenewables: A Practical Viewpoint, First Edition. Edited by Pablo Domínguez de María.

What if the persistence of plastics in natural ecosystems was no longer an ongoing environmental challenge? What if we could reduce greenhouse gas emissions as well as replace a significant amount of the world's plastic with a biodegradable substitute?

16.1 MOTIVATION: THE PROBLEMS WITH PLASTICS TODAY

Plastics are ubiquitous in everyday life. However, people rarely consider where they came from or what will happen to them when they are no longer needed. Traditional plastics are derived from petroleum, which is a nonrenewable resource. In 2013, the United States consumed almost 7 billion barrels (944 million tonnes) of petroleum products, or 2.6 million tonnes per day.[1] The manufacture of plastic products in the United States uses both petroleum and natural gas.[2]

Despite the development of alternative plastics made from biomass resources, such as corn or sugarcane, biobased plastics are often more costly than their mass-produced, petroleum-derived counterparts. Additionally, few plastics are actually reused or recycled at the end of their useful lives. As most plastics do not biodegrade, they continue to persist and accumulate in landfills, waste facilities, oceans, and other natural habitats. Nonbiodegradable plastics in the natural environment can harm ecosystems and all the species within them.

16.1.1 Plastics in the Natural Environment

Plastics can be detrimental to species living in oceans and other natural environments. To date, 136 different marine species have been found entangled in plastic debris.[3] Animals may also be harmed by ingesting plastic debris in natural environments.

Plastics can be used as a mode of transport for invasive species such as bryozoans, barnacles, polychaete worms, hydroids, and mollusks.[4] Chemicals can be transported by plastic pollution in oceans and ultimately harm ecosystems. It has been shown that hydrophobic pollutants (including polychlorinated biphenyl and dichlorodiphenyldichloroethylene) absorb onto the nonpolar surface of polypropylene pellets.[5] The concentration of the pollutants in the plastic debris was found to be 10,000–1,000,000 times higher than in surrounding waters.[6]

The Great Pacific Garbage Patch and Pacific Gyre are popular names used for a large swath of debris floating in the Pacific Ocean. According to the National Oceanic and Atmospheric Administration (NOAA), there are small pieces of floating plastic that cover a wide range of the Pacific Ocean due to mechanical mixing by wind and wave action.[7] The Pacific Gyre is just one of many ocean gyres where plastics are accumulating. There have been at least 5 large gyres noted made up of an estimated 33.6 thousand tonnes (37 thousand tons) of plastic.[8] The large quantity of small plastic fragments in the world's oceans is, therefore, a growing environmental concern.

Plastic has been found in plankton samples dating back to the 1960s.[9] If current nonbiodegradable plastic production and consumption persists, and if improper disposal endures, plastic waste will continue to end up in oceans. Without global intervention, it is hypothesized that garbage will continue to accumulate and not even hit its peak by 2021.[3]

Coastlines are also repositories for plastic pollution. In Orange County, California, plastic pellets are the most abundant debris on the coastline. In the waters off the coast of Long Beach, California, eight pieces of plastic per cubic meter have been reported, with seven times that amount measured after storms.[10]

16.1.2 Plastics in Landfills

Plastics typically end up in landfills, where they can persist for thousands of years. In 2012, the United States discarded about 29 million tonnes (32 million tons) of plastic, which comprised 13% (by weight) of the waste in municipal solid-waste landfills.[11] In California, plastics represent 9.5% of the waste discarded in landfills.[12] In the United States, almost 12.7 million tonnes (14 million tonnes) of the plastics disposed of was containers and packaging, about 10 million tonnes (11 million tonnes) was durable goods (i.e., appliances), and almost 6.3 million tonnes (7 million tonnes) was nondurable goods (i.e., plates and cups).[13]

16.1.3 Recycled Plastics

In 2012, only about 9% of discarded plastic in the United States was recycled, the majority of them being containers and packaging.[13] The most common type of plastic that is recycled is polyethylene terephthalate (PET), commonly known as #1 plastic.[14] In the past, municipalities generally had the ability to recycle only #1 and #2 (high-density polyethylene (HDPE)) plastics, but recycling other plastics is becoming more of the norm. In a survey of only the largest municipality per state, the number of municipalities recycling all plastics (#1–#7) grew from 60% in 2010 to 88% in 2012.[15]

California has instituted a “California Refund Value (CRV)” program where customers are paid when they recycle beverage containers, including plastic bottles, at certified recycling centers. The program began in 1987, and since then over 300 billion aluminum, glass, and plastic beverage containers have been recycled.[16] In 2009 alone, nearly 22 billion CRV-eligible containers were sold and more than 17 billion were recycled.[14]

Recycling plastics has clear benefits. First of all, recycled bottles and containers are used to produce new products from lumber for outdoor decking to carpeting, fleece jackets, and T-shirts.[17] Also, for every 4.5 kg (10 lb) of clear plastic that is recycled, carbon emissions are reduced by 1.5 kg (3.3 lb).[14]

16.2 THE BIOPLASTICS INDUSTRY: AN OVERVIEW

16.2.1 Regulations

Many communities are concerned that traditional plastics have a large carbon footprint and that they persist, rather than biodegrade, if improperly discarded. These concerns are increasingly being voiced through regulations. Bans of plastic with toxic additives are already in effect in Canada, France, Denmark, Minnesota, and New York.[18] Many cities in the United States, including San Francisco, Los Angeles, and Seattle, have policies that ban single-use plastic bags.[19] In Italy, a nationwide ban limits the use of plastic bags.[20] Ireland, China, and Washington D.C. have seen plastic bag use drop substantially after imposing fees or taxes on single-use plastic bags.[21] These trends are expected to significantly increase demand for safe, biodegradable plastics.

In addition to legislation restricting the use of nonbiodegradable or toxic plastics, some regulations now promote the use of biobased products. The 2002 US Federal Biobased Procurement Program requires governmental agencies to buy biobased products if these alternatives are not significantly more expensive or of lesser quality than fossil fuel-based materials.

16.2.2 Terminology

The bioplastics industry is awash with terminology that is used loosely and often improperly, leading to frequent misunderstandings. Confusion arises when organizations and analytical standard tests define words differently.[22]

The American Society for Testing and Materials (ASTM) and the International Standards Organization (ISO) have multiple tests and standards that materials must pass in order to be classified with certain designations. The commonly referenced ASTM D6400 proposes that a degradable plastic is one, "in which the degradation results from the action of naturally occurring microorganisms such as bacteria, fungi and algae."[23] This regulation regulates the labeling of plastics designed to be composted in municipal or industrial facilities. Similarly, ASTM D6866 describes a test that uses radiocarbon analysis to determine the biobased content of a material. This is to encourage products that are derived from rapidly renewable, biobased, and sustainable materials that result in a low environmental impact.[24] ASTM D6866 refers to the source where the materials originated, whereas ASTM D6400 involves the fate of the material at the end of its useful life. The state of California imposed new legislation in 2013 to clarify labeling of specific products, for example, "biodegradable" or "compostable."[25] According to California State Bill 567, a material such as plastic must pass specific ASTM tests to be marketed as compostable or with a similar label.[26]

The term "biobased" is completely unrelated to the term "biodegradable," but the two are often confused. "Biobased" carbon is from rapidly renewable, nonfossil sources.[27] Materials made from biobased sources such as sugar, starches, and cellulose may or may not be biodegradable. To confuse matters further, a bioplastic may be biobased, biodegradable, or both.

16.2.3 Bioplastics

Traditional (petroleum-based) plastics are the current industry standard and are produced worldwide. Their cost is currently about \$1.10/lb (\$2.44/k) when purchased in bulk quantities and is highly dependent on the cost of petroleum. Bioplastic production requires optimization at scale to achieve a similar price point. Due to the challenges facing the plastics industry, a number of alternatives related to biodegradable plastics have been developed.

16.2.3.1 Biobased, Nonbiodegradable Plastics Bioplastics that are not biodegradable but that are from rapidly renewable, nonfossil carbon comprise over 50% of the bioplastics market.[28] This includes polymers such as bio-polyethylene terephthalate (PET) and bio-polyethylene (PE). These polymers are structurally and functionally identical to their petroleum-based counterparts (i.e., PET and PE) but are derived from rapidly renewable biomass such as corn or sugarcane.

Materials such as bio-PET and bio-PE are considered "drop-in" replacements for petroleum-based plastics. An advantage of these materials is that their properties are already widely tailored and customized for many applications, and significant recycling infrastructure already exists to encourage material recapture and reuse. However, bio-PET and bio-PE are no more biodegradable than their petroleum-based counterparts and are subject to similar downcycling, or loss of material properties, upon recycling as PET or PE. Currently, a maximum fraction of 35% recycled PET can be used in plastic bottles before unacceptable effects of discoloration may be seen.[29]

Biobased PET and PE do not necessarily have better environmental footprints than their petroleum-based counterparts. A complete life cycle analysis must consider the carbon footprint of the feedstocks as well as the footprint of materials (e.g., fertilizer) required to produce those feedstocks. Biobased PET and PE may utilize agricultural lands and generate negative environmental impacts from fertilizers used to grow feedstock crops, such as on biodiversity and water quality.

PET is comprised of approximately 30% monoethylene glycol and 70% terephthalic acid, which comes from the precursor paraxylene. Bio-PET, such as that used in Coca Cola's "PlantBottle," that may be up to 30% biobased typically contains biobased monoethylene glycol and petroleum-based terephthalic acid. While it is possible to produce 100% biobased PET using biobased terephthalic acid in addition to ethylene – as demonstrated by PepsiCo in 2011 – this has not yet become commercially viable. Coca Cola is working with Virent to achieve the goal of 100% biobased PET.[30]

A potential downside of these biobased, nonbiodegradable materials is that they are often used to manufacture items such as packaging or water bottles, single-use items that could benefit from alternative end-of-life options. Additionally, the term "biobased" may be misleading to some general consumers who believe they are purchasing biodegradable options, so significant consumer education is ideal.

16.2.3.2 Biobased, Biodegradable Plastics Biobased, biodegradable plastics have the advantage of being both produced from rapidly renewable carbon and biodegradable. One of the most commonly known biobased, biodegradable polymers is polylactic acid (PLA). This biopolyester is generally made from corn or sugarcane and is manufactured by companies including NatureWorks and Purac (Corbion).

Polyhydroxyalkanoate (PHA) is a polyester synthesized by bacteria and stored as intracellular carbon and energy reserves.[31] PHAs are attracting considerable attention as environmentally friendly replacements for petroleum-based plastics due to numerous advantages, including that they (1) have mechanical properties similar to the properties of polypropylene, (2) are completely biodegradable, and (3) can be derived from renewable resources.[31]

Nearly a hundred different PHAs have been identified, and the copolymer composition influences the properties of the end polymer.[32] Poly-3-hydroxybutyrate (PHB) is the most common PHA and the most extensively studied.[31] Common copolymers include poly-hydroxyvalerate-co-butyrate and poly-hydroxyhexanoate-co-butyrate. Although PHB itself is relatively brittle, the addition of additives or other monomer units increases the polymer's flexibility. In addition to copolymerization, ecologically friendly additives may be added to PHB to alter its properties for specific applications.

Whereas PLA is only biodegradable under specific aerobic composting conditions, PHAs are biodegradable under a variety of conditions and will break down in aerobic and anaerobic composting as well as in marine and other natural environments.

16.2.4 Bioplastics Market

The global plastics market has shown strong growth despite the financial crisis of 2009. Due to existing market demand and population growth trends, global plastic production increased from 204 million tonnes in 2002 to 299 million tonnes in 2013.[33] Apparent consumption of goods produced by the US plastics industry grew to $267.3 billion in 2013, with plastic production reaching almost 108 billion pounds (49 million tonnes).[34]

Bioplastics occupy a small but rapidly growing segment of this market. The global bioplastics market is expected to almost triple in size from $7 billion (€6.13 billion) in 2014 to $20 billion (€17.45 billion) by 2018.[35] Similarly, the production capacity of the global bioplastics market is expected to exhibit explosive growth, with a predicted expansion of greater than four times from 3.6 billion pounds (1.6 million tonnes) in 2013 to almost 15 billion pounds (6.7 million tonnes) in 2018.[28] The North American bioplastics market is also estimated to grow significantly from 662 million pounds (300,000 t) in 2011 to 2.65 billion pounds (1.2 million tonnes) in 2016.[36]

Increased customer demand for biodegradable plastics has contributed to the rise of bioplastics in the market. The biodegradable plastics market made up 38%, or more than 1.3 billion pounds (600,000 tonnes), of the global bioplastics market in 2013 and is expected to almost double to 2.5 billion pounds (1.1 million tonnes) in 2018.[28] Out of the many biodegradable plastics on the market, PHA is the fastest growing type with its market size estimated to hit $1 billion by 2018 (Mango Materials forecast using expected compound annual growth rates (CAGRs) of 20–30%). At estimated CAGRs of 20–30%, PHA will see increased demand and command a larger share of the estimated $6 billion (€5 billion) biodegradable plastics market in 2018.[35]

Although the PHA market is large and growing, a number of challenges impede future growth. For PHA's market share to increase further, PHA production must improve in the following two critical areas: (1) scaling supply to meet demand and (2) cutting costs.

PHA is often found in high demand from customers who require minimum amounts of hundreds of tonnes each year. Companies currently producing PHA include Metabolix (United States), Tianan (China), Biocycle (Brazil), and Biomer (Germany). Metabolix, the largest US producer of PHA, markets its product at specialty materials prices of \$2.25–\$2.75/lb (\$4960–\$6063/t).[37] For PHA to compete with nonbiodegradable plastics and meet the growing demand for biodegradable plastics, PHA supply must increase significantly to reach economies of scale.

In addition to the need to scale up PHA production, PHA's high cost poses another challenge. PHA commercialization has been limited due to PHA's high cost compared to traditional petroleum-based plastics such as polyethylene and polypropylene. The increased cost of PHA stems largely from its most commonly used feedstock – sugar. Sugar not only has a high cost but also competes for land use with food crops; use of waste methane as a feedstock will greatly reduce this cost. Reliable production of PHA powder at a competitive price will contribute significantly to increasing bioplastics demand.

16.3 MANGO MATERIALS – A NOVEL PHA PRODUCTION PROCESS

16.3.1 Company Summary

Mango Materials produces biodegradable plastics from methane that are economically competitive with conventional, petroleum-based plastics. Mango Materials can use excess of methane gas from sources such as water resource recovery facilities (WRRFs) – these are also known as wastewater treatment plants – agricultural facilities, landfills, natural gas, or industrial processes to produce powder or pellets of PHA. PHA is a valuable polymer that is converted into a variety of high-margin or high-volume, eco-friendly plastic products such as children's toys, electronic casings, nonwovens, and packaging applications. Due to an increasing preference for green products from consumers, businesses, and government agencies, demand for biodegradable and nonpetroleum-based plastics is growing rapidly.

Mango Materials' competitors use either petroleum, which is low cost but produces nonbiodegradable plastic, or sugars, which are expensive but produce biodegradable plastic. In contrast, Mango Materials uses affordable methane gas and a process that competes favorably with petroleum-based plastics to produce low-cost, biodegradable plastics. This technology offers methane producers another profitable use for their biogas while transforming a greenhouse gas (GHG) into a valuable commodity.

Mango Materials' bioplastic can have a closed-loop life cycle without the significant cost and resources required to separate specific plastics from the waste stream

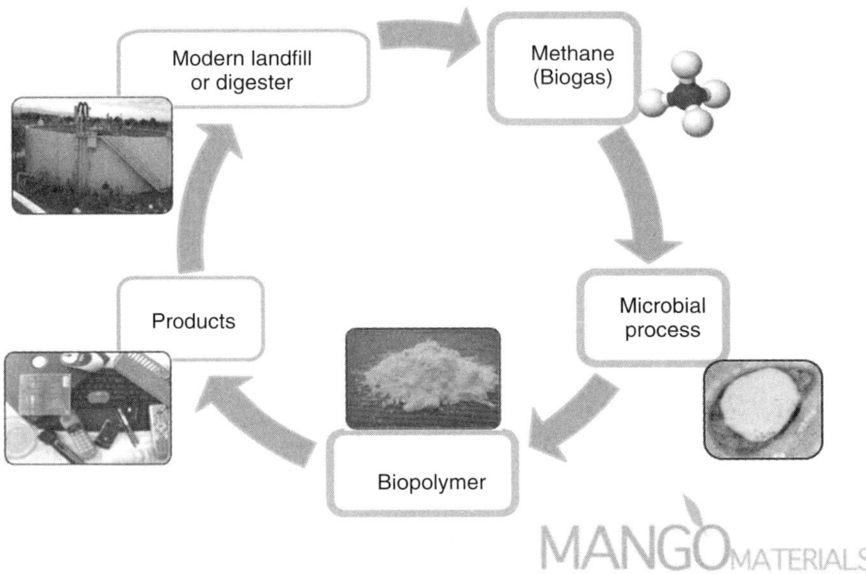

Figure 16.1 The closed-loop system of Mango Materials. (*See color plate section for the color representation of this figure.*)

through recycling. Bacteria can transform a renewable resource, methane biogas, into high-grade plastics that biodegrade back to methane gas when disposed in a modern landfill or in an anaerobic digester at a WRRF. The methane produced can be cycled back and reenter the Mango Materials process as feedstock to produce more PHA. Thus, a truly "cradle to cradle" system is created (Figure 16.1).

On the supply side, methane (Mango Materials' feedstock) is a GHG significantly more potent – in terms of global warming potential – than carbon dioxide. Waste methane, or biogas, from a variety of anthropogenic sources is often flared or simply vented to the atmosphere because there is rarely sufficient economic incentive to capture it. Mango Materials' technology can accommodate biogas that contains impurities and may be considered marginal quality by other users. This gives methane producers an environmentally beneficial method of disposal for their waste biogas.

On the demand side, manufacturing companies ("plastic converters" or "plastic formers") typically blend bulk plastic materials to obtain the desired end characteristics, including stiffness, temperature tolerance, ductility, and ease of processing. When plastic converters receive an order to produce biodegradable products, Mango Materials' PHA can be blended to produce a variety of eco-friendly plastic products.

Mango Materials' use of methane as feedstock is significant for several reasons:

1. It eliminates the need to first convert raw biomass into sugar, thereby significantly reducing both capital equipment and operating expenses compared to current practices.

2. Anthropological biogas sources, including WRRFs and landfills, provide an ideal source of methane (the primary feedstock).
3. Production plants could be widely deployed anywhere there is a large source of methane, often in urban centers. Facilities that dispose of biogas from anthropological sources are increasingly seeking reliable methods to prevent or minimize atmospheric venting.

In summary, the bioplastic produced via this process appears to be *exceptionally economical* and could *displace a significant portion of petroleum-based plastics by achieving a cost-competitive price* without requiring government subsidies or regulations.

16.3.2 Product Differentiation

Mango Materials' process is a transformative technology with numerous environmental benefits, including (1) reduction of GHG emissions, (2) sequestration of methane as bioplastic, (3) utilization of a potentially energy-independent process, and (4) production of a biobased, biocompatible, and completely biodegradable product. Thus, Mango Materials will reduce GHG emissions and increase the amount of environmentally friendly, nontoxic polymer on the market. Additionally, many bioplastic companies use genetically modified organisms (GMOs) in their process, through either a genetically modified feedstock (e.g., corn, sugarcane) or by using genetically modified bacteria to produce their biopolymer. Mango Materials can use rapidly renewable waste methane and naturally selected bacteria to produce GMO-free biopolymer.

Compared to other bioplastics such as PLA, Mango Materials' PHA biopolymer has superior heat, moisture, and end-of-line properties. It has advantages over competitors' PHA because at scale it can be cost competitive with petroleum-based plastics and because it uses a feedstock that is not in competition with food production.

In summary, advantages of the Mango Materials plastic include that it is:

- Completely biodegradable and does not accumulate in natural environments, thus reducing pollution.
- Biocompatible and safe for use in food packaging and biomedical applications.
- Produced from waste methane gas that does not compete with food crops.
- Capable of sequestering methane, a potent GHG, and promoting methane capture.
- Manufactured using a less energy-intensive process than other plastics and without using harmful solvents or chemicals.
- Not dependent on genetically modified feedstocks or bacteria.
- Processed on standard plastics manufacturing equipment.

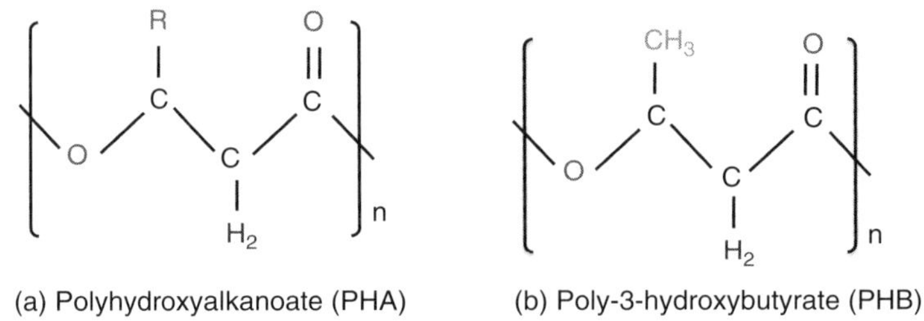

Figure 16.2 (a) Chemical structure of a generic PHA. (b) Chemical structure of PHB.

- Superior to other bioplastics in performance characteristics.
- Not produced with harmful chemicals, such as bisphenol A (BPA), and therefore will not leach them.

16.3.3 Technical Background: Production of PHA from Methane

16.3.3.1 PHA Production in Bacteria Mango Materials produces PHA, polyesters with the general formula as shown in Figure 16.2a. PHAs are a storage polymer synthesized by prokaryotes and stored as intracellular energy reserves.[31] Of all PHAs, PHB is the most common and most extensively studied (Figure 16.2b).[31] Since the original identification of PHB by Lemoigne in 1927, nearly a hundred different PHAs have been identified.[32, 38] That PHAs are so diverse makes them attractive for use as biodegradable plastics.

Several pathways for PHA synthesis and genes encoding these pathways have been identified.[31] There is one pathway that has been identified in many species, including the methylotroph *Methylobacterium extorquens*. This pathway has three key enzymes encoded by the genes *phaA*, *phaB*, and *phaC*.[31] Using a polymerase chain reaction (PCR) assay, the well-conserved *phaC* gene has been used to screen bacteria that produce PHA from environmental samples.[39]

The first step in the PHB synthesis pathway is the condensation of two acetyl-CoA molecules into one acetoacetyl-CoA molecule.[31] This step is mediated by a 3-ketothiolase, encoded by phaA. The next step in PHB biosynthesis is the reduction of acetoacetyl-CoA to 3-hydroxybutyryl-CoA; this is catalyzed by acetoacetyl-CoA reductase, encoded by *phaB*. The final step in PHB production is catalyzed by PHB polymerase and encoded by *phaC*.[31]

16.3.3.2 Methanotrophs Methanotrophic bacteria are a subset of methylotrophs and are unique in that they use only methane as a source of both carbon and energy. Methanotrophs are classified as either γ-proteobacteria (known as type I) or α-proteobacteria (known as type II). The two classes can be distinguished based primarily on their carbon assimilation pathways and internal membrane structures. Type I methanotrophs utilize the ribulose monophosphate (RuMP) pathway for carbon assimilation, whereas type II methanotrophs utilize the serine pathway.[40]

The methane monooxygenase enzyme, which catalyzes the oxidation of methane to methanol, is vital to methanotrophic metabolism. The genes encoding the methane monooxygenase are frequently used to differentiate and identify methanotrophs. All known methanotrophs, with the exception of some strains of *Methylocella*, possess particulate, or membrane-bound, monooxygenase (pMMO). The pmoA gene encodes a subunit of pMMO and is frequently used to detect and differentiate methanotrophs in environmental samples.[41] However, pMMO is copper dependent, and, in the absence of copper, many type II and some type I methanotrophs express another monooxygenase known as soluble methane monooxygenase (sMMO).[42] sMMO is unique in that it has a much broader substrate specificity than pMMO and other oxygenases.[40] This has made it attractive for use in remediation of toxic C1 compounds.

Another method of distinguishing type I and II methanotrophs is by internal membrane structure. Membranes in type I species are arranged in vesicular disks across the cell, while membranes in type II species are often found in stacks along the periphery of the cell wall.[40, 43]

Since they require both oxygen and methane, methanotrophs are typically found in environments where both oxygen and methane are readily available. They are often found at the interface between aerobic and anaerobic zones. Methanotrophs are ubiquitous in nature and are common in environments such as rice paddies, swamps and marshes, surface pond and lake sediments, activated sludge, and meadow and deciduous forest soils.[40] In addition to these, methanotrophs have also been isolated from more unique environments including deserts, landfills, coal mine surfaces, and oceans.[40]

16.3.3.3 PHA Production in Methanotrophs To date, research published on PHA production in methanotrophic bacteria is not extensive and is focused mainly on pure cultures. Type II methanotrophs produce PHA from methane under conditions of nutrient limitation or unbalanced growth.[44] The effects of limiting different nutrients on levels of PHA production have been studied. Limiting nutrients tested include nitrogen, magnesium, phosphate, sulfate, potassium, iron, manganese, copper, and combinations thereof.[44, 45]

PHB levels up to 67% of cell dry weight have been documented by Asenjo and Suk in a pure culture of *Methylocystis parvus* OBBP.[44] However, this research was conducted at low cell densities, and only 1.36 g PHB/L was produced. To date, this is believed to be the highest percentage of cell dry weight as PHB reported in a methanotrophic culture, although significantly higher overall concentrations of PHB have been documented. A concentration of 9.3 g PHB/L, or approximately 50% of the cell dry weight, has been reported.[45] Working with a nonsterile culture comprised mainly of *Methylocystis* sp. GB 25, Wendlandt *et al.* reported the production of approximately 55% of the cell dry weight as PHB, or approximately 30 g PHB/L.[44]

To date, the reported concentrations of PHB in methanotrophs are lower than those found in bacteria that feed on sugars; however, they are still significant and show promise for industrial applications. Part of the allure in using methanotrophs is their ability to use a "waste" as feedstock.

16.3.4 Methane Feedstock: Suitability for Widespread Market Adoption

16.3.4.1 Methane as Feedstock PHA commercialization has been limited to date because of its high cost compared to traditional petroleum-based plastics such as polyethylene and polypropylene. A major fraction of the cost of PHAs (30%) is due to feedstocks, typically glucose and sucrose, which require significant energy to grow, harvest, and ship to polymer production facilities.[46] The use of sugar as feedstock also necessitates a precise manufacturing process that requires significant energy to maintain pure cultures of specialized bacteria under sterile conditions. In contrast, Mango Materials' novel process to produce PHA uses methane gas as feedstock (which can be produced, i.e., through anaerobic waste digestion) and utilizes robust microbes.

Biogas is produced from the decomposition of organic matter in the absence of oxygen and is readily available from multiple sources, including anaerobic digesters at municipal WRRFs, landfills, and agricultural facilities. Bioplastic production plants can be sited at any facility where methane is produced, including WRRFs, landfills, agricultural facilities, coal beds, and other industrial sites. This enables Mango Materials to partner with established businesses and municipalities to leverage considerable infrastructure that is already in place.

16.3.4.2 Differences in Sources of Methane Biogas from WRRFs and agricultural wastes primarily contain methane and carbon dioxide. Methane composition typically ranges from 55% to 70%, while carbon dioxide is reported from 33% to 40%.[47, 48] However, agriculturally derived ("ag") biogas potentially has higher levels of trace contaminants, such as hydrogen sulfide.[47] Bothi reported a wide variation in hydrogen sulfide concentration of about 600 to over 7000 ppm when analyzing the biogas from four dairies.[47] Siloxanes are another trace contaminant found in biogas. Concentrations of siloxanes in biogas are reported to be rising due to the increase in silicon-containing compounds in personal care products, soaps, silicone oils, and even the production of food.[49] Raich found a range of siloxane concentrations from 10.3 mg Si/Nm^3 to 54 mg Si/Nm^3 when testing the biogas from six different WRRFs, a landfill, and a solid-waste anaerobic digester.[49]

Mango Materials enjoys flexibility in the grades and sources of methane that can be used as feedstock. Although biogas methane from waste sources is preferred due to environmental considerations, natural gas from a pipeline or a stranded natural gas well may also be used. Thus, Mango Materials could use the available biogas at a given site and then supplement bioplastic production with natural gas as necessary to meet peak product demand. In addition, Mango Materials could also use natural gas

Figure 16.3 The location of methane production at landfill point sources in the continental United States.

wells in a remote location (stranded gas), where it would be uneconomical to bring a pipeline.

Mango Materials' process has been tested on various types of biogas and appears to be sufficiently robust that the type of biogas does not affect the process.

16.3.4.3 Methane Availability Biogas is an underutilized, local resource that should be captured and utilized to not only prevent the release of methane emissions but also to produce valuable products at decentralized facilities. In reality, there is a significant quantity of landfill gas produced that is either collected and flared or not collected. Figure 16.3 illustrates the distribution of point sources of methane at landfills across the United States. The volume of landfill-derived methane in the United States is estimated to be 466 billion cubic feet (13.2 billion cubic meters) per year.[50] If the collected and unused methane from the landfills in the United States was used for bioplastic production, over 3 billion pounds (1.4 million tonnes) of bioplastic could be produced each year.

California has the highest overall collection and emission of landfill methane, collecting 45 billion cubic feet (1.3 billion cubic meters) per year and emitting 13 billion cubic feet (0.4 billion cubic meters) per year. There are 42 landfills in California and 532 in the United States that could support Mango Materials standard commercial bioplastic production plants – "MangoStandard" commercial plants – capable of producing 1000 tonnes of bioplastic/year. If the unused, vented methane from California landfills alone was collected and used for PHA feedstock, over 100 million pounds (45,000 tonnes) of plastic would be produced each year, enough to manufacture over 3 billion plastic water bottles.[51] In addition to landfills, California dairies could potentially generate nearly 13 billion cubic feet (0.4 billion cubic meters) of biogas each year.[52]

16.3.5 GHG Reduction Potential

16.3.5.1 GHG Reduction Mechanisms Mango Materials' process of converting biogas to bioplastic has the potential to significantly reduce GHG emissions through two methods. First, it directly utilizes a GHG, methane, as its feedstock; methane is consumed by bacteria and sequestered in bioplastic. Second, bioplastic production from methane displaces petroleum-based plastic production with a clean, energy-efficient process.

The GHG methane, one of the two major components in biogas, is 34 times more potent than carbon dioxide over a 100-year period.[53] It is often flared (burned to release carbon dioxide) or vented (released into the atmosphere) because there is rarely sufficient economic incentive to capture it. Mango Materials' process creates an incentive for biogas-producing municipalities and/or businesses to use their biogas to produce a valuable bioplastic, thus promoting methane capture. Mango Materials has estimated that the value of converting methane into plastic is over five times greater than that of generating electricity (Figure 16.4; see also Section 16.3.6). The resulting high-performance plastic can easily be shipped, if necessary, to markets near and far.

Additionally, Mango Materials' production process is more energy efficient and releases fewer GHG emissions compared to the production of petroleum-based plastics. Per unit, biobased plastic production releases only about 8% to 14% of the emissions released by petroleum-based plastic production.[54] Mango Materials' production process is energy efficient, and its production by-products decompose back to the original feedstock (methane); therefore, the production process inherently optimizes energy efficiency and minimizes pollutants.

In addition to being more energy efficient, the production of PHA-based plastics releases 7–12 times fewer GHG emissions (in terms of carbon dioxide equivalents) than petroleum-based plastics.[54] This will reduce overall GHG emissions as

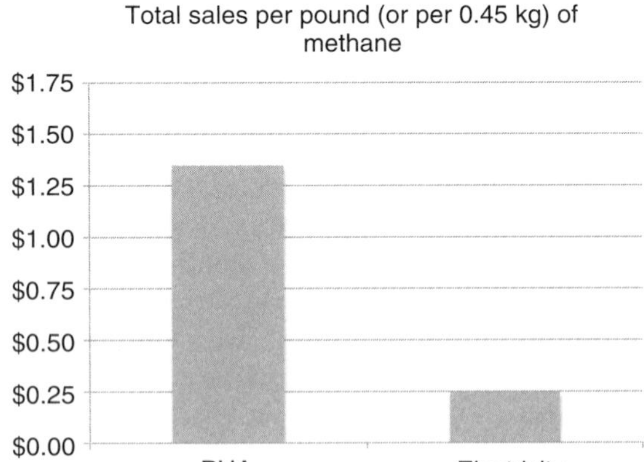

Figure 16.4 Comparison of revenue generated using methane for plastic production versus using methane to generate electricity.

PHA plastics replace petroleum-based plastics and may qualify the process to receive future GHG credits in the state of California. In the Mango Materials process, a gas handling system will be used to ensure that methane is not released from the production process.

16.3.5.2 Quantifying GHG Reduction Potential As the technology is commercialized and more, "MangoStandards" commercial bioplastic production plants, are in operation, sequestering methane and producing bioplastic, significant and quantifiable GHG reductions will be realized. The number of MangoStandards that could be in operation is limited only by the amount of available methane, and there is no shortage of methane in North America or the rest of the world.

Mango Materials' process consumes 2–4 t of methane per tonne of bioplastic produced. If one MangoStandard were sited at a typical WRRF treating 57 million liters of waste per day and producing 3 million cubic meters of methane annually, it would produce over 1100 t of plastic and would sequester 1700–3500 t of methane, reducing emissions by 58,000–1,200,000 t of CO_2 equivalents per year. The US EPA calculated that there was 7.6 million tonnes of methane emissions in 2011 from US WRRFs alone; theoretically, the Mango Materials process could be implemented to sequester this methane as bioplastic and thus reduce methane emissions by over 250 million tonnes of CO_2 equivalents each year.

The potential positive impacts on GHG emission reductions and carbon sequestration are huge; practical impacts are limited only by the extent of rollout of Mango Materials' commercial plants.

16.3.6 Bioplastic versus Other Uses of Methane

Although there are multiple options for biogas producers to convert or upgrade their methane using different technologies, these options are often impractical; they may be extremely costly to implement (cogeneration) or expensive to maintain (turbines require frequent cleaning), or else they may have low value (heat generation at digesters) or only narrow uses that may not be in high demand at the site of methane production (transportation fuels). Some facilities use their biogas for heating, electricity production, or renewable natural gas, but these options are not as economically favorable as converting biogas into a high-value, biodegradable plastic using Mango Materials' biological process.

The capital needed for energy-generation equipment is costly, and the payback period can be significant (Kennedy/Jenks Consultants, Personal communication). Energy-generation equipment can also be fouled quickly with unconditioned biogas (due to the presence of siloxanes and hydrogen sulfide), and, therefore, additional equipment is required to scrub biogas before it enters the energy recovery system.

Bioplastic production can utilize unfiltered biogas and, therefore, does not require scrubbing equipment, which could save facilities significant expenses. In addition to electricity, there are other potential uses for biogas, including production of compressed natural gas for vehicle fuels, upgrading for injection to the natural gas grid, and conversion into chemicals such as methanol.

Considering only the sale price of the product and the yield of the product on methane, bioplastic could generate five times as much revenue per unit methane as electricity. This estimate includes various assumptions including a sale price of $0.11 per kWh for electricity, and it does not account for initial capital expenditures or personnel requirements.

16.4 MANGO MATERIALS, THE STORY

16.4.1 Starting a Start-Up

The growth of Mango Materials is an example of how to build a start-up company on a shoestring. The company grew out of research from Stanford University. Two of the founders, Allison Pieja and Molly Morse, investigated the production of PHA bioplastic from methane and its biodegradation under anaerobic conditions, under the guidance of Professors Craig Criddle and Sarah Billington in the Department of Civil and Environmental Engineering. After Molly completed her Ph.D. work, she consulted briefly in the venture capital industry, which motivated her to incorporate Mango Materials. Allison and a third cofounder, Anne Schauer-Gimenez, joined to try to turn their academic research into a reality. Molly had met Anne years earlier at an anaerobic digestion conference hosted by Marquette University. Even before Molly started work on Mango Materials, she was used to the bootstrapping lifestyle. Rather than pay for a hotel, she asked if any other graduate student at the conference was willing to host in order to leverage resources. Anne volunteered, thus sowing the seeds for a future partnership.

Since the cofounders are all technical Ph.D.'s who at the time had little knowledge of the business world, they recruited longtime friend and fellow Stanford alum Bill Shelander to help navigate the ins and outs of starting a company. In addition to his MBA, Bill had thirty years of experience both in the venture capital field and in growing early-stage businesses. His advice was invaluable in getting the company off the ground.

Before initial funding was awarded or commercial lab-scale validation was begun, Mango Materials and its founders participated in several business plan competitions and business "boot camps." This included the Cleantech Open and the UC Davis Green Technology Entrepreneurship Academy. The idea of turning methane into bioplastic was met with generally positive feedback, if some initial skepticism. Mango Materials fared well at the Cleantech Open and used the competition as an opportunity to extensively network and to create an in-depth business plan.

Two key elements that helped Mango Materials secure initial funding were partnerships and intellectual property (IP). Before receiving funding, Mango Materials met with several local WRRFs and other methane producers that were generating an ample supply of methane through anaerobic digestion. The treatment plants were all excited at the prospect of putting their methane to a valuable use through bioplastic manufacturing. Therefore, Mango Materials developed a partnership with the WRRF in Redwood City, CA, now known as Silicon Valley Clean Water, or SVCW (formerly South Bayside Systems Authority). Through its partnership with SVCW, Mango Materials gained access to a field supply of biogas from anaerobic digestion and a site for preliminary testing.

Mango Materials were also selected for a partnership with the United States Department of Agriculture (USDA) in Albany, CA, through a trust agreement. While looking for a potential laboratory partner in the Bay Area, Allison and Molly remembered that the USDA had done some work relating to PHA in the past. They reached out to Dr. William Orts (whom Molly had met at a conference many years earlier) and were pleasantly surprised to receive a positive response within days. At their initial meeting, they brainstormed ways they could collaborate together to transform waste methane into affordable, biodegradable plastics and were eventually awarded a partnership with the USDA. This partnership with the USDA allowed Mango Materials to collaborate closely with USDA researchers interested in both anaerobic digestion and bioplastic processing. Additionally, it provided the company with a world-class polymer characterization facility and site where it could perform laboratory validation and optimization of its process.

In addition to partnerships, the other key element crucial to early success was access to IP created by the founders and their colleagues while at Stanford. Molly and Bill negotiated an agreement with the Office of Technology Licensing at Stanford where Mango Materials paid Stanford a sum of money in exchange for the exclusive option to license the technologies.

Mango Materials explored many opportunities for initial funding, including venture capital. However, funding through governmental grants held significant appeal for various reasons. Grant funding is often awarded based on the scientific merit of the technology and its chance of becoming a valuable product. It is completely nondilutive, that is, the granting agencies do not take equity or any other control in the company. This is advantageous because significant time for validation and development is required for early-stage technologies, especially in biotechnology. The founders sought out the Small Business Innovation and Research (SBIR) program and applied for a "Phase I" grant through the National Science Foundation (NSF). Coming from academic backgrounds, they were equipped to compile the rigorous technical proposals required, and Mango Materials was awarded a Phase I NSF SBIR in January 2012.

The receipt of this award kick-started the growth of the company to where it is today. Leveraging its partnerships, Mango Materials set out to make the most of the resources it had and to prove that its process would be successful with biogas, ignoring the advice that "you don't bootstrap a biotech start-up!." Early tests in the field and in controlled laboratory settings yielded positive results. Mango Materials built on this success and secured additional funding through a Phase II NSF SBIR award to commercialize its process. A contact recommended Mango Materials apply for the 2012 Postcode Lottery Green Challenge, an international clean-tech competition for start-ups, and Mango Materials applied. After being named a finalist and traveling to Amsterdam and New York to pitch, Mango Materials was declared the winner of the competition, taking home a check for €500,000.

The funding from these awards has helped Mango Materials grow substantially – at the time of publication, the company employs over fifteen people and has worked with ten student interns in conjunction with programs including the NSF Research Experience for Undergrads and for High School Students and Biomanufacturing to Markets. Working with student interns has provided a great experience for both the company and the interns; several of the interns have gone on to accept full-time positions at Mango Materials and have become invaluable members of the team, while others have stayed with Mango Materials for multiple years in a row. In addition to employees and interns, Mango Materials also works with several consultants to assist with both business and process development. Many of these are contacts gained through the founders' original network at Stanford.

Over time, the founders' roles within the company have evolved. Despite all having technical Ph.D.'s, they have divided up the responsibilities of running the company based on their respective strengths and expertise. From the beginning, Molly has been the company's CEO, and she has spearheaded the business development and dealt with details ranging from the intricacies of incorporating a company to insurance and payroll to hiring and taxes. Allison and Anne joined in on several of these tasks but were focused on the technical rather than the business aspect of the company from the beginning. Anne's focus has shifted to concentrate on methane sourcing; among other things, she is responsible for establishing relationships with potential suppliers of methane for Mango Materials' pilot- and commercial-scale plants. Allison has remained focused on the technology and heads the process scale-up and optimization, leading projects ranging from fermentation to downstream processing of the bioplastic to scaling the entire process. Molly remains involved in the technical work, whenever she can find time. The shift in roles has occurred as the company has grown and has helped to solidify an efficient team that is working together toward a common goal.

The company's initial goals were to demonstrate that biogas to plastic was a viable and potentially successful process; current goals are focused on scaling the process. Mango Materials has come a long way since its initial inception, and the path so far has been extremely rewarding for the entire team (Figure 16.5).

16.4.2 Growing a Start-Up: Next Steps for Mango Materials

Mango Materials' long-term goal is to commercialize its process successfully. Its current focus is to scale the entire bioplastic production process; now that the technology has been validated at small scales, the next challenge is to validate it at a larger scale. Scaling the process will also enable production of samples that can be given to potential customers. Customers need samples that they can test on their own manufacturing equipment to determine conditions required for processing. While customers might only need small quantities of certain specialty bioproducts to test, they require large samples of plastics to really understand the plastic properties and how to process it.

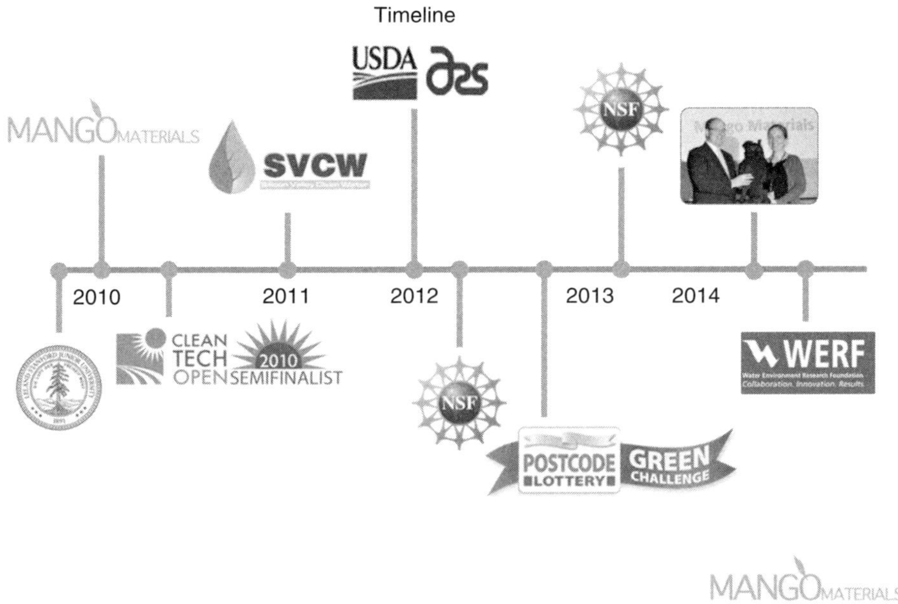

Figure 16.5 Timeline of significant events in the growth of Mango Materials.

Thus, once Mango Materials has produced large-scale samples for customers, it will verify the biodegradation of these samples in accordance with various standards such as ASTM regulations.

Mango Materials anticipates initially being able to sell its bioplastics at a premium due to their biodegradability and sustainable production profile. However, ultimately, Mango Materials aims to produce biodegradable plastics from methane at a price on parity with petroleum-based plastics in order to be competitive in the marketplace.

16.4.3 Biotech and Bioproducts: What Matters

Many factors are important in determining the success of a biotech start-up process and product. It has been found that it is important to optimize the entire production process rather than focus solely on a single part of the process with blinders. Ultimately, it is how well all the factors fit together that determines success. If the many different technical processes do not fit together successfully, the overall process may fail – even if each individual step works in a stand-alone unit operation. Additionally, attacking the "low-hanging fruit," or areas where there is the most room for improvement, may yield higher gains than more incremental improvements in other parts of the process.

Two important measures for predicting success in a bacteria-based process are productivity and yield. Used correctly, these can provide invaluable information about the efficiency of a process. For example, in the case of companies producing a biopolymer via a fermentation process, the yield is a measure of how efficiently the feedstock is used to produce the product; specifically, it refers to units of product created per unit of feedstock. For Mango Materials, that translates to grams PHA per gram methane, or how much methane is required to produce a given amount of plastic as an end product. This is important because feedstocks can be expensive to produce (e.g., sugars) or purchase. Even a feedstock that is currently considered a waste (e.g., methane) may have to be purchased, and, therefore, it is important to maximize yields to get the most product possible per unit of feedstock. In cost projections, Mango Materials always anticipates that it will have to purchase the methane it will require.

Productivity is a measure of how quickly product can be made for a given volume – it is reported as unit of product produced per unit volume per unit time. In other words, in fermentation, productivity is how quickly the microorganisms generate bioproduct for a specific fermentation volume. When considering capital costs, it is important to know how quickly product can be manufactured, and, by accounting for volume, productivity provides a measure of the efficiency of the process.

Although significant emphasis is often placed on the improvement of individual metrics (including productivity and yield), the key challenge in scaling is to optimize each step of the process. The context of the process, as a whole, must be considered so that all steps fit together to achieve a scalable, commercializable process. If all of the individual steps cannot be made to meld together into one cohesive process, it makes no difference if specific microorganisms grow quickly enough, or produce product rapidly, or uptake feedstock efficiently. In order for the bioeconomy to truly hit commercial scale, an eye must be kept not only on productivity and yield but also on overall capital equipment required for the entire process at scale and operating expenses per unit of product produced.

16.5 THE FUTURE – NEW IDEAS FOR POTENTIAL RESEARCH

Key challenges in achieving commercialization of biodegradable plastics from rapidly renewable resources include sufficient optimization of the microbial process and maintaining current cost-competitive yields and rates at scale; lower yields and rates would increase the cost of the polymer.

One frontier meriting investigation is achieving controlled rates of biodegradation of plastic products. For example, agricultural films used as crop covers might be used for several months, after which the season changes and they are no longer needed. Plastic microbeads in personal care products and cleansers are used very briefly prior to being flushed down a drain. Fishing nets and other plastic fishing equipment are often used in fragile ecosystems and can cause great harm if inadvertently discarded in aquatic environment.

One potential way to control biodegradation is by processing the polymer under very specific conditions; these may affect the polymer structure and, in turn, how rapidly it biodegrades. Another method of controlling biodegradation might be to add environmentally friendly additives that prevent biodegradation for a specified period of time. Yet another way might be to induce the polymers to break down upon certain triggers; for example, once microbeads are flushed down a drain and enter a WRRF, they are exposed to a much different environment than when they are present in a personal care product. This change in environmental conditions could be used to induce biodegradation.

In addition to focusing on biobased, biodegradable products, future research should also focus on how to establish an efficient infrastructure to accommodate biodegradable products at the end of their useful lives. While some municipalities collect "green waste" for composting, these can be few and far between. Green waste is commonly mingled with the waste stream headed for the landfill. Even when the infrastructure is available to collect green waste separately, consumer awareness must be increased so that biodegradable plastics are disposed in green waste rather than in the recycling or landfill streams.

In addition to composting, anaerobic digestion is an end-of-life option that will enable recovery of methane, a valuable resource. An infrastructure for establishing collection of waste and subsequent anaerobic digestion should be pursued. Dry fermentation enables the breakdown of organic matter in the absence of oxygen and is another option that may be less energy and water intensive than standard, aerobic, composting and should be explored in further detail.

The process presented in this chapter – to transform a waste, GHG into a valuable product – is just one way in which we can move toward a sustainable bioeconomy. This new bioeconomy would be one where "waste" is considered a resource. Innovative technologies will be required to address many of the current environmental issues we face today. Technologies must be taken all the way from nascent ideas to commercialization, which will require a full-system approach that considers the process as a whole during the scale-up period. Mango Materials is excited to be a part of this move toward a sustainable bioeconomy.

ACKNOWLEDGMENTS

The authors would like to thank Andrew Cal for help in preparing Figure 16.3 and Tze Wei U for research on market statistics. We would also like to thank our collaborators at the United States Agricultural Facility – Agricultural Research Service in Albany, CA, and Silicon Valley Clean Water in Redwood City, CA, for their ongoing partnerships. This work was funded by Phase I and Phase II Small Business Innovation Research grants from the National Science Foundation (NSF 1142566 and NSF IIP-1256623, respectively) and the Postcode Lottery Green Challenge (Project Number 120563).

REFERENCES

[1] U.S. Energy Information Administration, "Petroleum & other liquids." http://www.eia.gov/dnav/pet/pet_cons_psup_dc_nus_mbbl_a.htm, **2015**, accessed Mar 2015.

[2] U.S. Energy Information Administration, "How much oil is used to make plastic?" http://www.eia.gov/tools/faqs/faq.cfm?id=34&t=6, **2014**, accessed Mar 2015.

[3] L. Parker, "With millions of tons of plastics in oceans, more scientists studying impact." http://news.nationalgeographic.com/news/2014/06/140613-ocean-trash-garbage-patch-plastic-science-kerry-marine-debris/, **2014**, accessed Mar 2015.

[4] (a) D. Barnes, *Nature (London.* **2002**, *416*, 808–809; (b) J. Derraik, *Mar. Pollut. Bull.* **2002**, *44*, 842–852.

[5] Y. Mato, T. Isobe, H. Takada, H. Kanehiro, C. Ohtake, T. Kaminuma, *Environ. Sci. Technol.* **2001**, *35*, 318–324.

[6] NOAA, "What we know about: Plastic Marine Debris" http://marinedebris.noaa.gov/info/plastic.html, **2011**, accessed Mar 2015.

[7] NOAA, "What is the Great Pacific Garbage Patch?" http://oceanservice.noaa.gov/facts/garbagepatch.html, **2014**, accessed Mar 2015.

[8] M. Sesini, "The Garbage Patch in the Oceans: The Problem and Possible Solutions", http://www.seas.columbia.edu/earth/wtert/sofos/sesini_thesis.pdf, **2011**, accessed Mar 2015.

[9] R. Thompson, Y. Olsen, R. Mitchell, A. Davis, S. Rowland, A. John, D. McGonigle, A. Russell, *Science* **2004**, *304*, 838.

[10] (a) C. Moore, S. Moore, S. Weisberg, G. Lattin, A. Zellers, *Mar. Pollut. Bull.* **2002**, *44*, 1035–1038; (b) S. Moore, D. Gregorio, M. Carreon, S. Weisberg, M. Leecaster, *Mar. Pollut. Bull.* **2001**, *42*, 241–245.

[11] EPA, "Plastics." http://www.epa.gov/osw/conserve/materials/plastics.htm, **2014**, accessed Mar 2015.

[12] CalRecycle, "Plastics Information and Resources." http://www.calrecycle.ca.gov/plastics/, **2010**, accessed Mar 2015.

[13] EPA, "Municipal Solid Waste Generation, Recycling, and Disposal in the United States: Facts and Figures for 2012", http://www.epa.gov/waste/nonhaz/municipal/pubs/2012_msw_fs.pdf, **2014**, accessed Mar 2015.

[14] CalRecycle, "Beverage Container Recycling: It's in Your Hands." http://www.calrecycle.ca.gov/BevContainer/Consumers/Facts.htm, **2011**, accessed Mar 2015.

[15] Association of Postconsumer Plastic Recyclers, "2012 Survey – State's Largest Municipalities Plastic Collection Programs Overview", http://www.plasticsrecycling.org/images/pdf/resources/reports/States-Largest-Cities-Plastics-Recycling/2012_states_largest_munis_overview.pdf, **2012**, accessed Mar 2015.

[16] CalRecycle, "Beverage Container Recycling." http://www.calrecycle.ca.gov/bevcontainer/, **2014**, accessed Mar 2015.

[17] SPI, "Recycling." http://www.plasticsindustry.org/AboutPlastics/content.cfm?ItemNumber=1271&navItemNumber=1125, **2014**, accessed Mar 2015.

[18] Plastics Europe, "Plastics Europe Market Research Group." www.plasticseurope.org, **2008**, accessed Mar 2015.

[19] K. Deamer, "LA Bans Plastic Bags- Is Your Town Next?" http://www.motherjones.com/blue-marble/2012/05/plastic-bag-bans-usa, **2012**, accessed Mar 2015.

[20] J. Bardelline, "Italy Dumps Plastic Bags with Europe's First Nationwide Ban." http://www.greenbiz.com/news/2011/01/05/italy-dumps-plastic-bags-world-first-nationwide-ban, **2011**, accessed Mar 2015.

[21] (a) E. Rosenthal, "By 'bagging it,' Ireland Rids itself of a plastic nuisance." http://www.nytimes.com/2008/01/31/world/europe/31iht-bags.4.9650382.html?pagewanted=all, **2008**, accessed Mar 2015; (b) O. Zaleski, "China's Plastic Bag Ban Will Save 37 Million Barrels of Oil." http://www.thedailygreen.com/environmental-news/latest/china-plastic-bags-47010907, **2008**, accessed Mar 2015.

[22] A. Calmon-Decriaud, V. Bellon-Maurel, F. Silvestre, *Adv. Polym. Sci.* **1998**. 135, 207–226.

[23] ASTM D6400-12, *Standard Specification for Labeling of Plastics Designed to be Aerobically Composted in Municipal or Industrial Facilities*, ASTM International, West Conshohocken, PA, **2012**.

[24] ASTM D6866-12, Standard Test Methods for Determining the Biobased Content of Solid, Liquid, and Gaseous Samples Using Radiocarbon Analysis, ASTM International, West Conshohocken, PA, **2012**.

[25] State of California Department of Justice, "Quick reference guide to "biodegradable," "compostable," and related claims on plastic products in California." http://oag.ca.gov/sites/all/files/agweb/pdfs/environment/ag_website_environmental_claims.pdf, **2012**, accessed Mar 2015.

[26] California Legislative Information, "SB-567 Recycling: plastic products." http://leginfo.legislature.ca.gov/faces/billNavClient.xhtml?bill_id=201120120SB567, **2011–2012**, accessed Mar 2015.

[27] R. Narayan, "Biobased & Biodegradable Polymer Materials: Rationale, Drivers, and Technology Exemplars." http://standards.nsf.org/apps/group_public/download.php/5554/Narayan%20ACS%20BK%20Chapter%20Publication%202005%20biobased%20%26%20biodegradable%20FINAL%20REVISED.pdf, **2005**, accessed Mar 2015.

[28] European Bioplastics, "Bioplastics Facts and Figures." http://en.european-bioplastics.org/wp-content/uploads/2013/publications/EuBP_FactsFigures_bioplastics_2013.pdf, **2013**, accessed Mar 2015.

[29] L. Shen, E. Worrell, M. Patel, *Biofuel. Bioprod. Bior.* **2012**, *6*, 625–639.

[30] Virent, "Virent welcomes the Coca-Cola Company as a Partner." http://www.virent.com/partners/corporate/virent-welcomes-the-coca-cola-company-as-a-partner/, **2015**, accessed Mar 2015.

[31] (a) A. Anderson, E. Dawes, *Microbiol. Rev.* **1990**, *54*, 450–472; (b) L. Madison, G. Huisman, *Microbiol. Mol. Biol. R.* **1999**, *63*, 21–53.

[32] A. Steinbuchel, H. Valentin, *FEMS Microbiol. Lett.* **1995**, *128*, 219–228.

[33] Plastics Europe, "Plastics – the Facts 2014/2015." http://issuu.com/plasticseuropeebook/docs/final_plastics_the_facts_2014_19122, **2015**, accessed Mar 2015.

[34] (a) M. Taylor, *The SPI Magazine,* **2014**, *3*, 13;b)American Chemistry Council, "U.S. Resin Production & Sales 2013 vs. 2012." http://www.americanchemistry.com/Jobs/EconomicStatistics/Plastics-Statistics/Production-and-Sales-Data-by-Resin.pdf, **2014**, accessed Mar 2015.

[35] Institute for Bioplastics and Biocomposites, "Global biopolymer market size." http://ifbb.wp.hs-hannover.de/downloads/content/Statistics/Market%20statistics/Market%20share/Global%20biopolymer%20market%20size.png, **2015**, accessed Marc 2015.

[36] SPI Bioplastics Connect, "Bioplastic market update – fast facts." http://www.plasticsindustry.org/files/about/BPC/BC%20Newsletter%20-%20September%202012%20-%20Final.pdf, **2012**, accessed Mar 2015.

[37] R. Eno, J. Hill, "Metabolix." http://files.shareholder.com/downloads/MBLX/1295953128x0x447909/89b07ea2-d245-4f43-9d14 1c4928614388/MBLX_Jefferies_11th_Global_Clean_Tech_Conf_Feb_24_2011.pdf, **2011**, accessed Mar 2015.

[38] G. Braunegg, B. Sonnleitner, R. Lafferty, *Eur. J. Appl. Microbiol.* **1978**, *6*, 29–37.

[39] (a) D. Sheu, Y. Wang, C. Lee, *Microbiology.* **2000**, *146*, 2019–2025; (b) A. Pieja, E. Sundstrom, C. Criddle, *Appl. Environ. Microb.* **2011b**, *67*, 6012–6019.

[40] (a) J. Bowman, *The Methanotrophs – The Families Methylococcaceae and Methylocystaceae. (The Prokaryotes: A Handbook on the Biology of Bacteria),* Springer, New York, NY, **2006**, pp. 266–289; (b) R. Hanson, T. Hanson, *Microbiol. Rev.* **1996**, *60*, 439–471.

[41] I. McDonald, L. Bodrossy, Y. Chen, J. Murrell, *Appl. Environ. Microb.* **2008**, *74*, 1305–1315.

[42] J. Bowman, *Family I. Methylococcaceae and Family V. Methylocystaceae. (Bergey's Manual of Systematic Bacteriology*), Williams & Wilkins, **2001**, pp. 256–270; 411–420.

[43] I. Higgins, D. Best, R. Hammond, D. Scott, *Microbiol. Rev.* **1981**, *45*, 556–590.

[44] (a) J. Asenjo, J. Suk, *J. Ferment. Technol.* **1986**, *64*, 271–278; (b) M. Vecherskaya, C. Dijkema, A. Stams, *J. Ind. Microbiol. Biot.* **2001**, *26*, 15–21; (c) K. Wendlandt, M. Jechorek, J. Helm, U. Stottmeister, *J. Biotechnol.* **2001**, *86*, 127–133; (d) A. Pieja, PhD Thesis, Stanford University, **2011**; (e) A. Pieja, K. Rostkowski, C. Criddle, *Microbial. Ecol.* **2011a**, *62*, 564–57.

[45] (a) M. Daniel, J. Choi, J. Kim, *J. Lebeault. Appl. Microbiol. Biot.* **1992**, *37*, 702–706; (b) N. Shah, M. Hanna, K. Jackson, R. Taylor, *Biotechnol. Bioeng.* **1995**, *45*, 229–238.

[46] (a) S. Lee, *Biotechnol. Bioeng.* **1996**, *49*, 1–14; (b) M. Choi, S. Yoon, R. Lenz, *Appl. Environ. Microb.* **1999**, *65*, 1570–1577.

[47] (a) C. Sawyer, P. McCarty, *Chemistry for Environmental Engineering*, 3rd Edition, McGraw-Hill, Inc., New York, NY, **1978**; (b) A. Ghaly, *Energ. Source* **1989**, *11*, 221–235; (c) S. Hoenig, *Resource* **1998**, *53*, 7–8; (d) A. Schauer, MS Thesis, University of Wisconsin – Green Bay, **2002**; (e) K. Bothi, *MS Thesis*, Cornell University, **2007**

[48] AgSTAR, "Anaerobic Digestion." http://www.epa.gov/agstar/anaerobic/index.html, **2010**, accessed Mar 2015.

[49] J. Raich, *WERF Final Report – Evaluation of the Efficiency of Biogas Treatment for the Removal of Siloxanes*, IWA Publishing, **2013**.

[50] A. Schauer-Gimenez, A. Cal, M. Morse, A. Pieja, K. Holtman, W. Orts, *BioCycle*, **2014**, *55*, 43–47.

[51] EPA, "Landfill Methane Outreach Program (LMOP), Energy Projects and Candidate Landfills." http://www.epa.gov/lmop/projects-candidates/index.html, **2014**, accessed Mar 2015.

[52] K. Krich, D. Augenstein, J. Batmale, J. Benemann, B. Rutledge, D. Salour, "Biomethane from Dairy Waste: A Sourcebook for the Production and Use of Renewable Natural Gas in California." http://www.calstart.org/Libraries/Publications/Biomethane_from_Dairy_Waste_Full_Report.sflb.ashx, **2005**, accessed Mar 2015.

[53] G. Myhre, D. Shindell, F. Bréon, W. Collins, J. Fuglestvedt, J. Huang, D. Koch, J. Lamarque, D. Lee, B. Mendoza, T. Nakajima, A. Robock, G. Stephens, T. Takemura and H. Zhang, "Anthropogenic and natural radiative forcing", In *Climate Change 2013: The*

Physical Science Basis. Contribution of Working Group I to the Fifth Assessment Report of the Intergovernmental Panel on Climate Change (Eds. T.F. Stocker, D. Qin, G.-K. Plattner, M. Tignor, S.K. Allen, J. Boschung, A. Nauels, Y. Xia, V. Bex, P.M. Midgley). Cambridge University Press, Cambridge, United Kingdom and New York, NY, USA, **2013**, 714.

[54] M. Akiyama, T. Tsuge, Y. Doi, *Polym. Degrad. Stabil.* **2003**, *80*, 183–194.

17

NOVAMONT: PERSPECTIVES ON INDUSTRIAL BIORENEWABLES AND PUBLIC-PRIVATE NEEDS

STEFANO FACCO
Novamont SpA, Novara, Italy

17.1 STATE OF THE ART AND CHALLENGES FACED BY BIOBASED INDUSTRIES

In the tough economic climate that the EU currently faces, fostering the continuation of the development of the manufacturing industry is paramount. Global economic competition, just based on endless growth and based on pure financial values – following the "business as usual" scenario – simply will not work

Industrial Biorenewables: A Practical Viewpoint, First Edition. Edited by Pablo Domínguez de María.

anymore. The transition from our current resource-intensive use growth model to a resource-efficient growth model – toward a circular economy – is fundamental and would offer an opportunity to find out a large number of local, competitive niche sectors in the area of biobased products. Thus, it is of paramount importance for Europe to focus and invest on such models, which will enable sustainable economic growth as well as the provision of jobs at the local level. Public authorities, industry, researchers, and citizens should be aware that there is a different way to manufacture the products that we are using in everyday life, by minimizing waste of energy and natural resources. The economic recovery – needed worldwide – can benefit from new models based on sustainable development, which are able to create innovation and new jobs, improve production processes, and put local resources for their proper use in a "short supply chain situation." If we want to retain and further grow a dynamic industrial base, Europe must start to innovate, invest, and specialize in those activities, which will have a comparative advantage in the global competition for resources and markets. Only with this forward-looking approach Europe will protect its local biodiversity and ecosystems and will enable biobased industries to lead the transition toward a postpetroleum society. An approach inspired to a bioeconomy model, where resources will be addressed to technologies requiring less quantity of locally available feedstock. In this way Europe will specialize on its own niches based on high added value products, without competing with other models of countries having much larger surface of arable land and availability of feedstocks, which thus require different technologies compared to the European ones. As stated earlier, this strategy will enable Europe to be competitive and focus on resources and investments related to sustainable, local products. In this respect, the bioeconomy focused on the promotion of added value products, such as biodegradable bioplastics – through a cascading use of biomass – representing a powerful catalyst for recovery: spurring growth and jobs through innovation – particularly in rural areas – benefiting European farmers and regions, enabling the creation of local agroindustrial value chains and profuse innovation and opportunities for growth to traditional sectors, and leading the production of biochemicals and other added value products like bioplastics in synergy with the local production of food and feed.

The potential of this strategy may be expressed in several market projections. The global biobased market and revenue are estimated to reach €250 billion by 2020 by the World Economic Forum. Moreover the volume growth of EU biobased chemical products up to 2020, including bioplastics, biolubricants, biosolvents, and biosurfactants, is estimated at 5.3% p.a., resulting in a market worth € 40 billion and providing over 90,000 jobs within the biochemical industry alone.

However these promising opportunities risk being lost by Europe, despite the fact that the continent leads in the cutting-edge development of industrial biotech and green chemistry, as investments and technologies are deployed instead overseas. This shift is the result of competition from other world regions where countries are rapidly putting in place mechanisms to leverage private investment in technologies with higher financial risks such as the building of new demonstration and flagship biorefineries taking excellence in science and research through the commercialization of biobased products. Europe must therefore act swiftly and decisively to reverse the trend of deployment and commercialization of biobased technologies in other leading economies – namely, the United States (helped by the "shale gas revolution," although recent studies are presenting a less positive outcome), Brazil, and China. Moreover, European's competitor countries (e.g., United States) offer R&D incentives, demo- and pilot-scale incentives, and product incentives (e.g., green procurement programs, advanced biofuel subsidies). Some of these elements are now also emerging in China, Russia, and Brazil (e.g., biobased chemicals) and Thailand (in the Asian Pacific area competing with Malaysia and Indonesia). It is evident that these nations domestic policies are being developed to tackle barriers that exist in the creation of new value chains within the bioeconomy (i.e., with a strong focus on bioethanol in the United States and Brazil and on bioplastics in Thailand) as they aim to leverage the local potential and availability of natural resources.

The US focus on supporting primarily the production of biobased ethanol has created a model largely focused on funding commercial-scale biorefineries through loans from the Department of Energy (DOE) that can go up to 250 million dollars. This important support is clearly boosting private investment in cellulosic ethanol by industry players: according to a recent survey by the University of Iowa, at least ten projects on cellulosic ethanol exceeding 20 million gallons per year are expected to begin operations by 2014 with conglomerate project costs of over 2 billion euros. Needless to say that this trend and investments in the field are set to increase, given that cheap gas – at least at first glance – from shale gas is fueling a US industrial renaissance, which is drawing billions of dollars in investment from multinational corporations. Moreover, the strong support of the deployment in the United States is complemented by a scheme dedicated to the market uptake of innovative products: the BioPreferred program.

Thailand, which is also seeing increased attention and investments from EU companies, approved a framework for promoting investments in the bioplastics industry in 2010. A budget of approximately 10 million euro was set up to support the establishment and operation of small pilot plant (1 K – 10 kt/year). The government also offers a series of incentives in the forms of corporate tax exemption for up to 8 years and an additional 50% reduction in corporate income tax for 5 years.

China is even more ambitious and is increasingly strengthening the support of biobased industries and biotechnology. In the nation's 12th Five-Year Plan on National Emerging Industries of Strategic Importance, the need to accelerate China's bioeconomy to serve major needs in health, agriculture, and environmental protection is clearly stated.

Last but not least Brazil, with its vast resources of biomass and feedstock and a long tradition in the use of biofuels, aims now to become N°1 Global Bioeconomy. In terms of specific policy, the main item is the Plan for Supporting Industrial

Technological Innovation in the Sugar-based Energy and Chemical Sectors (PAISS). These policy trends worldwide testify how governments are acting decisively to push the most prominent sectors of the bioeconomy leaving an attractive and high-potential niche for Europe in the area of added value products, where several member states like Italy have a solid base of R&D&I and ongoing investments, which testify how even at crisis times the sector is set to boost growth and jobs in areas critically affected by the crisis, providing tangible opportunities for reconverting noncompetitive petrochemical sites into sustainable biorefineries.

17.2 WISDOM IN THE USE OF RENEWABLE RAW MATERIALS: THE CASCADING USE OF BIOMASS

As the first-generation biofuels have shown, when substituting fossil energy with biomass in any part of the economy, care should be taken to ensure that it actually helps to reach those goals of a more sustainable society. Government policy, through subsidies or obligations, can influence value; obligations in the transport fuel sector may divert biomass away from sectors where the biomass itself can be cascaded, and thus the whole chain would be more sustainable and valuable. A "sustainable cascading strategy" should help in creating transparency, when it comes to choosing an application for biomass, and will help policy makers in creating legislative frameworks able to minimize environmental impact. The main goal of applying cascading is to achieve the lowest environmental impact. In this respect, the goals one wants to achieve with cascading do not differ from general optimization principles for the biobased economy; thus, cascading is a tool, not a goal. It is, however, important to keep in mind that relative environmental impact and costs may change over time. Technological development in the reference system, (depreciation of) CO_2 mitigation costs, market prices, and yields of biomass production can actually influence results.[1]

If we take a look at the impacts of using biomass as an energy source or a material source, we can see some different results affecting the value generation. The analysis of recent studies, carried out by *nova-Institut* on the macroeconomic effects of the nonfood use of biomass, showed that the potential benefits of the material (nonenergetic) use in terms of employment and added value are significantly higher than those arising from the use of biomass for energy. Material use can directly support 5–10 times more employment and 4–9 times the added value compared to energy use. These comparisons relate, respectively, to the same raw material or the same farmed area. This is due to the significantly more complex and longer supply chain arising from material uses.[2] This is even true for traditional applications of wood: Using wood for particle boards or pulp and paper supports greater employment and added value compared to the production of energy pellets.[3] High resource efficiency in the use of renewable resources can only be achieved with biobased materials (higher input–output efficiency than biofuels) and strengthened through "cascading utilization." This starts with single or multiple uses (recycling economy) followed by energy use at the end of life.

17.3 CASE STUDY: BIOPLASTICS IN ITALY: GOING FOR GROWTH DESPITE THE CRISIS

17.3.1 Bioeconomy and Bioplastics Trigger to Unleash Growth of the European Chemical Industry

Biodegradable bioplastics and biobased chemical intermediates are an example of a systemic economy, as they enable the redesign of the entire sectors of application, with an impact on how raw materials are produced, the verticalization of entire agroindustrial chains, and the way in which products are used and disposed of. They are therefore a powerful and demonstrative example of sustainable development and cultural growth that can act as an example for other sectors.

Waste is often seen as such a big problem. But we have to start to see it as such a big opportunity. We have started to move in the right direction using European legislation, but it is the economic rationale for treating our waste as a resource, which will be the real driver leading us toward zero waste. The EU already estimates that just implementing existing waste legislation properly would create 400,000 jobs in the EU. But the potential of moving toward a truly circular economy – where materials are used again and again instead of going on a one-way trip – is truly phenomenal. In particular, the construction of integrated third-generation biorefineries, aimed mainly at innovative products with a higher added value than energy and biofuels, such as biochemicals and bioplastics, will enable the creation of a new socioeconomic and cultural model, more than an industrial one, by giving the correct priority to the use of biomasses with respect for local biodiversity and the creation of new jobs. The chemical industry – which had grown steadily until 2008 – suffered a double-figure contraction in 2009, forcing multinationals in the sector to shut down their older sites with heavy consequences for Italy.

However, regardless of the crisis, the petrochemical industry cannot justify any new large plants in Western countries, much less in Italy. For example, in 2010, according to ChemSystems data, the cash cost of ethylene from naphtha was €761 per ton in Italy and 93 per ton from ethane in the Middle East. This is the reason why the expansion of capacity for plastics is being concentrated in the Middle East; some plants have a capacity of more than 1 million tons per year. On the basis of these figures, there is no hope for their Italian or any other EU member's counterparts unless they are integrated or replaced with new technologies and ready to make the quantum leap based on local raw materials that can draw on the locally available resources, exploiting them in an environmentally friendly, efficient way.

Italy experienced major growth in the chemical industry between the 1930s and 1980s, particularly thanks to Montedison and the results of the great innovations of Giacomo Fauser who developed ammonia synthesis process and Giulio Natta who discovered polypropylene. During the 1990s, most of the Montedison plants were acquired by foreign multinationals with a view to short-term profit, and no provision was made for future growth and investment. Today, there are just 13 large chemical sites monitored by the National Chemical Observatory. All of them are essentially in an irreversible crisis. In particular, all the traditional plastic manufacturing plants in Italy are small scale, are obsolete, and have no upstream integration, therefore no prospects for profitability even in the short term. Therefore, if we take a look at the plastics sector, Italy has gone from being a leader in manufacturing and exports to being a net importer (7% of internal production against 19% imports). The latest Federchimica data also show that Italy has slipped from being Europe's second largest plastics manufacturer to fourth place behind Germany, Belgium, and France. It is still second in the plastics processing sector, after Germany, but its decline as a manufacturer of raw materials can only weaken this sector, as is already evident in other fields such as the transformation of bioriented polypropylene films. As a consequence, Italy's chemical industry is mostly made up of companies that are much smaller than they were in the past and are also very highly specialized. The country needs, more than other areas worldwide, immediate technological innovation not dependent of fossil fuels that can justify new investments and possibly enable the conversion of existing plants, leading the way to a postpetroleum society. Low-impact productions, renewable resources, and the search for new systemic economy models can be appropriately exploited in order to revitalize the Italian chemical industry and enable new investments capable of leading the way toward a circular economic model.

In this challenging context, bioplastics offer great potential for the following reasons:

- There have been significant investments (private and public) over the years in terms of research and also in industrial plants, so technologies are now ready for further industrial scale-up. Other related technologies are also at an advanced stage of development. There are, therefore, positions of technological leadership that can be immediately exploited, covered by patents, and recognized at European level, with production plants newly built or pending completion, faced with the phenomenon of deindustrialization in traditional chemistry.
- There is a network of companies in the machinery and plastics processing sector at risk of downsizing, without appropriate impulses from innovation in raw materials.
- In Italy there is a well-established system of biowaste and separate collection, by using biodegradable bags, which is able to generate quality compost (humus for the soil).

17.3.2 Small and Medium-Sized Enterprises: Leading an Agroindustrial Revolution

In Europe industrial systems heavily rely on SMEs. According to the European Commission, the 20.8 million European small and medium-sized enterprises (SMEs) create 85% of all new jobs in Europe, they employ 2/3 of the workforce in the EU, and they contribute significantly to innovation and growth. In particular, in Italy the world of small and medium-sized chemical companies expanded rapidly during the 1960s to the 1990s, driven by the large national industries. Nowadays, SMEs, without the strategic driving force of national raw material producers, are scaling down their innovation and risk and collapsing by dealing with low-cost competition. This situation is weakening the country with an abnormal use of "social shock absorbers" that end up subsidizing their lack of competitiveness.

The construction of a new industry, based on locally sourced agricultural materials in synergy with food industries and on agricultural and general (municipal/industrial) waste, represents a great opportunity that should not be missed. At Porto Torres, as in other parts of Italy, there is a great opportunity to cultivate plurianual dry-farming crops without irrigation, with respect for the territory and in collaboration with local livestock owners/farmers and SMEs. The challenges of today require increasingly sophisticated SMEs that are able to operate as part of a network, training young people with a systemic vision and competing at international level. The objective should be to breathe life into a knowledge-based systemic economy, which, starting with local resources and territorial biodiversity, is able to boost competitiveness at international level.

17.3.3 A Case Study on Incremental Innovation to Learn from: Key Data

In this scenario, Italy has implemented a case study that testifies how environmental policies addressed for leveraging market pull measures for biobased products play a pivotal role, since they trigger both new investments in innovative biorefineries and social benefits in terms of waste prevention and sustainable consumption habits among citizens. This strategy focuses on reducing the use of single-use plastic carrier bags and favoring the use of either reusable, durable bags or biodegradable, compostable ones. This measure builds upon the evolution of research and innovation in the biodegradable bioplastics sector on the one hand and the virtuous development of the quality compost industry and separated municipal waste collection on the other hand. The connections between these two developments over the years have activated a whole series of virtuous models of action and collaboration initiatives between various stakeholders (enterprises, institutions, research bodies, trade associations, consultancy companies, and regional authorities) generating an interconnected web that is ideal to promote a change in the development model, putting the efficient use of resources at its center. Leveraging on these solid pillars, the Italian government promoted in 2011 a visionary "market pull" measures aimed to boost the production and use of biodegradable products from renewable raw materials in niche markets, which are particularly critical from the environmental point of view. The measure hence enabled a scale-up of production of innovative technologies for bioplastics, which were already tested and economically viable.

A new publication launched at European Parliament and more recently in Italy by Kyoto Club "Bioplastics: a Case Study of Bioeconomy in Italy" goes deep into analyzing all the benefits and cascade effects triggered by this legislative measure providing sound evidence and data. It shows clearly how the new regulation, conceived primarily to tackle the serious problem of waste and landfill management, has already provided a number of benefits for the environment: an overall decrease in the consumption of disposable bags in supermarkets of around 50%, with the ensuing raising of public awareness (over 90% of Italian citizens consider the law to be a step forward in safeguarding the environment; source: ISPO 2012); a 20.7% reduction in waste sent to landfills with resulting annual savings of around 5.1 million euro; a 29% cut in CO_2 emissions; and a 39% reduction in oil used. In addition, around 50% of the bags used for the collection of organic waste are biodegradable and compostable shopping bags and provide an opportunity to extend separated collection of organic waste to municipalities that have not yet started it or generate savings for those virtuous communities that have already been implementing it for years.

As a result bioplastics have become a driver for spurring national growth triggering development of related biobased products such as tire additives, biolubricants, and biochemical building blocks. Confidence levels of potential in investors increased, given the supportive legislative framework in place, and for 2013 investments for about 1 billion euro have been planned in the field of biobased products in Italy, many of which are based on reconversion of abandoned and not operational plants into biorefineries with plans to reemploy local workers with little job prospects. Foreign companies such as DSM and Roquette (through the JV Reverdia) given the dynamic growth potential of the sectors decided to establish in Italy new plants for the production of biochemicals and bioplastics, proving that Italy can still be an attractive environment for foreign investors when supportive framework conditions are in place. In terms of reconversion projects, it is worth mentioning that Matrica is a 50/50 joint venture between Novamont and Eni Versalis, whose aim is to convert in Porto Torres an old petrochemical plant into 7 new plants, 3 flagships, for the production of bioplastics and building blocks as well as for bioadditives for rubber, with an investment of around 500 million euros. Another cutting-edge project is ongoing in Adria. Work is underway by Mater-Biotech, the biotechnological platform of Novamont, on the modification and redevelopment of an old fermentation plant, formerly owned by Ajinomoto. The aim is to produce a biobased chemical building block used in bioplastics (bio-BDO) through a technology that will be implemented on an industrial scale for the first time worldwide in Italy. This is the first major industrial biotechnology project in Italy and marks the beginning of an important biotechnology platform that will involve large and small players in Italian industry in several aspects and absorb personnel who would otherwise be unemployed. Moreover the government recently launched a cluster dedicated to green chemistry whose aim will be research and innovation on added value products gathering critical mass and building on the key know-how and skills available in the country.

In conclusion the Italian legislation on plastic bags is acting as a catalyst for change, promoting a process of incremental innovation by capitalizing and enhancing existing knowledge and competences. Opportunities are being provided to test and assess the effective repercussions on the local area, creating competences and bridges between otherwise divergent sectors: chemistry, agriculture, biotechnology, the petrochemical industry, the processing industry, the waste industry, public authorities, research centers, associations, compulsory and voluntary consortia, environmentalist forces, and the voluntary work sector. These developments will allow us to tackle the economic crisis with greater determination and the phenomenon of the deindustrialization of national chemical sites due to the loss of competitiveness of the oil commodity sector within Europe. It will also provide an answer to the problem of the desertification of some areas in Italy, through industries connected to biorefineries integrated in the local area, with due respect for local biodiversity and the greater availability of quality compost.

Remarkably, the Italian case study is followed closely by one of the major EU member states: France. On February 4, 2013, the French minister for economic development Arnaud Montebourg outlined to the press the intention and willingness to follow the Italian model and promote the use of biodegradable and compostable bags given the potential in terms of growth that a measure similar to the Italian one would have in triggering the creation of local value chain dedicated to the production of bioplastics and added value products.

The Italian case study could, if replicated in other EU member states, be an example capable of generating multiple direct and indirect benefits. In this way, with the right investment and support, Europe would stand a real chance at becoming the undisputed leader in the field of the bioeconomy, showing that leveraging and triggering private investments, even at times of crisis, can be easier than we think.

17.4 THE EU POLICY FRAMEWORK AND RELATED POLICY GAPS: THE EU STRATEGY ON BIOECONOMY AND THE ROLE OF INDUSTRIAL POLICIES

Despite the clear opportunity for Europe offered by biochemicals and bioplastics, there is currently no European policy framework to directly support biobased materials, in contrast to biofuels, which are incentivized via subsidies, quotas and mandates, and all measures, which clearly promote their use in the final market. So far this framework prevented the establishment of a level playing field for all sectors of biobased economy – bioenergy, biofuels, and industrial material use. This in turn has created barriers and limited potential in the industrial material to promote added value products and enable Europe to benefit from it; policies should therefore be focused more on supporting innovation and scale-up of new technologies, which can create a solid EU added value and are capable of responding to the societal challenges faced by our planet, together with clear measures aimed at triggering market uptake of those products.

Countries are in fact acting individually and collectively to develop the bioeconomy and reap the anticipated benefits, though efforts to support the specific development of bioplastics are less common. Generally, biofuel policy support is much greater than it is for either biobased plastics or biobased chemicals. This is likely to make the development of the bioeconomy uneven and may disfavor the use of biomass for bioplastics and biobased chemicals. It may also constrain the development and operation of integrated biorefineries.

In terms of stimulating innovation, besides individual company investments in biobased products, related research in Europe is today mainly funded via various public sources, in a very fragmented way. The main instruments at EU level is the European Framework Programme for Research and Technological Development, while at member states or regional level, specific public research funding for bioplastics and biochemicals is very limited in the EU member states. In fact, only a few countries are currently running dedicated research programs. The recently launched PPP on Biobased Industries is a great step toward more coordinated actions aimed to trigger the deployment of disruptive innovation in Europe. The focus of the PPP is to strengthen the development of EU-based value chains by connecting industries such as farming, chemicals, biotech and pulp and paper that would not be linked otherwise.

Specifically, the PPP will focus on:

- Building new value chains based on the development of sustainable biomass collection and supply systems with increased productivity and improved utilization of biomass feedstock (including co- and by-products) while unlocking utilization and valorization of waste and lignocellulosic biomass.
- Bringing existing value chains to new levels, through optimized uses of feedstock and industrial sidestreams, and offering innovative added value products to the market, thus creating a market pull and reinforcing the competitiveness of EU agriculture and forest-based industries.
- Bringing technology to maturity through research and innovation and through upgrading and building demonstration and flagship biorefineries, which will process the biomass into a range of innovative biobased products.

The European Commission will invest almost EUR 1 billion, while industrial partners have made a commitment to invest over EUR 2.8 billion in research and innovation efforts between 2014 and 2020 as a contribution to this initiative. If we take a look at policies aimed to boost market uptake, we see that subsidies and mandates for biofuels are dominating in Europe and worldwide (i.e., Brazil, United States). The focus on stronger and solid market pull measures for added value products is an essential element because the much needed change of paradigm leading to a post-petroleum and resource-efficient society cannot be just science or technology pushed and needs to be complemented by market-driven mechanisms. New knowledge and innovation are obviously essential, but this standing alone will not make Europe gain new markets nor will provide new high-quality jobs in rural areas, growth, and prosperity. Without a market, no new product or process can survive. Demand side policies are key to support the introduction of biobased products, and the implementation of the recommendations of the Expert Group on Biobased products of the Lead Market Initiative[4] should be a priority that needs to be implemented.

A positive signal in this direction was the recent Communication on the Update Industrial Policy for Europe[5] and the European Bioeconomy Strategy, where it can be seen that biobased products are considered a top priority. That said, much more needs to be done. Wisdom in the way we use renewable resources is key, and the EU would benefit from the creation of a framework, which favors the allocation of renewable resources to the highest-value use. The bioeconomy, *per se*, is not a panacea to all challenges faced by our planet and its inhabitants, but if we want to enable the bioeconomy to contribute at its best to some of those challenges, we need to think wisely on how to focus the use of natural resources toward productions capable of triggering sustainable economic recovery through local and regional growth without stressing the ecosystems. The aim is to leverage on the intrinsic capabilities and specializations of local areas and find collaborative and inclusive ways to valorize them.

A holistic approach, based on case studies, which have proven to be already economically viable and sustainable, is therefore essential when looking at the policy frameworks. Giving the cross-cutting nature of the bioeconomy, policies in the area of agriculture, innovation, environment, industry, regional development, etc., should all be focused on key sectors and converge toward a common central objective: enabling the European production of niche innovative added value products, which can themselves contribute to trigger more sustainable consumption patterns among citizens and society. This can only be achieved if support to innovation is coupled with measures that are able to trigger market uptake.

REFERENCES

[1] V. Dornburg, "Multi-functional Biomass Systems." PhD Thesis. University of Utrecht. **2004**. http://igitur-archive.library.uu.nl/dissertations/2004-1207-114735/full.pdf.

[2] M. Carus (nova-Institut GmbH). "The development of instruments to support the material use of renewable materials in Germany – Market volumes, structure and trends. Policy instruments to support the industrial use of renewable raw materials." **2010**.

[3] Pövry Forest Industry Consulting. "Value added and employment in PPI and energy alternative." Client: CEPI (Confederation of European Paper Industries). **2006**.

[4] European Commission. "Taking biobased from promise to market: A report from the ad Hoc advisory group for biobased products in the framework of the European Commission lead market initiative." **2009**.

[5] European Commission. "A stronger European Industry for Growth and Economic Recovery-Industrial Policy Communication Update" COM 582 Final. **2012**.

18

NOVOZYMES: HOW NOVOZYMES THINKS ABOUT BIOMASS

Brandon Emme[1] and Alex Berlin[2]

[1]*Novozymes North America, Inc., Franklinton, NC, USA*
[2]*Novozymes, Inc., Davis, CA, USA*

ABBREVIATION LIST

2G: Second generation, ref to use of biomass feedstocks

Industrial Biorenewables: A Practical Viewpoint, First Edition. Edited by Pablo Domínguez de María.

AXE: Acetyl Xylan Esterases
C5: Pentose sugar
C6: Hexose sugar
CAZy: Carbohydrate-Active enZymes
CBH I: CelloBioHydrolase type 1 for reducing ends
CBH II: CelloBioHydrolase type 2 for nonreducing ends
CBM: Cellulose-Binding Module
CHP: Combined Heat and Power
CIPs: Cellulose-Induced Proteins
COD: Carbon Oxygen Demand
DDGS: Distiller's Dry Grains and Solubles
EC: EndoCellulase
EG: EndoGlucanase
EUC: Enzyme Use Cost
FE: Feruloyl Esterases
GHG: Green House Gases
HHV: Higher Heating Value
KPI: Key Performance Indicator
LPMO: Lytic Polysaccharide MonoOxygenases
MMGY: Million Gallons per Year
NPV: Net Present Value
NR: NonReducing end of the cellulose polymer chain
NREL: National Renewable Energy Laboratory
OPEX: OPerating EXpenses
R: Reducing end of the cellulose polymer chain
R&D: Research & Development
US: United States
US DOE: United States Department of Energy
WWT: WasteWater Treatment
XUR: Xylose Uptake Rate
XYN: XYlaN polymer enzyme (xylanases)
α-AF: α-L-Arabinofuranosidases
α-GAL: α-D-Galactosidases
α-GLU: α-D-Glucuronidases
β-G: beta Glucanase
β-XYL: beta Xylosidase

18.1 THE COMPANY

The industrial enzyme leader Novozymes has developed expertise in biotechnology for more than half a century. Headquartered in Bagsvaerd, Denmark, Novozymes was founded in 2000 in a demerger from the pharmaceutical company Novo Nordisk. Novozymes' roots date back to the 1920s when Novo Terapeutisk Laboratorium and Nordisk Insulinlaboratorium were established in Copenhagen, Denmark. Enzyme production began in 1941. From its early work with insulin and protease extraction from swine pancreases in the 1940s, the company expanded into industrial fermentation-based production in the 1960s. Today, Novozymes produces more than 700 enzyme and microbial products used in more than 40 different industries and shipped to 130 countries, reducing energy requirements and raw materials needed at customers' manufacturing facilities. By 2013, Novozymes had produced more than 7000 pending and issued patents in 165 new patent families and introduced more than 50 new products between 2008 and 2013. By 2014, Novozymes had more than 6000 employees globally, with more than 1100 in North America including enzyme and microbial R&D (Davis, CA; Franklinton, NC) and production (Franklinton, NC; Blair, NE; Ottawa, Canada) along with industrial microbial R&D (Salem, VA; Houston, TX; Saskatoon, Canada; Research Triangle Park; NC) and production

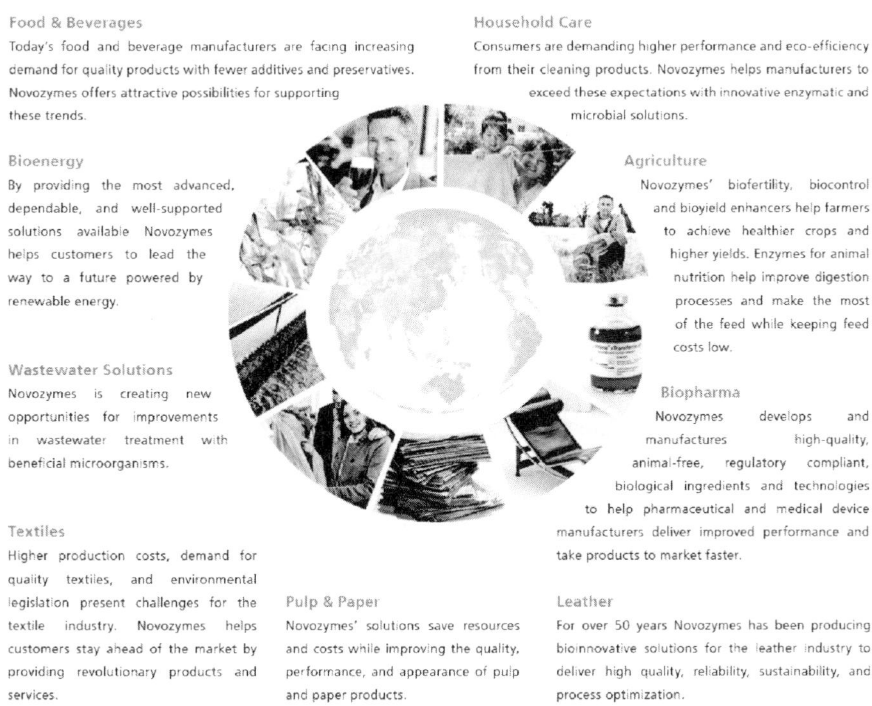

Figure 18.1 Novozymes overview. (*See color plate section for the color representation of this figure.*)

(Salem, VA; Saskatoon, Canada; Houston, TX). Novozymes also has R&D and production sites in Denmark, China, Brazil, United Kingdom, India, and Japan.

18.2 CASE STUDY: THE TRANSFORMATION OF CELLULOSE TO ETHANOL

In 2000 Novozymes initiated one of its largest R&D endeavor to date: commercializing the transformation of cellulose to ethanol via the biochemical process. More than 150 Novozymes researchers began work to reduce the cost of enzymes in the process by more than 90% over 5 years. Starting from the commercial enzymes used for the pulp and paper and textile industries used to improve fiber quality, Novozymes identified several new enzyme classes to efficiently break down plant fiber polymers to their base unit sugars. It worked: in 2010, Novozymes launched the first commercial product for the nascent industry with their line of Cellic products (Cellic CTec2®, Cellic HTec2®). Novozymes has continued to improve on its commercial offerings, each time bringing significant cost reductions to biofuel manufacturers with a consistent stream of new products, enabling a lower enzyme use cost (EUC) and reductions throughout the rest of the cellulosic ethanol process using integrated process optimization methods.

In the late 1990s, the National Renewable Energy Laboratory (NREL) began a concerted initial engagement with industry and technology developers through US DOE-funded research projects. These projects, including installation of their own pilot plant processes, produced the initial process designs and relevant data to form a basis for many commercial research studies into cellulosic biofuels, both through the thermochemical and the biochemical routes. Since that time, cellulosic biofuel development has been accelerating toward commercialization. Following those initial efforts by NREL, process development at technology providers and cellulosic ethanol producers has made great advances as well, improving pretreatment equipment design and chemistry and process controls to realize large reductions to both capital cost and yield losses due to inhibition of enzymes and microorganisms. Meanwhile, significant advancements to enzymes and fermenting organisms reached cost/yield levels to catalyze demonstration plant projects funded by the US DOE and private entities. Those factors allowed for smaller-scale demonstration of promising cost performance, such that by 2012 a handful of the first commercial-scale projects around the globe using the biochemical route had broken ground (Beta Renewables in Crescentino, Italy; POET in Emmetsburg, Iowa; Abengoa in Hugoton, Kansas; etc.).

Novozymes is the world's largest industrial enzyme producer and developer. Since the turn of the century, the company has put forth an unprecedented effort to provide cost-effective enzyme products for the biomass deconstruction (second generation, also called "2G") technology.

18.2.1 Background: Enzymatic Deconstruction of Plant Biomass to Fermentable Carbohydrates

The enzymatic deconstruction pathway – the so-called "enzymatic hydrolysis" *process step* – in biomass biorefining requires the action of glycoside hydrolases

(EC 3.2.1). However, these enzymes, as individual components, are not efficient enough and are not capable of achieving high degrees of biomass conversion under industrially relevant conditions. Biomass enzymes have to work in synergy in order to be efficient (Figure 18.2). The synergy between glycoside hydrolases and the so-called accessory proteins, which can be of catalytic or of noncatalytic nature, is necessary for the efficient depolymerization of major biomass polysaccharides, cellulose and hemicellulose, into fermentable hexoses and pentoses.[1,2] Classic (Figures 18.2a and b) and novel types of enzymatic synergisms (Figure 18.2c–g) combined with advanced protein engineering technologies and wild-type enzyme screening are the basis for the development of commercial biomass enzymes, such as, Cellic® CTec3 or Cellic® HTec3 (Novozymes, Denmark), Accellerase® TRIO™ (DuPont, NC, United States), and others.

The enzymatic cellulose degradation includes the concerted action of mainly exoglucanases or cellobiohydrolases (EC 3.2.1.176 and EC 3.2.1.91; reducing-end acting, R-CBHs, and nonreducing-end acting, NR-CBHs, respectively) or endoglucanases (EC 3.2.1.4; EGs) and cellobiases or β-glucosidases (EC 3.2.1.21; β-Gs) (Figure 18.2a). Cellobiohydrolases cleave cellulose, in a processive manner,[3] releasing cellobiose from the ends of the cellulose chain. Processive CBHs can be divided into two classes: reducing-end (R) and nonreducing-end (NR) acting enzymes. For instance, in the case of one of the most studied enzymatic cellulolytic systems, the soft rot fungus *Hypocrea jecorina* (formerly *Trichoderma reesei*) cellulase (*Hj* cellulases), the *Hj*Cel7A (formerly CBH I), is an R-acting CBH, while the *Hj*Cel6A (formerly CBH II) cleaves cellulose from the NR end.[4] In the *Hj* cellulase system *Hj*Cel7A and *Hj*Cel6A account for more than half of the total protein in the extracellular cellulase mix of the fungus indicating that these two enzymes play a dominant role in the cellulose deconstruction. Fungal processive CBHs show primarily modular structure with a catalytic domain connected through a peptidic linker with a *c*arbohydrate-*b*inding *m*odule (CBM).[5] This characteristic modular structure of fungal CBHs is critical for their processive action on crystalline cellulose, the preferred substrate, but it becomes less relevant if CBHs act on amorphous cellulose or soluble substrates. Partially degraded CBHs, where the peptidic linker had been proteolytically cleaved, have been shown to preserve activity on soluble cellodextrins.[6]

Unlike CBHs, most known endoglucanases randomly cleave the cellulose chain into cellodextrins, which are good substrates for CBHs. This classic vision of EGs as being enzymes randomly cleaving cellulose chains has been challenged in the past years by the identification of EGs, which show processive activity on cellulose, similarly to CBHs.[7] Another particularity of many EGs is that they tend to present broad substrate specificity. This fact is particularly well-illustrated by the existence of bifunctional endoglucanases with activity on xylan and cellulose[8] or mannan and cellulose.[9] In addition, it is known that "true xylanases" (lacking cellulase activity), such as, the xylanases from the family GH11,[10] can synergize with cellulases and boost the cellulose hydrolysis in pretreated biomass independently of the nature of the substrate or the total residual xylan content in the biomass substrate. It is believed that this boost in hydrolysis happens as a result of increasing

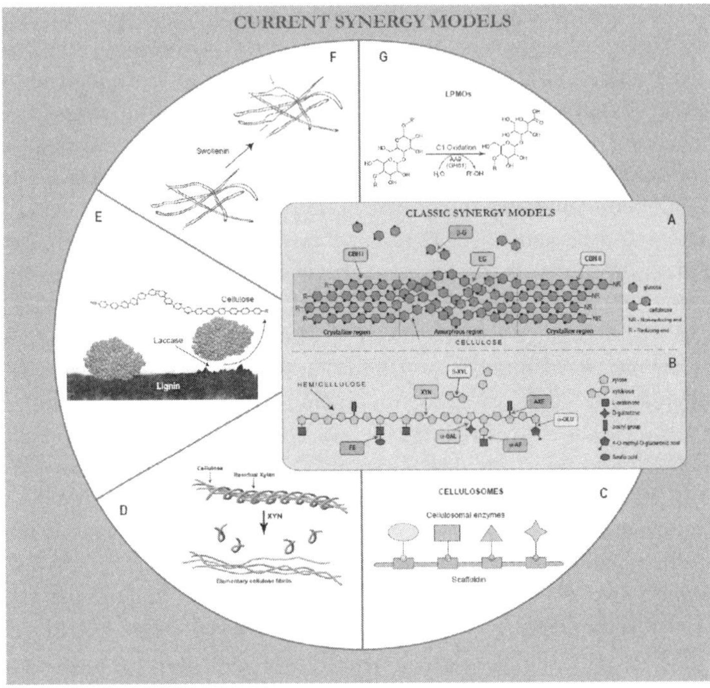

Figure 18.2 Enzymatic synergies in plant biomass deconstruction. Panel a – schematic representation of the enzymatic hydrolysis of cellulose by synergizing cellobiohydrolases (CBH I and CBH II), an endoglucanase, and a β-glucosidase; Panel b – schematic representation of the enzymatic hydrolysis of hemicellulose by synergizing xylanase, β-xylosidase, and xylan-debranching enzymes; Panel c – illustration of a typical bacterial cellulosome; Panel d – illustration of the increase of the accessible cellulose surface area in biomass as a result of hemicellulose degradation by an endo-β-1,4-xylanase; Panel e – illustration of the release of a nonproductively bound to lignin cellulase by a laccase; Panel f – effect of a swollenin on cellulose microfibril bundles; Panel g – oxidation of the anomeric carbon by LPMOs. List of abbreviations used in the figure: NR – nonreducing cellulose end; R – reducing cellulose end; CBH I – an R-end-acting exo-β-1,4-glucanase (EC 3.2.1.176), the *Hypocrea jecorina* (formerly *Trichoderma reesei*) Cel7A (formerly cellobiohydrolase I or CBH I); CBH II – an NR-end-acting exo-β-1,4-glucanase (EC 3.2.1.91), the *H. jecorina* Cel6A (formerly cellobiohydrolase II or CBH II); EG – an endo-β-1,4-glucanase (EC 3.2.1.4); β-G – a β-glucosidase (EC 3.2.1.21); XYN – an endo-β-1,4-xylanase (EC 3.2.1.8), which cleaves the main xylan backbone; β-XYL – a β-D-xylosidase (EC 3.2.1.37), which hydrolyses the released xylooligosaccharides to xylose; α-AF – an α-L-arabinofuranosidase (EC 3.2.1.55), active on arabinans; α-GAL – an α-D-galactosidase (EC 3.2.1.22), active on nonreducing α-D-galactose residues; α-GLU – an α-D-glucuronidase (EC 3.2.1.131), active on α-D-1,2-(4-O-methyl) glucuronosyl links; FE – a feruloyl esterase (EC 3.1.1.73), which releases ferulic acid by cleaving ester bonds; AXE – an acetylxylan esterase (EC 3.1.1.72), which catalyzes the release of acetyl groups; LPMOs – lytic polysaccharide monooxygenases (formerly, GH61); AA9 – CAZy auxiliary activity family 9 comprising LPMOs; GH61 – CAZy glycoside hydrolase family 61; R′, R – glycosyl groups; C1 – anomeric carbon. (Illustration by Vera M. Gutman.) (*See color plate section for the color representation of this figure.*)

the cellulose surface area accessible to cellulases, which was demonstrated by documenting an increase in the fiber width after treating pretreated biomass samples with a "true xylanase" (Figure 18.2d).[2] The bifunctional cellulase–xylanase substrate specificity created by nature and observed for certain EGs once again hints at the need of synergism of different activities as a requirement for efficient biomass enzymatic degradation. In this regard, the cooperative degradation of cellulose by the concerted action of CBHs and EGs has been one of the most studied examples of classic cellulolytic enzymatic synergisms and has become the backbone of today's commercial cellulase products. The final step in the enzymatic cellulose depolymerization is the cellobiose hydrolysis by β-glucosidases, which cleave the disaccharide into two molecules of glucose. β-Glucosidases are relatively minor but essential components in the naturally occurring cellulase systems, as well as, in the commercial biomass enzyme products. Most β-Gs do not have a CBM and generally are thought to act only upon soluble cellobiose, and, to a lesser degree, on other soluble cello- and xylooligosaccharides.[11]

The plant biomass composition is complex, and it comprises, besides cellulose, two other major polymers – hemicellulose and lignin, both considered natural barriers for biomass enzymes, which add to the biomass recalcitrance. Unlike cellulose, hemicelluloses and lignin lack a recalcitrant crystalline structure, but both are chemically very diverse and vary dramatically with plant source, as well as, with plants parts. Most often, a pretreatment of the biomass with heat and water/chemical catalysts is employed to open up the complex structure and provide access for enzymes. The equipment and chemistry applied to pretreatment also have a strong influence on the type and state of the deconstructed material by partially or fully solubilizing the different polymers based upon the melting temperatures and the degree of shear applied during the step. "Acid" (using acid during the heat treatment), "alkaline" (using base during the heat treatment), and "autohydrolysis" (only using water/steam during the heat treatment) are the main types of commercially applied pretreatment concepts in the first wave of 2G plants. The combined diversity gives rise to a similarly diverse enzyme requirement for the downstream enzyme reaction step.

Hemicellulose enzymatic depolymerization is achieved via the synergistic action of enzymes with broad substrate specificities, including those that degrade exclusively carbohydrate–carbohydrate bonds, such as, endo-β-1,4-xylanases (EC 3.2.1.8; XYNs), which cleave the main xylan backbone, or β-D-xylosidases (EC 3.2.1.37; β-XYL), which hydrolyze the released xylooligosaccharides to xylose, and a number of "*debranching enzymes*," [5] which cleave xylan branched side chains, including α-L-arabinofuranosidases (EC 3.2.1.55, α-AF), active on arabinans; α-D-galactosidases (EC 3.2.1.22, α-GAL), active on nonreducing α-D-galactose residues; α-D-glucuronidases (EC 3.2.1.131, α-GLU), active on α-D-1,2-(4-O-methyl) glucuronosyl links in certain xylans; feruloyl esterases (EC 3.1.1.73, FE), which release ferulic acid by cleaving ester bonds in branched xylans; and acetylxylan esterases (EC 3.1.1.72, AXE), which catalyze the release of acetyl groups. The ultimately released xylobiose is cleaved into two xylose molecules by β-D-xylosidases (Figure 18.2b). In conclusion, the combination of cellulases and hemicellulases dramatically improves the ability of cellulases to degrade cellulose as it was anticipated by the existence of naturally occurring bifunctional biomass enzymes, such as, cellulase–xylanases or cellulase–mannanases. The action of hemicellulases "activates" the biomass substrate by increasing the accessibility of cellulases to the target cellulosic elementary fibrils originally "wrapped up" in hemicellulose.[2]

18.2.2 Enzyme Commercial Development for Plant Biomass Deconstruction

Since the mid 1990s, the catalytic enhancement of the classically synergizing hydrolases, cellulases and hemicellulases, has been the basis for the development of advanced industrial enzymes. However, in the past decade novel types of enzymatic synergies have been discovered, which have accelerated the development of even more efficient industrial biomass enzymes. As much as hemicellulases, certain oxidoreductases have come to play an important role in the enzymatic degradation of cellulose, in particular, the copper-containing laccases. Both cellulases and hemicellulases have the tendency to bind nonproductively to lignin.[12] The oxidative action of laccases on biomass lignin seems to alleviate, in some cases, this undesirable behavior of glycoside hydrolases, which results in their increased active concentration in solution and, therefore, in better cellulose hydrolysis (Figure 18.2e). Another class of oxidative enzymes that plays an important role in biomass degradation, given their strong synergy with cellulases, is the copper-dependent lytic polysaccharide monooxygenases (LPMOs; CAZy family AA9, formerly GH61; Figure 18.2g)[13] and other oxidative auxiliary enzymes, such as, the family AA10.[14] LPMOs degrade cellulose in the presence of electron donors, water, and oxygen and yield oxidized cellodextrins, primarily aldonic acids, and unoxidized products.[15]

Most commercial biomass enzymes are of fungal origin driven primarily by the fact that fungal enzymes can be produced in higher yields compared to enzymes from other sources, and, the fact that they are very diverse in nature. A typical fungal enzymatic system can contain over 100 different enzymes with broad substrate specificity active on almost every type of plant polysaccharide and lignin. Recent advancements in the study of synergies between fungal cellulases and complexed cellulosomal enzymes might change the role that bacterial cellulases will play in future industrial biomass enzymes. Resch et al.[16] showed that a fungal enzyme cocktail and a cellulosome preparation display dramatic synergistic enzyme activity on cellulose. Cellulosomes are discrete multienzyme complexes of glycoside hydrolases anchored to noncatalytic subunits called scaffoldins (Figure 18.2c).[17] Cellulosomes are known to have higher specific activity on cellulose compared to free fungal cellulases. In this regard, during the past several years, major efforts have been dedicated to the production of the so-called *designed cellulosomes*. In the structure of designed cellulosomes, each enzyme is equipped with a docking module that interacts specifically with one of the cohesin modules of the chimeric scaffoldin. Artificial scaffoldins serve as docking backbones and contain a cellulose-specific carbohydrate-binding module that directs the enzymatic complex to the cellulosic substrate and one or more cohesin modules from different natural cellulosomal species, each exhibiting a different specificity, which allows for the specific incorporation of the desired matching docking-bearing enzymes.[18]

More recently, certain proteins lacking catalytic activity have been shown to boost cellulase activity presumably by either reducing the nonproductive binding of cellulases[19] or by increasing their accessibility to cellulose via reduction of cellulose crystallinity with the help of plant expansin-like proteins, such as, swollenins (Figure 18.2f),[20] loosenins,[21] cellulose-binding modules (CBMs), or cellulose-induced proteins (CIPs).[22]

The discovery of new insights on the mechanisms of action of biomass-degrading enzymes, as well as, the discovery of new types of synergisms has made possible the accelerated commercialization of cellulosic liquid biofuels via the enzymatic deconstruction pathway. These efforts have come a long way since the late 1990s; however there is still room for further improvement as new classes of enzymes, new synergistic modes of actions, and optimally engineered enzymes are developed or discovered. In the years to come, we should expect to witness continued improvements that will yield biomass enzymes even more robust and better adapted to the specific needs of the emerging global carbohydrate economy.

18.2.3 Industry Enablement Efforts

The cellulosic biofuel industry enablement in the early 2000s and onward was driven by the development of process cost models by a team at NREL[23] using advanced process simulation software (Aspen Plus). These models took into consideration many unique details related to biomass as a feedstock, in particular the merger of technologies from the existing pulp and paper industry with starch-based biofuels. The man-hours to generate such comprehensive models were considerable, but NREL generated believable output based upon copious amounts of laboratory data. An industry could envision a road map to cellulosic ethanol using the lower capital enzymatic hydrolysis and fermentation processes (Figure 18.3). Realizing the potential of process cost analysis, Novozymes leveraged process cost analysis into its own development program.

In order to create insights about the interplay between process factors, mass and energy balances, financial components, and commodity supplies, a sophisticated model was developed, using the 2002 NREL cost model as a basis. This so-called 2G cost model is dynamic, allowing for rapid and flexible changes to well over 400 different inputs, which then allow for tracking of a myriad of outputs of interest (installed capacity cost, minimum ethanol selling price, greenhouse gases (GHGs), individual raw material use costs, etc.) in a variety of units, such as, USD gal-$EtOH^{-1}$, kg/kg-$feedstock^{-1}$, etc. The design of the model allows for changes in process configuration, which in conjunction with lab- and pilot-scale data gives Novozymes the ability to predict winning concepts. Excel macro interfaces give the modeler the ability to simultaneously vary all of the input factors, thus making

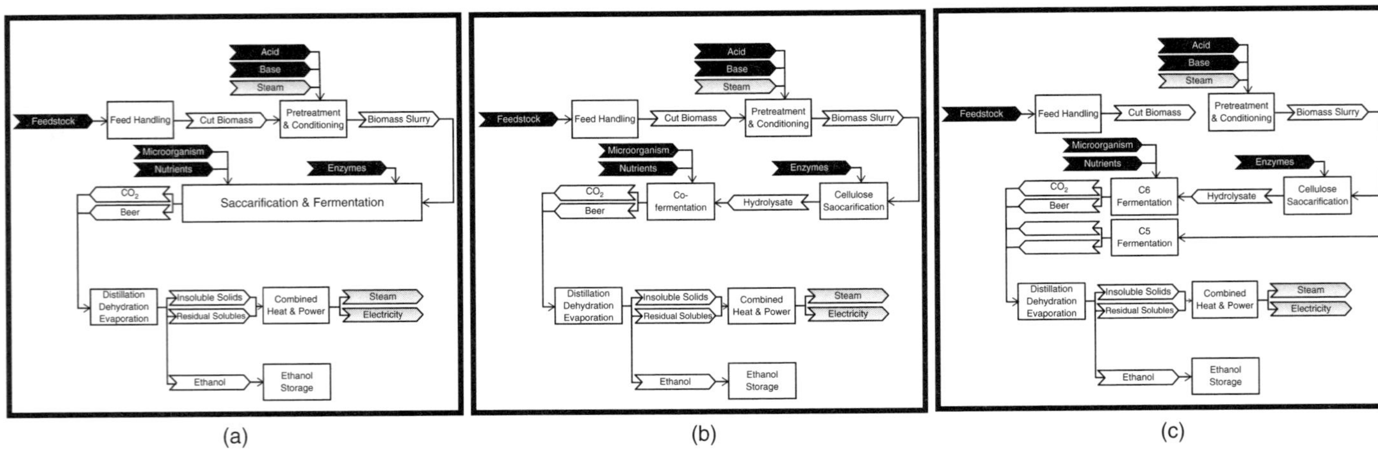

Figure 18.3 Common flow diagrams for a biomass to ethanol process, utilizing dilute acid in the pretreatment with a "whole slurry" concept. (a) Simultaneous saccharification and fermentation (SSF); (b) Separate hydrolysis and cofermentation (SHF); (c) Separate hydrolysis and parallel fermentation (SHPF). Alternative processes can separate and/or wash liquid streams generated during the pretreatment and conditioning step to remove impurities or create parallel process streams.

the tool a very powerful "brush" to "paint a picture" of the nascent cellulosic sugar/biofuel industry, both for current State of Technology and how decisions will be made as the technology moves toward *n*th plant performance.

Building up a model required some basis for the process. Bits and pieces of biomass process information were gleaned from public resources (NREL 2002 TEA by Aden, et al.)[23] and combined with internal knowledge of aerobic fermentation and biorefining from being the world's largest producer of industrial enzymes and microorganisms. Unit operation blocks were constructed using matrix arrays for chemical stoichiometry, as well as, performance factors (yield, capacity, etc.), followed by directed formulas and macro routines to "move" the mass and energy usage around the virtual plant. After deriving the mass and energy balance output for a given case, economic data (capital costs; net present value (NPV)) is calculated and key performance indicators (KPIs; see Figure 18.4) tracked to score each scenario considered. Use of simple Excel macros also facilitates rapid case studying, allowing for a quite vast field of candidate processes to be assessed in only a matter of minutes with a single laptop computer.

A key challenge to business development in technology-intensive industries is gaining a realistic grasp of the issues that drive the decision-making process. In addition to knowing the market potential, it is of great importance to have a system that allows cost sensitivities of a business venture to be studied and quantified. In the case of the biomass to sugar (and to biofuel, etc.) platform, many factors like raw material and capital costs are stacked up against uncertain product values (in the case of cellulosic ethanol, residue-based coproducts, and animal feed), an unchartered field of

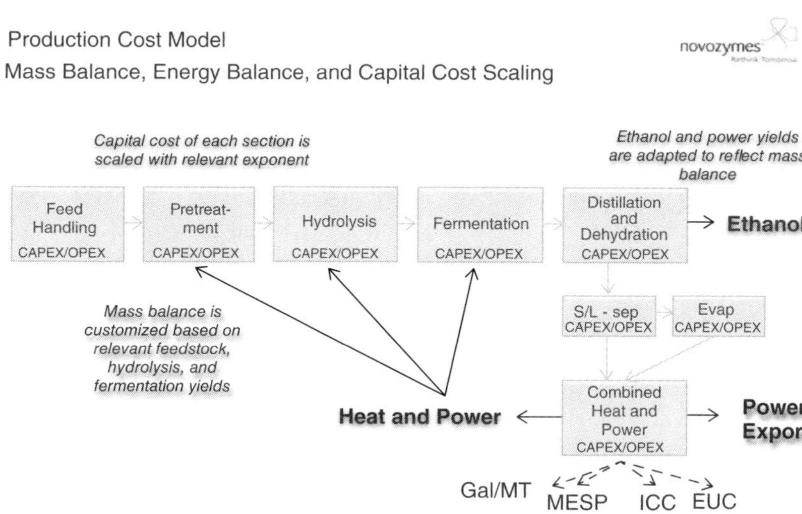

Figure 18.4 Diagram illustrating the block construction technoeconomic model used by Novozymes to estimate current and future state B2E concepts. (MESP = minimum ethanol selling price, ICC = installed capacity cost, EUC = enzyme use cost.)

regulatory issues, and a diverse set of potential stakeholders (the least of which are the oil companies and politicians) beyond the producers themselves. Simple models can be developed to conduct "back-of-the-envelope" scenario analysis, but many factors get ignored that might actually be significant. Conversely, complex modeling can be done using available software tools like Aspen Plus and Chemstation, but in order to get an output at all, a significant amount of information has to be known in order to specify the system. Additionally, such a complex model loses its flexibility such that diverse scenarios cannot be easily analyzed for comparison. In order to not be presumptuous about what is the best solution to a problem, it is important to both be accurate (enough) and visionary, lest the business model languish around a local optimum and fail to become the winning technology.

After analyzing the needs for such a complex, diverse, and uncertain industry as biomass sugar production, Novozymes adopted the approach of system analysis using a series of minimodels nested within a network of mass and energy balances. This dynamic method allows a highly multivariate system to integrate internal data along with both engineering- and business-derived correlations, such that all variables of interest are simultaneously acting on one another. By employing a block structure to the unit operations of the process, performance data can be easily linked with different process designs in order to be able to assess the holistic impact of changes to the process configuration. Additionally, regional aspects on things like feedstock type, composition, and logistics can also be included along with relevant market, finance, and governmental aspects. When all of these factors are taken into consideration, a global cost minimum – the so-called sweet spot – is visible and can be used to make best guesses of current state and its impact on decision-making, as well as, the future state – the so-called *n*th plant assumption – which can take into consideration relative impacts of technology and performance improvement and thus direct research and development efforts both within Novozymes and with their commercialization partners.

The Novozymes approach is relevant for cases where there is incomplete information available about a true commercial-scale system; hence no validation is possible. The use of the model output therefore is specifically to assess overall concepts: technology synergies, capex/finance-driven decision-making, basic process design, etc.; it is not for making detailed engineering design – more advanced process modeling would be done after the basic process design has been studied and confirmed by the dynamic model.

18.2.4 Important Sensitivities

Integration of lab data into process cost models, as correlations, allows for systematic study of performance factors impacting process design decisions. Figure 18.5 illustrates such an example. The enzymes used drive the conversion of lignocellulose into glucose and xylose. By linking the performance cost of enzymes with tank space–time cost, it is possible to understand the most reasonable conversion target for a process.

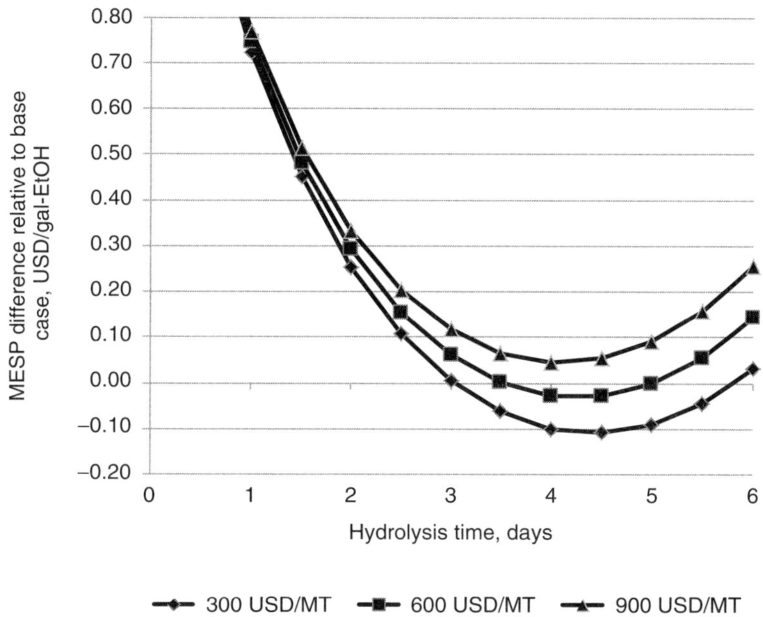

Figure 18.5 Example relationship of hydrolysis time to minimum ethanol selling price for several different assumed hydrolysis tank costs, using an enzyme performance correlation based upon laboratory data.

Likewise, it is also possible to assume the impact of performance improvements and determine if and how further process development might change the conversion target for any given process configuration (that is, if step "A" improves significantly, how would that change how one might configure/operate steps "B", "C", "D", etc.). When many factors are brought together, "global" cost sensitivities can be identified and a decision can be made and understood whether something can be done about it. Generally speaking, the model confirmed that the major cost "pieces" - feedstock, enzyme, yeast, and capital/financing - were key to driving down the overall cost of the process, and hence were (and still are) important areas of process development (Figure 18.6).

18.2.5 Enzyme Use Cost

Reducing the so-called "EUC" for a given process is a matter of several factors: enzyme dose, time, agitation, total solid content, temperature, and pH. Managing the interplay between the factors will allow for a commercially relevant unit operation

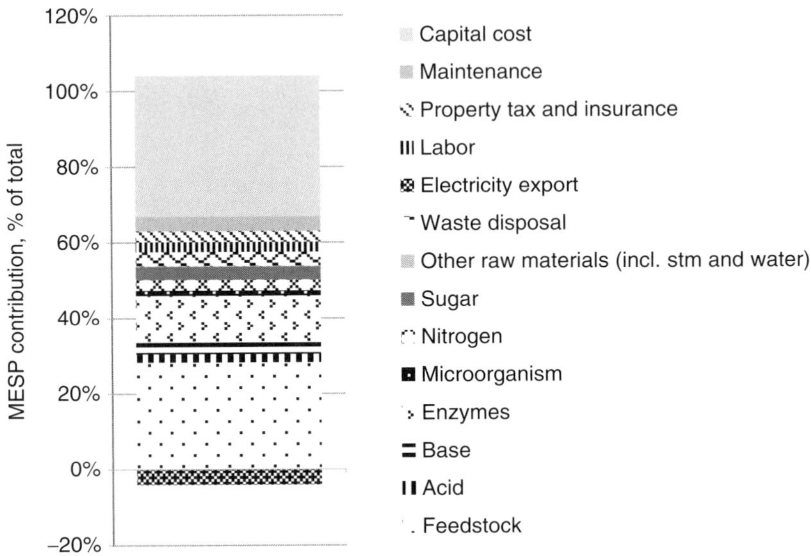

Figure 18.6 Common production cost breakdown for a dilute acid pretreatment process for biomass to ethanol. Note the negative cost associated with the electricity export.

step; failure to do so will drive optimization in the wrong direction, ending up with a more expensive overall process in the end.

At the outset of the current wave of 2G process development, much emphasis was put on the need to reduce the enzyme cost in the process. Although it is true that the enzymes did present a more significant component to the production cost (often referred to as operating expenses or OPEX) than they do for ethanol from corn (also called first-generation ("1G") ethanol), to consider biomass enzymes an unacceptable cost without also considering the value addition they give to the sugar/cellulose ethanol producer can be counterproductive to achieving a truly cost competitive process/technology. The substrates for corn and cellulosic ethanol, even though they are both primarily glucose polymers, are vastly different and "mingle" with rather different compounds inside the plant. Starch in corn ethanol exists in a highly branched, partially amorphous state – at the microscopic level a starch molecule looks much like a tumbleweed – making starch molecules rather accessible to enzymes (functionally, starch serves as energy storage for cells). Cellulose, on the other hand, exists in a much more crystalline and compact form – more like a tangle of ropes – with other complex polymers providing the branching (hemicellulose), as well as, antimicrobial, partially hydrophobic, and structural stability (lignin) characteristics. This distinct difference in physical and chemical structure for cellulosic biomass relative to starch makes it very stable and makes it orders of magnitude more complex and energetically strenuous for enzymes to depolymerize. As a result, a variety of enzyme activities has to be used to facilitate accessibility to each cellulose elemental fibril and polymeric chain.

When the EUC for a process is very high, data-driven cost optimization will suggest lowering the EUC by targeting a lower cellulose conversion to glucose (current cases; Figure 18.7). However, as the performance cost of enzymes begins to fall as a result of improvements in enzyme efficiency and selection (i.e., the Novozymes Cellic and HTec product lines) and in process development at pretreatment, hydrolysis, and fermentation steps, the cost optimization pathway changes. This can be modeled using tools like the Novozymes 2G model (Figure 18.7; the right side of optimum curves illustrates this point). When the enzyme performance cost reaches a critical point, the cost of increasing the sugar yield becomes less than the cost of the additional enzyme needed to reach that yield; hence enzyme dose should remain at a higher level in order to trade in for greater cost savings in other areas of the process design. Similarly, the enzyme cost becomes enabling for many other aspects of the process: increasing the solid loading in the process (which reduces water volume, hence equipment size and energy requirements), reducing biomass pretreatment catalyst (reducing one of the most expensive capital costs and saving some complexity in wastewater treatment (WWT)), and increasing product output (profits), among many others. At some point, correct use of generations of improved enzymes will have maximized the cost reductions of the other parts of the process; ethanol yields will be high, and lower EUC will occur, but the capital intensity of 2G processes in the cost models (2–4x that of 1G expressed in cost per gallon ethanol produced) suggests this will be in a mature biomass to sugar world, not what exists in the nascent cellulosic ethanol industry.

For any given enzyme mixture, there will be an optimal temperature for that mixture, below which the enzyme "machine" slows down and above which the most thermolabile enzyme proteins will begin to unfold or degrade as a result of induced surrounding media chemical changes or proteolytic degradation. Similar sensitivities are present in the hydrolysis for the pH: too high or too low pH and the enzymes will slow down, and pH extremes (especially at the site of acid or base addition for pH control) can completely denature the enzymes. High agitation can also impact the reaction rate by increasing the diffusion of enzymes on and off the surface of the lignocellulosic substrate, but other artifacts of high agitation can prevent the hydrolysis from reaching the same final substrate conversion.

The solids loading of the enzymatic hydrolysis step can also be adjusted to maximize the value of tank time cost, where choices of tank design (metallurgy, size, agitation, process controls, filling scheme, etc.) and capital financing play significant roles. In most cases, lower total solids will result in faster and more complete cellulose and hemicellulose hydrolysis. However, this must be balanced against the need for some realistic minimum ethanol concentration coming from fermentation in order to mitigate the cost associated with the distillation of a too diluted beer (a concentration step can be introduced after the enzyme hydrolysis to gain the best of both worlds, but the capital cost and energy use for that added step must then be managed).

Faster kinetics will save capital cost, but the kinetics of cellulose hydrolysis by cellulases is not straightforward to control. Many of the enzymes necessary to completely depolymerize the crystalline cellulose polymer into fermentable sugars are

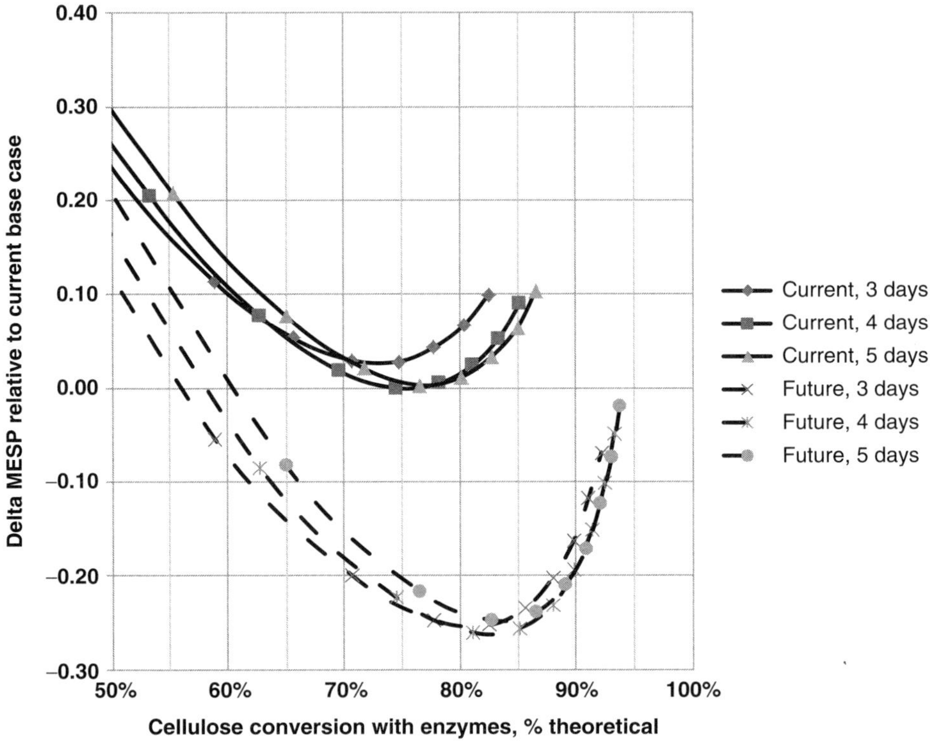

Figure 18.7 Illustration of the impact of improvements in enzyme efficiency (current Cellic CTec3 vs future) on the optimal reaction time and conversion targets for a whole slurry acid pretreated-type enzyme hydrolysis step. Enzyme improvement can be leveraged to increase the target cellulose conversion (75%/Current 4 days to 83%/Future 4 days). Changes in hydrolysis time can also be made (4 days Current to 3 days Future), but for the same tank type, the MESP benefit, if any, is small. The right side of the lowest delta MESP points for each curve illustrates the cost impact of targeting a higher conversion than the current technology will provide.

processive, meaning they move along the polymer similar to an inchworm, through a series of different reactions that "peel" the cellulose polymer chain from the crystal matrix. Thus, an improved mode of action would need to be developed in order to significantly impact the enzymatic cellulose kinetics (processivity is required to disrupt the cellulose crystalline structure of cellulose; most enzymes involved in the hydrolysis of amorphous cellulose are not processive). These advanced cellulase mixtures will need to have a similar or better supply cost to maintain commercial relevance. A high enzyme dose can be used to overcome limited enzyme hydrolytic rates, but the cost of enzymes under this scenario would make the process economically noncompetitive.

Because the enzymes can have such a pronounced impact not only on the process targets but also on the process design, often it is valuable to invest some capital to support optimal enzyme performance. Enzyme activity is a function of both chemical and physical structures; the protein molecules bend and move around in space as part of the reactions each carries out, so temperature, pH, time, and agitation can all play a significant role in reaching the best performance from the enzymes. A mixture of different enzymes work synergistically and sequentially to facilitate the breakdown of biomass to sugars (a sort of "de"-assembly line, if you will); thus it is important that the process operate at the point where the optimal performance curves of each enzyme component overlap. The more enzymes are needed to make the complete reaction, the narrower the operating range may be. Cellulose hydrolysis with cellulases tends to follow a kinetic curve similar to a Michaelis–Menten curve, starting with a high initial rate that tapers off after several days. A cost robust enzyme hydrolysis process has an increased residence time at the beginning of the hydrolysis (sometimes referred to as a preconditioning or prehydrolysis) to ensure temperature, pH, and enzyme distribution are all done such that enzyme kinetics run optimally. In order to reach the optimal hydrolysis space–time cost, the enzymatic hydrolysis should be allowed to proceed until the cost of the additional reaction time becomes the same/greater than the value of the additional sugar produced in that period.

Through a series of lab experiments, data can be generated to simulate a correlation between all of these factors in order to find the best possible hydrolysis in the context of the rest of the process. One might imagine that creating an optimal enzyme hydrolysis step in isolation of the other factors (feedstock, pretreatment, fermentation, distillation, and WWT) might risk settling on a more expensive process with a local optimum and a limited potential at large scale to capitalize on technology and process optimization.

18.2.6 Sugar Platform

Novozymes also used the aforementioned dynamic cost analysis tool to understand what a potential "pure" sugar platform process would look like. In order to conduct this study, the model was adapted to withdraw sugar streams from the process at different stages such that it would be possible to compare both the processes used as well as their impact on the resulting coproducts (in this case, ethanol). Biomass sugars were then compared against potential competitive carbohydrate sources to determine the commercial viability of a plant producing both sugars and ethanol in a competitive cellulosic ethanol industry, as well as, one for "green sugars." The Novozymes models suggest that biomass sugars in the United States will be competitive with sugarcane and beet molasses at historical world market prices and that a plant coproducing biomass sugars, as well as, cellulosic ethanol can have a strategic advantage over a dedicated cellulosic ethanol plant (Figure 18.8).

18.2.7 Financing

Biomass plants will be capital-intensive projects due to the size needed to capture the economy of scale,[23] with the debt burden likely being the single largest operating cost item. The financing model for a biomass plant as such then drives a lot of the

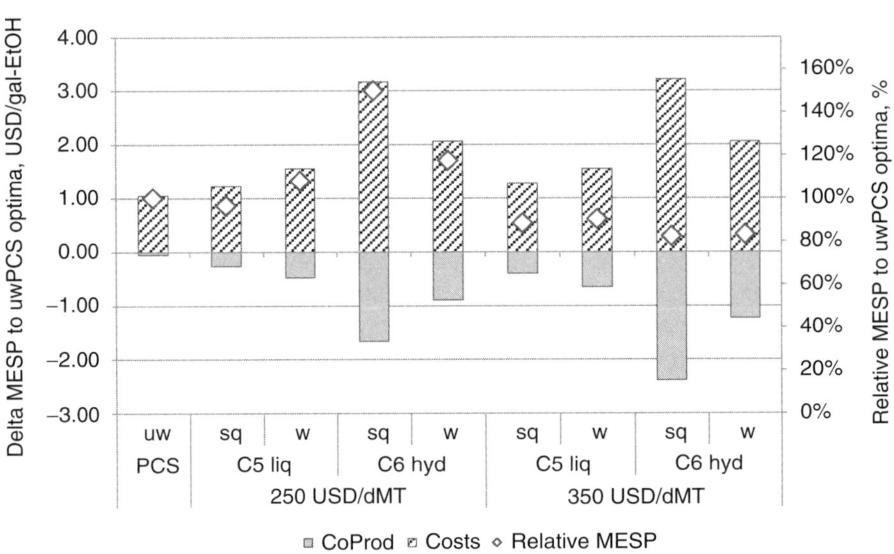

Figure 18.8 Comparison of several sugar platforms with ethanol processes to an ethanol only biomass process for varying sugar price in USD per metric ton dry weight sugars. (dMT = metric ton dry weight sugars, uw = unwashed substrate, sq = S/L separation process without washing, w = S/L separation process with washing.)

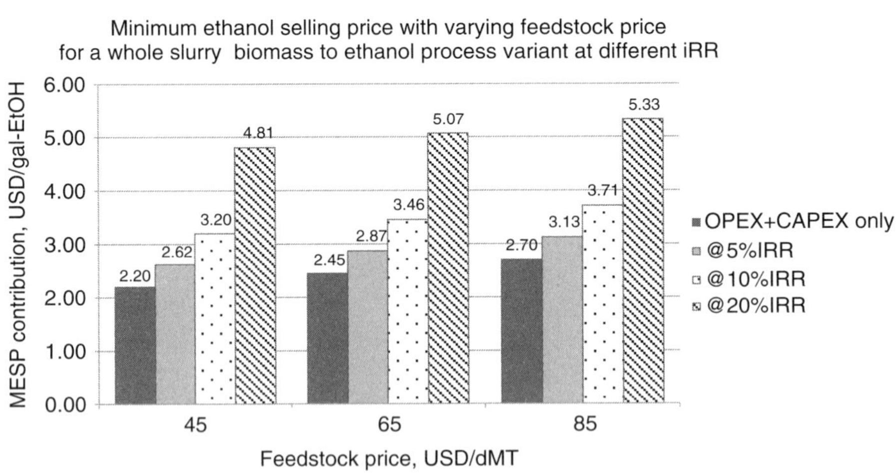

Figure 18.9 Impact of feedstock cost and internal rate of return (IRR) on the minimum ethanol selling price (MESP) for a Novozymes benchmark process model case with whole slurry converting biomass into ethanol. Other process layouts can lead to changes in the absolute and relative values. For high IRR requirements, the interest payments can be as large as the rest of the production costs combined.

process design decisions; varying the internal rate of return (IRR) for a project will impact the target process yields and reaction times to reach the targeted yields. For a given case shown in Figure 18.9, every percent increase in IRR will increase the MESP by roughly 5%.

18.2.8 Fermentation

18.2.8.1 Utilization of Pentoses For most lignocellulosic (biomass) feedstocks, roughly one third of the total carbohydrate is pentose sugars. Cost models will clearly demonstrate that processes that do not convert the main C5 (pentose) sugars (typically xylose but in some cases also arabinose) into cellulosic ethanol (or something else at least as valuable) will struggle to be commercially viable. There is and will continue to be much research both in academia and industry to develop pentose-fermenting microbes, so-called C5 microbes, but models will suggest there are some critical performance components that have to be in place to reach viable economy, namely, the propagation efficiency, xylose (pentose) uptake rate, and the ethanol/inhibitor tolerance.

18.2.8.2 Propagation Efficiency In most delivery strategies with an ethanologen in the fermentation step, the cost of the sugars used to grow up the cells (whether on-site at an ethanol producer or off-site at a microorganism supplier) is most often the dominating factor in the price. As such, the propagation efficiency – the weight of cells produced per weight of sugars fed to the cell growth phase – can directly set limits on what the cell loading can be in the ethanol fermentation step. Low cell growth efficiency will drive for processes with washed/fractionated hydrolysate, low cell loading, and/or cell recycling, while cost-efficient cell production will promote single-use fermentations with dirtier (cheaper) hydrolysate, making for an overall cheaper process.

18.2.8.3 Xylose Uptake Rate Xylose uptake rate (XUR) is a term that describes the realized consumption rate of xylose in the ethanol fermentation step, typically expressed in grams-xylose consumed per grams-broth per hour (g-xyl/g-broth/hr). Generally speaking, microorganisms will preferentially consume glucose before they metabolize any other carbohydrate, even in the case of genetically modified organisms (GMOs) containing specific C5 metabolic pathways. These microorganisms – when working on nonglucose carbohydrates – will generally ferment these "other" compounds at a much lower consumption rate than the glucose. Not every microorganism behaves the same way, so each case has to be scrutinized using empirical data correlations in the cost model to understand the impact on the optimal process design configuration. In a pure batch mixed-sugar fermentation with a C5 microorganism (i.e., everything is added at time zero), the glucose and other hexoses will normally be consumed at some fast rate, and then the pentoses will be fermented at much slower rate – as much as 3–4x slower – once the glucose and other hexoses are completely depleted, resulting in an inordinately long [expensive] fermentation step (Figure 18.10). There are several process decisions that can be

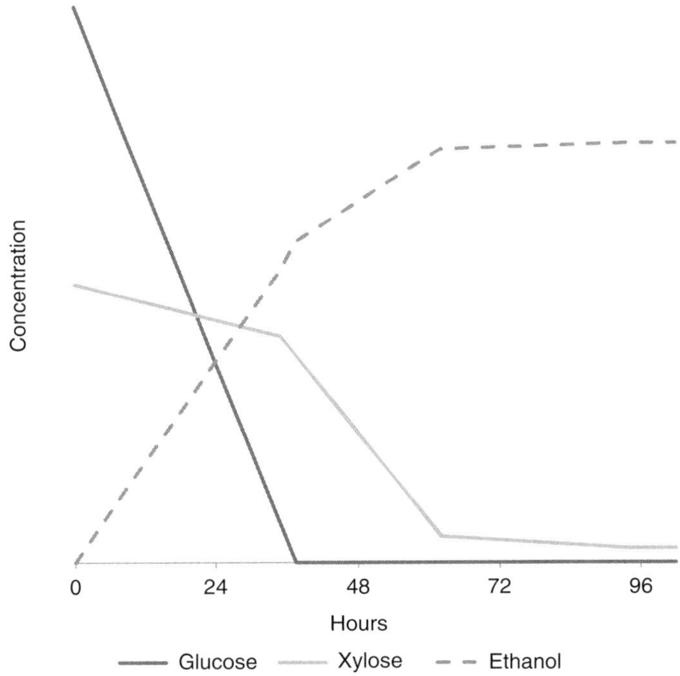

Figure 18.10 Typical concentration diagram for a pentose-fermenting ethanologen in a batch fermentation. The physical (sugar transporters on cell wall) as well as energetic/metabolic traits (intercellular enzyme pathways for sugar conversion to ethanol) often give considerable preferential activity on glucose over other sugars, including other hexoses (shown as phase lasting to about 36 h). After the glucose is consumed, the uptake rate of other sugars increases (36–60 h), until a critical point where the cells begin to be stressed by the ethanol concentration and/or the cells struggle as a result of reduced available maintenance energy.

employed to mitigate this problem, as well as, the inclusion of improved strain development.

18.2.8.4 Ethanol/Inhibitor Tolerance The ethanol and inhibitor tolerance of a microorganism will set limits on several other aspects of the process design. Stressed microorganisms will underperform, in many cases irreparably, so process limits have to be strictly maintained. Setting limits on the ethanol titer will impact both the hydrolysis (set points for conversion and total solids) and the distillation cost (water volume to be removed).

Some microorganisms will endure adverse conditions better in larger populations, but most will also suffer from overcrowding. A high cell loading (high pitch) in the fermentation can have several beneficial impacts (shorter fermentation time, inhibitor tolerance), but elevated cell counts can be costly: in the case of high pitch for use in a single fermentation, the cells are more expensive due to larger purchase volumes and/or greater on-site propagation requirements. In the case of high pitch for multi-batch use, the cells have to be successfully recycled from batch to batch (with preferential discharge of dead/low-performing cells and bacterial decontamination using, for instance, pH control). Novozymes developed and successfully demonstrated that a high pitch "fast fermentation" approach can be applied for mixed-sugar biomass hydrolysates (sugar slurry after enzyme hydrolysis has occurred) in the presence of elevated fermentation inhibitors from biomass (acetate, hydroxymethylfurfural, furfural, etc.).[24] In order to make such a high cell pitch process function effectively from a cost perspective (i.e., a sufficiently high number of recycles out of the microorganism population to make it inexpensive), the hydrolysate likely has to be clarified. This requires a solid/liquid separation system to be added to the process, which can be a significant cost to both installation and operation. Without the separation however, the residual solids from the enzyme hydrolysis will be mixed/mingled with the microorganism cells; thus when recycling cells from one batch to the next, the inert solids will build up and take up precious fermenter volume and hence increase the necessary fermenter size. Additionally, the solids – which often are also rather dark in color – will effectively "contaminate" the cell waste, making them less valuable as a protein-rich animal feed coproduct.

18.2.8.5 WWT Aspects The claim that production plants open and close based upon their water management will perhaps never be as true as it may be in the case of cellulosic sugar and ethanol plants. The impact of water on the process is both pervasive and dominant, as illustrated in Table 18.1.

As with all chemical production processes, there has to be a purge of the "mother liquor" stream (the residual liquid compounds after the manufacturing process has completed) else waste compounds will build up in the process. For many solid residual compounds from the process, the fate is combustion to CO_2, water, and ash. For many of the soluble residual compounds (unfermented sugars, weak acids, salt, etc.) however, their fate is likely going to be as aerobic and anaerobic digester feed, thus biogas (dirty methane), biosludge, and liquid effluent.

There is much discussion about the impact of the acid catalyst used in pretreatment on the cost of WWT. Acids can dramatically improve the enzyme accessibility from pretreatment.[25] However, when acid (or base) is used to catalyze the pretreatment, base (or acid) has to be added to neutralize the treated biomass to a pH range the enzymes can operate efficiently in. This generates what can be in some cases a significant amount of additional salt in the beer stream. NREL has commissioned several studies into the cost of WWT for a cellulosic ethanol/bioproducts.[26] Such studies have maintained the cost drivers to be the carbon oxygen demand (COD). However, the strong impact of salts (from pretreatment) can be indirect; elevated concentrations in wastewater streams can slow down the metabolism by the microbes in the digesters, thus dramatically increasing the residence time in the bioreactors. To minimize the COD, the solid/liquid separation that removes the lignin from the beer after distillation will need to be very efficient.

Fresh and clean water will be needed in the plant to displace the mother liquor purge, as well as, to supplement steam vapor losses in pretreatment. Some of the

TABLE 18.1 Some Impacts of Water on Process Steps for B2E.

Process Step	Water Impact on Process	Cost Impact/Mitigation
Pretreatment	Used to evenly heat and swell biomass to open cell wall structure	Water dramatically increases heat load needed; avoid excess
	Water/steam takes up volume	Minimize moisture content to keep reactors as small as possible
	Water/steam flashing facilitates rapid cooling of biomass to prevent degradation to enzyme inhibitors	Flashing is the cheapest way to drop temperature; alternatives are both capital and time-intensive
Enzyme hydrolysis	Used to facilitate enzyme diffusion to biomass	Optimize solid loading to get best enzyme performance at lowest processing cost due to dilution
	Serves as a "sink" for sugars after they have been made	
	Water is consumed in a hydrolysis reaction	
Fermentation	Microorganisms have a max ethanol concentration; too high sugar concentration and cells will die before fermenting all of the sugar	Optimize solid loading to get the best microorganism performance at lowest processing cost due to dilution
	Water is a key part of the osmotic pressure gradient for microorganisms; too little will stress the cells	Manage salt types and concentrations
Distillation	The more water per ethanol, the more energy is needed to distill each gallon	Maximize ethanol concentration (minimize water) in beer
Solid/liquid separation	Acts as rinsing solution; otherwise leave excessive sugars, etc., stuck in solids	Select technologies that can make the driest solid streams; use multiple stages and rinses where cost-effective
Wastewater treatment	Water volume determines size of bioreactors/lagoons	Maximize dry matter of process
	Capital largely determined by COD	Remove as much residual solids prior to WWT

makeup water will come in with the feedstock; a moisture content of the incoming feed of 12–15% will be common. For most pretreatment designs, direct steam injection will be used due to the faster heat transfer such that the flash condensate from the pretreatment will contain levels of volatile compounds like furfural, which are not compatible with condensate recycle to the boiler feed. Municipal water will therefore need to be brought in and/or additional water cleanup steps, such as, reverse osmosis (RO) will need to be added to the process design. The capital cost to the project is not large, but filtration systems can have a significant operating and cleaning cost that will have to be built into the cost models.

18.2.9 Colocation

Colocating a biomass plant can be one of the easiest ways to reduce the capital burden and increase the coproduct value, but it also can have some downsides that should be considered to evaluate commercial relevance. Partnering the project with relevant capacity allows for many cost downsides to be mitigated for both parties. If, for example, a biomass plant is located next door to a coal-fired power plant, some or all of the combined heat and power section capital of a 2G plant should be cut from the purchase price (savings around 30%), as well as, some of the back-end evaporation, as up to 15% of a coal feed stream can be displaced with the biomass and/or feedstock. If carbon credits are applied, additional benefits can be realized. However, it is also important to understand and appreciate the level of risk sharing involved with a colocation strategy, especially in cases of different entities controlling various parts of the value chain. Planning and logistics can impose unacceptable complication to operations that overall could make colocation in some cases in reality more expensive than a single stand-alone facility.

Lacking this best benefit option, it is also advantageous to merely be located nearby a sufficiently large electricity consumer. Many biomass plants will include some electricity export to the grid, but transmission line loss, as well as, power company profits have to be included in the valuation of the exported electricity. These aspects tend to reduce the value of the power significantly. If, however, the 2G plant's neighbor can take all of the excess power – with little or no transmission loss – the power may be sold to the neighbor at an elevated price compared to the grid power price while still allowing the neighbor to realize a discount over the grid power price. In the rural Midwest, there are many areas not readily accessible to a grid; in those cases, the 2G power export can be a cost savings to other rural projects by being the local power supply (noting that this would also bring the requirement for 365 days-per-year highly reliable operation).

A third colocation option could be with another industrial processing plant; one that would allow for more flow integration than just power. This scenario is modeled throughout the world in "biorefinery collectives." A good working example of this is the industrial park located in Kalundborg, Denmark, where more than nine companies and municipalities along with Novozymes have strategically located themselves such that they synergistically utilize each other's output streams to maximize value addition while also minimizing environmental impact from chemical discharge.[27]

Modeling such a case can be cumbersome, as correct valuation of flows is not always readily available; often the easiest way to assess these cases is not to find the actual best point (as it may be dependent on a false valuation) but instead to determine when the process will hit some satisfactory range and specify those limit values to sidestreams in the assumptions.

Colocation can also have downsides that need to be considered in the business plan. One of the most notable issues is road traffic in those cases when the biomass needs to be transported. Biomass is in general a low-density raw material, especially agricultural residual biomass, such as, wheat straw, corn stover, and rice straw, meaning that trucks loaded far below the weight limit loads will be driven on the roads collecting and delivering feedstock to the production plant. A rule of thumb suggests that a 2G biomass ethanol plant collocated with a 1G corn ethanol plant reaches synergy at roughly 1:2 scale; hence the biomass plant size should be 50 MMGY beside a 100 MMGY corn ethanol plant. Consideration of truck volume would suggest this would at least double the road traffic in the area. Municipal governments might push back against this, forcing the 2G plant to reduce its size – and its economy of scale benefit – or give some other economic requirement that will act on the economics unfavorably. Potentially viable alternatives to the use of low-density feedstocks "as is" might include, for instance, the use of denser biomass feedstocks, such as, wood chips, densification of low-density feedstock, or the use of other transportation means, when possible, such as, navigable waterways or the railway.

18.2.10 Coproducts

The ultimate goal of a biomass process is to create the highest value addition for the cost of a given quantity of biomass feedstock; thus the most cost competitive processes will have robust methods for maximizing the coproduct values while at the same time reducing the feedstock cost. Early biomass to ethanol process designs only created a few streams from the biomass: ethanol, (green) electricity, and gypsum. Value maximization efforts since then have generated several more viable candidates, shown in Table 18.2.

Animal feed is often regarded as the highest-value coproduct that also has a sufficiently large (and growing) market. Historically, the value of Dried Distillers Grains

TABLE 18.2 Various Forms of Coproducts from a Biomass to Ethanol Process.

Coproduct	Made From	Comment
Animal feed	Residual sugars, spent cells	Regulatory hurdle
Fuel pellets	Residual lignin solids	Balance between pelleting and transportation cost
Xylitol	Xylose	Small global market
Fertilizer (struvite)	Residual salts	Process yields?
Biogas	Residual carbon	Expensive to clean up for export

with Solubles (DDGS), a by-product of the 1G corn ethanol plants, has tracked well with the ethanol price (roughly 80% on dry weight basis of corn) up until the recent years. With the onset of the 2G industry, there will be diversification of the feed market based upon protein (both from feedstock and ethanologen biomass), fat, and other parameters. Animal feed will be an important feedstock, but valuation can be a challenge, and cost analysis should consider a range of values to understand the pinch points, utilizing relevant tools to develop parameter correlations. It is important to keep in mind the potential challenges and opportunities related to the regulatory and commercial acceptance of genetically modified organisms in the animal feed. In all cases, the timeline for general acceptance of genetically modified yeast as animal feed is much longer than for biofuels and other technical chemicals, so these factors will influence the NPV of a project.

The production of fuel pellets is an interesting way of bringing down capital cost of a 2G plant, especially in cases where the export electricity price locally is not attractive, perhaps due to remote location or other regional competitions for customers. The fuel pellet industry is thriving, primarily outside the United States. Pellets produced from residual hydrolysis solids will have a higher energy density than wood pellets and similar low higher heating value (HHV) because components (carbohydrates) are largely removed by the ethanol process, leaving the energy-rich lignin solids. Note, however, that the pellets made of residual hydrolysis solids of pretreated agricultural residues will contain in most cases a higher ash content compared to those made of wood. Some models might include some or all of the residual soluble solids, which will reduce the energy density but may reduce the overall ethanol plant production costs. In addition to reducing the initial investment (combined heat and power (CHP), etc.), when considering fuel pellets it is also important to properly estimate the secondary processing costs, namely, drying and pelletization.

Xylitol is an interesting coproduct as it utilizes the main C5 sugar, xylose. The main issue with a business case based upon xylitol production today is that the market is quite small; thus after only a handful of large 2G producers produce it, the world xylitol market price will become saturated. As other products made from xylitol reach the market, the outlook for xylitol processes will improve.

Another key challenge for 2G processes is how to deal with the native and process salts generated. In some cases, the majority of the salt comes with the feedstock into the plant, while in other processes the use of acid in pretreatment will significantly contribute to this issue. Directing of the salts to value-added fertilizer can dramatically improve not only the economic situation (favorable/reduced waste treatment costs), but it can also positively impact the logistics (return trip shipping for feedstock trucks) and the carbon footprint (restoration of nutrients to soil when possible). Fractionation of the salts, etc., can be a cost burden, and where and how they are removed can make or break a business case. Generally speaking, removal of salts prior to enzyme hydrolysis will be cheaper, but it will result in lower fertilizer yields (i.e., ammonium sulfate, in the case of the NREL-type process using sulfuric acid in pretreatment and ammonia for pH adjustment but depends upon acid and base choices for the plant process).

Biogas is a proven way to create value from the residual soluble COD. Although biogas operations are often predictably simple and therefore easy to cost estimate, the value is quite low due to the low quality of the produced biogas. To export biogas, expensive off-gas cleaning is needed, and with the vast natural gas reserves in the United States, it is unlikely there will be a significant export value for biogas from a 2G plant. It is therefore more reasonable to use biogas as a means to remove COD from the WWT feed and burn it for heat and, in the case of CHP, utilize it as extra export power.

REFERENCES

[1] M.J. Selig, E.P. Knoshaug, W.S. Adney, M.E. Himmel, S.R. Decker, *Bioresour. Technol.* **2008**, *99*, 4997.

[2] A. Berlin, J. Saddler, In "29th Symposium on Biotechnology for Fuels and Chemicals", (SIM-NREL, Denver, CO, USA), **2007**, 120.

[3] M. Kurašin, P. Väljamäe, *J. Biol. Chem.* **2001**, *286*, 169.

[4] B.L. Cantarel et al., *Nucl. Acids Res.* **2009**, *37*, D233.

[5] M. Silveira Buckeridge, G.H. Goldman, "*Routes to Cellulosic Ethanol*," Springer Science+Business Media, LLCC, New York, USA, **2011**.

[6] H. Van Tilbeurgh, P. Tomme, M. Claeyssens, R. Bhikhabhai, G. Pettersson, *FEBS Lett.* **1986**, *204*, 223.

[7] B.J. Watson, H. Zhang, A.G. Longmire, Y.H. Moon, S.W. Hutcheson, *J. Bacteriol.* **2009**, *191*, 5697.

[8] L. Chang et al., *Appl. Microbiol. Biotechnol.* **2011**, *90*, 1933.

[9] Z. Chen et al., *J. Biol. Chem.* **2013**, *287*, 25335.

[10] G. Paës, J.-G. Berrin, J. Beaugrand, *Biotechnol. Adv.* **2012**, *30*, 564.

[11] I. Kimura, S. Tajima, *J. Biosci. Bioeng.* **1999**, *87*, 572.

[12] A. Berlin et al., *J. Biotechnol.* **2006**, *125*, 198.

[13] P.V. Harris et al., *Biochemistry* **2010**, *49*, 3305.

[14] A. Levasseur, E. Drula, V. Lombard, P. Coutinho, B. Henrissat, *Biotech. Biofuel.* **2013**, *6*, 41.

[15] M. Bey et al., *Appl. Environ. Microbiol.* **2013**, *79*, 488.

[16] M. G. Resch et al., *Energ. Environ. Sci.* **2013**, *6*, 1858.

[17] E.A. Bayer, J.-P. Belaich, Y. Shoham, R. Lamed, *Annu. Rev. Microbiol.* **2004**, *58*, 521.

[18] Y. Vazana, S. Moraïs, Y. Barak, R. Lamed, E.A. Bayer, In "*Methods in Enzymology*" (Ed. J.G. Harry) Academic Press, vol. 510, **2012**, 429–452.

[19] S. Brethauer, M.H. Studer, B. Yang, C.E. Wyman, *Bioresour. Technol.* **2012**, *102*, 6295.

[20] M. Saloheimo et al., *Eur. J. Biochem.* **2002**, *269*, 4202.

[21] R. Quiroz-Castaneda, C. Martinez-Anaya, L. Cuervo-Soto, L. Segovia, J. Folch-Mallol, *Microb. Cell Fact.* **2011**, *10*, 8.

[22] F. Jacobson et al., *PLoS One* **2013**, *8*, e70562.

[23] A. Aden, et al., "Lignocellulosic Biomass to Ethanol Process Design and Economics Utilizing Co-Current Dilute Acid Prehydrolysis and Enzymatic Hydrolysis for Corn Stover" NREL/TP-510-32438, **2002**, 12–13.

[24] B. Emme, et al., "Pilot Scaleup of Enzymatic Hydrolysis and C5 Fermentation to a commercially relevant SHF Process for production of ethanol from biomass." Poster, 33rd Symposium on Biotechnology for Fuels and Chemicals. **2011**.

[25] M. Tucker, et al., "Effects of Temperature and Moisture on Dilute-Acid Steam Explosion Pretreatment of Corn Stover and Cellulase Enzyme Digestibility," *Appl. Biochem. Biotechnol.* (Biotechnology for Fuels and Chemicals The Twenty-Fourth Symposium), Vol. 105, No. 1–3, ISSN: 0273–2289, **2003**, 168–175.

[26] T. Steinwinder, et al., "Process Design of Wastewater Treatment for the NREL Cellulosic Ethanol Model," NREL/SR-5100-51838. **2011**.

[27] The Kalundborg case: http://www.iisd.org/business/viewcasestudy.aspx?id=77.

19

ORGANOCLICK: APPLIED ECO-FRIENDLY AND METAL-FREE CATALYSIS FOR WOOD AND FIBER MODIFICATIONS

Jonas Hafrén[1] and Armando Córdova[2,3]

[1]*Department of Forest Products, Swedish University of Agricultural Sciences, Uppsala, Sweden*

[2]*Department of Natural Sciences, Engineering and Mathematics, Mid Sweden University, Sundsvall, Sweden*

[3]*Berzelii Center EXSELENT on Porous Materials, Arrhenius Laboratory, Stockholm University, Stockholm, Sweden*

19.1 INTRODUCTION

Cellulose and hemicelluloses are polysaccharides and the main constituents of the plant cell wall, natural fibers, and wood. Moreover, cellulose is the world's most abundant organic polymer. Another prominent component of wood is lignin, which is the

Industrial Biorenewables: A Practical Viewpoint, First Edition. Edited by Pablo Domínguez de María.

most abundant natural aromatic polymer on Earth. Therefore, to use plant-derived natural polymers as renewable and environmentally friendly raw materials is highly interesting and desirable for sustainable chemistry. However, derivatization and treatments of wood and plant fiber material using environmentally friendly processes and chemicals bring certain challenges due to low reactivity of solid polysaccharides along heterogeneous reactions to deliver products. Herein, metal-free catalysis using organic catalysts (organocatalysis) provides for an environmentally friendly way for modification and derivatization of various heterogeneous and/or homogeneous substrates under mild, selective, eco-friendly, and nontoxic conditions. In this chapter, we provide a brief background to the research area of organocatalysis with respect to wood and fiber modification with focus on some of examples of industrially produced materials from the chemical company OrganoClick AB, which employ organocatalytic reactions for modification, and to add new or improved functionalities to wood and natural fiber materials.

At the moment the forest industry sector is facing several challenges. Our ways of using paper is changing and the most remarkable changes are the use of newsprint and graphic paper. The areas of packaging and hygiene products are increasing globally, but the challenge remains to create new fiber-based materials that can compete with other types of materials. Therefore, there is an increasing interest in finding new products and materials based on cellulose fibers, which is included in the aim of, for example, large-scale biorefineries. The forest industry in Northern Europe is today overall relatively well supplied by raw material (forest growth and logging), with some variations due to geographical and industrial factors. However, a significantly increased industrial use of wood would also increase the need for more raw material and energy, depending on the process. High-yield puling, for example, is raw material efficient but very electric energy demanding. One general approach that would allow for increased paper production, with unchanged demand for raw material or energy input per square meter paper, is to improve the product's properties such as strength and flexibility. A stronger paper may lead to a reduction of amount needed starting material by allowing for the use of thinner paper and/or to replace thicker paper with this thinner paper in products when possible. Hence, the employment of less material per finished product results in cost saving for a company and thereby potentially increased profits. Thus, it is important to find technologies for these properties that are in line with the "Twelve Principles of Green Chemistry" for a sustainable economy and environmentally friendly production.[1] In this context, the development of, for example, polyelectrolytes (PE)[2] has in recent years been found to offer great potential for improvement of the mechanical properties of papers made of treated fibers.[3] However, most of the PEs used in multilayers have been from oil-based starting materials, and expensive, thus unsuitable for standard paper grades. The use of PE from renewable sources (e.g., polysaccharides) exhibit great potential, but there is still place for significant further improvements in their production and applications.

One of the major limitations for new strategies toward utilizing the most abundant natural resource – cellulose – in sustainable technologies is the lack of environmentally friendly and efficient paths toward functionalization and modification of this natural material. For such purpose both homogeneous and heterogeneous approaches have been investigated. Cellulose is insoluble in water and most common organic solvents. Thus, with respect to the former approach, special solvents were developed to dissolve cellulose both for shaping and homogeneous chemical modification of the biopolymer.[4] The most important solvents NaOH–CS_2 (viscose process) and *N*-methylmorpholine *N*-oxide (NMO, Lyocell process) have serious limitation concerning the chemical stability, environmental impact, difficult recycling, and application for chemical reactions. Nonaqueous solvents like *N,N*-dimethylacetamide-lithium chloride (DMAC-LiCl) and ionic liquids show the potential of homogeneous cellulose modification in lab scale; however, these compounds still seem improper for large-scale transformations. Thus, new aqueous solvents were investigated to combine sustainability with a low environmental impact.[5] However, they require generally basic conditions (>pH = 9), which limits the transformations to nonbase sensitive reactants. Moreover, in the case of homogeneous cellulose chemistry, problems arise if reagents or the reaction products are highly hydrophobic, for example, if esterification, silylation, or tosylation reactions are intended to be carried out. It is also noteworthy that these transformations are not direct and often performed with lachrymatory reagents (e.g., acid chloride, tosyl chloride) both on solid and homogeneous cellulose materials. In addition, subsequent nucleophilic displacement reactions (S_N) on tosyl-modified cellulose are performed in DMSO or DMF.[6] As it can be concluded, there is a strong need for developing direct environmentally friendly modification methods of cellulose such as the use of organic acids instead of acid chlorides that will in the long run lead to new innovative products from renewable resources (e.g., forests, cellulose-based materials).

At the interface of enzyme and transition metal catalysis is the fast-moving field of metal-free catalysis employing organic catalysts (organocatalysis), which is an interesting and valuable research area due to the high selectivity of the transformations that even rivals that of enzymes.[7] The research that formed the scientific foundation to the company OrganoClick covers several bioorganic approaches to catalytic synthesis; small molecules for selective reactions with unmodified carbonyl compounds that utilize the same mechanisms as some natural enzymes do (enamine activation, iminium activation, hydrogen bond activation). For example, simple amino acids perform catalytic asymmetric carbon—carbon bond forming reactions similar to natural aldolase enzymes, which have the advantages of mild reaction conditions, high stereoselectivity, and the reduction of the use of protective group chemistry.[8] In addition, chiral small organic catalysts are inexpensive, nontoxic, may work both in organic or water solvents, and have the advantage to exist in both enantiomeric forms. To this date, several novel organocatalytic enantioselective reactions such as aldol, nitro-Michael, aza-Diels–Alder, α-oxidation, epoxidation, Mannich, and domino reactions have been developed.[7] The reactions may proceed with remarkably high enantioselectivities. For example, by employing environmentally benign catalytic reactions, a short synthesis of valuable heterocycles and amino acids have been developed (Scheme 19.1).

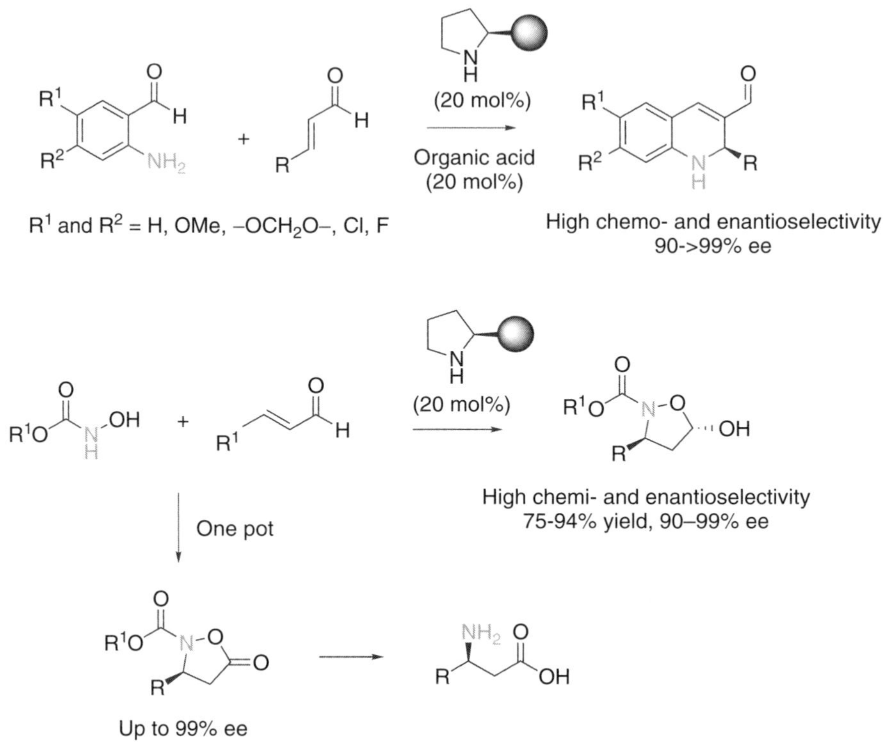

Scheme 19.1 Examples of organocatalytic reactions for the synthesis of heterocycles.

Inspired by our early accomplishments within the research field of asymmetric organocatalysis – the use of organic catalysis in cationic polymerizations of lactones and our studies on cellulose fibers – we embarked in 2003 on the task of creating a sustainable and green chemistry platform for the derivatization and functionalization of cellulose-based materials and lignocellulose. Thus, by applying the concept of organocatalysis for the modification and transformation of natural polymers we have developed catalytic transformations for direct modifications, which are useful and valuable for industrial applications. In our pursuit of developing biomimetic and eco-friendly industrially applicable catalytic reactions, we to some degree also mimic nature's highly atom-economic and selective biosynthesis and assembly of biomolecules. Moreover, organocatalysis may also be combined with metal, Lewis acid, and enzyme catalysis for the synthesis of biological active molecules and modification of heterogeneous polysaccharides with high economy and selectivity.

19.2 ECO-FRIENDLY AND ORGANOCATALYTIC SURFACE MODIFICATION OF LIGNOCELLULOSE

The biomimetic activation of organic compounds via fundamental activation mechanisms is key in our efforts to create new functionally modified wood or fiber materials. Importantly, the use of selective nontoxic catalysts will enable the setup of environmentally friendly and sustainable chemistry without the generation of toxic waste products. In addition, the use of starting materials and catalysts from renewable resources enables the development of sustainable catalytic processes in which wood or paper (i.e., cellulose, hemicellulose, and lignin) or other natural fibers are very suitable substrates for selective catalytic reaction and green chemistry. Our organocatalytic approach to wood and fiber modifications is based on the use of small catalytic molecules for selective reactions with unmodified compounds that utilize the same mechanisms as natural enzymes (e.g., hydrogen bond and Brønsted acid activation).

Focusing on these research lines, in 2004 we disclosed the ability to use highly selective nontoxic catalysts for the metal-free and solvent-free direct surface modification of heterogeneous polysaccharides.[9] This initial organocatalytic approach to solid carbohydrate modification is based on the use of small catalytic molecules such as fruit acids and amino acids for selective reactions with unmodified carbonyl compounds. For example, small organic acids such as tartaric acid and citric acid catalyze selective polymerization reactions, which have the advantages of mild reaction conditions, high selectivity, and the reduction of the use of protective group chemistry.[10] To this date we have developed several novel organocatalytic reactions such as α-oxidation, epoxidation, etherification, and esterification using the same type of activation mechanisms as nature's enzymes do. For example, we employed this metal-free environmentally benign catalysis for grafting of cellulose by surface-initiated ring-opening polymerizations (ROPs) from cellulose and paper products without the use of solvents (Scheme 19.2).[9] When the cellulose fibers are modified, a new material based on renewable starting materials is obtained, and by altering the catalysts or substrates, different functionalities can be added to the cellulose surface (as an example, in Scheme 19.2 hydrophobic polyester is formed onto the cellulose surface). Importantly, the use of highly selective nontoxic catalysts allows for environmentally friendly and sustainable chemistry without the generation of toxic waste products. This early work on surface-initiated grafting has subsequently been followed by other derivatizations of cellulose by ROPs as recently reviewed by Malmström and coworkers.[11]

Using the aforementioned organocatalytic approach, we have a platform of sustainable and green chemistry for environmentally benign wood and fiber modifications on industrial scale. Products made from renewable resources are often used based on their own merits (e.g., a specific property like appearance). But, also

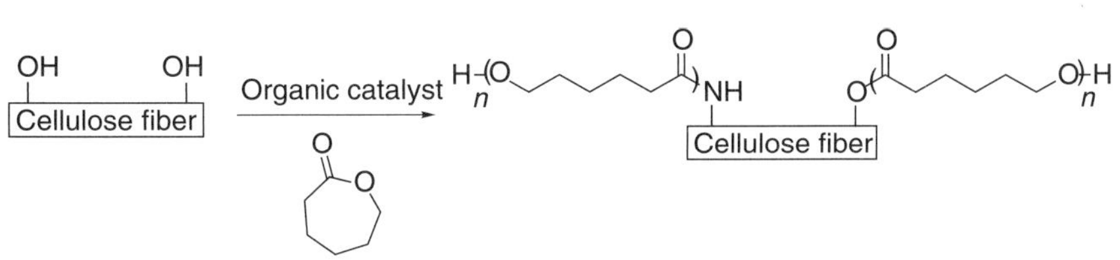

(3)

Scheme 19.2 Direct poly(caprolactone) derivatization of solid cellulose.

lignocellulosic products are desirable as substitutes for products that today contain nonrenewable petroleum or metal-based materials. The wood fibers in pulp can not only be selectively modified using hydrolytic enzymes and specific chemical treatments but also by addition of extraneous molecules. There are many reports of functionalization of cellulose and biopolymers; however, there are few available techniques for chemically controlled wood polymer covalent derivatization without using metal-containing catalysts. When using metal-containing catalysts, traces of metal are often found attached to the chain end on the macromolecular product. Of the various aims for cellulose derivatization, cellulose surface hydrophobicity is one important aim for modifications.[12] Especially aliphatic polyesters are interesting as hydrophobic polymers for cellulose derivatization due to their biodegradability and biocompatibility properties.[13] A common strategy for preparing these polyesters is by ROP of lactones.[14] However, later it was discovered that also small organic acids are able to catalyze ROP of cyclic lactones (for example, caprolactone)[15] and catalyzed the esterification of plant fiber hydroxy groups by acids.[16] Therefore, by adapting and applying this method on plant fibrous materials, the fiber-derived hydroxy groups have been utilized as initiators for ε-caprolactone polymerization, using an α-hydroxy Brønsted acid (tartaric acid) as the organic acid catalyst in bulk, plant fiber polyester product was afforded (Scheme 19.3).[16] A fiber-derived product produced by this method is biodegradable and biocompatible. After the initial discovery of organocatalytic direct esterification of solid cellulose, we have subsequently developed a number of systems for different cellulose functionalizations. These recent discoveries of organocatalytic modifications of cellulose fibers encouraged a commercial exploration of the full technical potential of finding environmental friendly means of developing new fiber-based products from paper.

The development of biopolymer-based materials with tailored surface properties is important for several industrial applications, and many products contain different

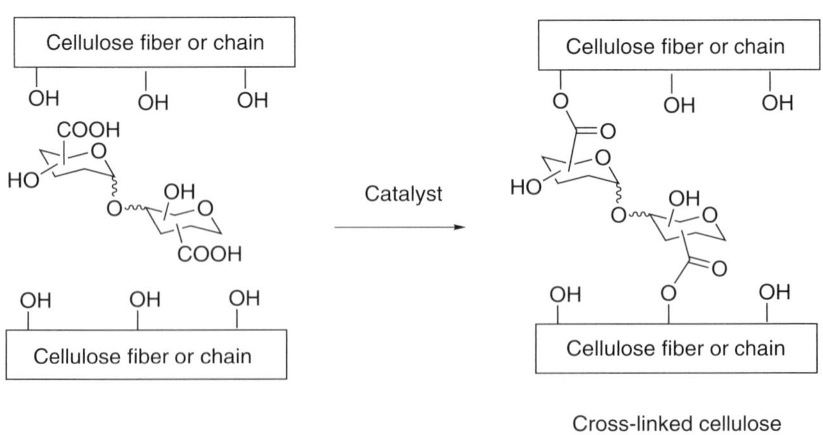

Scheme 19.3 Organocatalytic cross-linking of cellulose fibers.

materials that have to be compatible. Likewise, there is a political will and future ecological necessity to develop "green" chemistry based on renewable resources. Since cellulose and cellulose-derived products are of such huge importance to the global society, as a true renewable resource and sink for atmospheric CO_2, any possibilities of expanding plant fiber applicability by developing new products and substituting petroleum and metal-based materials are of great importance to the global economy and ecology. This, taken together with the interest for material recycling, has sparked interest for research on using cellulose as, for example, renewable, biodegradable, and environmental friendly additive and reinforcing fiber in plastic composites and fabrics.[17] This has led to the desire of controlling properties such as plant fiber hydrophobicity and adhesion of biofibers. Biofiber plastic composite failures are often due to weak boundary layers between the fiber and the matrix. This can be overcome by surface modification of the biofiber to improve the adhesion to the more hydrophobic plastic matrix.[18,19] In this context, polyesters have been used for plant fiber surface modifications. Aliphatic polyesters such as poly(ε-caprolactone) (PCL) and poly(lactic acid) are part of an important class of macromolecules for applications in biological and biomedical areas thanks to their desirable properties of biodegradability, biocompatibility, and permeability.[13,20] The functionality of organic acid catalysts in modifying a solid heterogeneous surface, the plant fiber, has just started to be explored, but the area shows great potential. In our studies, solid plant fibers have been derivatized using different cyclic lactones, and benzylic and aliphatic acids, and in all cases, also when polyesters are grafted onto the fibrous structure, the overall biological macrostructures, which in turn are given by the cellulose microfibril network,[21] are conserved.[16] Thus, the resulting polyester-derivatized plant fiber forms a multilevel structured material, a biopolymer–polyester biomimic composite on nanolevel, formed on the cellulose chain template. This may open up possibilities of new ways of producing plant fiber-based multicomponent biomimicking materials of, for example, defined nanoporosity (as fibers has pores, depending on plant origin, tissue, and treatments) of plausible interest for drug delivery as both microcellulose and biodegradable aliphatic esters are used and studied for this purpose. First, cellulose modification may be performed by derivatizations using organic acid catalysts, as outlined previously, since it is an environmentally friendly method, without toxic, hazardous catalysts, or organic solvents. In addition, a biodegradable and biocompatible composite, produced entirely of renewable material, can be made when using monomers such as ε-caprolactone or lactide together with plant fibers.

19.3 ORGANOCATALYTIC CROSS-LINKING BETWEEN POLYSACCHARIDES

There has been extensive research on the cross-linking of polysaccharides. The most common methods apply inorganic acids as catalysts for the cross-linking of the polysaccharide chains with low-molecular cross-linking agents, which are actually not eco-friendly. Here hazardous reactants such epichlorohydrin, dialdehydes, or formaldehyde are employed as the cross-linkers. With respect to our biomimetic approach, we began to investigate in 2006 about the possibility of using organocatalysis for the cross-linking between different polysaccharides using only simple organic acids as catalysts. Here we started by cross-linking different polysaccharides using citric acid, which is a well-known cross-linking agent, in combination with another organic acid catalyst or the autocatalysts.[22] This organocatalytic technology was next applied for cellulose modification by adding natural cross-linking molecules such as pectines, carboxymethylcellulose (CMC), chitosan, and oxidized cationic starch molecules covalently to the paper in the presence of organic catalysts (Scheme 19.3).[22] Next, the tensile strengths were measured and a significant improvement of it was observed for the resulting cellulose-based products. In fact, these catalytic reactions have been scaled-up to industrial level by OrganoClick AB, which have performed them on 100 t scale.

19.4 OC MODIFICATION OF LIGNOCELLULOSE

Based on the aforementioned primary organocatalytic modification, a second "click chemistry" step may be introduced in the wood or fiber modifications. In this context, we have recently disclosed catalytic direct esterification/dipolar cycloaddition[23a] relay sequences (Scheme 19.4a) and direct esterification/thiolene–[23b] relay sequences (Scheme 19.4b) for the functionalization of heterogeneous cellulose. The copper-catalyzed Huisgen 1,3-dipolar reaction[24] and the catalytic thiol–ene reaction[25,26] are both part of the concept of click chemistry[27] and can be performed in water. Moreover, by using "click chemistry" a multitude of functionalities could potentially be introduced onto the cellulose fiber surface in a modular fashion. Here we have explored and introduced the concept of "OrganoClick" modification of solid fibers and, as a proof of concept, added a fluorescent functionality onto paper.[23a] Moreover, dyes could be covalently and specifically linked to cellulose as a nonbleeding ink on paper or textiles. Many organic molecules of interest could be specifically linked to cellulose by employing the Huisgen 1,3-dipolar cycloaddition or the thiol–ene reaction, respectively, if successfully adapted for plant fibers. The novel materials could be of interest for the printing- or cellulose-based packaging industries. Other interesting products that might be explored using the "OrganoClick" route is dendrimeric derivatization of cellulose fibers, which will provide new biomaterials that may open up new frontiers in drug delivery and bionanomaterial.

19.4.1 Case Studies: Products Developed by OrganoClick

19.4.1.1 Background to the Company OrganoClick ab OrganoClick's business is to provide environmentally friendly fiber-based processes, formulations, materials, and chemicals for biofiber-based materials. By offering innovative and custom-designed products, the company works toward complete solutions for the manufacture of renewable and functional cellulose-based materials. OrganoClick

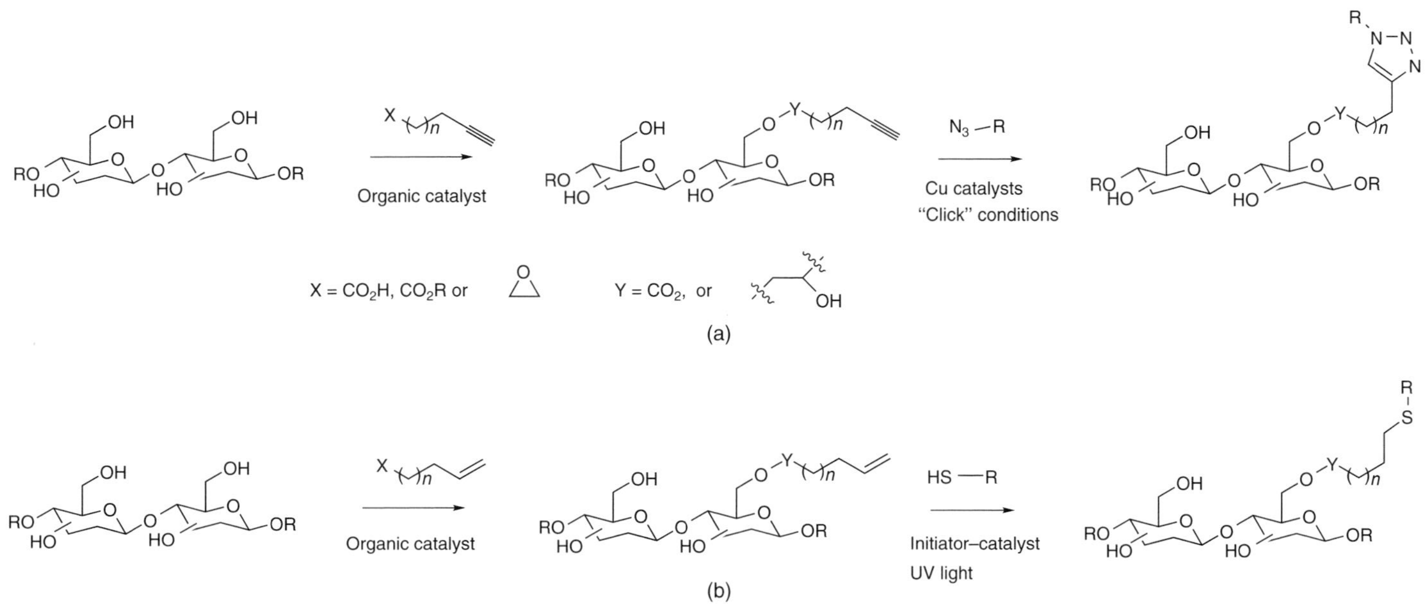

Scheme 19.4 "OrganoClick" surface modification of heterogeneous polysaccharides.

AB was appointed "Climate Solver" by the World Wildlife Foundation – WWF (www.climatesolver.org) in 2010 – and was in 2011 selected to be among Sweden's top 20 most promising companies by the Swedish Institute (SI). The research results described *vide supra* and derived from Swedish University for Agricultural Sciences and Stockholm University, and the subsequent patents thereof lay to the foundation to OrganoClick AB in 2006. The company headquarters and production facilities are located in Täby, and also a daughter company that handles sales and marketing of OrganoWood™ resides at the same location, just north of Stockholm.

The OrganoClick® technology is highly modular and applicable to a broad array of cellulosic products. To date the product portfolio include four major technical elements; OrganoTex™, a novel environmentally friendly technology improving textile's water resistance, and OrganoWood, a biofriendly chemical solution that delivers resistance to fungal decay and fire. OC-aquasil™ is an eco-friendly formula that produces extreme fiber hydrophobicity, and OC-biobinder™ for improving strengths for fiber-based materials.

19.4.1.2 Pulp and Paper The paper and pulp industry is one of the largest business areas in the world with a global estimated market of 35,000 billion SEK (2005). The pulp and paper industry is a large consumer of chemicals and energy, and with ever increasing costs for raw materials, energy, and pollution, there exists a demand for reducing chemical, fiber, and energy consumptions and a need for more environmentally friendly chemicals. The market for paper and pulp chemicals was estimated to 29.5 billion USD in 2010 (or ca. 84.2 million ton chemicals). The OrganoClick technology aims to provide customers with additional value to their paper-based products by adding new functionalities and/or to optimize the efficiency of raw materials usage. Paper additives are often used to improve strength properties. In this context, different formulations have been developed by OC including OC-biobinder. With respect to hydrophobization or sizing of paper, OC has technology that is applied on a variety of different paper grades. For example, OrganoClick has developed – by mimicking the structure of the Lotus flower – a new mean to render paper with superhydrophobic surfaces with contact angles in the range of 150° and Cobb (60 s) values below 5. Here the product OC-aquasil™ Paper is composed of a coating formulation for use on the dry end of the paper machine in coaters or size presses. Applications include specialty papers where the demand of hydrophobicity is high but vapor and air shall be able to flow freely through the paper. Moreover, the product can be customized into various viscosities and concentrations.

19.4.1.3 Textiles The global textile market is estimated to about 4000 billion US dollar, and man-made fibers and functional textiles are constantly gaining ground on natural fibers. Also in the textile processing, large amount of extraneous chemicals are added to produce desired material properties. These chemicals are often water leachable and a source of pollution, and consequently there is governmental and consumer pressure on textile producers to select more organic and eco-friendly products and processes. OrganoClick provides chemicals that can lift textile materials to a higher level of biocompatibility by substituting currently used chemicals with more environmentally friendly bioadditives. Moreover, even enhancement of textile functionality may be reached. OrganoClick has developed OC-biobinder, which is a new chemical binding system. OC-biobinder completely consists of renewable raw materials that are nontoxic for humans and are biodegradable. In comparison, conventional binders are predominantly made of petroleum-based substances (e.g., poly(acrylic acids) or poly(vinyl alcohols)), even though the nonwoven fibers are entirely made from renewable cellulose.

It is noteworthy that the OC-*biobinder* system can be used on woven textiles to increase stiffness and prevent fraying making it suitable for the manufacturing of sun blinds. On nonwoven or air-laid materials, OC-*biobinder* can increase tensile strength, decrease lining, and improve dimensional stability. Moreover, the system is compatible with most types of fibers, including cellulose-based fibers (e.g., cotton, paper pulp), synthetic fibers (e.g., polyester, PLA), mineral fibers (e.g., glass fibers), and mixed fibers.

OrganoTex is marketed as an environmentally friendly substitute for hydrofluorocarbons. Thus, it is a fluorocarbon-free water-repellent technology developed with inspiration of the Lotus flowers 3D structure (Figure 19.1). The technology provides functional fabrics with durable protection against water, snow, and water-based soiling such as stains from red wine, coffee, ketchup, etc. The water repellence is highly durable and withstands numerous cycles of home laundry (Table 19.1).

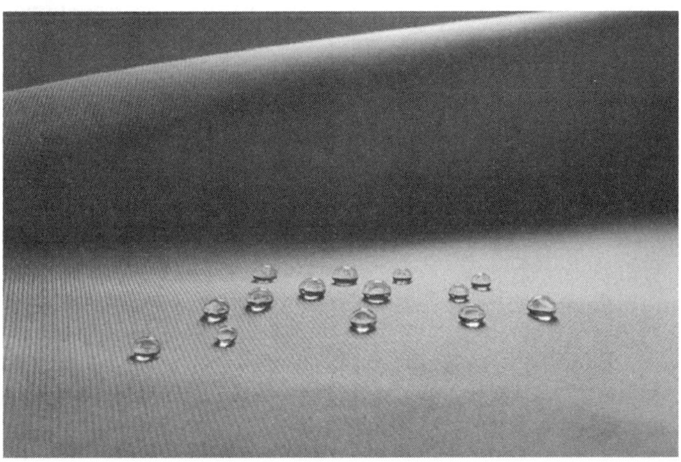

Figure 19.1 Water drops on a polyester-based textile treated with OrganoTex™.

TABLE 19.1 Results from Spray Tests.

	Before Home Laundry	After 20 Home Laundry Cycles (ISO 6330)
Spray test (ISO 4920)	5 of 5	4 of 5
Spray test (AATCC 22)	100 of 100	90 of 100

As stated previously, the OrganoTex® technology is free from fluorocarbons (e.g., PFOS, PFOA, etc) and other accumulative chemicals that negatively affect ecosystems. It is also free from hazardous chemicals, such as isocyanates that may cause respiratory problems. In addition OrganoTex is readily biodegradable in accordance with OECD 301. Fabrics treated with the OrganoTex technology are covered by an extensive quality assurance program comprising fabric performance, durability, and various sustainability parameters. Quality assured textiles are offered an attractive branding platform that can be used to promote the environmentally friendly characteristics of OrganoTex-treated textiles to consumers and retailers. The OrganoTex technology is based on the previously described biomimetic technologies. Thus, using plant-based catalysts to bind water-repellent "fatty" molecules directly to the textile fibers. The plant-based catalysts create a 3D-structured fabric surface of "fatty" molecules, resulting in the OrganoTex technology's excellent performance. The 3D-structured fabric surface still feels smooth and its roughness is invisible to the eye, much like the surface of the Lotus flower's leaves.

19.4.1.4 Functional Wood The construction wood sector generates about 20 billion euro of market value. Wood preservation treatments of wood include all processing that make wood, planks, and timber more resistant to biological or photochemical decay. The wood preservation industry is growing 2% annually and relatively more in North Africa and Asia. Traditionally wood preservatives have contained hazardous metal salts; however, the number of metals legally eligible for wood impregnation has been reduced and new national and international regulations enforce the use of more environmentally friendly chemicals in wood preservation. For example, chromated copper arsenate (CCA) is a highly effective wood preservative and it is universal in providing good protection against decay fungi and bacteria, termites, marine borers, and weathering. However, considering the environmental and health effects of CCA, its use has been restricted in Europe. A number of preservatives free of arsenic and chromium have been introduced to the market. These preservatives are based mainly on copper as main active component. Typical examples of these preservatives are amine copper quat (ACQ) and copper boron azole (CBA).

In this area, OrganoClick has developed a new formula to protect wood from fire and fungal decay, which is an environmentally friendly and completely nontoxic product called OrganoWood. It is based on the addition of silica to the wood matrix. Silicon is the second most abundant element on earth after oxygen. It is naturally found in the form of mineral silicates (salt of silicic acid), which mainly consist of polymeric SiO_2 ($mSiO_2.nH_2O$ or $[Si(OH)_4]$. Silicates and most other silicon compounds are classified as nontoxic. A wide variety of organic and inorganic silicon compounds can be applied to the wood modification process. Inorganic silicon compounds are mainly based on condensation products of silicic acid (colloidal silicic acids, silicates, "water glass") or tetraalkoxysilanes, which undergo hydrolysis and condensation steps to form sols and finally gels (sol–gel technology). When the sol–gel process is completed, an inorganic silicate free of organic groups is formed. Many silicon compounds have been described earlier.[25,26] Various silicofluorides, that is, salts of hexafluorosilicic acid (hydrofluorosilicic acid, H_2SiF_6), have been proposed as wood preservatives.

The OrganoWood process is based on mimicking the natural fossilization process in which wood is slowly transformed into mineral (cf. fossilization). As a consequence, a highly durable and fire-resistant wooden material is created. Through the use of organocatalysis for modification of biofibers, the speed of the fossilization process has been effectively increased. The silicon-based compounds used in the products are also strongly attached to the wood fibers so that the modified wood can be placed in wet outdoor environments with good durability of the material and its improved properties. In addition to modified wood, the *OrganoWood formula* is also available as a do-it-yourself surface treatment product formula. It is noteworthy that OrganoWood is an eco-friendly and highly fire-resistant material. It has been tested by SP Technical Research Institute of Sweden. Here the fire resistance was tested in the standard test EN 13501-1 and obtained the highest class of fire protection (Bfl-s1) for floor constructions. Moreover, in standard test EN 113 against rot, OrganoWood has shown good durability against biological decay.

REFERENCES

[1] (a) P. Anastas, J. Warner, *Green Chemistry: Theory and Practice*, Oxford University Press, New York, **1998**; (b) P.A. Clarke, S. Santos, W.H. Martin, *Green Chem.* **2007**, *9*, 438–440.

[2] G. Decher, *Science*, **1997**, *277*, 1232–1237.

[3] (a) L. Wågberg, S. Forsberg, A. Johansson, P. Juntii, *J. Pulp Pap. Sci.* **2002**, *28*, 222–228; (b) R. Lingström, L. Wågberg, P.T. Larsson, *J. Coll. Interf. Sci.* **2006**, *296*, 396–408; (c) L. Lundström Hämälä, J. Lindgren, E. Svensson-Rundlöf, T. Sennerfors, L. Wågberg, *Nord. Pulp Pap. Res. J.* **2009**, *24*, 459–468.

[4] T. Heinze, T. Liebert, "Polymer science: a comprehensive reference", In "*Celluloses and Polyoses/Hemicelluloses*",, Vol. 10 (Eds. K. Matyjaszewski, M. Möller), Elsevier BV, Amsterdam, **2012**, 83–152.

[5] (a) H. Ke, J. Zhou, L. Zhang, *Polym. Bull.* **2006**, *56*, 349–357; (b) J. Zhou, Y. Qin, S. Liu, L. Zhang, *Macromol. Biosci.* **2006**, *6*, 84–89; (c) S. Schmidt, T. Liebert, T. Heinze, *Green. Chem.* **2014**, *16*, 1941–1946.

[6] T. Liebert, C. Hänsch, T. Heinze, *Macromol. Rapid. Commun.* **2006**, *27*, 208–213.

[7] (a) J. von Liebig, *Ann. der Chem.Pharm.* **1860**, *113*, 246–247; (b) W. Langenbeck, *Liebigs Ann.* **1929**, *469*, 16–25; (c)A. Berkessel, H. Groeger Ed. *Asymmetric Organocatalysis*. Weinheim, Wiley-VCH, **2005**; (d)Special Issue: "Organocatalysis". B. List, Ed. *Chem. Rev.* **2007**, *107*, 5413–5415; (e) K.A. Ahrendt, C.J. Borths, D.W.C. MacMillan, *J. Am. Chem. Soc.* **2000**, *122*, 4243–4244.

[8] (a) Z.G. Hajos, D.R. Parrish, German Patent DE 2102623 **1971**; (b) Z.G. Hajos, D.R. Parrish, *J. Org. Chem.* **1974**; *39*; 1615–1621; (c) G. Zhong, T. Hoffmann, R.A. Lerner, S. Danishefsky, C.F. Barbas III, *J. Am. Chem.* **1997**, *119*, 8131–8132; (d) B. List, *J. Am. Chem. Soc.* **2000**, *122*, 9336–9337; (e) A. Córdova, W. Zou, I. Ibrahem, E. Reyes, M. Engqvist, W. -W. Liao, *Chem. Commun.* **2005**, 3586–3588.

[9] J. Hafrén, A. Córdova, *Macromol. Rapid Commun.* **2005**, *26*, 82–86.

[10] P.V. Persson, J. Casas, T. Iversen, A. Córdova, *Macromolecules* **2006**, *39*, 2819–2822.

[11] A. Calmark, E. Larsson, E. Malmström, *Eur. Pol. J.* **2012**, *48*, 1646–1648.

[12] D. Klemm, B. Philip, T. Heinze, U. Heinze, W. Wagenknecht, "*Comprehensive Cellulose Chemistry*", Vol. *2*, Wiley, New York, **1998**.

[13] S. Slomkowski, *Macromol. Symp.* **1996**, *103*, 213–228.

[14] P. Dubois, C. Jacobs, R. Jerome, P. Teyssie, *Macromolecules* **1991**, *24*, 2266–2270.

[15] (a) P.V. Persson, J. Schröder, K. Wickholm, H. Hedenström, T. Iversen, *Macromolecules* **2004**, *37*, 5889–5893; (b) J. Casas, P.V. Persson, T. Iversen, A. Córdova, *Adv. Synth. Cat.* **2004**, *346*, 1087–1089.

[16] (a) A. Córdova, J. Hafrén, *Nord. Pulp Pap. Res. J.* **2005**, *20*, 480–485; (b) J. Hafrén, A. Córdova, *Nord. Pulp Pap. Res. J.* **2007**, *22*, 184–188.

[17] B. Singh, M. Gupta, A. Verma, O.S. Tyagi, *Polym. Int. Ed.* **2000**, *49*, 1444–1451.

[18] Y. Li, Y.-W. Mai, L. Ye, *Compos. Sci. Technol.* **2000**, *60*, 2037–2182, and references therein.

[19] A. Carlmark, E. Malmström, *J. Am. Chem. Soc.* **2002**, *124*, 900–901.

[20] (a) J.-M. Vion, R. Jerome, P. Teyssié, M. Aubin, R.E. Prud'homme, *Macromolecules* **1986**, *19*, 1828–1838; (b) E. Chielleni, R. Solaro, *Adv. Mater.* **1996**, *8*, 305–313; (c) T. Fujisato, Y. Ikada, *Macromol. Symp.* **1996**, *103*, 73–83.

[21] J. Hafrén, F. Fujino, T. Itoh, *Plant Cell Physiol.* **1999**, *40*, 532–541.

[22] A. Córdova, J. Hafrén, J. Samec, K. Enquist, WO/2009/121926, **2009**.

[23] (a) J. Hafrén, W. Zou, A. Córdova, *Macromol. Rapid Commun.* **2006**, *27*, 1362–1366; (b) G.-L. Zhao, J. Hafrén, L. Deiana, A. Córdova, *Macromol. Rapid Commun.* **2010**, *31*, 740–744.

[24] (a) V.V. Rostovtsev, L.G. Green, V.V. Fokin, K.B. Sharpless, *Angew. Chem. Int. Ed.* **2002**, *41*, 2596–2599; (b) C.W. Tornøe, C. Christensen, M. Meldal, *J. Org. Chem.* **2002**, *67*, 3057–3064.

[25] (a) C.E. Hoyle, C.N. Bowman, *Angew. Chem. Int. Ed.* **2010**, *49*, 1540–1573; (b) J. Mergy, A. Fournier, E. Hachet, R. Auzély-Velty, *J. Polym. Sci. Part A. Polym. Chem.* **2012**, *50*, 4019–4028.

[26] H.C. Kolb, M.G. Finn, K.B. Sharpless, *Angew. Chem. Int. Ed.* **2004**, *43*, 5138–5175.

[27] (a) T. Furuno, T. Uehara, S. Jodai, *Mokuzai Gakkaishi* **1993**, *39*, 462–467; (b) C. Mai, H. Militz, *Wood Sci. Technol.* **2004**, *37*, 453–461; (c) C. Mai, H. Militz, *Wood Sci. Technol.* **2004**, *37*, 339–348; (d) S. Saka, H. Miyafuji, F. Tanno, *J. Sol–gel Sci. Techn.* **2001**, *20*, 213–217.

20

PETROBRAS: THE CONCEPT OF INTEGRATED BIOREFINERIES APPLIED TO THE OLEOCHEMISTRY INDUSTRY: RATIONAL UTILIZATION OF PRODUCTS AND RESIDUES VIA CATALYTIC ROUTES

EDUARDO FALABELLA SOUSA-AGUIAR[1,2], JOÃO MONNERAT ARAUJO RIBEIRO DE ALMEIDA[2], PEDRO NOTHAFT ROMANO[2], AND YURI CARVALHO[2]

[1]*Petrobras Research Centre (CENPES), Ilha do Fundão, Q7, Cidade Universitária, Rio de Janeiro, Brazil*

[2]*School of Chemistry, Department of Organic Processes, Centro de Tecnologia, Bloco E, Ilha do Fundão, Federal University of Rio de Janeiro (UFRJ), Rio de Janeiro, Brazil*

Industrial Biorenewables: A Practical Viewpoint, First Edition. Edited by Pablo Domínguez de María.

20.1 INTRODUCTION

Petrobras is a publicly traded corporation, the majority stockholder of which being the Government of Brazil. While having been traditionally an oil company, Petrobras is nowadays performing as an integrated energy company. Hence, Petrobras has activities comprising the following sectors: exploration and production, refining, marketing, transportation, petrochemistry, oil product distribution, natural gas, electricity, chemical gas, and last but not least biofuels.

Petrobras is surely a leader in the Brazilian oil industry. However, the company has expanded the operations thereof aiming at becoming one of the top five integrated energy companies in the world by 2030. Therefore, the company is currently present in 17 countries. Petrobras is expecting to invest R$ 104.41 billion, displaying a net revenue of R$ 304.89 billion and a net profit of R$ 23.57 billion. There are 798,596 shareholders and 86,111 employees. A daily output of 2,539,000 barrels of oil equivalent a day, proven reserves of 16.57 billion barrels of oil equivalent (boe), 15 refineries, and an oil product production of 2,124,000 barrels per day make Petrobras one of the most important energy companies of the world. As part of Petrobras' philosophy to become an integrated energy company, the area of green chemistry and biofuels has received much attention. Indeed, Petrobras operates currently 5 biodiesel plants (3 owned and 2 in partnerships); also, 10 ethanol plants in partnerships are an important new asset of the company.

The refining industry is currently facing a number of challenges whose solution is not easy to achieve. Indeed, more stringent environmental regulations resulting from the growing demand for cleaner fuels are increasing everywhere. Furthermore, the quality of today's crude oil is inferior to the crude oil produced several years ago. In fact, the crude oil nowadays may be rather heavy, acidic, and more impure, implying new steps in the refining process, such as the setup of extra desulphurization units. It must be borne in mind, however, that these new units are very energy consuming and reduce the overall thermal efficiency of the refinery. Additionally, traditional desulphurization processes usually require hydrogen, whose production via shift reaction also produces CO_2, a well-known greenhouse gas. Hence, meeting the society's demand by improving the air quality due to the use of cleaner fuels may come at the expense of higher greenhouse gas emissions during the production of such fuels. In addition to the growing pressure to reduce emissions, globalization, and oscillating customer's choice, it is worth noticing that, like any ordinary industry, the refining industry must be obviously profitable. Clearly, the aforementioned requirements are changing the traditional goals of petroleum refineries, frequently imposing a conundrum. For that reason, intelligent solutions must be searched to meet all needs. In this respect, the use of alternative raw materials such as natural gas and biomass is playing an important role in the refinery of the future. Moreover, the production of synthetic fuels from synthesis gas is an option that must not be disregarded. Nevertheless, regardless of the chemical route chosen, the role of catalysis is paramount, since it is easier to change the catalyst, rather than changing the entire production systems to generate a new drop-in fuel.

As previously mentioned, the refining of the future will encompass the concept of biorefineries, which, according to the 2008 Farm Act,[1–5] may be defined as "a facility – including equipment and processes – that converts renewable biomass into biofuels and biobased products, and may produce electricity." More recently, the term *integrated biorefinery* has been coined.[6] An integrated biorefinery is capable of efficiently converting a broad range of biomass feedstocks into affordable biofuels, biopower, and other biobased products. By definition, the integrated biorefinery must cope with the problem of residues. Integrated biorefineries are similar to conventional refineries, being forced to generate a variety of products to optimize both the use of the feedstock and the whole production economics. However, integrated biorefineries still require much research, that is to say, new processes must be developed and practically implemented in order to reduce production costs, thereby improving competitiveness. Above all, the potential of generated residues must be explored, and innovative chemical routes for their valorization are to be proposed.

The concept of integrated biorefinery may be applied to several types of industries, using different raw materials. In this area, vegetable oil producing facilities (or FAME industries) seem to be an excellent example of how different by-products and residues may undergo chemical transformations aiming at increasing the overall profitability of the process. Figure 20.1 depicts the potential use of vegetable oils as raw material in nonconventional chemical routes. The oleaginous biomass must be firstly harvested and thus undergo (bio)chemical processing. These two steps will generate the oil and the first residue, namely, the lignocellulosic fraction. Such fraction contains mainly cellulose, hemicellulose, and lignin. All three of them are potential raw materials for chemical transformations.

As far as the main product (the oil fraction) is concerned, several chemical transformations may be proposed aiming at adding value to this important feedstock. In this text, one will broach those considered the most promising ones, which are the following:

- Transesterification
- Hydrotreating (HDT)
- Decarboxylation

These aforementioned routes will be discussed in details later on in this chapter, highlighting the mechanism of reaction, the main catalysts deployed, as well as the

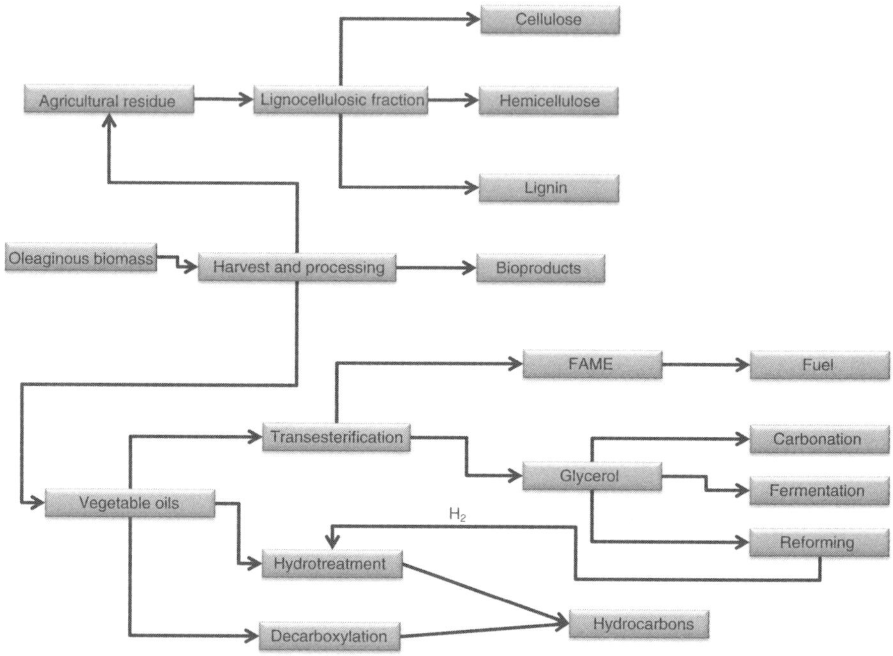

Figure 20.1 Potential use of vegetable oils as raw material in nonconventional chemical routes.

advantages and drawbacks thereof. In addition, the concept of integrated biorefinery will also be debated, being the feasibility of the routes presented.

20.2 GLYCEROL FERMENTATION

As mentioned in the previous section, Petrobras is operating 5 biodiesel plants, generating huge quantities of glycerol in different regions of Brazil. Since the birth of the biodiesel industry, a large amount of crude glycerol is available at low cost, and many efforts have been made trying to add value to this feedstock (waste). A particular interest is given to some microorganisms capable of transforming glycerol into more valuable chemicals through fermentation. Since glycerol can be converted into a series of metabolites with yields similar to the ones obtained from sugars, it is interesting to study this kind of fermentation. Some of these products with related biochemical intermediates are shown in Figure 20.2. The main species that can produce more reduced chemicals from glycerol includes *Citrobacter*, *Clostridium*, *Enterobacter*, *Klebsiella*, and *Lactobacillus*. Currently, the commercial availability of such fermentations is going toward recombinant species by overexpressing those pathways to optimize products. In this section the main products that can be achieved from glycerol fermentation procedures will be discussed, highlighting their importance in chemical industry and for Petrobras.

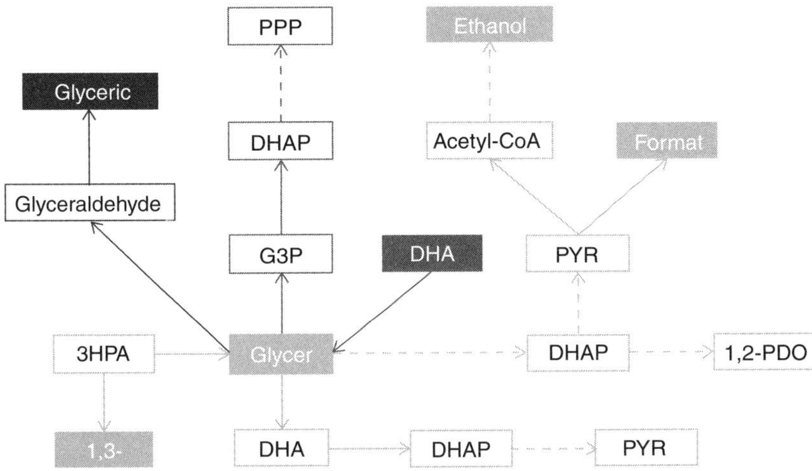

Figure 20.2 Main products from glycerol.

20.2.1 1,3-Propanediol

1,3-Propanediol (1–3 PDO) is one of the oldest known fermentation products and the most studied derivative of glycerol fermentation. However, its commercial production starting from glycerol as carbon source has not been reported yet. 1,3-PDO has several important applications, deserving special attention in its use as raw material for polytrimethylene terephthalate (PTT) production. PTT is a modern biodegradable polymer patented by Shell and DuPont for carpets and special textile fibers manufacturing. PTT can also be used as engineering thermoplastic. Polybutylene terephthalate (PBT) is another important derivative from 1–3 PDO. PBT main applications remains in automotive and electronic sectors.[7] Main bacteria strains that can produce 1,3-PDO from glycerol are *Lactobacillus sp.*, *Klebsiella sp.*, and *Clostridium sp.*[8] For the biosynthesis, there are two main biochemical pathways to 3-PDO, the oxidative one and the reductive one. The first route, which is an ATP-dependent route, passes through dihydroxyacetone (DHA) by dehydrogenation. The second (reductive) pathway undergoes a dehydration step to 3-hydroxypropionaldehyde (3HPA), which is then reduced to 3-PDO. Due to the low concentrations obtained from native producers by these pathways, genetic engineering has been widely studied aiming at overexpressing the enzymes involved in each step. This technique enables not only increased concentrations and productivity from native producers but also the manipulation of nonnative producers to recreate these pathways, for example, *Escherichia coli* and *Klebsiella pneumoniae*.[9]

20.2.2 2,3-Butanediol

2,3-Butanediol (BDO) is used to produce 1,3-butadiene (synthetic rubber production), methyl ethyl ketone (fuel additive), and diacetyl (flavoring). It can also be used in solvent preparations and as antifreeze agent. BDO is traditionally obtained from petroleum. In glycerol fermentation by *Klebsiella sp.*, which produces mainly 1–3 PDO, BDO represents a minor product along with acetate, lactate, succinate, and ethanol.[10]

20.2.3 Ethanol

Ethanol is mainly used as biofuel for transportation and can also be blended with gasoline. In fact, commercial gasoline in Brazil represents a blend containing up to 25% ethanol. Moreover, recently the Brazilian government has announced that the concentration of ethanol in the commercial gasoline may be increased, reaching 27% of blending, and thereby reducing gasoline imports.

Ethanol is traditionally produced from yeast fermentation of sugarcane sucrose and corn starch. Many bacteria from *Enterobacteriaceae* family are able to produce ethanol from glycerol, yet in low concentrations as a by-product. Production concentrations vary between 8 and 17 g/L.[11] *Enterobacter aerogenes* can also produce ethanol from biodiesel-derived crude glycerol at 0.40 g ethanol g glycerol^{-1}. The presence of higher concentrations of glycerol in the culture broth may inhibit fermentation.[12]

20.2.4 Butanol

Butanol has been produced by *Clostridium sp.* fermentation of sugars for years. The most known bioprocess for butanol production is ABE fermentation, which produces also acetone and ethanol. Butanol has been identified as an alternative fuel and also an important platform for other chemicals as acrylates, ethers, and butyl acetate. It can be produced simultaneously with 1,3-PDO from glycerol fermentation by *Clostridium pasteurianum*.[13]

20.2.5 DHA

DHA is mostly used in cosmetic industry as the main active substance in tanning products. Other applications includes the production of chemicals such as 1,2-propylene glycol, lactic acid, methotrexate, surfactants, etc. It is produced in a chemical route using glycerol and formaldehyde as feedstock. In the industry, *Gluconobacter oxydans* can produce DHA from glycerol fermentation. However, some problems may complicate the microbial DHA production. High glycerol concentrations and by-production of glyceric acid may reduce or inhibit DHA formation.[14]

20.2.6 GA

The glyceric acid main uses can be found in chemical and pharmaceutical industries as building block, making benefit of the fact that the molecule is highly functionalized, bearing three functional groups in its structure. Moreover, it is also used for the production of polymers and surfactants. GA can be biosynthesized from glycerol as carbon source by both *Acetobacter sp.* and *Gluconobacter sp.*[15]

20.2.7 LA

Its main application still remains in food industry, but it has several other applications as disinfectant and major component of intravenous fluids. Lactic acid can also be used to produce polylactic acid (PLA), acrylic acid, 1,2-propanediol, and polyurethane. Lactate esters are used as green solvents for coating and in the cleaning industry.[10] Traditionally, lactic acid is produced from sugars fermentation by *Lactobacillus sp.* bacteria. Alternatively, lactic acid can also be produced from glycerol as carbon source by a series of microorganisms such as *E. coli*, *Klebsiella*, *Clostridia*, *Bacillus*, etc. Due to low yields achieved with these natural strains, genetic engineering becomes essential to develop a competitive bioprocess.[10]

20.2.8 Succinic Acid

It has many applications in food, pharmaceutical, and cosmetic industry. It is an important building block to produce adipic acid, surfactants, green solvents, antibiotics, biopolymers, and vitamins. Natural succinate-producing rumen bacteria, such as *Anaerobiospirillum succiniciproducens*, can produce succinate from glycerol as carbon source. Due to the high complexity of this process in terms of medium composition and purification, engineered *E. coli* has been studied for this case.[16]

20.2.9 Citric Acid

It is an important commodity, having several applications specifically in food industry as preservative, acidifier, and antioxidant. Other important applications include cosmetics and washing detergents.[17] Citric acid is industrially produced from sugars fermentation by the filamentous fungi *Aspergillus niger*. There are also some yeasts capable of producing citric acid, including *Yarrowia lipolytica*, *Candida oleophila*, and *Candida guilliermondii*. Among these species, *Y. lipolytica* is the most promising for glycerol conversion to citric acid, giving maximal yield of product in this case. However, fermentation seems to be sensitive to impurities and also shows product inhibition problems.[18]

20.2.10 Oxalic Acid

Oxalic acid is mainly used as cleaning agent. Most of applications concern paper and detergent manufacture. Its conjugate base oxalate is used as a chelating agent. In industry it is produced by catalytic oxidation of carbohydrates. The production of oxalic acid from glycerol can be carried out via fermentation by *Aspergillus spp.*[19]

20.3 HYDROTREATING

HDT processes have been employed for decades in the petroleum refining industries with different purposes, such as fuel contaminant removal or upgrading of heavier streams. More recently, the increasing global demand for renewable fuels has put a spotlight on hydroconversion technologies. This occurred due to the possibility of generating green fuels – such as jet fuel – via hydroprocessing renewable oils and their mixtures with petroleum fractions.[20]

The traditional HDT processes employ tungsten/molybdenum sulfide catalysts supported on alumina and promoted with Ni and Co. The reactor is a trickle-bed type and the process is carried out at high temperature and pressure. There are several reactions taking place inside a HDT reactor. However, the most relevant are hydrodesulfurization (I, HDS), hydrodenitrogenation (II, HDN), hydrodemetallization (III, HDM), and hydrodeoxigenation (IV, HDO). These reactions involve the hydrogenolysis of carbon–heteroatom bond. Also, there are other important reactions, such as the hydrogenation of olefins (HO) and aromatics (HA) and hydrocracking.[20]

Despite the similarities between conventional HDT process and the hydroconversion of natural oil, there are many new challenges in processing or coprocessing renewable oils via HDT.

20.3.1 Coprocessing Renewable Oils in a Conventional HDT Fossil Diesel Unit

From the economics viewpoint, coprocessing triglyceride-based feedstock with petroleum fractions in existing refinery units is a relatively low-cost method to produce renewable fuels.[21] In this area, Petrobras has developed a new chemical route in which vegetable oils are coprocessed in an existing HDT facility, generating the so-called HBIO, a new green diesel.[22,23] Figure 20.3 shows the general coprocessing flow sheet. In a first step the natural oil is mixed with the petroleum fraction resulting in a stream containing up to 20% of renewable oil. After that, hydrogen is added to this stream to obtain the first reaction stream, which is heated and injected in the first catalytic bed at temperatures ranging from 220 to 350 °C. Due to the exothermicity of the reactions, a hydrogen stream is injected between beds. Following the reaction session, the stream is cooled down, gaseous products are separated, and the liquid hydrocarbons stream recuperated and rectified for dissolved gas removal, obtaining a hydrotreated product containing *n*-paraffins.

One of the major issues of coprocessing biomass oil in conventional HDT units is the high exothermal hydroconversion reactions of the triglyceride and the generation of water, CO, and CO_2 gases due to decarboxylation of fatty acids. Because of that, dilution of renewable oil with mineral stream should contain less than 20% of natural oil to be adequate for coprocessing in existing refinery HDT units.

As previously mentioned, sulfided bimetallic catalysts made of metal oxides (Ni–Mo, Co–Mo, and Ni–W) supported on alumina are the conventional HDT catalysts. The selectivity of natural oil HDT is very dependent on the catalyst

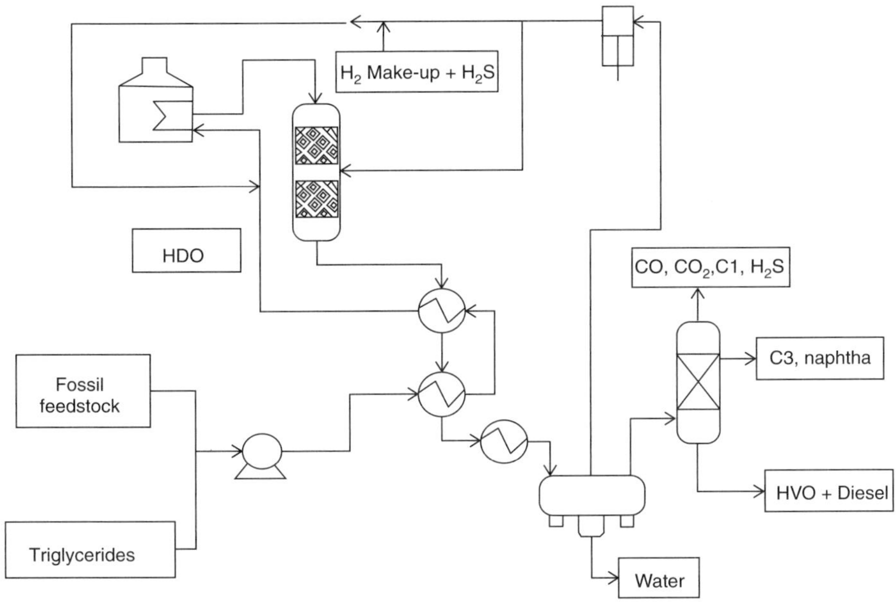

Figure 20.3 Coprocessing flow sheet. *Source*: Adapted from [24].

composition. The use of bimetallic catalyst containing metals from VIB and VIIIB results in high gas yield, mainly CO_2. In contrast, catalysts containing only VIB metals lead to a very low selectivity to gas yield and the production of *n*-paraffins, with the same carbon number of the oil's fatty acid.[24]

In conclusion, coprocessing triglyceride-based feedstock in petroleum refinery catalytic hydrotreating units is an advantageous alternative for producing diesel from renewable sources. Moreover, it is clear that the more important variables for coprocessing natural oil are catalysts characteristics and molecular structure of the feedstock components.

20.3.2 Hydroprocessing in Stand-Alone Mode

From the economic point of view, processing renewable oils in stand-alone mode may not seem as advantageous as coprocessing. On the other hand, renewable fuel products derived from this route are higher quality green fuels. Because of that, the aviation industry has chosen this process to fulfill the demand for biojet fuel in a short-term horizon.

Several and concurrent reactions occur during the hydroprocessing of triglyceride-based raw materials. These reactions can be classified as saturation of olefinic bonds, degradation of triglycerides, and hydrogenation of intermediates produced in this degradation. This reaction network was illustrated in the work by Kubica and Kaluza using rapeseed,[25] demonstrating that in the presence of hydrogen at elevated temperatures saturation/degradation of triglycerides occurs followed by decarboxylation, decarbonylation, and hydrogenation reactions to produce alkanes. In addition, the hydrogenolysis of fatty alcohols yields alkanes and fatty esters.

Besides conventional hydrogenation catalysts, such as Ni–Mo/Al_2O_3, Co–Mo/Al_2O_3, and Ni–W/Al_2O_3, many other catalytic systems have been proposed for hydroprocessing renewable oil, as shown in Table 20.1. In an interesting piece of work, Sousa et al.[37] used β-Mo_2C/Al_2O_3 as catalyst for HDT sunflower oil. They have not only shown that β-Mo_2C promotes the deoxygenation of free fatty acids forming alkanes with the same carbon number than the FFA but also that employing these catalysts decarbonylation and decarboxylation do not play an important role, contrary to what is commonly observed with conventional HDT catalysts.

From results reported in Table 20.1, it appears clear that there is a lot of work to be done in the development of better catalysts for natural oil HDT. However, existing systems already fulfill the requirements for industrial production and because of that stand-alone hydroconversion of natural oil is nowadays the best way to convert natural oils into truly renewable drop-in fuels.

20.4 DECARBOXYLATION

Indeed, the most common way to produce biofuels from vegetable oils nowadays is through the transesterification pathway. However, there are several disadvantages concerning this particular route such as engine compatibility, storage problems, and poor low temperature properties, among others. In a nutshell, the produced alkyl esters are not drop-in fuels. Therefore, novel catalytic routes are being studied in order to overcome these limitations.

As mentioned in the previous section, hydrotreatment (hydrodeoxygenation) is currently the best way to produce selectively long-chain alkanes derived from triglycerides/fatty acids.[26] Although having several unquestionable advantages, the hydrotreatment process has some drawbacks, mainly concerning the high consumption of H_2. Currently, this feedstock is primarily produced from fossil fuels by routes such as steam reforming of natural gas. Such routes, as previously mentioned, generate CO_2, a well-known greenhouse gas whose abatement is searched when biofuels are deployed. Moreover, H_2 has a high cost that may challenge economics of the process further.

In that context, decarboxylation reactions appear as a novel approach for producing clean, sustainable, and alternative biofuels. Decarboxylation removes the oxygen atoms in fatty acids as CO_2 rather than H_2O. Thus, no stoichiometric H_2 is required for the reaction.[36] In addition, the loss of a CO_2 molecule does not lead to a loss in any of the chemical energy in the molecule. In fact, it produces a hydrocarbon molecule with an even higher energy density.[36] A simplified scheme showing different reactions pathways starting from an ordinary triglyceride is depicted in Figure 20.4.

As can be seen in Figure 20.4, there are several different paths for the production of straight-chain hydrocarbons. Decarboxylation removes the carboxyl group producing carbon dioxide and a paraffinic hydrocarbon, as illustrated by Equation 20.1.

TABLE 20.1 Different Catalytic Systems for Hydroprocessing Vegetable Oils/Fatty Acids.

Reference	Catalyst	Oil/Fatty Acid	Temperature (°C)	Pressure	Reaction Time (h)	Liquid Phase	Gas Phase	Alkane Yield
[26]	5% Pt/C; 5% Pd/C	Palmitic acid	380	280 bar	±2	Water	Air	>80%
[27]	5% Pt/C	Stearic/palmitic/ lauric acid	330	–	2.5	Water	Air	>80%
[28]	5% Pt/C	Oleic/linoleic acid	330	–	2.5	Water	Air	<20%
[29]	1% Pd/carvão sintético	Palmitic + stearic acid	260-300	17.5 bar	5	Dodecane	5% H_2/ 95% Ar (vol)	47.5%
[30]	5% Pt/ nanotubo C	Stearic acid	330	–	1	Water	Air	78.7%
[30]	5% Pd/C	Stearic acid	330	–	0.5	Water	Air	29.9%
[30]	5% Ru/C	Stearic acid	330	–	0.5	Water	Air	18.1%
[31]	5% Pd/C	Stearic acid	300	17 bar	6	Dodecane	5% H_2/95% Ar (vol)	55.8%
[32]	0,3% Rh/ZrO_2	Heptanoic acid	250	80 bar	6	Water	H_2	6.9%
[32]	0,3% Rh/ZrO_2	Heptanoic acid	250	80 bar	6	Water	N_2	8.4%
[32]	0,3% Rh/ZrO_2	Methyl heptanoate	340	80 bar	6	Water	H_2	74.3%
[33]	5% Pd/C	Stearic acid	300	15 bar	1	6% wt Dodecane	H_2/Ar (1/9 vol)	90.4%
[33]	5% Pd/C	Oleic acid	300	15 bar	6	6% wt Dodecane	H_2/Ar (1/9 vol)	93.7%
[33]	5% Pd/C	Oleic acid	300	15 bar	6	6% wt Dodecane	H_2/Ar (1/1 vol)	95.1%
[33]	5% Pd/C	Linoleic acid	300	15 bar	6	6% wt Dodecane	H_2/Ar (1/9 vol)	61.7%
[33]	5% Pd/C	10% Stearic/ 65% oleic/ 25% linoleic Acid	300	15 bar	6	6% wt Dodecane	H_2/Ar (1/9 vol)	81.7%
[33]	5% Pd/C	"Brown grease"	300	15 bar	6	6% wt Dodecane	H_2/Ar (1/9 vol)	24.1%
[34]	5% Pd/C	Stearic acid	300	135 psi (9.3 bar)	1.5	Dodecane	10% H_2/N_2	96.0%
[34]	20% Ni/C	Tristearin	360	580 psi (40 bar)	6	Dodecane	10% H_2/N_3	66.0%
[35]	Pt/SAPO-11-Al_2O_3	Soy oil	380	30 atm	–	–	–	–
[36]	Pt/C	Jatropha oil	350	±17 MPa	1	Water	N_2	40.8% (C_{17})
[36]	NiC	Jatropha oil	350	±17 MPa	1	Water	N_2	54.5% (C_{1-4})

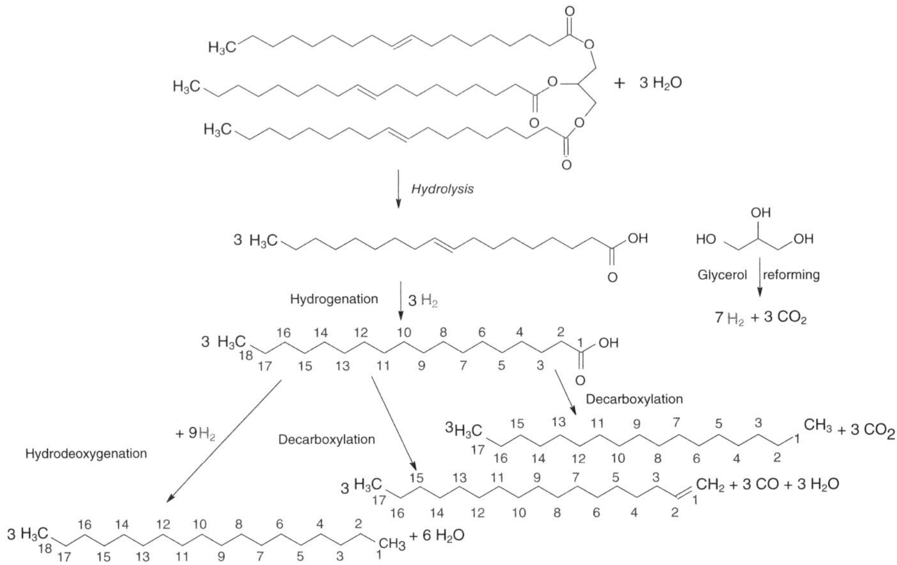

Figure 20.4 Reaction pathways starting from an ordinary triglyceride.

It is worth noticing that the produced hydrocarbon has one carbon atom less than the original fatty acid, and no H_2 is involved in the case of a saturated fatty acid.

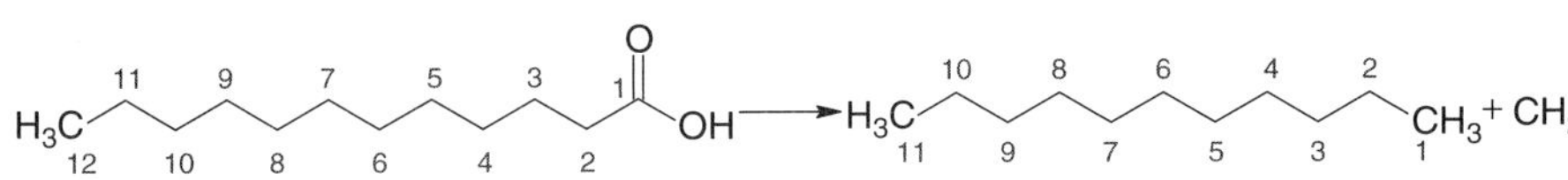

(20.1)

In what concerns decarbonylation, on the other hand, a terminal alkene is produced by removing the carbonyl group forming carbon monoxide and water, as depicted in Equation 20.2.

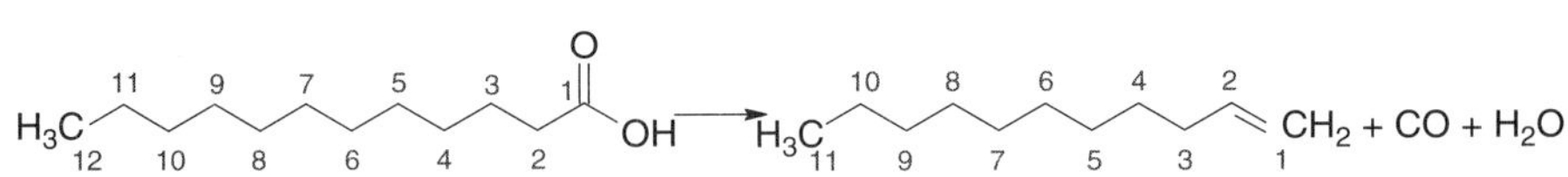

(20.2)

Nevertheless, CO can be rapidly converted into CO_2 by the water–gas shift (WGS) reaction and terminal alkenes can be readily hydrogenated,[27] as illustrated by Equations 20.3 and 20.4.

$$CO + H_2O \longrightarrow CO + H_2$$

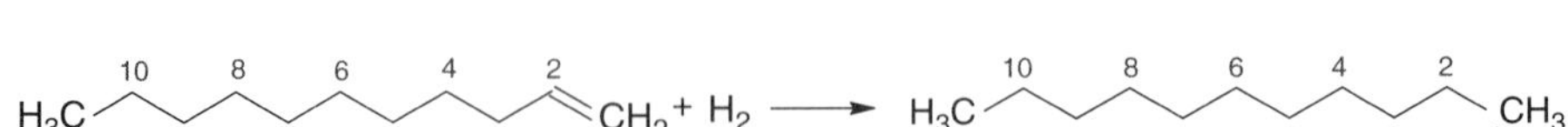

(20.3)

Several catalysts, such as supported Ni, Ru, Pd, Pt, Ir, Os, and Rh, were evaluated in the deoxygenation of fatty acids, and the most promising one is a carbon-supported palladium catalyst.[38] However, for these processes there are manifold new types of catalysts being studied, among which carbides are worth highlighting.

Fu et al.[39] proposed a two-step process for producing n-alkanes from low-value feedstocks such as cooking oils. The first one is hydrolysis of the triglycerides at elevated temperatures and pressures in aqueous medium producing free fatty acids. Posteriorly, deoxygenation via decarboxylation would occur in the same hydrothermal medium as the hydrolysis step producing fuel range hydrocarbons. The final products can be easily separated since organic compounds are not soluble in ambient water. Furthermore, no hydrogen is needed in neither hydrolysis nor decarboxylation process.[26]

Regarding the hydrolysis step, the produced glycerol may play a substantial role when dealing with unsaturated fatty acids. Glycerol, through hydrothermal catalytic reforming or aqueous phase reforming (APR), can be catalytically gasified producing H_2, facilitating the unsaturated fatty acids hydrogenation, thereby allowing subsequent decarboxylation, as illustrated in Equations 20.5 and 20.6.[27,36] In addition, glycerol removal from the reaction medium may also accelerate hydrolysis, which is restricted by the chemical equilibrium.[36] Traditional decarboxylation catalysts such as Pt/C and Pd/C are also active for glycerol gasification at the reaction temperatures (300–380 °C). In short, glycerol reforming contributes to the hastening of hydrolysis and *in situ* hydrogenation of unsaturated fatty acids, leading to more n-alkane products.[36]

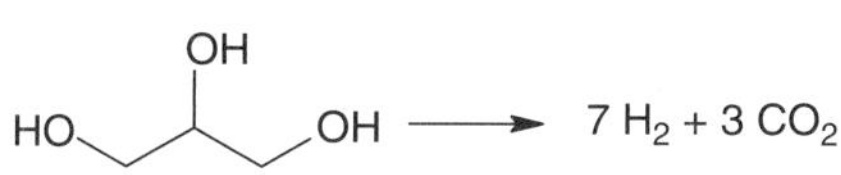

(20.4)

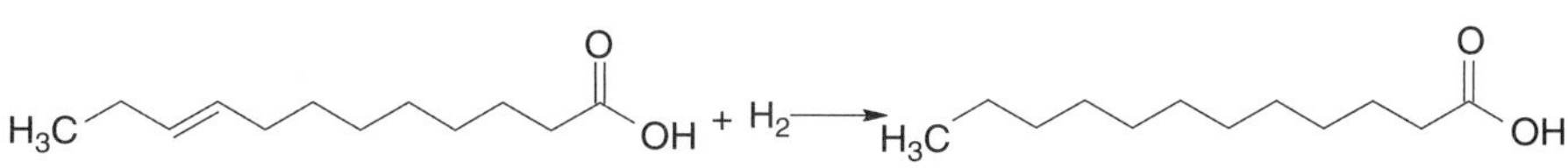

(20.5)

A key point that must be mentioned in this process is that the final product will be a function of the feedstock composition, that is to say, depending on the number of carbon atoms in the fatty acid chain, different types of fuel will be obtained. Long-chain fatty acids shall result in hydrocarbons in the diesel range whereas shorter fatty acids would produce jet fuel or gasoline range hydrocarbons.

Eventually, deoxygenation of vegetable oils via decarboxylation is a promising pathway to produce high quality, clean, and alternative drop-in biofuels. Although significant progress has already been done in this research area, several challenges still remain. In this sense it is worth mentioning the constant need for the development of more active and selective catalysts for the decarboxylation products.

20.5 CONCLUSIONS

As previously mentioned in this chapter, Petrobras is an energy company in which green chemistry is studied and developed, in order to cope with society's concerns regarding sustainability. Bearing that in mind, some important chemical routes developed using the concept of integrated biorefinery have been presented. As a case study, the aforementioned concept has been applied to an oil producing facility. Potential transformation of the oil and some by-products, such as glycerol, are discussed. A careful analysis of the routes presented indicates that indeed oleochemistry is an excellent example of how value-added products may be generated, thereby increasing the overall profitability of a green refinery. Conscious of the importance of the green philosophy, Petrobras is investing a considerable amount of resources in R&D activities aiming at developing innovative transformation routes from biomass.

REFERENCES

[1] C. Briens, J. Piskors, F. Berruti, *Int. J. Chem. React. Eng.* **2008**, *6*, R2.

[2] G. Huber, S. Iborra, A. Corma, *Chem. Rev.* **2006**, *106*, 4044–4098.

[3] D.J. Hayes, State of Play in the Biorefining Industry. **2007**. Available on http://www.carbolea.ul.ie/files/State_of_Play.pdf, accessed 2014.

[4] J.C. Erickson, *Int. Sugar J.* **2007**, *109*, 163–173.

[5] E. Falabella Sousa-Aguiar, N. Pereira Jr., A. Aranda, A.M.S. Antunes, "Biofuels and Biochemicals in Brazil", In "*Renewable Resources and Renewable Energy*" (Eds. P. Fornasiero, M. Grazianni) 2nd Edition CRC Press – Taylor & Francis Group, Boca Raton, **2012**, 391–455.

[6] E. Falabella Sousa-Aguiar, L.G. Appel, P. Zonetti, A.C. Fraga, A. Bicudo, I. Fonseca, *Cat.Today* **2014**, *234*, 13–23.

[7] NNDCC: Biochemical Opportunities in the United Kingdom. **2008**. www.nnfcc.co.uk/publications.

[8] S.S. Yazdani, R. González, *Curr. Opin. Biotech.* **2007**, *18*, 213–219.

[9] V. Kumar, S. Ashok, S. Park, *Biotechnol. Adv.* **2013**, *31*, 945–961.

[10] J.R.M. Almeida L.C.L. Fávaro, B.F. Quirino, *Biotechnol. Biofuels* **2012**, *5*, 48.

[11] K.K. Cheng, J.A. Zhang, D.H. Liu, Y. Sun, H.J. Liu, M.D. Yang, J.M. Xu, *Process Biochem.* **2007**, *42*, 740–744.

[12] T. Ito, Y. Nakashimada, K. Senba, T. Matsui, N. Nishio, *J. Biosci. Bioeng.* **2005**, *100*, 260–265.

[13] H. Biebl, *J. Ind. Microbiol. Biotechnol.* **2001**, *27*, 18–26.

[14] D. Hekmat, R. Bauer, J. Fricke,. *Bioprocess Biosyst. Eng.* **2003**, *26*, 109–116.

[15] H. Habe, T. Fukuoka, D. Kitamoto, K. Sakaki, *Appl. Microbiol. Biotechnol.* **2009**, *84*, 445–452.

[16] P.C. Lee, W.G. Lee, S.Y. Lee, H.N. Chang, *Biotechnol. Bioeng.* **2001**, *72*, 41–48.

[17] C.R. Soccol, L.P.S. Vandenberghe, C. Rodrigues, A. Pandey, *Food Technol. Biotechnol.* **2006**, *44*, 141–149.

[18] W. Rymowicz, A. Rywińska, B. Źarowska, P. Juszczyk, *Chem. Pap.* **2006**, *60*, 391–394.

[19] I. Musial, E. Cibis, W. Rymowicz, *Appl. Clay Sci.* **2011**, *52*, 277–284.

[20] M. Al-Sabawi, J. Chen, *Energy Fuel* **2012**, *26*, 5373–5399.

[21] T.V. Choudhary, C.B. Phillips, *Appl. Catal. A: Gen.* **2011**, *397*, 1–12.

[22] J.R. Gomes, US20060186020 A1, **2006**.

[23] J.R. Gomes, US Patent 2010/0270207, **2010**.

[24] J.R. Gomes, S. Bezergianni, J.L. Zotin, E.F. Sousa-Aguiar, "Biofuels generation by hydroconversion of vegetable oils", In "*Catalytic Hydrogenation for Biomass Conversion*", (Ed. R. Rinaldi). RSC, **2014**. Chapter 11, http://pubs.rsc.org/en/content/chapter/bk9781849738019-00204/978-1-84973-801-9#!divabstract.

[25] I. Kubickova, L. Kaluza, *Appl. Catal. A: Gen.* **2010**, *372*, 199–208.

[26] J. Fu, X. Lu, P.E. Savage, *Energy Environ. Sci.* **2010**, *3*, 311–317.

[27] J. Fu, X. Lu, P.E. Savage, *ChemSusChem* **2011**, *4*, 481–486.

[28] J. Fu, F. Shi, L.T. Thompson, X. Lu, P.E. Savage, *ACS Catal.* **2011**, *1*, 227–231.

[29] I. Simakova, O. Simakova, P. Mäki-Arvela, A. Simakov, M. Estrada, D.Y. Murzin, *Appl. Catal. A: Gen.* **2009**, *355*, 100–108.

[30] C. Yang, R. Nie, J. Fu, Z. Hou, X. Lu, *Bioresour. Technol.* **2013**, *146*, 569–573.

[31] I. Kubickovà, M. Snare, K. Eränen, P. Mäki-Arvela, D.Y. Murzin, *Catal. Today* **2005**, *106*, 197–200.

[32] Y. Bie, A. Gutierrez, T.R. Viljava, J.M. Kanervo, J. Lehtonen, *Ind. Eng. Chem. Res.* **2013**, *52*, 11544–11551.

[33] E. Sari, C. DiMaggio, M. Kim, S.O. Salley, K.Y. Simon Ng, *Ind. Eng. Chem. Res.* **2013**, *52*, 11527–11536.

[34] E. Santillan-Jimenez, T. Morgan, J. Lacny, S. Mohapatra, M. Crocker, *Fuel* **2013**, *103*, 1010–1017.

[35] M. Herskowitz, M.V. Landau, Y. Reizner, D. Berger, *Fuel* **2013**, *111*, 157–164.

[36] S. Idesh, S. Kudo, K. Norinaga, J.-I. Hayashi, *Energy Fuel* **2013**, *27*, 4796–4803.

[37] L.A. Sousa, J.L. Zotin, V. Teixeira da Silva, *Appl. Catal. A: Gen.* **2010**, *449*, 105–111.

[38] D.R. Vardon, B.K. Sharma, H. Jaramillo, D. Kim, J.K. Choe, P.N. Ciesielski, T.J. Strathmann, *Green Chem.* **2014**, *16*, 1507–1520.

[39] H. Bernas, K. Eranen, I. Simakova, A.-R. Leino, K. Kordás, J. Mylloyaja, P. Maki-Arvela, T. Salmi, D.Y. Murizni, *Fuel* **2010**, *89*, 2033–2039.

21

PHYTONIX: CYANOBACTERIA FOR BIOBASED PRODUCTION USING CO_2

BRUCE DANNENBERG[1], PETER LINDBLAD,[2] AND GARY ANDERSON[3]

[1]*Phytonix Corporation, Asheville, NC, USA*
[2]*Department of Chemistry – Angstrom Laboratory, Uppsala University, Uppsala, Sweden*
[3]*Department of Agricultural and Biosystems Engineering, South Dakota State University, Brookings, SD, USA*

Industrial Biorenewables: A Practical Viewpoint, First Edition. Edited by Pablo Domínguez de María.

21.1 BACKGROUND: THE COMING CO_2 ECONOMY AND CIRCULAR ECONOMY PRINCIPLES

The CO_2 economy is developing rapidly with research, technology, and new routes for the production and subsequent commercialization of renewable chemicals and biofuels. Carbon dioxide utilized for a carbon-based feedstock holds great promise to become a permanent solution for our civilization's crucial problems: solar power, CO_2 and water together could actually supply all of humanity with a future of abundant energy and raw materials while at the same time stabilizing anthropogenic, GHG-induced climate change and other effects of human activity on Earth's biophysical environments, biodiversity, and natural resources. According to this vision, CO_2 needs no longer to be seen as a waste product with dangerous environmental effects but increasingly as a valuable feedstock for chemicals, fuels, or polymers. This positive vision has been gaining momentum and is nowadays emerging from research laboratories and new commercial ventures as a realistic alternative path to securing the constant supply of carbon atoms for the industrial chemistry sector, continuing its need for their production cycles, even in a future world where fossil resources may be completely depleted. Assuming a planet with about 9 billion expected by 2030 (with increasing middle-class consumers as well), the challenge of meeting future's demand cannot be simply tackled by relying on efficiency gains. Therefore, systemic changes and new paradigms are needed, such as the circular economy, which may decouple economic growth from resource consumption, while providing innovation, job creation, and economic development.[1]

21.2 TECHNOLOGY FOR CYANOBACTERIA AND DIRECT PHOTOBIOLOGICAL PRODUCTION

Phytonix and its organism development partners are using a synthetic biology approach to develop efficient photosynthetic microbial cell factories for the direct generation of *n*-butanol – a well-known biofuel and renewable chemical – from solar energy, utilizing carbon dioxide as the sole, direct feedstock. There are no wild-type organisms existing in nature that photosynthetically/photobiologically produce butanol. There are, however, other organisms, such as several bacteria, that do that efficiently. In our minds, the production of biofuels (e.g., ethanol/butanol) via fermentation has fundamental environmental and sustainability flaws, as over one third of the biomass feedstock processed – feedstocks that in some cases are, or may be, in competition with food crops – converts into waste carbon dioxide rather than into the desired biofuel. Moreover, sourcing and processing the biomass also adds significant cost to the fermentation production process. Actually, the provision of an affordable, available, and sustainable carbon source has been one of the greatest barriers to the production of economically viable, affordable biofuels and renewable chemicals.

Phytonix's photobiological process uses carbon dioxide as the sole, nonbiomass feedstock and along with solar energy and water produces the desired biofuel/chemical with only oxygen and residual biomass as by-products. This is a significantly carbon-negative and sustainable process. The Phytonix technology/IP essentially takes the key genetic code/genes for a butanol synthesis pathway from nonphotosynthetic organisms and engineers them into the genome of a photosynthetic cyanobacteria. Thus, this enables butanol to be generated via a photosynthetic/photobiological "engine," using carbon dioxide as the sole feedstock/carbon source.

21.2.1 Cyanobacterial "Chemical Production Platforms"

Cyanobacteria are prokaryotic microorganisms – covering unicellular to filamentous ones – displaying the same type of photosynthesis as higher plants. Cyanobacteria are found in largely diverse and even extreme environments, in every biome on Earth, with significant fluctuations in temperature, salinity, pH, and water availability. That adaptation is based on their remarkable ability of using solar energy and atmospheric CO_2 as carbon source, combined with their faster growth rates when compared to other plants. Furthermore, from a practical perspective, an important asset is that their genetic engineering is relatively straightforward when compared to algae. All these advantages make cyanobacteria outstanding from all other organisms used in biotechnological applications.[2–4]

Native and genetically engineered cyanobacteria have been used by Phytonix's Swedish organism development team at the Angstrom Laboratory of Uppsala University as model systems to examine, demonstrate, and develop photobiological hydrogen production. More recently, the production of carbon-containing solar fuels like ethanol, butanol, and isoprene has been demonstrated. In that respect, Phytonix and its Swedish organism development team have recently completed a working cyanobacterial prototype that is now growing in liquid cultures, directly converting carbon dioxide into *n*-butanol photobiologically, with oxygen as the main by-product. This organism was engineered with a modified branch of the butyrate synthesis pathway. This prototype organism is suited for growing in temperate climates. As is typical with new transgenic organisms, continuing development activities need to be performed to "turn on" and optimize the function of the newly engineered synthesis pathway(s), such as overexpression and underexpression testing and comparison of native and nonnative enzymes, elimination of synthesis pathway enzymatic/catalytic bottlenecks, and testing and comparison of various gene deletions (knockout genes) and combinations thereof, in order to find the genetic pathway structure that initiates and maximizes *n*-butanol production. This optimization work will be ongoing through the development of commercial-scale organisms with the goal of eliminating all chemical synthesis pathway bottlenecks.

21.2.2 Cyanobacterial Solar Efficiency: The Actual Photosynthetic Efficiency

As stated in the previous sections, the Phytonix photosynthetic process takes place in cyanobacteria as aquatic-based (planktonic) photosynthetic microorganism. Phytonix's proprietary cyanobacteria are genetically modified photosynthetic organisms with several redundant biosafety features (see following text). As ultimate outcome, the bacteria effectively act as tiny microrefineries. Importantly, cyanobacteria are considered to be the most energy-efficient organisms existing on Earth. They are characterized by a favorable energy balance, with photosynthetic efficiency of up to five times that of green algae and other plants.[5]

Phytonix initially estimated an average of 15% actual photosynthetic efficiency ("APE") for its initial target host organisms. The APE assumption is dependent on a number of factors that must be taken into account for the ultimate calculations:

- The cyanobacterial (CB) antenna structure.
- The proportion of the cell surface that is antenna versus the portion that is not.
- The cell shape.
- The reaction rate of the photosynthesis.
- The amount of reflected light off of the PBR/phytoconverter.
- The behavior of photons in the blue energy range. Photons in the blue range (~400 nm) must loose energy down to the approximate energy content of photons in the red photon range (~ 700 nm). The degree to which the energy of the blue photons is lost as thermal energy, or is utilized for other cell functions, is uncertain.

Furthermore, researchers at Arizona State University have assumed that as much as 50% of the photons are not used by the cell. Phytonix considers this as being the percentage of photons either not impinging on an antenna structure so they cannot take part in photosynthesis or impinging on an antenna structure that is already active in photosynthesis, thereby making it so that the photon cannot be used in photosynthesis.

Following further analyses of the related scientific literature and the structure of our target host organism, Phytonix considers that a realistic estimate of APE for the initial Phytonix organism is 9.1–12.2%, dependent on the host organism. This may be retrieved as follows: researchers suggest that cyanobacteria can have maximum photosynthetic efficiencies as high as 37.5%. The general photosynthesis equation can be depicted as

$$CO_2 + H_2O + 8 \text{ photons} \rightarrow CH_2O + O_2$$

According to the equation, theoretically 8 photons would be needed to fix 1 mol of CO_2. However, some researchers have suggested using 10 photons due to inefficiencies of the light-harvesting system. The energy required to fix a mole of CO_2 as CH_2O is estimated to be between 470 and 496 kJ/mol, with an average value of 482.5 kJ/mol often reported by researchers. Taking these figures, it has to be considered that a 700 nm photon has 170.9 kJ of energy. Thus, to fix a mole of CO_2, $482.5\ \text{kJ/mol}/170.9\ \text{kJ/mol} = 2.8$ photons, that is, a minimum of 3 photons are required. Given the maximum efficiency, $3\ \text{photons}/8\ \text{photons} \times 100\% = 37.5\%$. A maximum photosynthetic efficiency of 28% assumes it takes 10 photons to fix a mole of CO_2. The cell metabolism is often assumed to be negligible. However, we question that assumption and some have suggested that it is as high as 35% of the photosynthetic energy intake. We assume that 35% of this energy intake is utilized by the cell and accounts for basic life processes, cell mass increase for cell division, and repair of damaged/worn systems like the antenna structure, leaving a conservative estimate of utilizable photosynthetic energy at 65%:

Phytonix's APE projection, given 28% cyanobacteria maximum efficiency would be$28\% \times 50\% \times 65\% = 9.1\%$.

Phytonix's APE projection, given 37.5% cyanobacteria maximum efficiency = $37.5\% \times 50\% \times 65\% = 12.2\%$.

21.2.3 The Emerging Field of Synthetic Biology

The field known as "synthetic biology" (SB) can be defined as the design and construction of new biological parts, devices, and systems, as well as the redesign of existing biological systems for useful purposes.[2,6] SB is part of a continuum in the evolution of recombinant DNA technology, with the ultimate aim of integrating the practice of biotechnology with the principles of engineering. It comprises technical fields like metabolic engineering, genomics, proteomics, and bioinformatics. Generally speaking, synthetic biology can be built over several basic principles[2]:

- Using standardized and well-characterized building blocks (e.g., BioBricks).
- Hierarchical design of nature-inspired, artificial genetic circuits and proteins *in silico*.
- Use of chemical DNA synthesis, enabling the production of nonnaturally found DNA sequences.[7]

An important aspect to be carefully assessed in SB is the rational design of artificial genetic circuits. Herein, the biochemical behavior of each individual part has to be well characterized separately. In concrete, promoters are key parts for the heterologous expression of proteins, and in cyanobacteria they have been characterized in a standardized experimental setup.[7] The ultimate goal of cyanobacterial synthetic biology would be to set up a repository of standard genetic parts to be used in the hosts that are shared by the cyanobacterial community. This can certainly accelerate the successful genetic engineering of cyanobacterial strains for the sustainable production of valuable biobased products like *n*-butanol and related isomers (in the particular case of Phytonix).

21.2.4 Gene Optimization

As stated previously, the use of cyanobacteria for the generation of nonnatural bulk commodities or added value products via photobiological routes will definitely require the efficient expression of an array of heterologous genes. These can be originated from any organism naturally displaying the desired metabolic capabilities and biochemical pathways. Obviously, some technical hurdles will need to be sorted out, as the gene expression in the cyanobacterial host will demand specific systems for propagation and/or integration within the host genome, together with the control of transcription and translation.[2,4] Yet, as commonly occurring in metabolic engineering challenging projects, it is not sufficient to merely introduce the gene sequences of interest under control of suitable promoters for cyanobacteria. Different organisms display varied preferences for codon use, and thus the optimization of the heterologous gene sequences to be adequately fit in the host organism will be necessary to enhance protein expression of the desired enzymes. Herein, codon optimization is carried out by comparing the gene sequence with the codon use preferences of the intended host organism and then changing the sequence of the gene to create a better fit for the host.[1] To carry out codon optimization, given the current routinely available DNA synthesis technology, the entire gene sequence may be straightforwardly reconstructed starting from the amino acid sequence of the encoded protein. This strategy allows for optimization of all codons to identify whose one(s) fit better with the host's preferences. Moreover, the approach may be used to minimize problems caused by strong secondary structures in the mRNA at the end of the gene. Furthermore, repetitive sequences are then avoided, and new features such as protein tags and other modifications may also be introduced.[8]

21.2.5 Direct Solar Fuel and Chemical Organism Development

Dr. Peter Lindblad, Phytonix's organism development project leader, and his senior scientific team at the Angstrom Laboratory, Uppsala University (Sweden), are using a standard synthetic biology approach to genetically design, engineer, construct, and analyze cyanobacteria for targeted and improved biofuel and biochemical production (*n*-butanol for Phytonix and related examples) via photobiology. The experimental design is followed by the construction, characterization, and analyses of the cyanobacterial cells. In a detailed form, it comprises the following steps[2]:

- Bioinformatic research for codon optimization and experimental design.
- Synthesis of some crucial DNA parts – promoters, ribosomal binding sites, and genes – to fit and be optimized for the specific cyanobacterial host.

[1]A codon is a sequence of three DNA or RNA nucleotides that corresponds with a specific amino acid or stop signal during protein synthesis. DNA and RNA molecules are written in a language of four nucleotides; meanwhile, the language of proteins includes 20 amino acids. Codons provide the key that allows these two languages to be translated into each other.

- Assembly of the synthesized genetic parts into the genetic vehicle.
- Insertion of the novel genetic constructs into the cyanobacterial host.
- Controlled cultivation of the recombinant microorganism in *ad hoc* specially designed photobioreactors.
- Analysis of the culture broth to assess biofuel/renewable chemical production performance.
- High-throughput data acquisition (and interpretation) from multiple levels, mainly transcriptome, proteome, metabolome, and bioinformatic network analyses. The aim is to create gene association, regulatory, and metabolic networks.
- Identification/targeting of important process steps to further improve biofuel and biochemical production titers. Also important is to identify bottlenecks and suggest modifications and optimizations accordingly.
- Based on the new generated knowledge through the previously described process, start a new optimized cycle.

21.2.6 Analysis of Modified Cells via Proteomics: Bottleneck Identification and Further Steps Suggestion

The intended strain construction and envisaged processes – cyanobacterial-based photobiological biofuel and biochemical production – are rather complex and challenging, thus requiring systems-level strategies to exploit the full capacity of these recombinant microorganisms in efficient photobiological production processes. By carefully integrating growth analysis and enzyme activity with quantitative mass spectrometry-based proteomics, Dr. Lindblad and team have identified a number of potential targets for future genetic engineering, aiming at improving the cyanobacterial-based production of *n*-butanol. A system-wide analysis, on proteome level, of cyanobacterial cells genetically engineered to produce *n*-butanol is an important step in the development process. By following that strategy, potential limitations on the engineered cyanobacteria in large-scale cultivations in phytoconverters – both at demo or pilot scale – from differing sources of carbon dioxide waste streams photobiologically converted into *n*-butanol, or during varying cultivation conditions, could be successfully identified at an early stage.[2] A conclusion of this study was that quantitative large-scale proteomics is necessary to successfully identify targets for directed genetic/physiological modification to optimize the system for a biotechnological approach such as the production of sustainable *n*-butanol.

21.2.7 Design, Engineering, and Construction of Cyanobacterial Cells for Biofuel Production

For cyanobacteria, to Phytonix's knowledge there is only one example whereby codon optimization of an entire gene sequence has been demonstrated to enhance the overexpression of a heterologous protein. In that case the plant enzyme isoprene synthase was overexpressed in the unicellular cyanobacterium *Synechocystis* PCC 6803, and it was found that when the codon usage of the gene was adapted to the preferences of the host bacterium, the overexpression of the protein was 10-fold increased, compared to when the original plant gene sequence was *simply* expressed in the same cyanobacterium.[2,9] With this proof of principle in hands, it is clear that the cyanobacterial approach may hold high promise in developing novel strains for different photobiological production purposes utilizing CO_2 as the sole carbon source (feedstock).

Using the previously described outlined strategy to design, engineer, construct, and analyze cyanobacteria for improved *n*-butanol production, Dr. Lindblad and his scientific team have successfully developed a solid platform to design and engineer efficient photosynthetic microbial cell factories for the future large-scale photobiological production of *n*-butanol in utilizing carbon dioxide as the sole, direct feedstock, with oxygen and residual cyanobacterial biomass as the sole by-products. All modified biofuel and biochemical producing organisms developed by Phytonix are proprietary. In 2010 Phytonix purchased the exclusive global license, with full sublicensing rights, to its core technology: "Designer organisms for photobiological butanol production from carbon dioxide and water." Additionally, follow-on IP/national phase patent applications have been filed for the photobiological production of pentanol, hexanol, heptanol, octanol, and other higher alcohols.[10,11]

Once these first strains have been successfully developed, further continuous development and optimization activities need to be performed to "turn on" and improve the function of the newly engineered synthesis pathway(s), such as overexpression and underexpression testing and comparison of native and nonnative enzymes, elimination of synthesis pathway enzymatic/catalytic bottlenecks, and testing and comparison of various gene deletions (knockout genes) and combinations thereof, in order to find the genetic pathway structure that initiates and maximizes *n*-butanol production. This optimization work will be ongoing through the development of commercial-scale organisms with the goal of eliminating all chemical synthesis pathway bottlenecks.

21.2.8 Phytonix: Biosafety-Guarded Technology

Phytonix is also working to be a leader in biosafety and environmental stewardship. Cyanobacteria are not inherently dangerous organisms, but when dealing with recombinant microorganisms, it is always appropriate to err on the side of caution and safety. In this respect, Phytonix's proprietary biosafety-guarded technology was created to protect ecosystem health. It uses several redundant mechanisms to ensure that our engineered bacteria – or any other designer microorganism – would not survive and reproduce in the natural world. To that end, the first line of defense is using enclosed tanks rather than open ones. In case of a leak or undesired rupture, the organisms would not be able to survive because they are engineered to grow in water containing 3–4% of CO_2, a concentration much higher than what exists in rivers, lakes, or the Earth's atmosphere. Remarkably, the two cyanobacteria already engineered are freshwater species that would not be able to survive in the ocean, an example of a natural, rather than an engineered, biosafety guard. Another safety guard is that once they achieve a certain density, the engineered cyanobacteria stop replicating.

These designer biosafety features are conferred by a number of technical, SB-based mechanisms, including:

- Inducible insertion of designer proton channels into the cytoplasm membrane to permanently disable any cell division and mating capability.
- Selective application of designer cell division cycle interference RNA (iRNA) to permanently inhibit the cell division cycle and preferably keep the cell at the G1 phase or G0 state.
- Innovative use of a high carbon dioxide requiring host photosynthetic CB organism for expression of the designer biofuel production pathway(s).
- Expression of a Nai1 promoter-controlled designer proton channel gene prompting the insertion of proton channels into the cytoplasm membrane when exposed to atmospheric nitrogen gas levels (N_2 makes up 78% of the ambient atmosphere), thus killing the cell.

Any single application or combination of these biosafety-guarded design features can be utilized within the modified cyanobacterial organisms depending on the cell culturing and carbon dioxide feedstock strategy utilized. Furthermore, our biosafety-guarded technology can be used for other transgenic and modified organisms and is not limited in its application to biofuel production.

21.3 PHYTONIX: PATH TOWARD FULL COMMERCIALIZATION OF THE TECHNOLOGY

Phytonix's full production process requires both an organism (*n*-butanol producing microbial cell factory) as well as the equipment to grow the organisms for the direct secretion of liquid *n*-butanol into the culture media and a back-end process for the separation of the butanol from the water-based culture solution. Phytonix calls its proprietary equipment phytoconverters, generically known in industry as photobioreactors. Phytonix and its scientific partner, South Dakota State University, continue to progress the design of its low-cost, scalable, and modular phytoconverter/photobioreactor technology and to assess potentially competitive and/or complimentary solutions. Two technically detailed phytoconverter™ (PBR) prototype design concepts have been completed to address differing environmental and resource availability criteria. One of them is a rigid, thin, modular panel/tank system that can be configured vertically up to three units high. This design could reduce required land usage by up to 75% but will reduce the yield per phytoconverter as well. The second design is a soft, "bag-like" system that is not stackable and would require more land. There is a trade-off with these two designs between the cost of the phytoconverter – soft, bag-like system is lower cost – and the amount and cost of land available for production. Testing is ongoing with lab-scale phytoconverters. Larger, bench-scale prototypes of these designs will be built as a next step.

21.3.1 Phytoconverter Design Approach

Our phytoconverters, being developed by Dr. Anderson's team at SDSU, are modular systems that will be piped together into compartmentalized arrays to produce butanol/biofuel in a continuous batch process – contrary to individual batch approaches that many photobioreactors use – and are specifically designed for culturing fuel-secreting cyanobacteria (which will have a lower ultimate biomass concentration upon "harvesting" than typical PBRs are designed for) and with downstream membrane separation–filtration–pervaporation systems to separate the butanol at low cost, as well as removing any biological contaminants or potentially toxic metabolites. The back-end butanol separation process will likely be performed at the end of each 7-day growth cycle. Up to 97% of the water is then recycled to continue the process.

Air loaded with CO_2 will be bubbled vertically through a porous membrane into the medium volume. The target concentration of solubilized carbon dioxide in the culture media is 2–3%. The bubbles will exchange CO_2, O_2, and other compounds between the medium in the phytoconverter. The gas bubbles also cause flow patterns to be developed in the medium volume that mixes the medium, minerals, and cyanobacteria. Bicarbonate will be added to the media to increase the concentration of soluble carbon dioxide. The porous membrane tends to allow air more readily through the center of the phytoconverter medium volume. The phytoconverter width and height will be adjusted with carbon dioxide flow rates to obtain optimum mixing and gas diffusion. We expect that with the design parameters of our phytoconverters, such as bubble size, velocity, retention time, and carbon dioxide concentration, along with factors like medium pH and temperature, the amount of carbon dioxide feedstock transferred into solution and converted to *n*-butanol will approach 60% of the total amount input. Generally speaking, 25–60% of the CO_2 can be removed from the air stream in one pass through a phytoconverter. If a higher CO_2 removal rate is desired, a multistage system may be appropriate. The air would first be passed through several algal cultures (phytoconverters) until the desired amount of CO_2 was removed from the air. Culture density may be less in the units the air is passed through after the first stage. The air passed through the phytoconverter may also be mixed with new incoming air and passed through a second time.

21.3.2 Nutrient Demand

The phytoconverter must provide an environment in which the cell can perform two basic functions: cell growth and production of the product (*n*-butanol). Cell growth will populate the phytoconverter to the desired cyanobacteria density in the phytoconverter and to maintain the cell population once the desired population is obtained; that is, cell growth equals cell death rate. Nutrients are provided to support both cell growth and product production. Product production requires a supply of photons, carbon dioxide, and water, while cell growth will have the additional requirement of macronutrients (nitrogen, phosphorus, and potassium) and micronutrients like iron, magnesium, and cobalt. The cell nutrient demand will be different for the cell growth and production nutrient streams in the phytoconverter.

The cell growth nutrient requirement may be met with commercial salts/fertilizers or from waste streams from other processes that the phytoconverter is colocated with. The supply of these nutrients from waste streams from other facilities will reduce the cost of operating the phytoconverter and possibly increase the efficiency of the cyanobacteria. Cyanobacteria that are supplied organic nitrogen in the form of nitrate and/or ammonia do not need to develop heterocysts to fix aerial nitrogen. The heterocysts for nitrogen fixation do not take part in photosynthesis, and it has been estimated that photosynthetic production can be increased 10% in some species.

21.3.3 The Photosynthetic Factory

The cyanobacterial cell is a miniature factory that requires inputs to be delivered to the production line as demanded. Production of the product requires that photons (light), carbon dioxide, and water be brought to the cell when and as required. Also, it is essential that oxygen be removed from the vicinity of the cell for efficient production. The cells consume carbon dioxide, produce oxygen and product, and harness light energy. As photosynthesis progresses, the carbon dioxide and oxygen concentrations change in the reactor volume.

Carbon dioxide is brought into the phytoconverter with air that removes oxygen and mixes phytoconverter reactor volume. The photosynthetic process needs light, and that light should be provided where carbon dioxide concentrations are high and oxygen concentrations in the reactor volume are low. The process is further complicated by the fact that the cyanobacteria not only absorb light for photosynthesis but reflect and scatter light blocking cells deeper in the reactor volume from receiving light. This effect can restrict the width (light path length) of the phytoconverter requiring large surface area to volume ratio that may be detrimental to the thermal environment in the reactor and reduce phytoconverter efficiency. Light guides that project light from the surface of the phytoconverter deeper into the reactor volume will allow for greater phytoconverter depths and allow for the use of excess light on the surface of the phytoconverter deeper in the reactor volume (Figures 21.1–21.3).

21.3.4 Potential Butanol Recovery Bottleneck

n-Butanol may inhibit the cyanobacteria growth and production when it reaches a limiting concentration in the phytoconverter. Product recovery is enhanced when the *n*-butanol is at a high concentration in the reactor water. Therefore, it is desirable to have an organism that can tolerate high concentrations of product in the water and still grow and produce efficiently. Organism tolerance to product may limit or increase energy demand for product recovery.

21.3.5 Purity and Availability of Acceptable Carbon Dioxide "Feedstock"

Phytonix considers that the carbon dioxide required by its processes does not need to be pure. Actually, the carbon dioxide may contain contaminants including carbon monoxide, hydrogen, and nitrogen. Conversely, carbon dioxide produced through the

Figure 21.1 Shading in a photobioreactor. (*See color plate section for the color representation of this figure.*)

combustion of coal would not be suitable as feedstock to Phytonix's processes due to the high amounts of nitrogen oxides, sulfur oxides, and/or heavy metals in the off-gases that would be toxic to Phytonix organisms. Nevertheless, Phytonix understands that there are many potential sources of carbon dioxide that could be used in its processes. Based on the variety and availability of low-cost waste carbon dioxide sources, Phytonix does not anticipate any significant amount of supply or pricing risk by using carbon dioxide as carbon source for fermentations. On this basis, Phytonix considers that the colocation of one of its plants with an industrial source of waste carbon dioxide is preferable to a stand-alone plant, for reasons that include capital efficiency and operational complexity. Phytonix believes that there are many industrial facilities, in multiple countries and in various climatic zones, which produce carbon dioxide that is compatible with the feedstock requirements of its processes.

21.3.6 Potential Phytoconverter Limitations and Bottlenecks

The scale-up to large outdoor phytoconverters will require multiple compartments and support material. Likewise, the phytoconverters will be arranged in discrete arrays of multiple units where the environmental parameters (carbon dioxide concentration and oxygen levels, pH, temperature, salinity) will need to be maintained in a uniform manner. To that end, sensors will be strategically located across these arrays

Figure 21.2 Bubble flow in a reactor volume that mixes the reactor contents.

to help maintain our economic/CAPEX targets (not redundant in each phytoconverter unit). Controlling and maintaining the various environmental parameters is a crucial area that will be addressed in our design parameter testing and optimization work. Several technical approaches are being evaluated for low-cost, high-yield separation of *n*-butanol from the culture media – the back-end process – including combined advanced membrane separation and pervaporation using water as the sole solvent for the polydimethylsiloxane (PDMS) membrane process component and cloud point separation/extraction utilizing recyclable nonionic surfactants/copolymers with phase separation of the condensate. There are also proprietary *n*-butanol separation systems that have been developed by other companies that could be sublicensed.[12]

21.3.7 The Phytonix Business Model

In order to attain the aforementioned goals of significantly reducing fossil-based production processes and extractive fossil fuel production, slowing and stabilizing the progression of anthropogenic climate change, and rapidly transitioning to a bioeconomy and ultimately to a (circular) CO_2-based economy, Phytonix has implemented strategic alliances with academic institutions and companies. By working with its

Figure 21.3 Porous membrane used to mix PBR with air.

partners through its capital-light, distributed business model, Phytonix can build on the strengths of each when it comes to biofuel and renewable chemicals development and commercialization. This distributed and collaborative business model enables the firm to achieve breakthrough development and realistic commercialization results and improved time to market with limited capital, a clear necessity in the currently capital-constrained environment for earlier stage, environmentally focused industrial biotechnology companies.

Phytonix's innovative business model has already demonstrated its value as it has enabled us to bridge the *technological valley of death*, the stage at which a firm moves from R&D/invention to prototype and/or proof of concept. In this respect, working in collaboration with world-class laboratories and scientists (who have developed a depth of expertise and "economies of learning" in the highly specialized fields of photobiologically based microbial chemistry and synthetic biology) has allowed us to develop prototype organisms and a protected, powerful, and global core IP portfolio at a much lower capital outlay than a traditional firm with in-house R&D teams and laboratories. Thus, the Phytonix team believes that we will be successfully moving through the *commercialization valley of death*, the stage at which a firm moves from prototype to pilot and subsequently to full commercial production, due to multiple factors:

- Our "capital-light" business model – enabling a high "progress per dollar" ratio for development and commercialization.
- An existing and growing global industrial commodity chemical market for butanol with established distribution channels and higher-value, market-based pricing.
- A biobased, cost-advantaged, highly competitive, and significantly carbon-negative butanol production process versus the incumbent fossil-based, energy, and GHG-intensive butanol production process that we intend to ultimately replace.
- The increasing urgency for the adoption of renewable energy and biobased products that will help to mitigate GHG-induced climate change and ocean acidification, substituting for environmentally destructive fossil fuel and chemical production. Presently an increasing amount of people – involving technology adopters, customers, and consumers – are realizing that there is no "Planet B" for us and future generations. Phytonix believes that the capital required to develop and fully commercialize its technology will be materially lower than that experienced or required by competitive technologies as a result of its distributed and collaborative development and commercialization approach and "economies of learning" by its scientific teams.

A relevant example of this may be the Phytonix's leverage of economies of learning for organism development and commercialization through the Angstrom Laboratories, Uppsala University, Sweden. The Angstrom Laboratory is among the world's leading research institutions for renewable chemical and solar fuel production utilizing photosynthetic microorganisms. For over a decade, the photobiology and microbial chemistry teams' main focus was the investigation and improvement of hydrogen evolution from sunlight and water by unicellular cyanobacteria and the development of molecular tools. More recently, the Angstrom team has focused on designing and engineering strategies and organism development initiatives utilizing cyanobacteria as carbon-based liquid chemical and biofuel production platforms such as Phytonix's "butanol organism."

Through collaborative partnerships such as this, Phytonix eliminates the need to build in-house laboratories, hire and train scientists, and work through the learning curve associated with in-house R&D. As a result of utilizing a distributed and collaborative business model, Phytonix expects that it can achieve commercial yield organism and phytoconverter (photobioreactor) development to pilot-scale and then to large demonstration-scale production for 20% of the total capital expenditures of its nearest competitor for the achievement of these same milestones.

Phytonix and its partners will use distributed approaches and commercial-scale installations to reduce toxins while using existing pipe-to-pump infrastructure to deliver and store biofuel and valuable industrial chemicals. Through industrial biotechnology and SB, Phytonix and its partners will optimize cyanobacterial species and growth conditions that can reduce greenhouse gas emissions significantly.

21.4 *n*-BUTANOL: A VALUABLE INDUSTRIAL CHEMICAL AND POTENTIAL "DROP-IN" GASOLINE REPLACEMENT

Normal butanol – also known as *n*-butanol – is a four-carbon, straight-chain alcohol. It is traditionally produced through a chemical process from propylene, a petroleum derivative. Butanol is currently used as an industrial solvent and is a major component in paints, coatings, adhesives, and household cleaning products. Butanol can also be used to run spark-ignition engines, as a fuel additive, or as a "drop-in" replacement fuel for gasoline-powered vehicles. In 2011 the global *n*-butanol market was estimated at \$5.9 billion, and this market is expected to triple from \$3.0 billion in 2005 to \$9.2 billion in 2015.[13]

The butanol market represents a small fraction of the size of the global fuel ethanol market that is estimated at \$87 billion in 2011. In 2010 China was the largest consumer of *n*-butanol, with 34% market share and over 1 million metric tons. China imported 600,000 metric tons (57% of its consumption) in 2010. China is forecasted to be a material net importer of *n*-butanol for the foreseeable future.[13]

Currently the price of *n*-butanol is highly correlated with that of propylene, a petroleum derivative. Since biobutanol is produced from a renewable feedstock, and not propylene, biobutanol could potentially achieve a significantly lower price than propylene-derived *n*-butanol. *n*-Butanol can be dehydrated to produce butenes that have many industrial uses in the production of plastics, fibers, rubber, and other polymers. Moreover, *n*-butanol can also be converted into other valuable chemicals such as 2-ethylhexanol and 1-butene. Compared to other alternative biofuels (e.g., ethanol), *n*-butanol offers several key advantages over alternative biofuels such as ethanol:

- *High energy: n*-Butanol has 91%+ energy content of gasoline and has 50% greater energy content than ethanol.
- *Current infrastructure: n*-Butanol can be transported with existing infrastructure, while ethanol must be shipped by rail, truck, or ship.
- *Safe fuel: n*-Butanol is 13.5 times less evaporative than gasoline and 6 times less evaporative than ethanol.
- *No maximum blend: n*-Butanol can be used as a 100% gasoline substitute in an unmodified engine, while ethanol is a blended additive to gasoline, and high ethanol blends require significant engine modifications.
- *Clean combustion:* Spark-ignition engine combustion in ambient air produces no carbon monoxide, no particulates, and low NO*x* and SO*x*. Complete combustion produces only carbon dioxide. Fuel life cycle from production through combustion is carbon neutral.

The global wholesale gasoline market is projected to be $900 billion USD in 2013. The size of the market increased threefold between 2000 and 2010, primarily as a result of the new demand from emerging economies such as China and India. As a gasoline additive, *n*-butanol blends well with ethanol and provides enhanced performance over ethanol alone. The potential wholesale fuel additive market for *n*-butanol is up to $150 billion/year in 2016. As a drop-in gasoline replacement fuel, the potential wholesale market, assuming all fossil gasoline were replaced by *n*-butanol (and potentially other higher alcohols such as *n*-pentanol), approaches $1 trillion in 2020 (according to Phytonix projections). Furthermore, there are also opportunities for *n*-butanol to serve as a feedstock for jet fuel. *n*-Butanol can access a > $40 billion USD per year for jet fuel feedstock. An assortment of technologies exist and continue to be developed that dehydrate *n*-butanol to 1-butene and then go on to polymerize 1-butene into a suitable jet fuel.[14]

21.4.1 Phytonix Projected Production Cost versus Incumbent Fossil Producers

Phytonix projects its target commercial-scale production cost at $1.35/gallon versus the incumbent fossil producer cost (BASF, DOW, etc.) that is in excess of $4.00/gallon and with rising expectations for the coming years (Figure 21.4).[15]

Butanol production is dominated by marginal costs. More than 70% of the current petroleum-based *n*-butanol cash cost is related to propylene pricing, with approximately 25% of the remaining costs variable items such as energy (high-temperature processes) and catalysts (e.g., cobalt and rhodium) (Figure 21.4). Phytonix considers that its total production cost is materially lower than the marginal cost of incumbent butanol production facilities. A further supportive external data source for incumbent producer costs is Cobalt Biofuels. In an April 2011 interview, Rick Wilson, Cobalt CEO, noted that "The cost of making petroleum-based butanol is around $1230 per

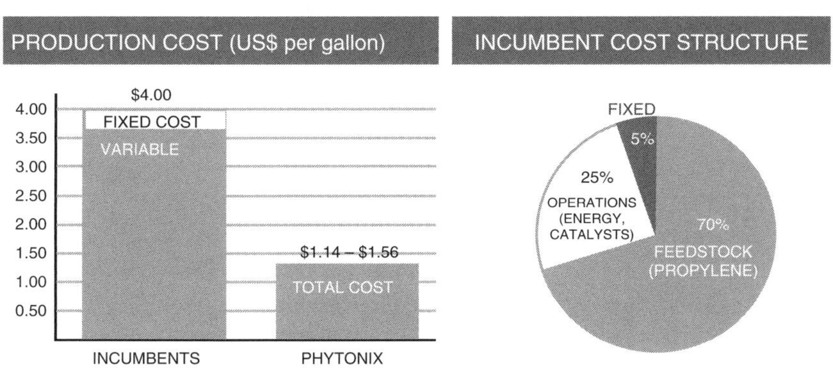

Figure 21.4 Production cost structure. Projected *n*-butanol production costs according to Phytonix estimations. See data in literature for comparison.[15]

metric ton." Given 2204.6 lb/t and 7.142 lb/gallon *n*-butanol, then $1230/t represents approximately $4/gallon.[16]

With regard to the GHG benefits of utilizing carbon dioxide as the direct feedstock, Phytonix's photobiological conversion process for the production of *n*-butanol and other higher alcohols from carbon dioxide is significantly carbon negative. We believe that commercial-scale production of our technology can contribute to stabilizing the effects of GHG-related climate change. Each gallon of *n*-butanol produced requires 17.86 lb of carbon dioxide feedstock to produce. The only by-products of this photosynthetic/photobiological production process are oxygen (17.54 lb of oxygen generated per gallon of *n*-butanol produced) and some residual and benign cyanobacterial biomass that could be used as fertilizer.

21.4.2 Carbon Dioxide: Phytonix Mass Balance and Other Technoeconomic Assessments

The overall mass balance concept of Phytonix is depicted in Figure 21.5.

At commercial scale, each 111.1 million gallons of *n*-butanol produced annually will result in the elimination of 1 megaton of carbon dioxide via bioconversion to *n*-butanol, oxygen, and residual biomass. Phytonix's current projections are for two to four 80 MGPY or 40 MGPY commercial-scale facilities, respectively, to come online between 2018 and 2025, which would create a net annual reduction in waste CO_2 emissions of 1.30 megatons per year by 2024/2025.

The Phytonix chemical reaction formula can be depicted as follows:

$$4CO_2 + 5H_2O + \text{organism} + \text{light} \rightarrow C_4H_9OH + 6O_2 + \text{organism}$$

Phytonix's understanding of reaction: 176 g of carbon dioxide plus 90 g of water plus organism plus light will yield 74 g of butanol plus 192 g of oxygen, neglecting cell mass growth and maintenance energy of the cell. Therefore, 1 gallon of *n*-butanol (3.0664 kg/gal) would require 7.2931 kg $((176 \times 3.0664)/74)$ of carbon dioxide.

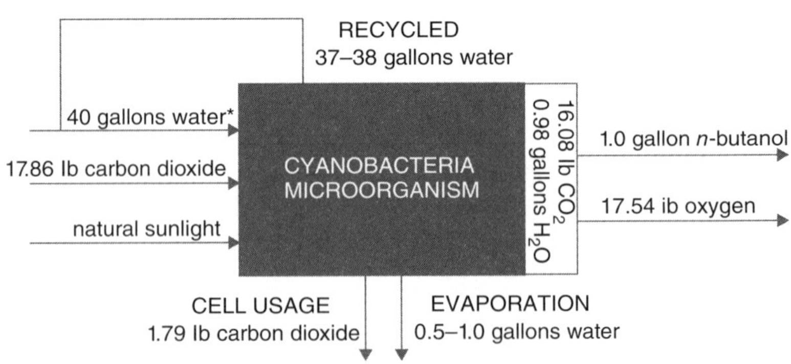

Figure 21.5 Mass balance for the Phytonix's *n*-butanol production.

The water required for one gallon of n-butanol is 3.7294 kg (8.2047 lb) or 0.9835 gallons of water ((90 × 3.0664)/74) and 3.7294 × 2.2 × 7.48/62.4). The oxygen yield is calculated to be: 192 g $O_2 \times 3.0664$ kg n-butanol/74 g n-butanol = 7.9561 kg O_2 7.9561 kg $O_2 \times 1000$ g O_2/kg O_2/453.6 g O_2/lb. O_2 = 17.5399 lb.

The assumptions for the inputs to Phytonix's intended production processes will be 17.86 lb carbon dioxide, being 16.08 lb for n-butanol production plus 1.79 lb for cellular usage, 40 gallons of water – fresh, brackish, or saline – being 0.98 gallons (8.2 lb) for the butanol synthesis reaction; 1–2 gallons for cellular usage and evaporation; and 37–38 gallons recycled back in the process, natural sunlight/photons.

Likewise, Phytonix's assumptions for the outputs from its intended production process will be 1.0 gallon of 100% n-butanol (6.76 lb), 17.54 lb oxygen and residual cyanobacterial biomass. 17.86 lb $CO_2 \times 40$ million gallons = 0.714 billion lbs. CO_2/2204.6 lb = 324.05 t CO_2 × 4 = 1.30 Mt/year of carbon dioxide eliminated (gross). Thus, Phytonix projects that its production process will generate 1.5 lb of carbon dioxide per gallon of n-butanol produced, offsetting the gross carbon dioxide reduction by 8.4%.

With regard to carbon dioxide feedstock cost, Phytonix technology is not dependent upon pure carbon dioxide feedstock, nor does it require zero contaminants. Phytonix assumed carbon capture cost is based on a 5% gas (feedstock) concentration. Our base case assumes carbon dioxide capture/feedstock costs of \$40/t, which we consider to be a reasonable and realistic carbon capture feedstock cost estimate based on its waste CO_2 sourcing strategy. Moreover, Phytonix assumes that production of one gallon of n-butanol will require 17.86 lb of carbon dioxide, of which 1.79 lb will be used by the microorganism culture and 16.08 lb will be directly used for n-butanol synthesis. Our base case CO_2 cost per gallon, assuming \$40/t for carbon dioxide feedstock, is 35 cents per gallon. By way of comparison, a company that is utilizing third-generation intracellular ethanol fermentation technology requires 16 lb of carbon dioxide per gallon of ethanol produced, resulting in a carbon dioxide cost per gallon, assuming \$40/t, of 32 cents per gallon. However, it must be noted that 1-butanol/n-butanol has a 49% greater energy density (in Btu's/gallon) than ethanol.

21.4.3 Carbon Dioxide: Colocation Options

With regard to carbon dioxide purity and availability, as stated previously, Phytonix considers that the carbon dioxide required by its processes does not need to be pure. The carbon dioxide may contain contaminants including carbon monoxide, hydrogen, and nitrogen. Though carbon dioxide produced through the combustion of coal would not be suitable as feedstock (due to high amounts of nitrogen oxides, sulfur oxides, and/or heavy metals in the off-gases), there are many other potential sources of carbon dioxide that could be used in its processes. In this regard, the colocation of the plants within an industrial source of waste carbon dioxide is preferable to a stand-alone plant, for reasons that include capital efficiency and operational complexity. There are many industrial facilities, in multiple countries and in various climatic zones, which produce carbon dioxide that is compatible with the feedstock requirements of its processes.

A series of illustrative examples are noted in the following. Phytonix has recently engaged in positive direct dialogue with one or more large corporate stakeholders in several of the industries that may offer sources of carbon dioxide:

- *Option 1: Industrial biofuels or renewable chemical fermentation facilities.* Examples of such sources are production facilities for the first-generation or second-generation production of ethanol, butanol, or pentanol. Phytonix understands up to 45% of the carbon from feedstock inputs, in related processes, is released as waste carbon dioxide emissions.
- *Option 2: Large breweries.* Beer and liquor breweries generate large volumes of carbon dioxide emissions. Phytonix production facilities could be colocated with large brewery operations to utilize their waste carbon dioxide for its feedstock. Many facilities of this type are located across North America and globally.
- *Option 3: Natural gas production facilities.* Extractive natural gas technologies can generate significant carbon dioxide emissions as a by-product to their day-to-day operations.
- *Option 4: Industrial cement manufacturing facilities.* Industrial cement manufacturing facilities generate significant volumes of carbon dioxide. Phytonix notes that industrial cement manufacturing facilities produce large amounts of carbon dioxide waste gases that are likely suitable for Phytonix's feedstock needs.
- *Option 5: Gas-fired production of electricity.* The off-gases generated by gas-fired power generation could provide waste carbon dioxide suitable for use as Phytonix's feedstock without any costly scrubbing or precleaning required. Phytonix could colocate its facilities with these power generation facilities. Many electric utilities are in the process of or planning to switch from coal to natural gas combustion for their electricity generation due to environmental concerns and the relatively low cost and high availability of natural gas. These types of facilities are numerous and globally located.

21.4.4 Carbon/GHG Life Cycle Analysis and Production Process Life Cycle Considerations and Implications

Phytonix has a dual market strategy. Initially, the target market is the $9 billion and growing industrial chemical *n*-butanol market. In the existing global *n*-butanol chemical market, we look to replace carbon and GHG-intensive petroleum-based *n*-butanol production output with Phytonix's carbon-negative biobutanol. Given sufficient production volume over time, we expect to penetrate and gain market share in the gasoline fuel additive and "drop-in" gasoline replacement market, potentially displacing both ethanol and gasoline.

Standard life cycle analysis (LCA) analysis of gasoline and biofuels typically only considers "well-to-pump" or "field-to-pump" analysis. Phytonix considers these incomplete LCAs as they do not include the full life cycle from sourcing the biomass feedstock (in Phytonix's case carbon dioxide is the feedstock) through combustion in the carbon-based liquid-fueled vehicle. We have attempted to bring the full life cycle into view with combustion included in our LCA analysis of biobutanol used as a biofuel: from the waste carbon dioxide emissions stream input to the bioconversion process through combustion and related emissions output, that is, "phytoconverter-through-combustion" (PTC) biofuel LCA. This is helpful for evaluating several scenarios and identifying potential opportunities for carbon intensity reductions across our fuel or chemical life cycle to ensure that carbon reductions are maximized. This is one benefit of conducting LCAs early in the design phase. There are opportunities to choose from various energy and processing technologies that will be used in our bioconversion process to manufacture cyanobacterial-based biobutanol. For instance, does it make sense to install cogeneration for electricity and steam if the region has low carbon intensity for grid electricity? If we were to utilize a cogeneration strategy to produce energy and heat, would we use woody biomass or locally sourced biomethane for Phytonix's back-end pervaporation or cloud point separation heating step and electricity needs?

In the Phytonix process there are a couple of production steps that will generate some amounts of carbon/GHGs:

- Small amounts of electricity will be needed for the on-site culturing facility and to run phytoconverter pumps and sensors.
- Any required heating or cooling of the culture media in the phytoconverters to maintain the organism's optimal temperature range.
- The back-end butanol separation process that will have a heating step involved in either a pervaporation process or a heating step involved to perform cloud point separation. Transportation and related emissions to get the product to the end-user market are also a consideration.

Depending on the location of the facility, some heating or cooling of the culture media within the phytoconverters may be required. Our process will heat or cool the culture media by adding either steam water or chill water into the water/media input piping of the phytoconverter arrays. It is important to note that Phytonix's organism and phytoconverter engineering approach and designs attempt to minimize heating and cooling requirements. Our organisms are designed from host organisms that naturally live in and can more easily tolerate and adapt to the climatic conditions and solar energy characteristics of the production facilities location.

Thus, Phytonix will generate electricity, heating, and/or cooling for our production facilities along a strategy based upon the climatic and resource availability characteristics of the facilities location. If our colocated carbon dioxide stream provides waste heat or there is waste steam available, we would plan on utilizing as much of this waste heat or steam heat as possible for our process. We would get any supplemental energy required for running the phytoconverter pumps, sensor equipment, and the back-end butanol separation process by employing a cogeneration strategy to produce supplemental electricity and heat on site. If the location had an abundance of woody biomass available for heat and power generation, we would use wood chips as the fuel source for an on-site cogeneration system to generate production process electricity and heat. The combustion of wood chips would emit some GHGs, but wood chip combustion is much less carbon intensive than that of other fuels. For example, per a recent study conducted by the BIOMASS Energy Center in the United Kingdom, approximate life cycle CO_2 emissions (carbon or carbon dioxide emissions from complete combustion) of wood chip fuel (including its production and transport chain) are 2 kg CO_2/GJ energy produced. This compares to 75 kg CO_2/GJ energy produced for the natural gas fuel life cycle, 90 kg CO_2/GJ energy produced for the LPG fuel life cycle, and 150 kg CO_2/GJ energy produced for the electricity generation life cycle based on the UK grid.[17] If we assume that our process energy requirements are 1/4 GJ for each gallon of biobutanol that Phytonix produces, our wood chip cogeneration fuel life cycle CO_2 emissions would be 1/2 kg CO_2 per gallon, or approximately a 1.1 lb. CO_2 carbon offset for production process energy generation to the 17.86 lb of CO_2 used as feedstock input into the bioconversion process, for a net CO_2 reduction of 16.76 lb carbon dioxide per gallon of *n*-butanol produced. To be conservative, we estimate a CO_2 offset of our production process at 1.5 lb, for a net CO_2 reduction of 16.36 lb carbon dioxide per gallon of *n*-butanol produced.

Our electrical and heating requirements are expected to be very low compared to the very GHG- and energy-intensive petroleum sourcing, extraction, and refining/cracking process required to produce the petrobutanol fossil feedstock propylene, as well as the GHG- and energy-intensive "oxo synthesis" production process for the production of petrobutanol from propylene, as described in the following sections. We also expect to be able to get some of the heat needed for culture media warming from the waste heat and steam generated from the colocated industrial process. We would utilize a state-of-the-art wood chip cogeneration combustion system to minimize GHG emissions for site electricity needs and the energy/heating required for the back-end butanol separation process. Our electrical and heating requirements are expected to be very low compared to the very GHG- and energy-intensive petroleum sourcing, extraction, and refining/cracking process required to produce the petrobutanol fossil feedstock propylene, as well as the GHG- and energy-intensive "oxo synthesis" production process for the production of petrobutanol from propylene.

Another important aspect is the land use requirements for the process. The land utilization of biofuel technologies varies materially. This is, in part, due to the comparative efficiencies of carbon conversion technologies. For example, in the production of corn ethanol then approximately half of the carbon is converted to ethanol but the remaining half is lost as carbon dioxide emissions.

To illustrate land utilization, were the United States to displace all motor gasoline with domestically produced biofuels, corn ethanol would require a land area multiple times the size of total corn crop acreage. Phytonix assumes that a commercial facility in 2019, with an annual production capacity of 25 million gallons *n*-butanol, would require approximately 600–1000 acres of land for biofuel production, depending on both the phytoconverter configuration and organism secretion yield. Phytonix assumes that its intended 2018 pilot facility, with an annual production capacity of 1.8 million gallons/year butanol, would require 150 acres of land.[18]

REFERENCES

[1] Circular Economy. World Economic Forum. http://www.weforum.org/issues/circular-economy, accessed Sept 2014.

[2] P. Lindblad, P. Lindberg, P. Oliveira, K. Stensjö, T. Heidorn, *AMBIO* **2012**, *41*, 163–168.

[3] S.A. Angermayr, K. Hellingwerf, P. Lindblad, M.J. Teixeira de Mattos, *Curr. Opin. Biotech.* **2009**, *20*, 257–263.

[4] T. Heidorn, D. Camsund, H.H. Huang, P. Lindberg, P. Oliveira, K. Stensjö, P. Lindblad, *Methods Enzymol.* **2011**, *497*, 540–579.

[5] L.R. Mur, O.M. Skulberg, H. Utkilen, "Cyanobacteria in the environment" (Chapter 2). In "*Toxic Cyanobacteria in Water: A Guide to their Public Health Consequences Monitoring and Management*", World Health Organization, **1999**.

[6] http://syntheticbiology.org, accessed Sept 2014.

[7] H.H. Huang, D. Camsund, P. Lindblad, T. Heidorn, *Nucl. Acids Res.* **2010**, *38*, 2577–2593.

[8] M. Welch, A. Villalobos, C. Gustafsson, J. Minshull, *Methods Enzymol.* **2011**, *498*, 43–66.

[9] P. Lindberg, S. Park, A. Melis, *Metab. Eng.* **2010**, *12*, 70–79.

[10] L.J. Weifu, WO 2009/105733, **2009**.

[11] L.J. Weifu, US8735651, **2014**.

[12] J. Huang, M.M. Meagher, *J. Membrane Sci.* **2001**, *192*, 231–242.

[13] *Prospectus, Biobutanol: The next big biofuel. A techno-economic and market evaluation.* Edited by Nexant, **2009**.

[14] http://elcriton.com/Markets.html, accessed Sept 2014.

[15] G. Duembgen, D. Neubauer, *Chem. Ing. Tech.* **1969**, *41*, 974–980.

[16] http://www.lowcarboneconomy.com/profile/green_power_conferences/_news_and_press_releases/a_2_fuel_vs_a_5_chemical/13982, accessed Sept 2014.

[17] http://www.biomassenergycentre.org.uk/portal/page?_pageid=75,163182&_dad=portal&_schema=PORTAL, accessed Sept 2014.

[18] http://www1.eere.energy.gov/bioenergy/pdfs/billion_ton_update.pdf, accessed Sept 2014.

22

PHYTOWELT GREEN TECHNOLOGIES: FERMENTATION PROCESSES AND PLANT BREEDING AS MODULES FOR ENHANCED BIOREFINERY SYSTEMS

PETER WELTERS, GUIDO JACH, KATRIN SCHULLEHNER, NADIA EVREMOVA, AND RENATE LUEHRS

Phytowelt Green Technologies GmbH, Nettetal, Germany

Industrial Biorenewables: A Practical Viewpoint, First Edition. Edited by Pablo Domínguez de María.

22.1 INTRODUCTION

The development of sustainable production systems, saving of energy and resources, and improvement of the carbon footprint clearly are very important issues in today's economies and societies. To address these topics it is imperative to find ways for integrating novel and innovative approaches and technologies into established ones. Their implementation, however, is not trivial and provides quite a challenge to currently existing technologies and even business models. In fact, the nationwide German initiative "*Energiewende*," which promotes the use of alternative energy sources to save resources and improve carbon footprints, already caused severe consequences for the power supply industry in Germany, due to increasing decentralized power generation by solar panels and wind turbines. Slow adaptation to the new situation is prone to cause heavy losses. For example, the balance of energy producer RWE in 2013 showed a loss for the first time in its history (local press media Handelsblatt, 05.03.2014). Nevertheless, in the long-term scenario, technology implementation and business adaptations are hardly avoidable due to the importance of sustainable energy production.

It is important to note that besides the already mentioned major alternative energy sources, light and wind, there is also an increasing interest to complement these sources with the use of renewable biomass (e.g., trees) for energy production. Norwegian power producer Vattenfall, for example, already exploits poplar plantations (short rotation coppice) for a biomass power plant in Germany. The overall planning for the plantation of poplars in Germany is expected to reach up to 100,000 ha of short rotation coppice on marginal soil. This development opens up new chances and opportunities ranging from the holistic exploitation of biomass and usage of existing waste streams to implementation of new businesses.

22.2 THE NEXT STEP: BEYOND ENERGY PRODUCTION

Savings of fossil resources and energy, as well as replacement of petrochemical feedstock by renewable alternatives were the major driving forces behind various individual approaches conducted by the industry and academia to develop feasible solutions during the last decade. In our view it is only logical to enter the next level without delay and combine gained knowledge and expertise on the various fields and disciplines to develop enhanced biorefineries driven by biorenewable resources. Although there is no strict definition, in analogy to current refineries producing fuels and feedstocks for chemical industry from petroleum, biorefineries are generally referred to as facilities housing processes to convert biomass into valuable goods such as fuel, energy (e.g., power, heat), and a spectrum of biobased products. Being a holistic approach biorefineries aim to take advantage of the various components in biomass and their intermediates to address the topics mentioned previously while maximizing the value derived. Of course, biomass used in such a biorefinery itself should be renewable. However, such a next-generation biorefinery has yet to be fully realized in a broad extent.

As basis for realization of a next-generation biorefinery, there is a need for novel concepts following holistic approaches combining energetic and material uses of biomass properly. With concept development being in progress, it instantly became obvious that this demands new technologies and tools broadening the range of disciplines involved. In this respect, plant breeding and fermentation processes are important modules for the development of biorefineries, which need to be evaluated and optimized in order to enable a range of new possibilities and applications beyond pure energetic use of renewable biomass.

In recent years Phytowelt Green Technologies became increasingly interested into biorefineries and put a strong focus on it. This led to the development of a new biorefinery concept coupling the breeding of poplar or rapeseed plants with their use for energy generation and production of materials. Briefly, this biorefinery concept combines cutting-edge breeding technologies for optimization of biomass yield and content of secondary metabolites with enzyme technology driving novel fermentation processes for production of high-value compounds, the extraction of high-value compounds already present in residues (see, for instance the grant of the German research ministry for the research project EcoTan), and biotransformation of extracted compounds to even higher-value products. More specifically, residues of poplar short rotation coppices (branches and bark) are planned to be used to extract valuable compounds, whereas cellulose and hemicelluloses converted to sugars will be used as feedstock for microorganisms in fermentation processes. Compounds to be produced include natural antioxidants – for example, carotenoids like astaxanthin and ß-carotene – or valuable fatty acids for polymer chemistry. For the first time, a concept was developed (Figure 22.1), which not only targets maximum exploitation of given biomass but also the generation and use of improved plant materials (e.g., higher biomass production per plant, improved metabolite composition, etc.).

This unique feature aims to implement an innovation circuit leading to constant optimization of the system. Moreover it will help to avoid possible shortcomings due to ever increasing biomass demands envisioned for the future. Additionally, a highly important feature of this biorefinery concept is its fully modular structure. Processing and production steps will be realized in the form of individual container systems with standardized interfaces. Scalability therefore is not restricted to the refinery as a whole, but conversely it can be applied at single steps on demand. In addition, such a modular system will also allow setting up a decentralized biorefinery thus reducing transportation costs and improving economics.

Being a spin-off of the Max Planck Institute for Plant Breeding Research (MPIPZ) in Cologne, Germany, Phytowelt Green Technologies has excellent skills and know-how in plant tissue culture and breeding and constantly works on the development of new breeding techniques (see in the following). Moreover, Phytowelt is equally well based in the state-of-the-art enzyme technologies. Its Phytozymes department is completely devoted to the identification, cloning, and use of (plant) enzymes in industrial biocatalysis and metabolic engineering aiming at creation of novel fermentative production systems and the improvement of existing ones. The gained expertise not only provides the basis for the biorefinery concept development but will also be equally important for its successful realization in combination with

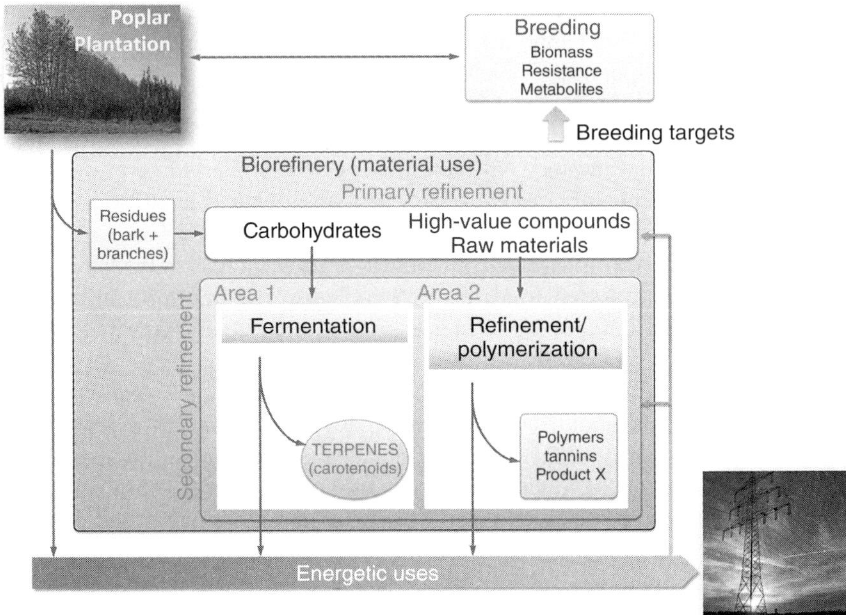

Figure 22.1 The biorefinery concept. A simplified graphical representation of Phytowelt's concept of a poplar-based biorefinery. Tree biomass produced in short rotation coppices serves a biomass source. Residues not suited for immediate energetic use like bark and branches will enter the primary refinement extracting carbohydrates, preformed high-value compounds, as well as raw materials. These intermediate products will serve as feedstock for fermentative production of high-value compounds such as carotenoids and for production of additional products for chemical or biochemical treatment. Wastes produced during all steps will again be used to generate energy, first used to cover the requirements of the biorefinery processes. While running the biorefinery desirable improvements of the input biomass will be revealed. This directly translates into new breeding targets to be realized using Phytowelt's time-lapse breeding techniques and closes the feedback loop.

other expertise related to process design and engineering, downstream processing, etc. Moreover, external expertise and collaborations will be of particular importance for the development of the envisioned container modules. An example of such a container is given by Evotrainer developed by Evonik Industries AG in the European research project Copiride, reflecting also the trend of shortened product cycles and shorter amortization time in the life cycle of a product.

22.3 MATERIAL USES OF RENEWABLE POPLAR BIOMASS

Already today poplar plants hold great promise and offer a number of interesting (potential) material uses. New technologies for the extraction and separation of natural sugar polymers are under development. These would enable the direct use of poplar-derived cellulosic polymers in fabrication processes and in particular, after being enhanced enzymatically to improve existing or develop new properties. Lignin with its high content of phenolic compounds might be another useful polymer out of poplar residues if the technology for its use can be optimized, and it would serve as a biobased source for aromatic chemicals. Besides useful biopolymers poplar also produces a number of valuable compounds of lower molecular weight, most abundantly in the bark. One prominent example is benzoic acid, which can be used as food preservatives or precursor for the synthesis of many other organic substances. Other compounds of interest are phytosterols (triterpenes), tannins, and flavonoids. Their usefulness and resulting economic potential is emphasized by the fact that bark extracts were used in former times as antiseptics against hemorrhages and wound infections.

These possibilities provide a first option to exploit poplar as a biorenewable resource and to combine energetic and material uses within a biorefinery, simply by implementing a module for extraction of valuable, previously deposited, plant-produced compounds prior to energy production (burning). In fact, branches and bark of poplar trees are poorly suited and are usually not used in energy production (due to increased ash production) and could be exploited immediately as feedstock in such approaches. After compound extraction remaining residues could still be used for energy production (for internal and external use).

Another promising process to be realized in an additional module is the use of these waste streams after breakdown of included polysaccharides as carbon source, thus allowing to couple fermentative production systems for high-value products with renewable plant resources. In this scenario, again leftovers can be used for bioenergy production.

22.4 FERMENTATIVE PRODUCTION OF HIGH-VALUE COMPOUNDS

Originally being defined as a yeast process converting sugar to carbon dioxide and alcohol, fermentation nowadays is generally used for biotechnological processes employing microbes to produce desired substances from organic materials. Today a wide range of compounds is already produced via microbial fermentations. Major advantages of fermentative production systems include easy host cell handling, ability to implement biosynthetic pathways through metabolic engineering, fast growth rates of the host cells, use of relatively cheap carbon sources (e.g., sugar), scalability, standardization abilities, and ease of downstream processing, as well as high product yields, purity, and quality. These factors directly translate into cost, making fermentative production very attractive. Thus, one of the best known examples is the production of riboflavin (vitamin B_2) by BASF. The complex chemical synthesis of vitamin B_2 requires eight steps and was replaced already in 1990 by a fermentative one step process using soybean oil and enzymes from the fungus *Ashbya gossypii*. This process had considerable advantages over the traditional, petrochemical method that led to a 95% reduction of waste, a 30% reduction of CO_2 emissions, and a 60% reduction of the total amount of resources required. In total, the costs for the production of vitamin B_2 could be reduced by 40% ("Biotechnology/Life Sciences in Baden-Württemberg Opportunities and Limitations of White Biotechnology in the Chemical Industry"). Nowadays, this system has been replaced by an even better fermentative production process in *Bacillus subtilis* developed by DSM.

22.4.1 Plantdustrial – Linking Plant and Industrial Biotechnology

The sustainability of chemical processes, especially those where chiral products are synthesized, can be increased by the use of fermentation processes or direct biocatalytic steps. Unlike most chemical reactions, biocatalytically driven reactions are highly stereoselective for the substrate. Due to the intrinsic properties of the involved enzymes and their activities, the resulting product has a particularly high substrate and reaction specificity (Figure 22.2). Moreover, in contrast to chemical synthesis enzymatic reactions may also work effectively even using raw materials of lower purity because of the precise selectivity of the enzyme for the substrate structures. Yet another advantage is the fact that natural enzyme-based processes operate at lower temperatures and produce less toxic waste and fewer emissions than complex conventional chemical processes,[1] for example, the conversion of elementary nitrogen to ammonium is done by plants in symbiosis with bacteria at room temperature and normal pressure (cf. nitrogen-fixing nodule in the background of Figure 22.2) whereas the chemical process demands high pressure and temperature and metal catalysts (Haber–Bosch process).

The symbiotic process of nitrogen fixation is one of the many examples where nature leads the way to a strong collaboration between plant and microbial processes solving challenging problems in an elegant way. This has been the trigger for our slogan "plantdustrial – linking plant and industrial biotechnology" and is intended to result in a strong interdisciplinary collaboration with a high synergistic potential, for example, to produce aromatic substances in a sustainable fermentation process.

22.4.1.1 Biocatalysts Availability of suitable biocatalysts is crucial for the construction of novel biosynthetic pathways and their successful implementation into the chosen host organism via metabolic engineering. It is most important to note that usage of plant enzymes and their biosynthetic abilities is not restricted to plant cells as host systems (e.g., as bioreactors). In fact, the exploitation of the plant gene pool and the use of plant enzymes for synthetic biology is one of the unique selling positions of Phytowelt Green Technologies. Through the targeted use of biotechnology and biodiversity, isolated plant genes and their optimized coding regions are employed to produce enzymes in microorganisms with specificities and stabilities tuned to particular industrial purposes.

To serve these needs Phytowelt Green Technologies utilizes the very rich diversity of plant enzymes through its gene identification and characterization platform phytomining. Plants produce a high number of compounds not directly involved in or essential for the primary metabolism of the cells. However, many of these natural products – the so-called secondary metabolites – commonly play important roles as signal compounds or defensive chemicals. The number of already described compounds exceeds 100,000, but the real number is certainly much higher as only a small number (20–30%) of the available plants have been analyzed yet. With more than 30,000 known substances, the terpenes form the biggest class of plant natural products. In fact, terpenes not only represent the largest class of natural compounds, but they also are a rich source for bioactive compounds of economic interest. Their biosynthesis is thoroughly investigated and Phytowelt has put a focus of its research on these plant secondary metabolites as well.

Plant secondary metabolites show a very broad range of structural variation and complexity. To be able to synthesize this huge biochemical portfolio, plants are well equipped with a very high number of enzymes and enzyme variants. They are outstandingly rich in oxygenases (mostly members of the cytochrome P450 monooxygenases but also dioxygenases) and terpene synthases. The latter are necessary and essential for the synthesis of the compounds of the different terpene classes forming multimers out of isoprene such as monoterpenes (e.g., menthol, see previous text),

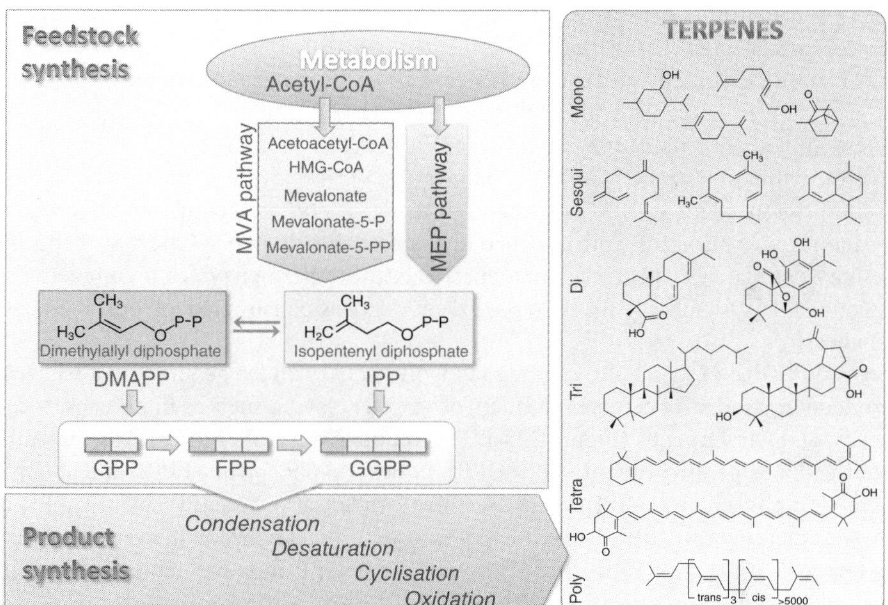

Figure 22.2 Schematic view of terpene biosynthesis. Essential feedstock for the synthesis of all terpenes are IPP and DMAPP and, depending on the terpene class, either the condensation products GPP, FPP, or GGPP. In most bacteria (like *E. coli*) IPP can only be synthesized via the MEP pathway and GGPP cannot be produced. The additional implementation of the MVA pathway increases IPP production capacities resulting in increased terpene yields. Final product formation requires a range of additional reactions such as condensations, desaturations, cyclisations, and oxidations. Examples of terpene structures belonging to the different classes are given.

sesquiterpenes (taxol, artemisinin), diterpenes (resins for glues), triterpenes (phytosterols, squalane), tetraterpenes (carotenoids, e.g., zeaxanthin and astaxanthin), and the polymeric forms known as natural rubber (Figure 22.2).

Due to the importance of terpenes, both scientifically and economically, Phytowelt follows several lines of work strongly focusing on different aspects of fermentative terpene production. Using its rich set of metabolic engineering tools (Figure 22.3) together with selected plant enzyme genes, Phytowelt has developed new bacterial platform production strains, optimizing the feedstock production for terpene synthesis. Irrespective of the organism all terpenes are produced via condensation of the same two precursors: isopentenyl pyrophosphate (IPP) and dimethylallyl diphosphate (DMAPP). For their biosynthesis two independent nonhomologous metabolic pathways are known: the methylerythritol phosphate (MEP) pathway in bacteria and photosynthetic eukaryotes and the mevalonate (MVA) pathway in eukaryotes and archaea. However, the MVA pathway is also found in a few bacteria. Higher plants use both pathways in parallel, though they take place in different cell compartments. The enzymes of the MVA pathway are located in the cytosol, whereas MEP pathway enzymes are found in the organelles.

The bacterium *Escherichia coli* is often used for metabolic engineering approaches, because its genome and metabolism is best understood. However *E. coli* is only capable of synthesizing IPP and DMAPP via the MEP pathway. Aiming at providing *E. coli* cells with additional IPP/DMAPP production capacities, we set out to implement a plant enzyme-based MVA pathway into this host. For each of the five biosynthetic steps of the MVA pathway, several enzyme variants were cloned from different plant species and analyzed using *E. coli* as recombinant expression system. Active enzymes were obtained in all cases showing up to 84× higher specific activity compared to bacterial analogues. Based on these enzymes a complete and functional MVA pathway for bacterial expression was constructed to boost precursor synthesis.[2]

Another line of metabolic engineering work focuses on the generation of bacterial production strains for terpene products of various classes such as diterpenes, triterpenes, and tetraterpenes (Figure 22.4). This requires the ability to build the essential intermediates geranyl diphosphate (GPP), farnesyl diphosphate (FPP), and geranylgeranyl diphosphate (GGPP) by consecutive condensations of IPP and DMAPP, as these compounds are the building blocks to form the backbones of the various classes of terpenes. As *E. coli* is not capable to synthesize GGPP, this enzymatic functionality was additionally implemented.

Currently, the fermentative production of tetraterpenes such as carotenoids is most advanced. Taking advantage of molecular tools and regulatory elements developed by Phytowelt, plant and bacterial genes were successfully combined in artificial operon constructs allowing simultaneous expression of suitable enzymes to perform the various condensation, desaturation, cyclization, and/or oxidation reactions required for product formation (Figure 22.2). Using these constructs high-level production of various carotenoids such as ß-carotene, lycopene, lutein, canthaxanthin, zeaxanthin, and astaxanthin in *E. coli* has been achieved (Figure 22.4a). Moreover, the combination

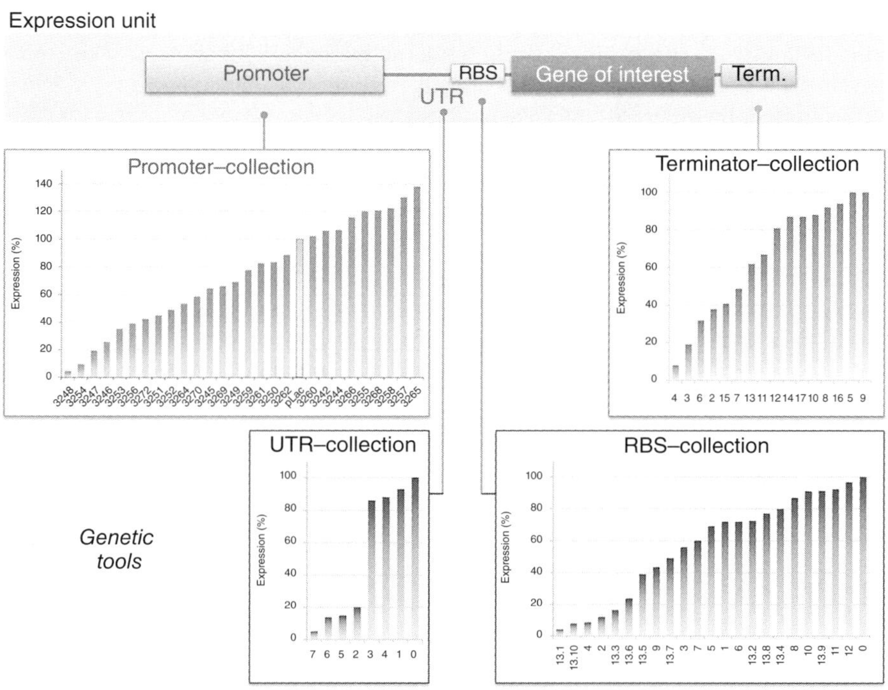

Figure 22.3 Tools for metabolic engineering of microbial hosts. Design and cloning of suitable individual expression units for a given gene of interest or operon-like multigene constructs require availability of various DNA elements such as promoters, ribosome binding site (RBS), and their flanking untranslated regions (UTR), as well as transcriptional termination signals (terminators). Phytowelt has generated collections of functional proprietary sequences for each of these elements as genetic toolbox for metabolic engineering.

of these operons with MVA pathway enzymes resulted in strains showing even higher carotenoid productivity (50–100% depending on the individual compound).

These achievements of Phytowelt not only provide an excellent basis for the implementation of fermentative production of high-value compounds such as carotenoids as a module within a biorefinery but also demonstrate the usefulness and potential of innovative technologies for the design and generation of eco-friendly production systems, holding great promise as part of a sustainable bioeconomy.

22.5 COOPERATIONS WITH CHEMICAL INDUSTRY

Today's chemical industry is still intensive in raw material and energy consumption. Renewable resources should be an increasingly important issue for the chemical industry over the coming years due to growing scarcity of fossil resources, steadily increasing oil price, and a need to reduce greenhouse gases. Germany, fourth in the world for production of chemicals, is due to take a leading role in these efforts.

Although the availability of cheap shale gas in the United States is currently causing some hesitation in (further) investing in alternative material resources, some major players such as Evonik increase their already high investments in R&D to find alternative production processes to replace mineral oil.[3,4]

Besides its usage in own product developments, Phytowelt offers its technologies as a service package to customers or in cooperations. For example, Phytowelt has been the source for the main enzymatic activity employed in a pilot process that Evonik has recently set up in its production plant in Slovakia. In such process, omega-amino carboxylic acid is produced from palm kernel oil.[5]

Another example is the use of plant oil as a renewable resource for the production of brassylic acid out of rape seed oil containing high amounts of erucic acid. This process was developed within the Project SynRg (granted by BMEL/FNR and coordinated by Phytowelt) aiming for the use of plant oils and polyphenols for polymer production. Phytowelt also acted as research partner in this project working on the identification, recombinant expression, and characterization of suitable plant enzyme activities to facilitate the production of long-chain dicarboxylic acids from fatty acids.[6] This is of particular interest for the wholesome use of plant oil-derived fatty

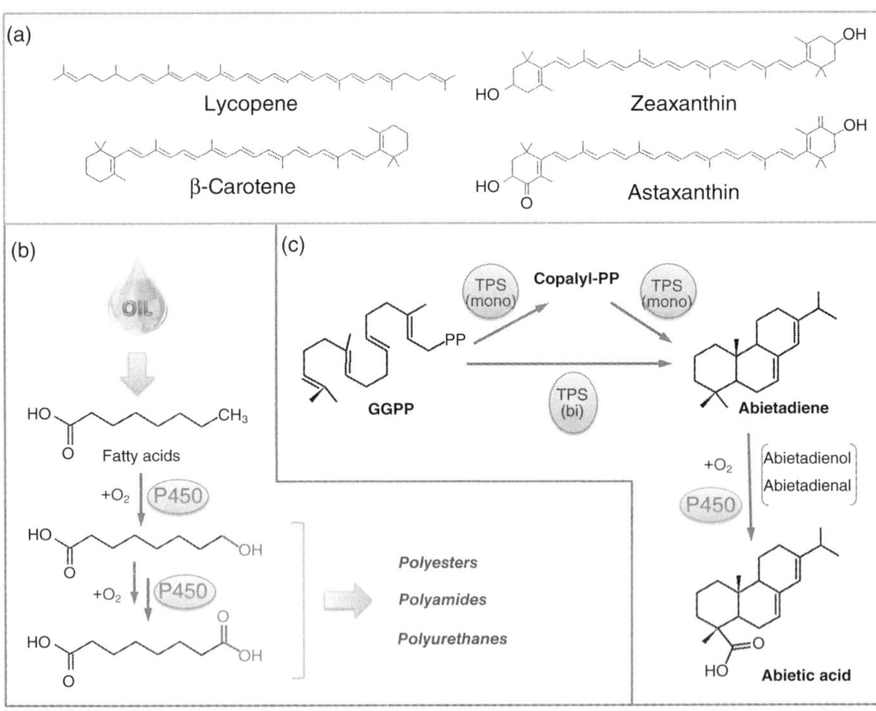

Figure 22.4 Fermentative production of high-value compounds. (a) Structures of some of the carotenoids produced by Phytowelt's metabolically engineered *E. coli* strains. (b) Omega oxidation of plant oil-derived fatty acids results in feedstocks for polymer chemistry. (c) Biosynthesis of diterpene compounds currently under investigation.

acids. Such enzyme activities would enable the production of dicarboxylic acids, omega-hydroxy carboxylic acids, and omega-amino carboxylic acids (Figure 22.4b) to serve as building blocks in polymer chemistry. Besides all oil-derived fatty acids, also breakdown products of fatty acid modifications could be used as substrate, as for example, pelargonic acid obtained from the ozonolysis of erucic or oleic acid. These reactions basically can be realized in two different ways. A one-enzyme approach driven by multifunctional cytochrome P450 monooxygenases (P450 enzymes) is capable of successively synthesizing fatty alcohols, fatty aldehydes, and diacids from fatty acids (Figure 22.4b). As an alternative to this, a cascade of three different highly specific enzymes can be implemented, each being responsible for individual oxidation steps. Though the latter approach appears to be more complex, it offers the benefit to allow for better reaction control, making it possible to stop the reaction at every desired oxidation state (Figure 22.4b). In any case, the enzymes belonging to the cytochrome P450 monooxygenases (P450 enzymes) play a major role to introduce the oxygen into the target molecules. However, P450 enzymes are not self-sufficient and ultimately require a cytochrome P450 reductase (CPR) as essential partner protein to be functional. Moreover, highest P450 enzyme activity can only be achieved upon proper choice of the CPR partner protein to maximize protein–protein interaction. Hence, the availability of suitable CPRs is key to success. Again using its phytomining technology, Phytowelt therefore isolated plant CPR enzymes and subsequently constructed a library of CPRs (Figure 22.5) modified and/or recombined for high-level functional expression in microbial cells (e.g., bacteria). This unique library enables the testing and identification of best-suited CPRs for a given P450 enzyme and, thereby, optimization of product yields leading to improved production strains for fermentation.

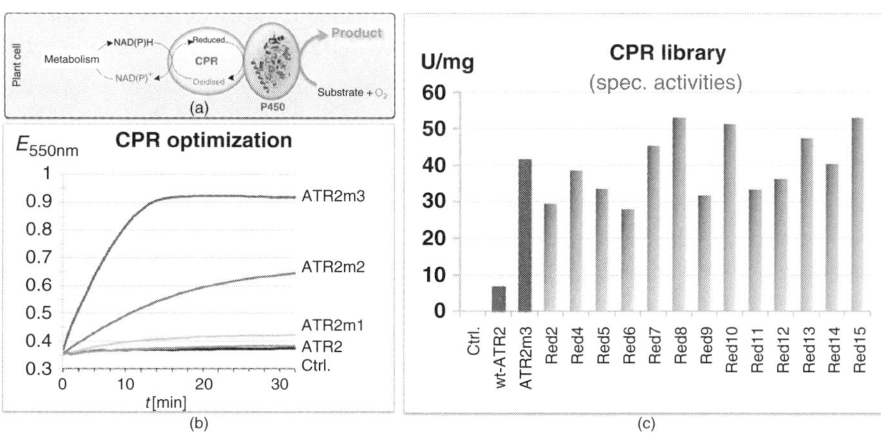

Figure 22.5 Cytochrome reductase (CPR) enables activity of P450 enzymes. (a) In the cell CPR mediates the transfer of electrons from cofactor NADPH to the P450 enzymes. (b) Expression and activity optimization of a CPR enzyme developed by Phytowelt. (c) Activities measured for a subset of Phytowelt's CPR library.

In the field of fermentative terpene production, Phytowelt also cooperates with the chemical industry, in particular for the production diterpenes for resin fabrication. In this regard, the BioHarz project (granted by FNR) aims at the development of a biotechnological/fermentative method for resin acid synthesis (Figure 22.4c). The work conducted includes identification, recombinant expression, and evaluation of suitable enzymes for the resin acid synthesis, with relevant enzyme classes being diterpene synthases (diTPS) and, again, specific cytochrome P450 monooxygenases. Suitable diTPS build the cyclic backbone of the resin acids in a two-step reaction process with copalyl diphosphate (CPP) as intermediate. Monofunctional diTPS can only specifically catalyze the second reaction step and are dependent on an additionally expressed copalyl pyrophosphate synthase (CPS). Bifunctional diTPS possess all protein domains to conduct both reactions on their own and CPP is not released. Subsequently a P450 enzyme is responsible for the oxidation of the build structure to result in the resin acid via alcohol and aldehyde intermediates (Figure 22.4c). Further project tasks address the implementation and testing of corresponding synthetic metabolic pathway in microbial hosts. Further tasks are yield optimization and evaluation of possibilities for a targeted synthesis of new product qualities through coexpression of further enzymes.

Furthermore, the production of triterpenes and derivatives in microorganisms, establishment of corresponding compound libraries, as well as library screening and identification of triterpenoids for medical applications are further goals of Phytowelt's research efforts. This is the topic of currently ongoing project conducted together by the group of Prof. Prüfer at the University of Münster, which has been also the partner in a project on enzymes for the production of polyterpenes (latex, caoutchouc). Likewise, drug testing and marketing will be done together with industrial partners.

All the described activities and processes developed for fermentative production of high-value compounds provide ideal modules to be coupled with the use of poplar as a biorenewable source for the carbohydrates to feed the compound producing microbes.

22.6 TOWARD OPTIMIZED BIORENEWABLES: TIME-LAPSE AND SMART BREEDING

In crop breeding, backcrossing processes are commonly used to generate new varieties via introducing and stabilizing desired agronomical traits. Yet, in tree breeding these processes are not applicable due to the long generation times of the plants, and thus hybrid breeding is used instead. In recent years Phytowelt developed a time-lapse breeding process based on its protoplast fusion technology. This started with a successful protoplast fusion-driven breeding program for poplar trees resulting in plant materials with increased biomass production in their juvenile phase (Figure 22.6).

This new process enables for the first time to perform backcrossing-like breeding procedures to trees or other plants with long generation times such as grapevines or nearly sterile plants like mint, banana, and *Miscanthus*. In addition, this technology is

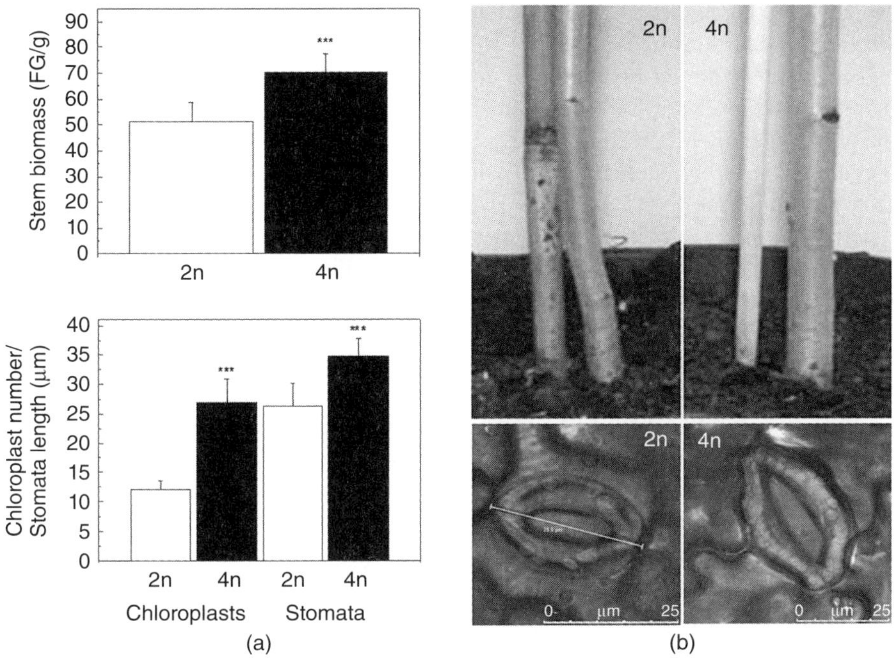

Figure 22.6 Enhanced biomass of tetraploid poplar plants derived from protoplast fusion. (a) New poplar line (tetraploid, $4n$) shows significant increases in stem biomass, chloroplast number, and stomata length compared to diploid ($2n$) poplar. (b) Corresponding pictures of the plant materials. *Source*: Pictures and data provided by Dr. T. Teichmann, University of Goettingen. (*See color plate section for the color representation of this figure.*)

suitable to increase the ploidy number (multiplication of chromosome sets) of a given target plant. The history of plant breeding proved that higher ploidy levels (tetraploid, hexaploid, etc.) are frequently coupled with improved growth behavior, increased yield, and high biomass production. As it can be rapidly envisaged, these properties may become also highly desirable for trees serving as feedstocks.

With respect to their further usage as feedstocks in a biorefinery, poplar trees are highly interesting because they can easily be grown in short rotation coppices, which immediately supports and fulfills the requirement of being a renewable resource thus ensuring sustainability. Moreover, poplars growing in short rotation coppice show the best efficiency in comparison to other energy plants such as triticale or rapeseed in terms of output to input energy ratio.[7] This is further supported by data showing that from the technoeconomic point of view, the use of hybrid poplar as feedstock is superior compared to wheat straw, corn stover, switchgrass, or loblolly pine.[8] In addition, latest evaluations show that ethanol production from poplar causes lower greenhouse gas emissions than corn, wheat, sugarcane, and *Miscanthus*.[9] Furthermore, depending on the species they even can be grown on marginal land. In fact, a number of the most important agronomical traits for the desired usage in short rotation coppice, such as drought tolerance, pathogen resistance, and biomass production, are already available in the poplar gene pool. Unfortunately, these traits occur dispersed over different poplar species and are not present in combination in a single variety/species. Hence available poplar plant species currently in use are far from being optimal, but there is a high potential for improvement by breeding. However, despite the fact that the poplar gene pool already offers various interesting traits, their recombination by breeding is severely hampered for mainly two reasons: (1) tree poplars have a very long juvenile stage (5–15 years) and, therefore, conventional breeding is extremely time consuming; (2) existing crossing barriers between poplar species belonging to different subgroups/sections make poplar breeding difficult. To overcome these problems Phytowelt developed an innovative breeding platform (*phytodiversity*) for new genome combinations employing their protoplast fusion technology resulting in a process best described as time-lapse breeding. Thus, *phytodiversity* enables the forest and agricultural industry to improve their raw material based on plants with long juvenile phase by a realistic breeding approach to generate results in reasonable time frames by shortening breeding cycles from 5–15 years to 1 year. This time-lapse breeding process offers the possibility to perform 1–2 "backfusions" per year. In combination with marker-assisted breeding, it opens up completely new opportunities and chances for breeding of trees. The technology, remarkably, is not restricted to plants with long generation times, and, of course, it can also be beneficial for breeding programs of important crops. For instance, in potatoes, it would also be possible to achieve 3–4 generations per year under ideal conditions. Using this *phytodiversity* platform Phytowelt already obtained several poplar lines showing an increased biomass production thus proving the suitability of this approach. The implementation of further important traits into poplar awaits its realization in future breeding work.

As already mentioned previously, the technology is not restricted to trees. An example for breeding of a nearly sterile plant species (only vegetative propagation of clones) is given by Phytowelt's protoplast fusion project with mint plants. Here, the profile of monoterpenes being produced, for example, carvone, menthol, and *iso*menthol, as well as eucalyptol, have been the goal for plant improvement as the development of a plant with a high eucalyptol oil profile could possibly lead to a replacement of eucalyptus-based aromatic oils by a mint-based European production. The technology was successfully applied to mint plants giving rise to numerous new mint lines with significantly altered metabolite profiles, including one showing a eucalyptol content of 40% in its essential oil. This emphasizes the power of protoplast fusion technology and demonstrates that not only basic agronomic traits but also biosynthetic pathways leading to valuable secondary plant metabolites can be modified and optimized.

22.6.1 Molecular Markers and Smart Breeding

Molecular markers have been established as indispensable and important tools to monitor and validate breeding approaches and to identify the best offspring of performed crosses long before the plants show the desired phenotype. Marker development and usage can be easily combined with the protoplast fusion technology and, in fact, is beneficial to gain fastest breeding progress.

The latest additions to the modern breeders' toolbox are the recently developed techniques TALEN and CRISPR, allowing improvement of plants via new smart breeding approaches without the need to introduce foreign DNA, thus resulting in nontransgenic plants lines. By identifying target genes and their respective promoters and by using these methods to specifically mutate them, one would be able to alter biosynthetic pathways with very high precision. Thus obtained enhanced production of secondary metabolites within the plant could make it economically feasible to produce and extract high-value compounds directly from the source instead of implementing and synthesizing them within a fermentative production system. Circumventing of (additional) fermentation step(s) improves the carbon footprint, energy demand, and efficiency of a given production process by up to 10-fold. Therefore the use of smart breeding technologies such as TALEN or CRISPR is recommended, whenever feasible.

22.7 NEXT-GENERATION POPLARS/PLANTS

In poplar breeding combining fundamental and important agronomical traits such as drought tolerance, pathogen resistance, and biomass production represents an obvious first step toward tailor-made renewable plant materials optimized for energetic or material uses as described previously. The medium-term use of such plants in the biorefinery will undoubtedly lead to the identification and definition of new desirable plant properties such as increased content of identified high-value compounds or improved wood quality (lignin content, energetic properties). This feedback will lead to the definition of novel breeding targets, which can be addressed employing the described time-lapse breeding technology. Thus the innovation circle aiming at continuous improvement of plant materials for combined energetic and material use is completed.

22.8 TOWARD NOVEL BIOREFINERIES: NETWORKING FOR SUCCESS

Phytowelt's envisioned biorefinery exploits poplar as biorenewable resource of energy, valuable plant metabolites, and carbon to drive fermentative microbial production systems for high-value compounds, as described. Moreover a feedback mechanism to poplar breeding is planned allowing to steadily improve the renewable plant resources used in the process. Phytowelt combines quite a number of expertise required for various aspects of the biorefinery such as expertise in poplar breeding, gene discovery, metabolic engineering and biocatalysis, development of microbial production strains, as well as in project planning and management. Nevertheless, the actual realization of the concept certainly requires a multidisciplinary approach, involving experts in biomass extraction, chemistry, downstream processing, and energy production to name only a few. However, cooperation only on the technical level is not sufficient to ensure that a critical mass is reached. Undoubtedly, the formation of diversified clusters with partners from academia, large international corporations, and SMEs are required and very important for a further exploitation of the hidden treasures of plants. Good examples for such kind of activities are the *Cl*uster *I*ndustrielle *B*iotechnologie *2021* (CLIB2021) and the European *P*ublic–*P*rivate *P*artnership on *B*io-*b*ased *I*ndustries (PPP BBI), being converted in 2014 into a joint undertaking (JU). These clusters with a broad array of technologies and the implementation of substantial information available from plants will enable partners to generate value over the whole supply chain, improving sustainability, and the impact of renewable resources, reducing cost and time to market. Phytowelt will participate in developments by its synergistic approach of microbial and plant science comparable to successful microbial–plant interactions in nature as in symbiotic nitrogen fixation of *Rhizobia* with *Fabaceae* (legumes). The cluster approach is crucial for the establishment of an increased number of biorefineries as intended by the BBI and a strong commitment of the European Commission (EC). Both are dedicating substantial amounts of money and resources to the implementation of new biorefineries (EC ~0.975 billion €, BBI ~2.7 billion €) through R&D, demo, and flagship projects over a period of 7 years until 2021. Phytowelt as technology provider is one of the few members of the BBI at the very beginning of the value chain as well as at the end by offering improvement of breeding technologies to improve biomass generation in quantity and quality in addition to discovery of new enzymatic activities and production strain optimization for the switch from fossil and high energy-demanding production processes to enzymatic reactions out of renewable resources.

REFERENCES

[1] H. Fischer, *Stoff-Wechsel: Auf dem Weg zu einer solaren Chemie für das 21. Jahrhundert.* Verlag Antje Kunstmann, München. **2012**.

[2] K. Schullehner, P. Welters, G. Jach, *BIOspektrum* **2013**, *19*, 214–215.

[3] Evonik startet Innovationsoffensive, available from: http://www.chemanager-online.com/news-opinions/unternehmen/evonik-startet-innovationsoffensive, **2013**.

[4] Evonik startet Innovationsoffensive | chemanager-online.com – Chemie und Life Science, available from: http://www.chemanager-online.com/news-opinions/unternehmen/evonik-startet-innovationsoffensive, **2013**.

[5] A. Karau, V. Sieber, T. Haas, H. Haeger, K. Grammann, B. Buehler, L. Blank, A. Schmid, G. Jach, B. Lalla, A. Mueller, K. Schullehner, P. Welters, T. Eggert, A. Weckbecker, US0100324257, **2009**.

[6] P. Welters. *BIOspektrum* **2012**, *5*, 550.

[7] L. Eltrop, A. Voß, R. Berger, K. Hein, M. Braun-Unkhoff, M. Aigner. *Konkurrenzanalyse der energetischen Nutzung von Biomasse*, available from: http://elib.dlr.de/58712/, **2005**.

[8] J. Han, S.M. Sen, D.M. Alonso, J.A. Dumesic, C.T. Maravelias, *Green Chem.* **2014**, *16*, 653.

[9] A. Azapagic, *Trends Biotechnol.* **2014**, *32*, 1–4.

23

BIORENEWABLES AT SHELL*: BIOFUELS

JEAN-PAUL LANGE, JOHAN WILLEM GOSSELINK, ROB LEE, EVERT VAN DER HEIDE, COLIN JOHN SCHAVERIEN, AND JOSEPH B. POWELL

Shell Technology Centre Amsterdam, Amsterdam, The Netherlands

Global energy demand could double between 2000 and 2050. Supply from lower-CO_2 energy sources, such as renewables and nuclear, is predicted to become a larger part of the energy mix in the future. The mobility/transport sector is a significant contributor

Industrial Biorenewables: A Practical Viewpoint, First Edition. Edited by Pablo Domínguez de María.

to global CO_2 emissions. New drive trains like the electric engine and hydrogen fuel cell have been introduced in the light-duty sector, but heavier-duty vehicle segments (heavy duty transport, marine vessels) and airplanes will continue to rely on liquid fuels in the long term. Commercially, the most realistic way to reduce transport fuel's carbon footprint is the blending with sustainable biofuels. Compared to conventional biofuels, the development of processes for advanced biofuels, derived from lignocellulosic biomass, is more complicated. Moreover the advanced biofuels should have characteristics enabling blending into crude oil-based transportation fuels up to levels beyond the current blend walls to allow for a smooth fuel transition. This double challenge should be tackled by developing new conversion processes in such a way that these biofuels have substantial greenhouse gas emission reductions qualifying them as advanced biofuels, are accepted in the society (for instance, by car drivers), are sustainable, and pose a sufficiently interesting commercial case for the biofuel manufacturers.

Shell was one of the first energy companies to invest in advanced biofuel research and has been active in this area for some 15 years. Our experience so far is that developing cost-effective production processes for advanced biofuels is clearly a challenging research area, both in terms of chemistry and process technology.

In the early years of this technical journey, many options were investigated. Via an active and continuous portfolio approach, they were narrowed down to a few that show economic potential. However, these choices still face significant challenges. Shell though remains committed to deliver economically viable processes for the production of advanced biofuels.

Shell has investigated various pathways from biomass to advanced biofuels:

1. Oxygenate platform (sugars and sugar derivatives): cellulosic ethanol, butanol/ABE, levulinic acid derivatives, and furfural derivatives.
2. Drop-in biofuels: algae to hydrocarbons, microbes producing hydrocarbons, aqueous thermal conversion, and Fischer–Tropsch synthesis.
3. Biomass coprocessing in crude oil refinery: via pyrolysis oil and (hydro)liquefaction oil.

The biggest challenge in the route from biomass to advanced biofuels via platform molecules, such as sugars and sugar derivatives, is the efficient production of these platform molecules. In developing process routes via intermediate biocrudes, the main challenge is to produce biocrudes with improved characteristics to avoid the need and cost of upgrading prior to coprocessing in existing refinery process units. Both conversion of platform molecules and biocrudes to fuels and fuel components are close to Shell core capabilities.

A lot of fundamental research will still be required to accelerate the development of the cost-effective production of advanced biofuels. Research in catalysis, process, and equipment engineering could focus on catalytic materials; catalytic deoxygenation; fermentation; separation; lignocellulose fractionation; handling of ash, solids, and lignin; and materials of construction, to mention a few.

23.1 INTRODUCTION

23.1.1 Energy and Mobility

Human society faces an overall energy challenge. Global energy demand could double between 2000 and 2050, as the number of people on our planet grows from 7 to 9 billion and economic development continues apace in non-OECD countries. This increasing prosperity drives increasing energy demand as societies industrialize, become mobile, and increase consumption. Yet the world needs to significantly reduce CO_2 emissions by 2050 to avoid the worst effects of climate change.[1] Whatever actions society will take to tackle climate change, they must be reconciled with population growth, economic growth, and extraordinary growth in demand for energy.

Supply from lower-CO_2 energy sources, such as renewables and nuclear, will become a larger part of the energy mix in the future. However, even with strong government support, it takes time for newer energy technologies to become affordable and available at scale. It takes 30 years to span the 10,000-fold growth needed to get from pilot-plant scale up to 1–2% of the world's total primary energy.[2] Therefore it is estimated that fossil fuels will still meet > 65% of world energy demand in 2050.

Mobility is fundamental for society (creating access to markets, jobs, education, etc.), enabling economic growth, trade, and social interaction and improving living standards. For oil companies, like Shell, mobility creates markets for their main products, transportation fuel, lubricants, bitumen, and petrochemicals for plastics production. The mobility/transport sector is a significant contributor to global CO_2 emissions. Demand is expected to grow rapidly as emerging nations continue through their most intensive phase of economic development. By 2050, the global car park could increase to 2 billion and the number of trucks could double. A more sustainable transport system must be further developed; otherwise transport CO_2 emissions could increase by up to 80% by 2050.

In their efforts to tackle CO_2 emissions, governments worldwide also target the transport sector. Policy makers in more than 65 countries have developed or are developing renewable fuel policies. These transport policies will drive the outcomes and influence the future fuel mix. New drive trains like the electric engine and hydrogen fuel cell have been introduced in the light-duty sector but are still facing many challenges in battery technology and roll out coordination between vehicle manufacturers, governments, and society. Heavier-duty vehicle segments (heavy duty transport, marine vessels, and airplanes) will continue relying on liquid fuels in the long term.

23.1.2 Biofuels

With liquid fuels remaining a dominant factor in the transport sector, the commercially most realistic way to reduce transport fuel's carbon footprint is the blending with sustainable biofuels. Sustainable biofuels start with the selection of sustainable feedstock that provides significant net CO_2 emission savings (even after considering eventual direct and indirect land use change (ILUC)) but also benefits local communities (e.g., in terms of food availability, economical benefit, labor conditions, human and property rights, etc.), preserve the quality of soil and water on/with which it is grown, and do not endanger local biodiversity.[3] However, sustainable biofuels also imply the use of manufacturing processes that are clean, safe, efficient, and affordable.[4]

23.2 SHELL AND BIOFUELS

Shell is one of the world's largest distributors and developer of biofuels. Capacity in biofuels is being built, which provides the best combination of performance and low "well-to-wheel" CO_2 performance from more sustainable feedstocks. Via the joint venture, Raizen, with the Brazilian firm Cosan, Shell is one of the largest manufacturers of ethanol. The JV was formed in 2011 for the production of ethanol, sugar, and power from sugarcane and the distribution and retail of transportation fuels in Brazil. The use of sugarcane ethanol, a conventional biofuel, results in significantly, up to ~70%, less greenhouse gas emissions than conventional crude oil-based gasoline.[5] Bagasse, a lignocellulosic by-product from sugarcane sugar/ethanol production, is used for power generation. The power is used for the production process and surplus electricity is sold to the grid. Furthermore, both filter cake (from filtration of cane juice) and ashes (from the boiler burning bagasse) are used in the fields as organic fertilizer. Total annual ethanol capacity is ~2.2 billion L (2014), which makes the JV one of the world's largest ethanol producers.

Shell champions sustainability standards in its biofuel supply chain. This is achieved by having systems, policies, and resources in place to assess potential sustainability risks in the biofuel supply chain, to implement controls, and to monitor and report progress. Since 2007, Shell has introduced environmental and social sustainability clauses into new and renewed contracts for the biofuels that it purchases for blending to ensure they are not knowingly linked to violation of human rights or cultivated, produced, or manufactured in areas of high biodiversity value. These clauses are used for all long-term contracts except where the product is already certified as sustainable by a robust third party (e.g., Bonsucro (sugarcane), the Roundtable on Responsible Soy, the Roundtable on Sustainable Palm Oil, and the Roundtable on Sustainable Biomaterials) or if the biofuel is certified sustainable by a European Commission-approved voluntary scheme and the feedstock was grown in the EU, Canada, or the United States. > 99% of the purchased biofuel volumes are either covered by these sustainability clauses or certified as sustainable by an independent auditor.

Shell also engages with industry, governments, intergovernmental agencies, and policy makers to encourage sustainability standards in the supply chain, raise standards, and improve practices across the feedstock industry. There is an active cooperation with environmental and social experts to develop projects that help address potential direct and indirect impacts of biofuel production. On issues as blend walls and ILUC, there is engagement with stakeholders to find workable solutions.

23.3 DEVELOPMENT OF ADVANCED BIOFUELS IN SHELL

23.3.1 Technical Introduction

The conversion of lignocellulose to transportation fuel typically proceeds via one of three major routes, namely, pyrolysis, gasification, or hydrolysis, all followed by upgrading (Figure 23.1).[3,6] Accordingly, lignocellulose is being converted to a crude liquid or gaseous stream and, subsequently, subjected to "deoxygenation" and rearrangement to a final fuel component. These three routes only represent archetype routes. Numerous variations and integration of these various steps have, and still are, being explored.

The processing is not limited to these two steps, however. The intermediate product stream often requires extensive cleaning to remove contaminants present in the biomass or eliminate undesired by-products of the cracking reaction. Moreover, the solid feedstock often requires a pretreatment that may consist of sizing, drying, predecomposition, etc. Finally, low-temperature processes may proceed in a liquid solvent that needs to be recycled through pumping, heating, cooling, recovery, and cleaning up. All these steps will contribute to the complexity and cost of lignocellulose conversion processes.

This chapter will briefly review the research programs that Shell has undertaken over the last 15 years on each of these process steps and variation or integration thereof. This includes:

- The hydrolysis to platform oxygenates with subsequent fermentation or chemical conversion to hydrocarbons and oxygenates.
- The pyrolysis or liquefaction to biocrude with subsequent upgrading using typical oil-refining processes.
- The gasification and subsequent Fischer–Tropsch synthesis.

Moreover, this chapter will also review research carried out on alternative biomass, namely, algae.

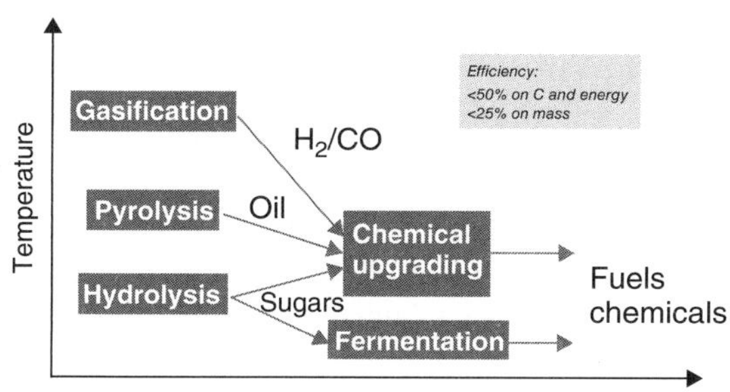

Figure 23.1 Routes for converting lignocellulose to fuels and chemicals. *Source*: Adapted from Ref. [3]. Reproduced with permission of Wiley.

23.3.2 Portfolio Approach

Shell was one of the first energy companies to invest in advanced biofuels and continues to innovate via dedicated teams, research agreements with leading academic institutes, participation in public–private cooperation, and JV partnerships. A portfolio approach to advanced biofuels is being applied, aiming to concentrate resources on the most promising competitive technology pathways.

Compared to conventional biofuels, the development of processes for advanced biofuels is more complicated. Lignocellulosic biomass, such as crop wastes or inedible crops (used so as not to compete with food), for example, corn stover, wood residues, and energy grasses, are more resistant to conversion than sugars and starch. Moreover the advanced biofuels should have characteristics enabling blending into the crude oil-based transportation fuels up to levels beyond the current blend walls. This implies that the molecules in advanced biofuels should have chemical structures similar to their crude oil-based analogues. This double challenge should be tackled by developing new conversion processes in such a way that these advanced biofuels have greenhouse emission reduction performance that would qualify them as advanced biofuels, are accepted in society (for instance, by car drivers), and pose a sufficient interesting commercial case for the biofuel manufacturers.

For more than a decade, Shell's laboratories have carried out research on advanced biofuels. In the early years of this technical journey, many options were investigated, such as biofuels derived from algae and microbes, cellulosic ethanol, and biofuel components derived from levulinic acid and furfural. As part of ongoing portfolio analyses, the various options are compared on a regular basis and decisions made to concentrate on the most promising ones. Such decisions are made not only on technical grounds (will the technology actually work and deliver on its promise?) but also on economic (will the biofuels be delivered at an acceptable price?), commercial (how does one share the risk, what is competition doing?), organizational (is the organization capable of successfully progressing and delivering the project?), and political (can governments create and support the market conditions required for the development of a sustainable biofuel market?) grounds. Today, the number of choices has been narrowed down to a few, which are best aligned with Shell's competences. However, these choices still face significant challenges.

23.3.3 Oxygenate Platform

23.3.3.1 Cellulosic Ethanol (Incl. Biomass Pretreatment) Ethanol is clearly the most popular biofuel today. Its global production of 73.4 Mt/a (2014)[7] is mainly produced by fermenting edible sugars extracted from cane or corn. The overall reaction shows theoretical efficiencies of 51 wt%, 67 C%, and 97% LHV when expressed in terms of mass, carbon, or energy retained in ethanol relative to monomeric sugar (Eq. 23.1)[3, 6]:

$$C_6H_{12}O_6 \rightarrow 2\ C_2H_5OH + 2\ CO_2 \tag{23.1}$$

The sugars contained in lignocellulosic feedstock are readily recognized as being promising feedstock for ethanol production. However, the sugars are not easily fermentable because they are present in polymeric form in the cellulose and are further inaccessible behind a layer of hemicellulose and lignin. Moreover, lignocellulose contains some 20 wt% of hemicellulose that generally contains C5 sugars, which are not readily fermentable by conventional ethanologen yeast. Consequently, the production of cellulosic ethanol requires the development of dedicated technologies to access and depolymerize the various sugars and, ideally, ferment all of them to ethanol.

Costs of lignocellulosic feedstock are generally low relative to corn and cane sugars, while intensification continues to increase yields per area. This provides an excellent starting point for conversion to transport fuels, competing with bulk applications like paper, power generation, and biogas generation. Meanwhile, technology maturation and commercialization require step-change improvements in biomass handling, pretreatment, hydrolysis, fermentation, waste treatment, and energy requirements, intended to reduce capital costs, operational costs, the carbon footprint of the process, and technical and nontechnical risks, including a public license to operate.

Numerous academic and private research institutions have been working on such technologies. In 2002, Shell joined this community by partnering with the enzyme specialist Iogen that had developed a proprietary technology.

Iogen's technology at that time consisted of a number of building blocks,[3,8] namely:

- A dilute-acid pretreatment step to open the lignocellulosic structure and increase its accessibility and reactivity.
- An enzymatic hydrolysis step to depolymerize the cellulose and the remainder of the hemicellulose, using a proprietary cocktail of enzymes.
- A conditioning step to remove the acetate and lignin components that are known to inhibit the fermentation microorganisms.
- A fermentation step that used a proprietary bacterium of the genotype of *Zymomonas mobilis* (AX101) to coferment the C5 and C6 sugars present in the hydrolysate.

Iogen has operated a demonstration facility of $\sim$6.7 kt/a biomass intake (20 t/d) since 2004 in Ottawa, Canada.[9] The technology is claimed to produce $\sim$0.22 t ethanol (75 gallons) per ton of wheat straw, which corresponds to a yield of 70–75% theory relative to carbohydrates contained. In 2009, Shell further partnered with Codexis to apply Codexis[10] gene-shuffling technology for further improving the biocatalysts used for hydrolysis and fermentation to enhance the Iogen process and shorten the timeline to commercialization.

In lignocellulosic pretreatment, the majority of the players apply dilute sulfuric acid. Alternatives considered by Shell include a dilute formic acid pretreatment, optionally combined with lignin extraction, avoiding the generation of sulfate containing wastewater and enabling anaerobic fermentation and biogas generation. Clean low-sulfur lignin enables burning in cheaper boilers or opens lignin outlets in higher-value applications. Pretreatment at low formic acid concentrations and higher pH values appeared to be effective,[11] enabling feedstock flexibility[12] and cheaper materials of construction.[13] Higher sugar purities and densities can be obtained after the extraction of lignin by water/ethanol at atmospheric conditions, prior to hydrolysis.[14] Enzymatic conversion of cellulose can be enhanced by the application of trace amounts of additives like polyvinylpyrrolidone.[15]

Further technology improvements are required, including cost reductions and efficiency improvements, enabling commercialization and building a solid biofuel business competitive with alternative lignocellulose applications as well as with fossil fuels and contributing to a sustainable future.

23.3.3.2 Butanol/ABE/Butanol Recovery *n*-Butanol also appeared promising for blending in gasoline but also for conversion to diesel components. However, the conventional acetone–butanol–ethanol (ABE) fermentation process required significant improvements in three specific areas to become economically attractive, namely, improvement in yields to *n*-butanol (lower coproduction of ethanol and acetone), in production rate and in product concentration in the fermentation brew. Shell joined the Dutch B-Basic consortium[16] to work on these challenges and contributed by assisting the Technical University of Delft in recovering of *n*-butanol from the fermentation brew[17] and by exploring the upgrading of butanol to diesel components such as dibutyl ether and 2-ethyl-hexanol. We will limit the coming discussion to the conversion of butanol to diesel component and refer to the literature[17,18] for progress made on other fronts.

The upgrading of butanol to diesel components followed two routes, an acid-catalyzed and a base-catalyzed one. The acid-catalyzed route was geared to produced dibutyl ether and achieved a yield of ~65% (at 80% conversion) and a productivity of 0.15 $g/(g * h)$ using gel-type ion-exchange resins that operate at a mild temperature (130°C). Similar systems were already reported for the etherification of other alcohols.[18] Promising alternative modes of operation, namely, reactive distillation and biphasic reaction, have been identified to shift the equilibrium toward ether formation. The reactive distillation exploits the presence of water-rich azeotropes with butanol, with dibutyl ether, and with their mixture, which are all boiling at 91–93 °C, to remove the water *in situ* and, thereby, shift the equilibrium toward ether formation. The biphasic reaction recognized the presence of liquid–liquid demixing when the butanol/water ratio drops below ~4 : 1 wt/wt (1 mol/mol). Accordingly, the reaction would be carried out at high pressure, to ensure liquid-full conditions, and would allow interstage removal of the water-rich liquid phase, to shift the equilibrium.

As for the base-catalyzed route, a mixture of C_{7+} components that is rich in 2-ethylhexanol and 2-ethylhexanal could be produced at 60% yield and 0.1 $g/(g * h)$ productivity using a ZnCr catalyst at 350°C. The reaction could be diverted toward 4-heptanone or toward butyl butyrate when carried out at higher or lower temperature, respectively. However, Ca-hydroxyapatites appeared to be more promising as alcohol condensation catalysts, consistent with reports in the literature.[19]

23.3.3.3 Levulinic Acid Derivatives The biotechnological routes discussed above require an expensive pretreatment and hydrolysis steps to unlock the sugars needed for fermentation. However, acid treatment of lignocellulose at higher severity is known to deliver furfural and levulinic acid in a single step, which are versatile precursors for oxygenated fuel components. This approach was extensively studied and covered the conversion of lignocellulose to levulinic acid and furfural, their upgrading to various oxygenates, and the evaluation of the resulting oxygenates for fuel applications.

A number of batch experiments confirmed the possibility to digest lignocellulose in the presence of strong mineral acid such as H_2SO_4 at 200 °C to release furfural and levulinic acids as well as significant amounts of water-insoluble humins. Levulinic acid was confirmed to be a stable end product. In contrast, furfural is a reactive intermediate that degrades at prolonged reaction time unless removed efficiently from the reaction medium, for example, by flashing.[20] The acid hydrolysis of lignocellulose is generally carried at low solid loading, which is economically unattractive. It was found that high solid loading gave reasonable yield of levulinic acid – ~55 mol% on contained hexose – when the concentration of H_2SO_4 was increased accordingly. In fact, numerous feedstocks each with varying content of hexose (including pure cellulose) appeared to deliver the same yield profile as long as the H_2SO_4/hexose concentration is kept constant (Figure 23.2).

Accordingly, high solid loading allowed the size of reactor and recovery unit to be minimized. However, several challenges were identified. Firstly, the resulting high H_2SO_4 concentration also raised the corrosiveness of the medium and the need for expensive material of construction. Secondly, the recovery of levulinic acid (and furfural) from the highly acidic reaction medium is expensive. Extraction technologies

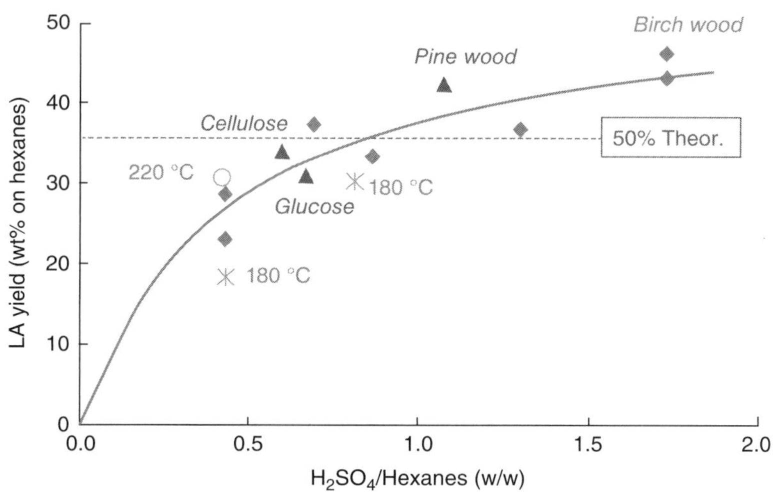

Figure 23.2 Acid hydrolysis of lignocellulose to levulinic acid (180 – 220°C, 5–20 wt% solid, 3–9 wt% H_2SO_4, ~30 min).

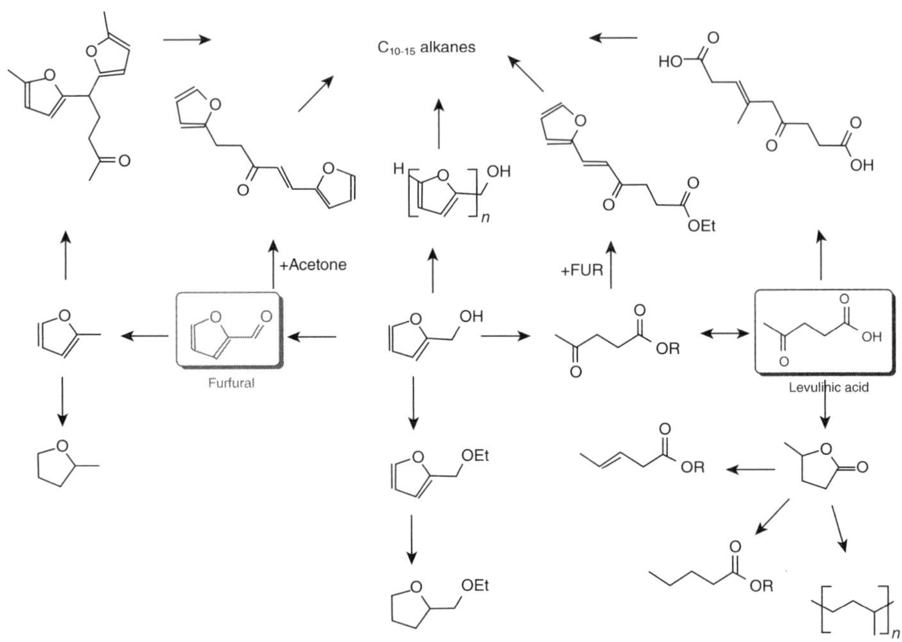

Figure 23.3 Fuel derivatives from furfural and levulinic acid.

were explored to recover furfural and levulinic acid from the aqueous phase.[21] The combination of modest yield, expensive material of construction, and difficult product recovery eventually present severe challenges to the economical viability of this approach.

Contrasting the technical challenges in producing levulinic acid, its upgrading to oxygenated fuel components offered a number of opportunities (Figure 23.3).

Levulinic acid can be converted to levulinate esters, γ-valerolactone, and methyl tetrahydrofuran, which are all claimed to be potential fuel components.[20,22] However, early studies on fuel quality in Shell laboratories (discussed later in this chapter) drove the development of less polar pentenoate and valerate esters for gasoline and diesel application.[23] The valerate esters were produced by combining a series of process steps, some of which offer room for process integration/intensification.[23]

- Hydrogenation to γ-valerolactone using Pt/ZrO_2 catalyst operating under gas-phase conditions.
- Hydrodeoxygenation (HDO) of γ-valerolactone to valeric acid using Pt/ZSM-5 catalysts.
- Esterification of valeric acid with ethanol for gasoline application but also with heavier alcohols (e.g., pentanol) or diols/triols (e.g., ethane/propane diol or glycerol) for diesel application.

Alternatively, the pentenoate esters could be produced by reacting γ-valerolactone with methanol or ethanol under reactive distillation conditions using a single process step.[24] Coupling products of levulinic acids were also explored as intermediates to diesel components. For instance, levulinic acid can be dimerized through an acid-catalyzed aldol-coupling reaction under hydrogenating conditions.[23] Levulinate esters can react with furfural through base-catalyzed aldol coupling reactions under reactive distillation conditions.[25] Other groups converted γ-valerolactone to butenes for subsequent oligomerization to diesel-range hydrocarbons[26] or dimerized valeric acid to 5-nonanone via ketonization reaction.[27]

23.3.3.4 Furfural Derivatives, Inclusive Oligomers Much like levulinic acid discussed above, furfural offers a variety of derivatives that are promising oxygenated fuel components (Figure 23.3).[25] Furfural can undergo HDO to methyl furan, a powerful octane booster. The HDO reaction can be performed either at high temperature in gas phase using Cu-based catalyst or at lower temperature under reactive distillation conditions using a combination of metal and acid catalysts.[28] Alternatively, furfural can be hydrogenated to furfuryl alcohol and, subsequently, rearranged to alkyl levulinate in the presence of acidic ion-exchange resins.[29] Alternatively, furfuryl alcohol can be converted to ethyl furfuryl ether, which is also a premium octane booster.[30] Furfural can also be converted to heavier diesel-range products.[25] Furfural can be reacted with levulinate esters via aldol coupling reaction and, subsequently, undergo HDO to C_{10} alkanes. However, furfuryl alcohol can also undergo acid-catalyzed oligomerization to produce tri- and tetramers with up to 60 wt% yields (Figure 23.4).[31] These oligomers spontaneously separate from the aqueous reaction medium, thereby offering the possibility to control the extent of oligomerization by withdrawing the desired short oligomers from the acidic medium by *in situ* extraction. Subsequent HDO leads to premium-quality diesel-range alkanes.

In the same period, other laboratories confirmed the production of (di)methyl furan[32] and explored the manufacture of diesel-range hydrocarbons by coupling furfural with ketones[33] or by trimerization of methylfuran.[34] Much as in the case of levulinic acid, however, the production of furfural appeared more challenging than its upgrading to oxygenated biofuels.[25] The coproduction of furfural and levulinic acid presented the challenges of yield, recovery, and material of construction, as discussed above. The more traditional manufacture of furfural leads to poor biomass valorization. Less than 15 wt% of the biomass or < 50 mol% of the contained pentoses are converted to furfural. The rest of the biomass is degraded to insoluble humins that are burned to raise the steam needed for stripping furfural from the reactor before it is degraded.

Furfural production may also be coupled with pulping or pretreatment processes that extract xylans from biomass and produce cellulose for fermentation processes or for the manufacture of paper pulp.[25] Accordingly, the extracted xylan can be dehydrated to furfural by acid hydrolysis. When carried out in water, such acid hydrolysis leads to a modest furfural yield of ~50 mol%. Acid hydrolysis can also be carried out in organic solvent. This allows much higher furfural yield (e.g., 80 mol%) but requires to switch solvent and, consequently, to recover the dilute xylose/xylan from

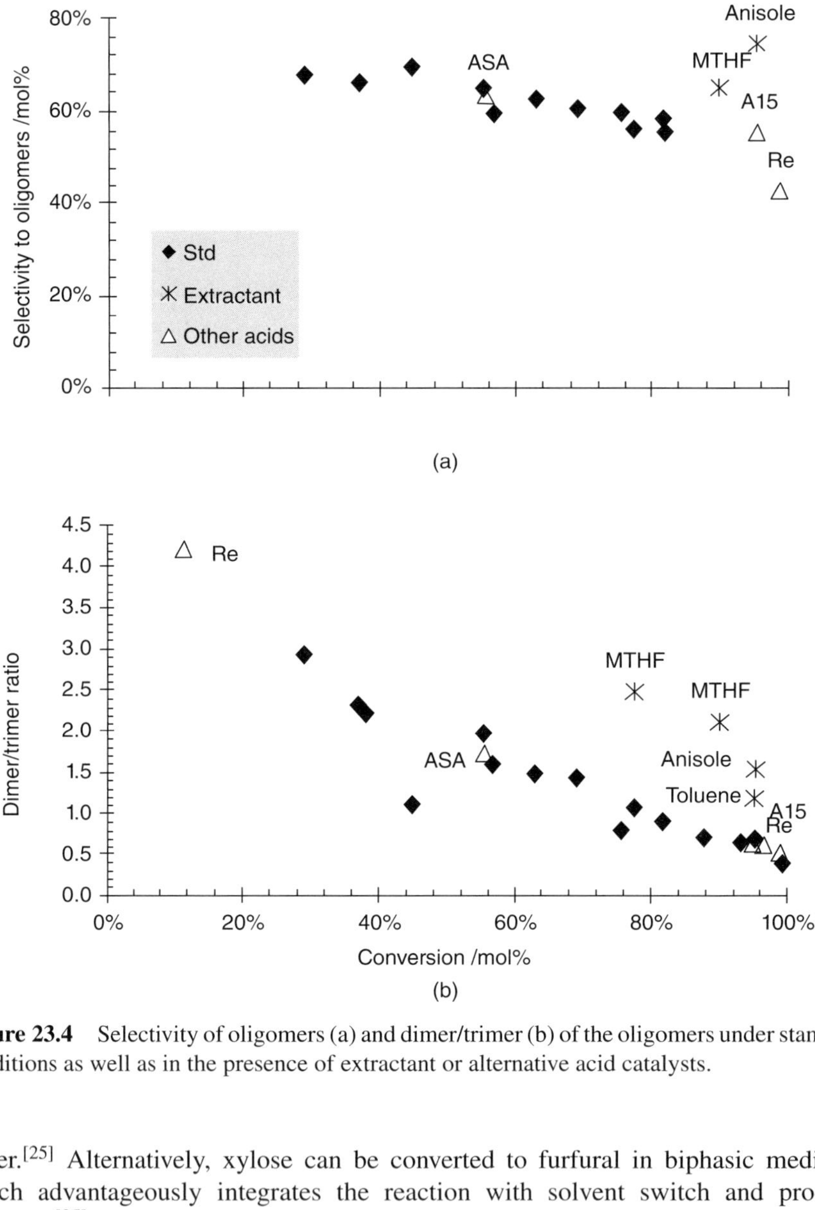

Figure 23.4 Selectivity of oligomers (a) and dimer/trimer (b) of the oligomers under standard conditions as well as in the presence of extractant or alternative acid catalysts.

water.[25] Alternatively, xylose can be converted to furfural in biphasic medium, which advantageously integrates the reaction with solvent switch and product recovery.[25]

However, a third route was reported: furfural could be stripped off from lignocellulose with > 95% yield upon passing a hot gaseous stream of HCl and steam.[35] A later study confirmed the production of furfural but could not push the yield beyond ~ 60 mol% (based on contained pentosan).[36] Most of the pentosan was degraded by such treatment, while most of the cellulose remained unaffected. Preliminary experiments confirmed the possibility of recycling the hot reaction effluent, which allowed

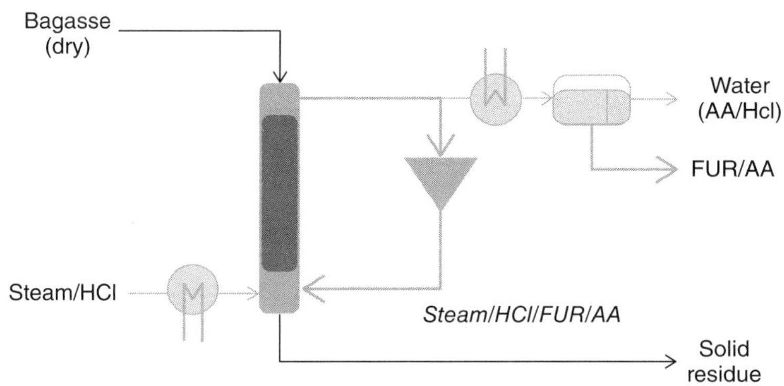

		Liq.-phase process	Acidic steam stripping
Furfural yield	mol%	30–50	50–60
Steam consumption	$t_{steam}/\ t_{furfural}$	20–30	5–10
	%LHV bagasse	40–60	10–20
Pressure of solid feeding	bar	15–25	1–5
Product recovery	–	Distillation of diluted stream	L/L enrichment + distillation

Figure 23.5 Acidic steam stripping of furfural from lignocellulose. *Source*: Adapted from Ref. [36]. Reproduced with permission of Wiley.

significant buildup of furfural with limited degradation. This recycle reduced the consumption of steam down to 5–10 t steam per ton furfural, compared to 20–30 t/t in conventional liquid-phase processes. Moreover, such buildup facilitated the recovery of furfural by enabling spontaneous liquid–liquid phase separation of the furfural/water/acetic acid product. All these improvements promise to make the furfural production process competitive (Figure 23.5), although numerous process issues still need to be resolved.

23.3.3.5 Fuel Properties of Oxygenates As discussed above, the fermentation or acid conversion of pentoses and hexoses leads to a wide variety of oxygenated components that could be valuable as blending components for gasoline or diesel. Owing to the large number of potential components to be considered, a preliminary framework was developed to evaluate the potential of oxygenated fuel additives. It considered four key physical properties (Figure 23.6), namely, volatility (atmospheric boiling point), solubility (Hildebrand solubility constant), energy density, and ignition properties (research octane or cetane numbers).[23,25]

Accordingly, promising fuel components should first exhibit a high energy density and high hydrocarbon solubility (low Hildebrand constant) to blend seamlessly in today's hydrocarbon fuels. They should then either combine a high research octane number and moderate boiling point for blending in gasoline or, conversely, combine a low octane number (i.e. a high cetane number) and high boiling point for blending in diesel. A quick inspection of Figure 23.6 reveals that oxygenates are promising candidates for gasoline but less for diesel. Their modest energy density is generally

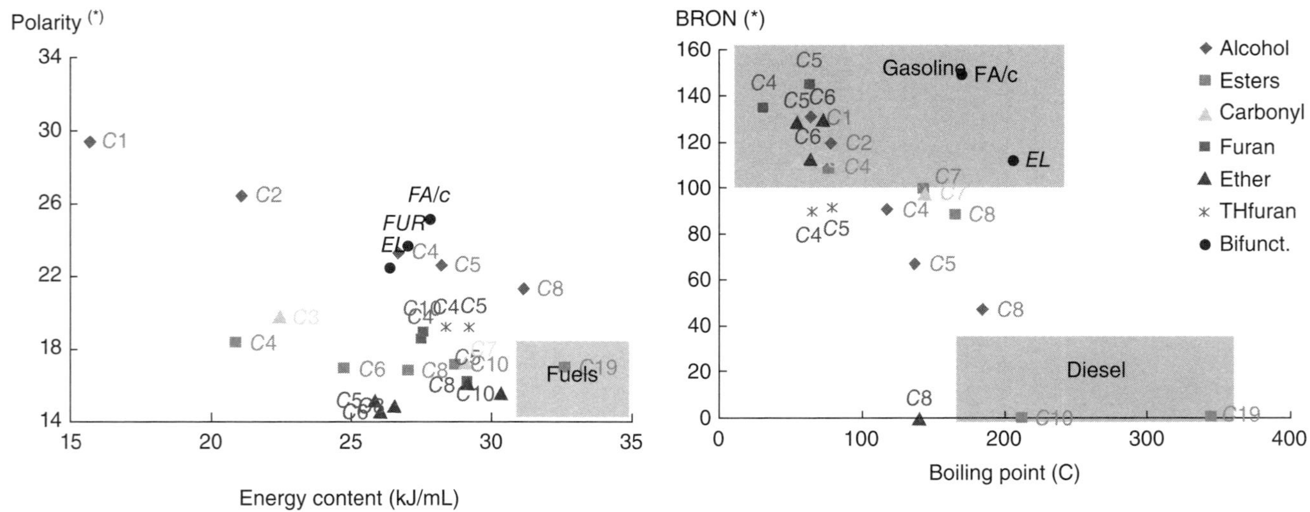

Figure 23.6 Fuel properties of bio-oxygenates. *Source*: Adapted from Refs. [23,25]. Reproduced with permission of Wiley.

compensated by good octane properties, particularly when combined with unsaturation, as exemplified by furans versus tetrahydrofuran components. Not all forms of oxygen are desirable, however. Hydroxyl and carbonyl groups generally depress the miscibility in hydrocarbons, much more than ether and ester functions. In contrast, diesel fuels are not improved by blending with heavier oxygenates. The loss of energy density is generally combined with loss in cetane number and reduced hydrocarbon solubility. The literature occasionally claims a favorable reduction of soot production with oxygenate blends. However, similar reduction can be achieved more cheaply by fine-tuning the engine.

Oxygenates may still be valuable for diesel. Recent developments show promise in running diesel engines with low cetane fuel and early injection.[37] Such operation allows for partial mixing of fuel and air prior to spontaneous ignition and, thereby, seems to combine the high efficiency of diesel engines with the low emissions of gasoline engines. Oxygenated fuel components may present new opportunities in this area.

23.3.4 Drop-In Biofuels

23.3.4.1 From Algae to Hydrocarbons Microalgae are an interesting source for bio-oils as they have high growth rates; they utilize a relatively large fraction of solar energy. The bio-oils, produced by microorganisms like algae, can be harvested by cell wall disruption and extraction. Harvesting the algal lipids from a dilute algae suspension in water in an economic manner is a significant challenge and was investigated by Shell.

For several years, Shell was a major stakeholder in Cellana. The Cellana joint venture operated a pilot facility in Hawaii to grow microalgae and produce algal lipids. The project was supported both by an academic research program to screen and select natural microalgae species on highest lipids yields and by research programs in Shell laboratories to develop efficient downstream processes. This latter research led to various inventions, which will be highlighted below. Shell and others have since realized that it will take 10+ years to develop a commercial process for the production of biofuels from algae that is economically viable and sustainable,[38] and the amount of research and development should not be underestimated.[39]

A new technique[40] for algae harvesting was derived from a dynamic filtering technique[41] developed for filtering biological mud, in particular liquid manure originating from pigs. The algae substance is filtered over a deformable filter. This technique has a reduced energy consumption and can be operated in a continuous way, and it avoids the damage of the cell walls. When the mixture containing the algae is preconcentrated by coagulation, flocculation, and/or sedimentation, the volume to be processed over the dynamic filter can be reduced. After filtering the algae mass can either be mildly heat treated (80–150 °C, 1–5 bar) to disrupt cell walls, after which the algae release the majority of the lipids,[42] and/or mechanically treated (e.g., ultrasound, high shear), and/or chemically treated to obtain the algal lipids. The algal lipids recovered, for instance, after solvent extraction, can then be further upgraded to hydrocarbon fuels by hydrotreating (see below) or enzymatic decarboxylation processes.

Various alternative schemes have been developed to obtain algal lipids from algae sources, among which is a method that circumvents handling of solids and sludge[43]: An algae-containing feedstock, after an optional centrifugal step to increase dry matter content to 20–25 wt%, is treated to disrupt algae cell walls and release lipids (via biochemical, (thermo)chemical, or physical routes). The resulting mixture, after an optional centrifugal step to increase dry matter content to 25–50 wt%, is extracted with a solvent and filtered to obtain and organic fraction, aqueous fraction and solids. The extracted lipids are then obtained by distilling off the solvent from the organic fraction. The water fraction can be recycled to the algae-growing system.

Microbial oils, produced by oleaginous microorganisms, including bacteria, algae, and yeasts, can be upgraded by a two-step hydrotreatment process[44]: First the microbial oil (triglycerides) undergoes HDO to produce paraffins and water. After (optionally) separating the water, the paraffin stream is hydroisomerized to improve the cold flow properties of the paraffinic middle distillate product. In order to better control the exothermicity of the HDO step, either a mineral oil can be coprocessed or part of the effluent stream can be recycled, diluting the feed to the hydrotreating reactor. By applying, in the second step, nonnoble metal-based hydroisomerization catalyst, like NiMo and NiW supported on an acidic heterogeneous carrier, the process can be a simple series flow operation. This means that gaseous impurities, like H_2S and NH_3, do not need to be removed after the first, HDO, reactor.[45,46]

Alternatively, microbial oils can be cofed to an FCC unit (as investigated in a small-scale laboratory FCC unit) to produce more gasoline. Extracted lipids from *Chlorella* microalgae have been blended with a vacuum gas oil (VGO, a routine feed to FCC) in 20/80 wt/wt algal lipids/VGO ratio[47] at 500 °C and 67% conversion. Figure 23.7 shows the yields of the main products, compared with those from cracking of pure VGO and a 20/80 blend with rapeseed oil. Although the product distribution is rather similar, more coke is produced when cofeeding algal lipids at the expense of mainly gasoline and LPG. This could mean that in practice less coke producing feeds should be cofed to the FCC. The advantage of the FCC process in this case is that CO_2 produced by the FCC regenerator can be used for the cultivation and propagation of the microalgae.

The process for extracting lipids from aquatic microbial biomass is a cost- and energy-intensive process. Therefore direct processing of the microbial biomass in, for instance, an FCC unit, has also been studied. This concept has been tested with dried and milled microalgae in a laboratory-scale FCC reactor at 500 °C.[48] The resulting liquid hydrocarbon yield was 20.2 wt% (5.1 wt% gasoline yield). Lots of water, coke, gas, and inorganics are coproduced.

Alternatively, microorganisms can be used that are able to secrete the lipids. Special care should be taken in separating microorganisms from the bio-oil to prevent damage to the cell walls. A special separation process has been developed for this making use of a cyclone.[49] A cyclone requires less energy and puts less local energy and shear to the microbes.

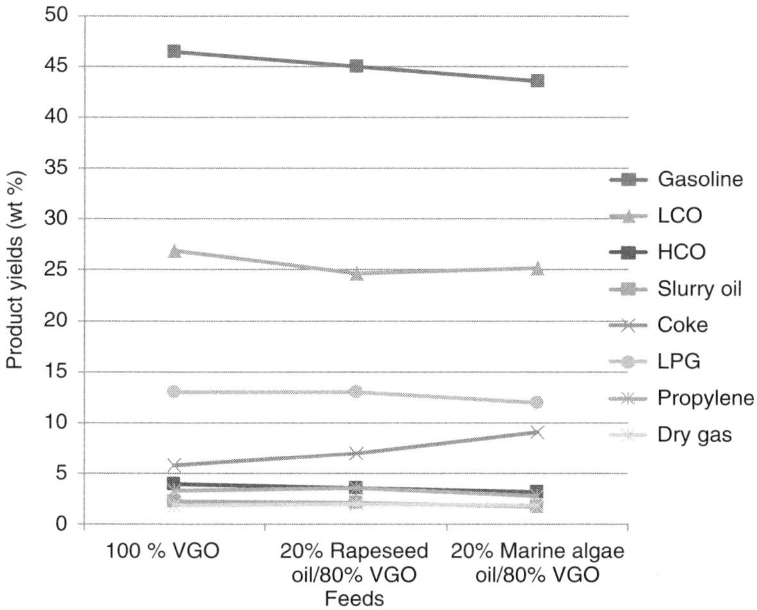

Figure 23.7 Comparison of FCC yields from vacuum gas oil and blends with rapeseed oil (20%) and marine algae oil (20%) as feed.

23.3.4.2 Microbes Producing Hydrocarbons From the start, it was recognized that to achieve a production host for a premium fuel component via genetically modified organisms would require a sizeable commitment of both people and cutting-edge expertise. A research program with Codexis gave ready access to both, with scientists from Shell and Codexis working to produce two different hydrocarbon component sets. Codexis was picked as a partner for its success at DNA shuffling and directed evolution of genes and metabolic paths to produce a number of pharmaceutical compounds. The two projects were focused on the two retail fuel types – gasoline and diesel.

An ideal gasoline biocomponent for power and acceleration would have a good blending octane number (BON) and a high energy density and improve the flame speed of any gasoline that it was blended with. One possible novel molecule was tricyclene, a saturated highly strained tricyclic monoterpene (Figure 23.8).

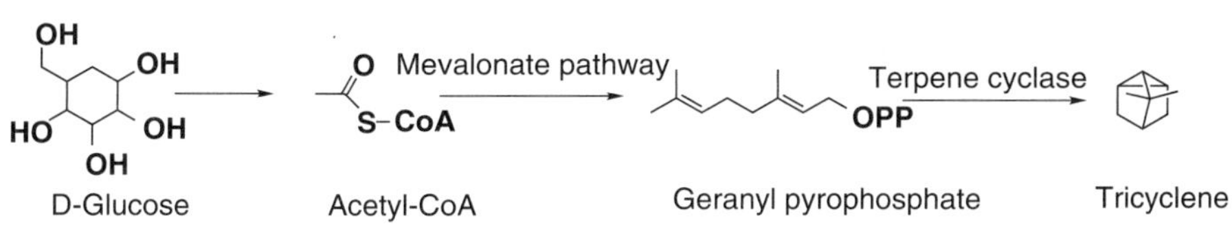

Figure 23.8 Schematic path showing the metabolic route to tricyclene.

Extensive testing of tricyclene confirmed that it was the almost ideal power component, having good blending motor and research octane numbers and excellent energy density, oxidative stability, and lubricity properties.[50] A screen of terpene cyclases identified two cyclases that showed promise: (1) bornyl diphosphate synthase from *Salvia officinalis* and (2) camphene synthase from *Pseudotsuga menziesii*. Unfortunately, the native pathways to tricyclene produced only trace amounts of the molecule, with bornyl diphosphate and camphene being the primary products, respectively. A short program of directed evolution to improve both the carbon flux to geranyl pyrophosphate[51] and then to tricyclene[52] led to mutants with 25–50X improvements in the production of tricyclene compared to the wild types, with tricyclene consisting of between 4% and 11% of the observed monoterpenes produced at titers of ca. 1 mg/L.[52]

For diesel the target was an alkane stream that could be added beyond the 8% FAME blend wall. At the time while there were many proposed paths to biologically derived alkanes, none of the genes had been identified.[53] However, a number of systems looked promising, including a putative alcohol reductase in *Vibrio furnissii* and a decarbonylase in cyanobacteria.[54–57] As a consequence, a two-pronged approach was undertaken: (1) an exploratory program to identify and characterize putative alkane synthases in *Vibrio furnissii* and cyanobacteria and (2) a directed evolution program to produce high titers of fatty alcohols. If the alkane synthase(s) could be identified, it was hoped that they could be "bolted" into the high titer fatty lipid host, and if not, then the fatty alcohols offered a couple of other options for diesel: (1) the formation of reverse esters and (2) hydrotreating to alkanes. The reverse esters offered some novel fuel lubricity properties and excellent cetane values[58] (fatty alcohols themselves have limited use as drop-in components due to their poor cold flow and cetane properties).

From a biological perspective to produce fatty alcohols, the two big questions were as follows: (1) Which genes to target (Figure 23.9)[59] and (2) what organism to use?[60]

For the scale required for fuel volumes, *Escherichia coli* was felt to be deficient in robustness, while *Saccharomyces cerevisiae* always produced an alcohol as a side product, limiting the potential carbon conversion efficiency to the fatty alcohols. The solution was to use *Yarrowia lipolytica*, which had a naturally high flux path to fatty acid-derived lipids. The downside was a minimal toolbox for genetically manipulating *Y. lipolytica*. So both tool development[61] and gene/pathway identification, proving and evolution were undertaken simultaneously. Initial directed evolution work focused on the carboxylic acid reductase (CAR) genes, which, with an alcohol dehydrogenase (ADH), converts fatty acid to fatty alcohol (Figure 23.9) and gave 7–10-fold improvements, while multiple gene disruptions to the metabolic paths gave improvements of up to 30X in C_{16}–C_{18} fatty alcohol production in *Y. lipolytica*.[62,63] However, the big improvements came with the identification of the fatty alcohol-forming acyl reductase (FAR) genes in *Marinobacter aquaeolei VT8* and *Oceanobacter RED65*[64] (which catalyze both the fatty acyl-CoA and fatty acyl-ACP paths (Figure 23.9) and which upon improvement gave titers of 45–50 g/L).[65] Unfortunately, the alkane synthase project was not as successful.

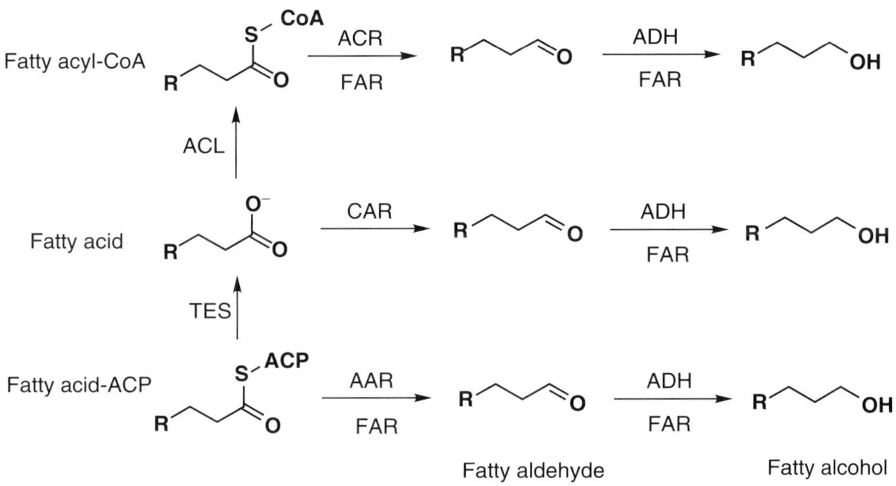

Figure 23.9 Schematic path showing the metabolic routes to fatty alcohols (AAR: acyl-ACP reductase; ADH alcohol dehydrogenase; FAR fatty alcohol-forming acyl reductase; CAR: carboxylic acid reductase; TES: thioesterase; ACL: fatty acid-CoA ligase; ACR: fatty acyl-CoA reductase).

No alkane-producing genes were found in *V. furnissii*[66,67] and the cyanobacterium alkane decarbonylase,[68] while shown to produce "drop-in diesel,"[69] has been found to be severely inhibited.[70]

23.3.4.3 Aqueous Thermal Conversion In 2007, Shell and Virent Inc. (Madison, Wisconsin, USA) entered a joint research and development effort to convert plant sugars directly into gasoline and gasoline blend components based on Virent's BioForming™ platform technology. This technology emerged from the discovery of Dumesic, Cortright, and coworkers[71–78] that hydrogen and alkanes can be produced from aqueous mixtures of oxygenated hydrocarbons by catalytic aqueous-phase reforming (APR) over a suitable metal catalyst. Carbohydrates have the generic structure $C(H_2O)$ and can be reformed to H_2 and CO over a metal surface to produce synthesis gas[76]:

$$C_nH_{2y}O_n \rightarrow n\text{CO} + y\text{H}_2 \tag{23.2}$$

An effective APR catalyst must also promote the water–gas shift reaction to release CO adsorbed on the catalyst surface:

$$\text{CO} + \text{H}_2\text{O} \leftrightarrow \text{CO}_2 + \text{H}_2 \tag{23.3}$$

For hydrogen generation to be optimized, reforming reactions including C—C bond scission and water gas shift must be favored, relative to HDO reactions entailing C—O bond scission and hydrogenation. Product distributions can be controlled by catalyst and reactor design including the use of multiple or bifunctional catalyst systems, whereby methanation and Fischer–Tropsch synthesis of methane and light alkanes from CO and H_2 are both minimized. For ethylene glycol as a model reactant, activity for CO_2 generation by *Group* 8–10 metals follows the order[76]:

$$Pt = Ni > Ru > Rh = Pd > Ir$$

Pt and Pd exhibit highest activity for H_2 formation via the avoidance of recombination to alkanes and methane. The modification of Ni with Sn enabled improved H_2 selectivity with a non-noble metal catalyst.[73]

In *aqueous-phase processing*,[74,78] the APR reaction can be combined with dehydration and HDO reaction with the generated hydrogen, to give alkanes. For example, upon the generation of H_2 via APR, C_6 sugar glucose can be hydrogenated to sorbitol and undergo dehydration and hydrogenation reactions to produce hexane, either as a second reaction step or in concerted reaction with careful design of catalyst and reactor configurations. However, the competition between APR and HDO reaction generally leads to some C—C cleavage and, thereby, shorter alkanes than C_6.

Building on this early discovery, Virent and Dumesic et al. found that the APR reaction could also be geared toward mono-oxygenates.[78–80] Subsequent condensation reactions over acid or base catalysts can be employed together with reduction by H_2 generated by APR to generate even longer-chain hydrocarbons desired for liquid biofuels in the gasoline or diesel range. Dehydration and condensation reactions of C_6 sorbitol followed by HDO can lead to the formation of a C_{12} alkane, for example. Virent's BioForming technology platform combines the conversion of sugars into light oxygenates via APR and their subsequent condensation to hydrocarbons in the boiling range of gasoline, kerosene, and/or diesel (Figure 23.10).

One of the unique features of APR is the potential to generate water-immiscible liquid products from biobased feedstocks without having to vaporize water. This offers the potential for significant improvement in energy efficiency and net energy yield, given that water vaporization or predrying can be energy intensive for competing process routes. Overall reforming of sugars or sugar alcohols to hydrogen via the APR reaction is exothermic, further improving the potential for high energy yields in biofuel production.

In 2010, Shell and Virent demonstrated the BioForming technology at a scale of 38,000 L annual capacity at Virent's facility in Madison. In 2013, Shell started a pilot unit at its Houston Technology Center to evaluate a thermocatalytic process technology that is licensed from Virent and is similar to the process being used at the Virent pilot plant in Madison, Wisconsin, USA. The Houston pilot plant explores the use of a range of feedstocks, starting with sugars and extending to nonfood cellulosic alternatives, leading to the production of a range of products, including gasoline, diesel, and jet fuel.

23.3.4.4 Syngas Platform Hydrocarbon fuels can also be produced by means of gasification and Fischer–Tropsch synthesis as applied for the upgrading of coal and natural gas to diesel-range hydrocarbons.

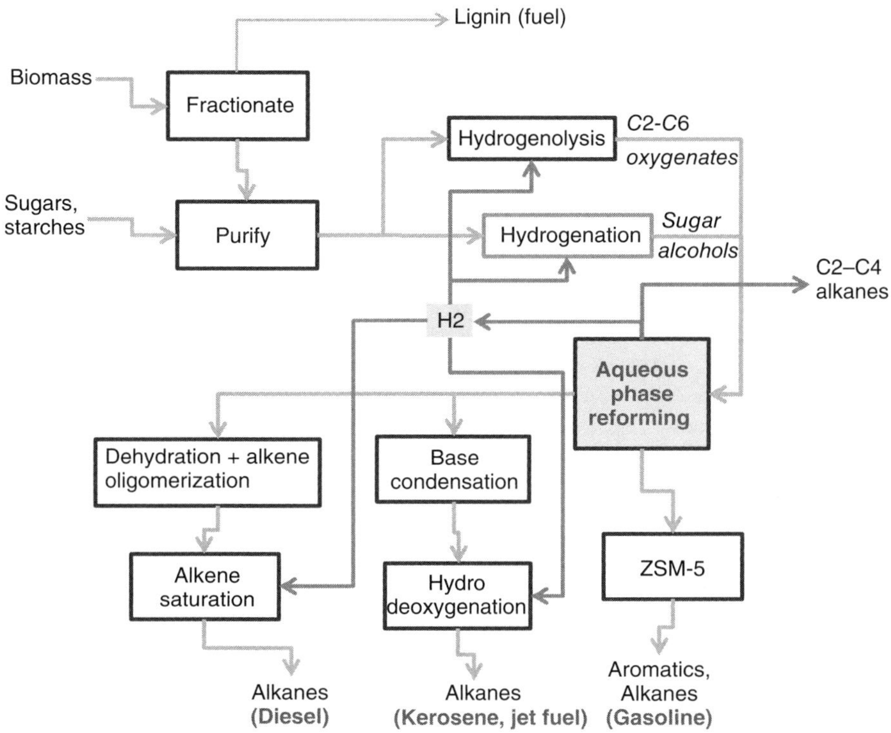

Figure 23.10 Virent's BioForming technology platform.[79]

As mentioned above, biomass can be converted to synthesis gas, a valuable mixture of CO and H_2, upon heating at high temperature in the presence of inert atmosphere or steam, optionally in the presence of catalyst. Such pyrolysis and steam reforming reaction are endothermic and, thereby, require large and costly heat input. Alternatively, one can conveniently cofeed some oxygen to burn a fraction ($\sim 1/4$) of the feed (Eq. 23.4) to reach the required temperature (>1000 °C) and provide the heat required by the endothermic formation of synthesis gas. The gasification reaction is mainly driven by the thermodynamics. Accordingly, the thermoneutral gasification of the carbohydrate follows the typical stoichiometry of Equation 23.4, which is characterized with > 95% energy efficiency (based on LHV).[3,6]

$$C_6(H_2O)_6 + 1.5O_2 \rightarrow 6CO + 3H_2 + 3H_2O \quad (23.4)$$

The resulting synthesis gas can subsequently be converted to diesel-range hydrocarbons via the Fischer–Tropsch synthesis according to Equation 23.5, which proceeds with theoretical energy and carbon efficiencies of ~76%LHV and 50 C%, respectively:[3,6]

$$6CO + 3H_2 \rightarrow 3/n\text{-}[CH_2]_n\text{-} + 3CO_2 \quad (23.5)$$

It should be noticed that Equation 23.5 applies to the ideal synthesis gas obtained from the carbohydrate fraction of the biomass. Since lignin fraction provides as slightly higher H_2/CO ratio, it should allow incorporating slightly more than 50% of the total carbon into the hydrocarbon product than suggested in Equation 23.5. It should indeed be realized here that lignin and carbohydrate are equally well utilized in such a process. In contrast to the gasification reaction, the Fischer–Tropsch synthesis proceeds at mild temperature (200–300 °C) in the presence of a suitable metal–base catalyst.[81] Proper design of the catalyst will determine the distribution of the hydrocarbons produced.

The crude Fischer–Tropsch product is not yet suitable for diesel application as it is spread over a too wide range of chain length, from C_1 to $C_{>100}$, and consists almost exclusively of linear alkanes that tend to solidify at low temperature. The crude Fischer–Tropsch product is, therefore, upgraded via a mild hydrocracking step that converts the heavy fraction into lighter diesel-range cut and slightly isomerizes the alkanes to confer them the desired low-temperature properties.[81]

Shell invested in Choren in 2005 to develop such a Biomass to Liquids process. The process design, based on Choren's gasification technology[82] and Shell's Fischer–Tropsch technology,[81] has been in demonstration in Freiberg, Germany. The demonstration unit is designed to convert 68 kt/a of wood chips into 15 kt/a of biofuel. A commercial unit would convert ~1 million t/a of biomass into 200 kt/a of biofuel and would require €800 million in capital investment.[82] Through a detailed energy flow chart, Choren reported to convert 53.3% of the energy of the feed into liquid fuels (39% to diesel and 14.3 to naphtha), 4.1% into exhaust gas (incl. 3.1% exhaust dryer), and 6.4% into electricity.[82] From the energy lost, 41% is lost as cooling water and 0.7% is lost in wastewater and slag. In summary, the Choren–Shell technology is claimed to operate with mass and energy efficiencies of 22% and 53%, respectively.

A simplified flow diagram of the Choren–Shell biomass to synfuel process consists of the following segments:

- Feed conditioning.
- Three-step gasification composed of a pyrolysis reactor followed by an O_2-blown burner to gasify the vaporized fraction of the biomass and, finally, a chemical quench that injects the char residue into the hot synthesis gas to cool it while being gasified.
- A synthesis gas treating train that consists of a cooler, wash vessels, and a water–gas shift reactor.
- A Fischer–Tropsch synthesis reactor followed by a mild hydrocracking step.

Despite the elegance and efficiency of this technology, several factors make a BtL process capital intensive, namely, a large number of process steps, high temperatures, large heat release (gasification and Fischer–Tropsch sections), and complex synthesis gas cleaning.[3] The conclusion for gasification processes is that while energy efficiency is very important, the biggest economic challenge is in reducing capital intensity while maintaining reasonable operability.

23.3.5 Biomass Coprocessing in Crude Oil Refinery/Crude Substitution

23.3.5.1 Pyrolysis Oil Shell wished to investigate the possibility of using existing refinery assets to (co)process biomass to fungible biofuels. These thermocatalytic conversion routes include the hydrogenation of vegetable oils (HVOs) but also the conversion of lignocellulosic biomass sources from forestry and industrial wastes,[83] as they do not compete with food production.[3]

Shell rationalized that pyrolysis oil, available from fast pyrolysis, would provide a convenient starting point. It is a pumpable liquid; hence catalysis, both heterogeneous and homogeneous, and conventional conversion processes could be applied. Many companies and institutes have actively investigated the production and applications of pyrolysis oil. Shell actively participated in BIOCOUP, a pan-European Consortium of 16 partners supported by the European Commission through the Sixth Framework Programme for Research and Technological Development (Figure 23.11).

Pyrolysis oil could not be coprocessed directly in an FCC riser pilot plant due to its intrinsic thermal instability at elevated temperatures, leading rapidly to coke formation. Hence one of the objectives of BIOCOUP was to develop upgrading technology such as HDO, including catalyst development, to give suitable upgraded pyrolysis oil (UPO). The upgrading, or stabilization, by mild hydrogenation reduces the MCRT or coke-forming tendency, total acid number (TAN), and oxygen and water content and increases the hydrogen to carbon ratio of the UPO. The TAN decreases with increasing upgrading severity, although it still remains high compared to fossil-based refinery feeds. This has consequences for materials of construction.

In complementary BIOCOUP research with CNRS in Lyon,[84–87] it appeared possible to coprocess up to 20 wt% of the UPOs in laboratory-scale FCC fluidized bed units at 500 °C or coprocess UPO with a straight run gas oil in continuous hydrotreating microflow units at 300 and 320 °C, 60 bar, WHSV = 1.[88]

A range of NiCu-based catalysts supported on δ-alumina, ZrO_2, and TiO_2 were developed by the BIOCOUP partners BTG, BIC, and RuG. These were shown to give UPO with superior properties compared to that obtained with conventional pyrolysis oil hydrotreating Ru/C catalysts.[89–91]

UPOs have been coprocessed with a typical VGO under realistic conditions in an FCC riser pilot plant; see Table 23.1. The beneficial effect of the mild

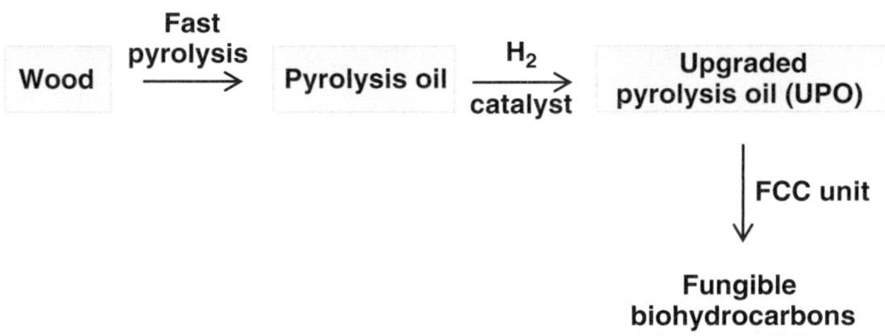

Figure 23.11 The pyrolysis oil pathway as investigated by BIOCOUP.

TABLE 23.1 Coprocessing of Upgraded Pyrolysis Oils in Laboratory-Scale FCC Unit (Product Distribution of 20 wt% HDO Oil Blends at Constant 60% Conversion. From HDO-1 to HDO-5, the Severity of Upgrading Increases).

Product Yields (wt%)	Long Residue	HDO 1+ LR	HDO 2 + LR	HDO 3 + LR	HDO 4 + LR	HDO 5 + LR
Water		7.9	6.7	5.7	3.7	3.9
Cat/oil ratio	3.1	3.3	2.5	2.6	3.2	3.3
LPG	8.7	9.3	8.2	9.0	8.4	8.7
Gasoline	44.1	41.4	42.8	43.6	44.1	43.6
LCO	24.9	24.9	25.5	25.1	25.6	25.9
Coke	5.6	6.9	6.8	5.6	5.5	5.6

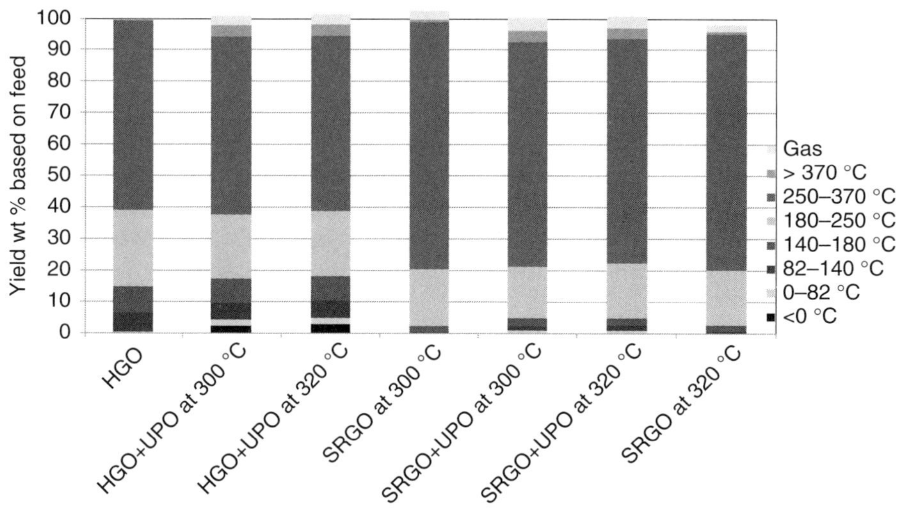

Figure 23.12 Coprocessing of upgraded pyrolysis oils in laboratory-scale hydrotreating unit (feed: blend of gas oil (hydrotreated gas oil), HGO), or straight-run gas oil (SRGO)) and upgraded pyrolysis (UPO; bars refer to boiling range fraction, e.g., 180 – 250 °C being kerosene and 250 – 370 °C being diesel).

hydrotreating (HDO) step results in more gasoline and less coke in the laboratory-scale FCC.

Shell has also coprocessed UPO in a laboratory-scale gas oil hydrotreater. The introduction of UPO results in a small decrease of diesel yield, while more higher boiling point components and a bit more gas are produced; see Figure 23.12.

23.3.5.2 (Hydro)liquefaction Oil Considering the poor quality of the bio-oil that is produced by flash pyrolysis and the consequent need to intermediate upgrading, alternative routes that could produce bio-oil with sufficient quality to be fed directly to existing oil refineries were explored, specifically focusing on liquefaction processes that attempt to depolymerize lignocellulose in a liquid medium.

One of the approaches exploited the knowledge gathered in acid hydrolysis to levulinic acid and furfural that is described above. Acid hydrolysis processes suffered from excessive formation of undesirable insoluble humins that were formed through degradation of sugars and repolymerization of the degradation products. However, the addition of water-soluble organic cosolvent to the acidic aqueous medium appeared to significantly reduce the formation of insoluble humins in favor of more desirable mono- and oligomeric components.[92] Interestingly, some effective cosolvents such as acetic acid and γ-valerolactone could be produced by the hydrolysis of lignocellulose itself, avoiding thereby the need to supplement expensive chemical-grade cosolvent. This finding was later confirmed by Dumesic's group.[93] Further improvements of the acid hydrolysis were achieved upon providing hydrogenation catalyst and hydrogen atmosphere digestion.[94] For instance, the addition of 80 bars of hydrogen and 0.3 wt% of hydrogenation metal, for example, Ru or Pd in the form of organic complex or supported metal catalyst, allowed to reduce the formation of insoluble humins to < 10 wt% in favor of ~ 50 wt% of mono-/oligomeric product (Figure 23.13).

The monomeric product consisted of a variety of oxygenates such as alcohols, ketones, carboxylates, furans, tetrahydrofurans, tetrahydropyrans, and phenolic components.[94,95] The oligomeric components showed a molecular weight distribution between ~ 200 and 2000 Da, as determined by size-exclusion chromatography. Interestingly, the hydroliquefaction bio-oil contained ~ 27 wt% of oxygen and exhibited an effective H/C ratio of ~ 0.8, which is much more favorable than the ~ 50 wt% oxygen and $H/C_{eff} \sim 0.4$ reported for flash pyrolysis. It also contains > 50 C% of paraffinic type, as determined by ^{13}C-NMR, compared to the < 20 C% observed for pyrolysis oil. Finally, the resulting hydroliquefaction bio-oil seemed to be more suitable for upgrading to biofuel by means of fluid catalytic cracking than is flash pyrolysis oil.[95] Indeed, preliminary cracking of a blend of 20 wt% of hydroliquefaction bio-oil in vacuum gas oil resulted in a modest decrease of −0.5 wt% of valuable products (namely, LPG, gasoline, and light cycle oil) and marginal increase of +0.2 wt% coke, which is to be compared with a larger loss of −3.2 wt% of desired product and +3.2 increase in coke with a blend of 20 wt% of pyrolysis oil.

The hydroliquefaction bio-oil was successfully recovered from the aqueous medium by means of liquid–liquid extraction or nanofiltration using TiO_2 membrane with 0.9 nm pores.[92] Interestingly, both methods allowed to separate the "heavy" component of the hydroliquefaction oil (e.g., > 250 Da) from the lighter components, water and the inorganic acid catalyst, which could be easily recycled to the hydroliquefaction reactor and used as a reaction medium.

However, the use of acidic conditions for liquefaction presents challenges in the metallurgy of reactors, heat exchangers, pipes, and separation units. On the other hand, processes that do not require strongly acidic medium required higher

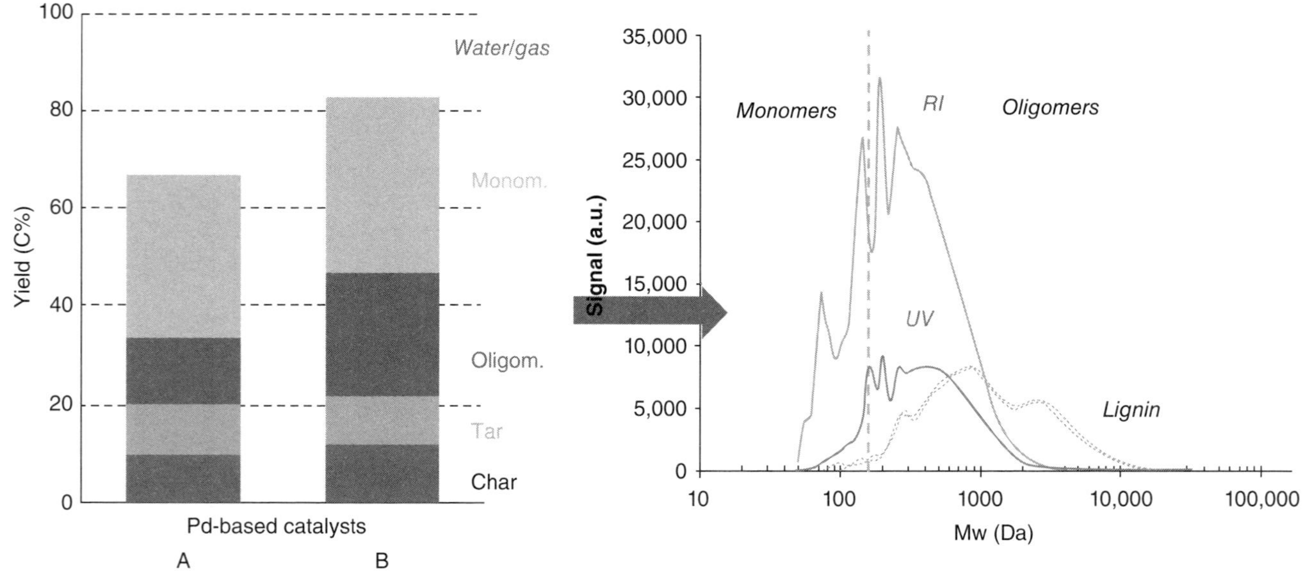

Figure 23.13 Acidic hydroliquefaction of lignocellulose (acetic acid as cosolvent, pH < 2, 2200–250 °C, 40–100 bar H_2). (*See color plate section for the color representation of this figure.*)

temperatures for depolymerizing lignocellulose. The move to higher temperatures also results in a move to heavier solvents to avoid buildup of excessive pressure as required, for example, for the hydrothermal upgrading (HTU) process that was developed at Shell and further by Biofuels BV that operates in water at 330 °C and 200 bar.[96]

Phenolic components appeared to be effective medium for dissolving wood with a low production of char (< 10 wt%) and gas (< 10 wt%) at temperature of 300 – 350 °C (Figure 23.14).[97]

Successive charges of wood in the autoclave were also converted to bio-oil, opening thereby the possibility to use the bio-oil as liquefaction medium. However, the bio-oil appeared to accumulate more heavy components than the acidic hydroliquefaction bio-oil. After a few consecutive wood charges, the bio-oil consisted of some 50 wt% of heavy product with molecular weight between 1 and 20 kDa. These are expected to fall within the range of refinery vacuum residue and, thereby, may likely require residue processes for upgrading to biofuel.

The resulting biocrudes could therefore be considered for coprocessing in a refinery FCC unit. This opportunity was evaluated by blending various biocrudes in a reference FCC feed at 20:80 weight ratio and, subsequently, injected in the fluidized catalyst bed of a MAT 5000 FCC laboratory unit that was operating at 520 °C with an equilibrium catalyst that contained Y-zeolite as active phase. Table 23.2 illustrates such test in the case of a blend with pyrolysis oil and a hydroliquefaction biocrude.

23.3.5.3 Alternative Routes The biomass conversion routes via either pyrolysis oil or hydroliquefaction oil as intermediate for coprocessing in refinery units have the disadvantage of resulting in additional unit operations (pyrolysis, upgrading, and liquefaction). Therefore we are also exploring direct cofeeding of solid biomass (lignocellulosic) to, for instance, FCC processes. Via such an approach, biomass pyrolysis and catalytic cracking of conventional fossil feeds may be combined in one step. A number of examples will be given.

Feeding 5–10 wt% (on total feed) solid (lignocellulosic) biomass to FCC riser reactor downstream of the hydrocarbon feed[98] will reduce coke production, as result of interaction of *in situ* produced hydrogen – from the hydrocarbon feed – with the bioproducts from the biomass feed. This is important as biomass will lead in principle to more coke formation as result of lower $(H/C)_{eff}$ ratio as compared typical hydrocarbon feeds. The FCC products containing biocarbon might be further upgraded via HDO. Alternatively, if more residence time is required for the solid biomass feed than for the typical hydrocarbon feed, feeding biomass upstream of feeding point for hydrocarbon feed to the FCC riser reactor[99] creates an interesting option for more efficient conversion. The produced pyrolysis oil vapors will be further catalytically cracked in the riser reactor. Similarly, feeding solid biomass to the bottom section of the FCC riser reactor,[100] with the bottom section having larger diameter than riser reactor, will create more residence time for the biomass feed and higher temperature, resulting in a more efficient biomass conversion.

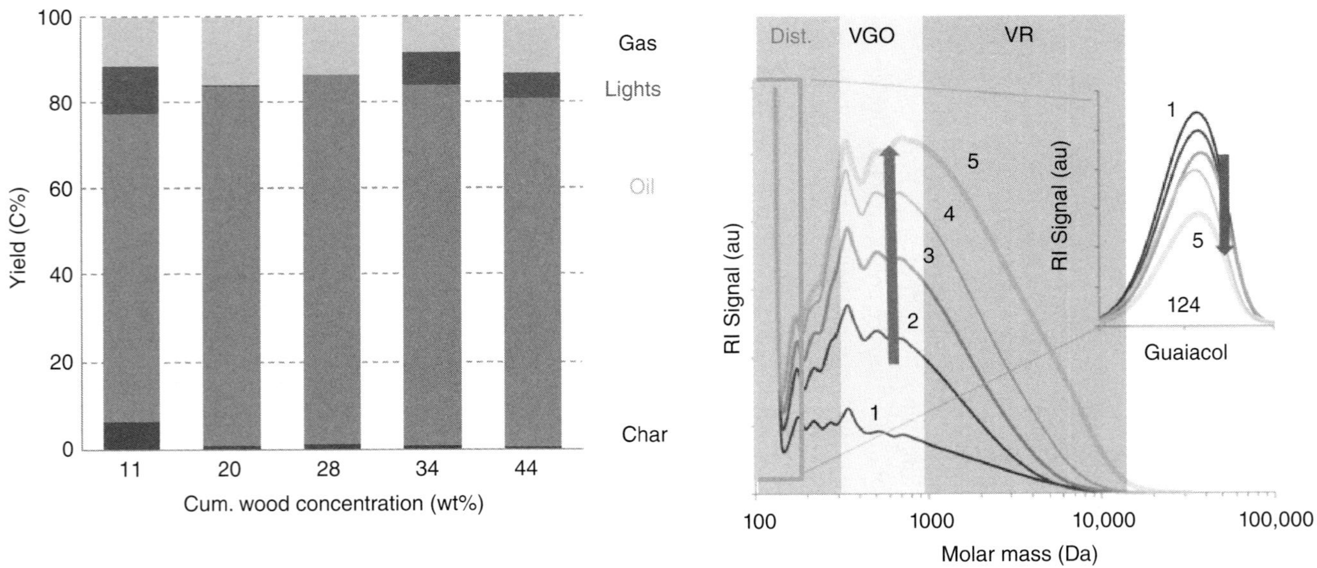

Figure 23.14 Liquefaction of lignocellulose in phenolic medium (30 min at 300 °C). *Source*: Adapted from [97]. Reproduced with permission of Wiley. (*See color plate section for the color representation of this figure.*)

TABLE 23.2 Product Distribution After Catalytic Cracking 100 wt% VGO Compared with 80 wt% VGO and 20 wt% Liquefied Biofeed and 80 wt% VGO and 20% Pyrolysis Oil (at a Constant Cat/Oil Ratio of 3.0 and a Temperature of 520 °C).

	100% VGO	20% Bio-Oil in VGO	20% Pyrolysis Oil in VGO
Oxygen wt%	–	6.5	10.4
Water wt%	–	6.9	11.9
Conversion wt%	60.8	+1.3	−0.2
Yield			
Dry gas wt%	1.5	+1.3	+0.7
CO_x wt% (C basis)	0	+0.5	+0.5
LPG wt%	9.0	+1.0	+1.0
Gasoline wt%	44.6	−0.9	−3.6
LCO wt%	25.1	−0.6	−0.8
HCO wt%	7.7	−0.3	−0.6
Slurry oil wt%	6.4	0	−0.3
Coke wt%	5.7	+0.2	+3.2

LCO = light cycle oil (221–370 °C); HCO = heavy cycle oil (370–425 °C), LPG = liquefied petroleum gas.

23.4 CHALLENGES LEADING TO MORE RESEARCH

23.4.1 Catalysis

23.4.1.1 Catalytic Materials In contrast with oil refineries, biorefineries operate in corrosive media. These are polar, often aqueous, and acidic as result of the introduction of biomass. This poses additional challenges for homogeneous and heterogeneous catalysts, the main topic of the Dutch consortium CatchBio.[101] For instance, the typical support applied in oil refining such as SiO_2 and Al_2O_3 and their mixtures may not be stable under hydrothermal conditions. Cheap nonnoble metals and numerous promoters may leach. Strong acid or base sites may leach and get inhibited by the polar medium or poisoned by undesired by-products. New catalyst materials and/or concepts are needed to face these challenges.

23.4.1.2 Catalytic Deoxygenation Lignocellulose and, particularly, its carbohydrate constituents contain more oxygen than desired in the final fuel. Hence, deep deoxygenation is generally required. This can be achieved in one of two ways: eliminating water or CO_2.

Water elimination using acidic or basic catalysts or catalytic hydrotreating with H_2 is known for decades. However, it is rarely performed in hot water, as often required in biorefineries. Dehydration also results in highly unsaturated products that have a moderate stability and, therefore, require hydrofinishing. However this increases the cost and carbon intensity of the product, as discussed elsewhere.[25] Many catalysts and technologies are available for the hydrogenation, hydrogenolysis, or hydrofinishing of model components. However, challenges are still encountered due to the presence of trace sulfur in the biomass. Sulfided catalysts, which are commonly used for such operations in refineries, are challenged in case of high oxygen concentrations in biofeeds, such as pyrolysis oil.

As for deoxygenation by CO_2 elimination, decarboxylation technologies are scarce. Typical examples are the decarboxylation and the ketonization of carboxylic acid. However, decarboxylation may require a rearrangement of the oxygenated feedstock into a carboxylic acid. Such rearrangement can be a complex task, as exemplified by the fermentation of sugar to ethanol, which proceeds via splitting of the C6 sugars into C3 sugars, rearrangement to pyruvic acid, decarboxylation to acetyl groups, and, finally, hydrogenation to ethanol. More technologies for such rearrangement and for the subsequent decarboxylation would be valuable. CO_2 elimination result in C loss. This can increase to carbon intensity of the remaining C that ends up in the fuel when biomass production has a significant C footprint.

23.4.2 Process

23.4.2.1 Biomass Fermentation Fermentation processes are generally limited to the utilization of sugars, preferably hexoses. Efficient fermentation of pentoses and, particularly, lignin remains challenging. The fermentative production of advance biofuels often requires complex metabolic pathways and, thereby, proceeds at low rate ($< 1\,g/L/h$). It also proceeds at high dilution ($< 1\,g/L$), either because the sugars are obtained at high dilution or because the final product is toxic for the microorganism beyond a critical concentration. Cost-effective solutions to these challenges are very desirable. They could include sugar reconcentration, advanced product recovery, or improved microorganisms. Fermentation organisms are also sensitive to various toxic substances present in the biomass or formed during its pretreatment. Cost-effective handling of these toxins, either by toxins removal or by improving the resistance of the organism, is very desirable. The latter seems imperative when one seeks to integrate the pretreatment, hydrolysis, and fermentation steps in a single vessel.

23.4.2.2 Separation Processes Biorefineries provide new separation challenges. They often involve intermediates and product that are highly functional, thermally unstable, and highly diluted. These conditions make distillation economically unattractive. Alternative separation and purification at milder conditions are therefore required, for example, based on extraction, absorption, or membrane. Such separation processes can optionally be integrated with distillation or, preferably, the reaction step. The integration of reaction and separation may be particularly desirable to minimize the degradation and polymerization of reactive intermediates or products, which leads to fouling of catalysts and equipments. However, the best approach is arguably to avoid them by applying smart reaction cascades. Such separations may also require improved thermodynamic model to describe such complex, polar, H bonding and/or ionic mixtures and model/predict the required thermodynamic properties and interactions.

23.4.2.3 Lignocellulose Fractionation Several conversion processes require chemically well-defined feedstock such as hexose, pentose, or lignin. They cannot process lignocellulosic feedstock without prior fraction into its main components, namely, cellulose, hemicellulose, and lignin. As most processes cannot digest cellulose either, there is an additional need for clean depolymerization into the constituting glucose. These treatment steps are costly for a number of reasons. The three top challenges are probably the following:

- Solid handling, generally under elevated pressure and temperature.
- Large volume of liquid, generally water but occasionally more expensive organic solvents or even ionic liquids. This results in large wastewater treatment or expensive solvent recovery, purification, and recycling steps.
- Either acidic conditions, with resulting challenges in metallurgy, or basic conditions, with challenges in regenerating and recycling the base.

These challenges explain why we have explored routes that do not require pretreatment or, more precisely, that integrate these challenges into the conversion unit. Nevertheless, some processes may still need to isolate components from lignocellulose remains. To be promising, future pretreatment should address the three main challenges mentioned above and preferably integrate them in a smart way into subsequent conversion steps.

23.4.2.4 Handling Ash Lignocellulose, particularly agricultural residue and grasses, contain significant amounts of inorganic contaminants. Some may present challenges in the conversion units. Some should ideally be recovered to allow recycling as nutrient to the field. Hence, proper attention should be paid to these contaminants. For instance, chlorine is a notorious corrosive element for common construction materials. Its corrosiveness is further exacerbated under hydrogenation conditions as it may be converted to HCl, which is volatile, spread throughout the unit, and will cause extensive corrosion at cold and aqueous locations.

Alkali and alkali earth are also essential to vegetal life. Such components will neutralize much of the acids that are eventually added to the process, for example, to depolymerize lignocellulose, or poison heterogeneous catalysts used in process. The alkali may form volatile salts that spread in hot sections of the units, while the alkali earth may lead to undesired precipitates in piping.

Residual proteins will likely contain S and N components. These may poison hydrogenation catalysts used in the process.

Cost-effective solutions to clean lignocellulose or intermediate process streams are highly desirable.

23.4.3 Equipment

23.4.3.1 Solid Handling Lignocellulosic processes generally require solid handling steps, which contribute to the complexity, energy demand, and operational sensitivity of the overall process. For instance, lignocellulose may need to be cut or milled into small pieces and eventually be dried. However, much more challenging is to feed the resulting solid particles or powder into high-pressure and high-temperature reactors. The pulp and paper industry has tackled the challenged and developed ways to feed solid biomass into a liquid-full reactor operating at ~20 bar and ~200 °C. This is generally performed by slurrying the biomass into a liquid and, subsequently, pumping the slurry into the hot pressurized reactor. Similar solutions are indeed considered for biomass pretreatment processes. Other processes may require different feeding options, however. Solid may have to be fed against higher pressure (e.g., 50–150 bar) and higher temperature (e.g., 300 – 400 °C) and as solid particles rather than slurry. Numerous designs have been investigated, but few have yet been properly demonstrated.[102] Other solid handling steps include filtering and washing of solid residue such as pulp or humins. Innovation in this area is highly needed.

Beyond improved equipment design, solid biomass feeding also presents more fundamental challenges. For instance, the hydrodynamic behavior of fibrous material in gas or as slurry needs more accurate understanding and description. The same applies to the wetting and flotation of lignocellulose with/in reaction liquid.

23.4.3.2 ***Lignin*** Lignin represents about 20–25 wt% of lignocellulose. Unless it ends up in the main product, for example, in bio-oil, lignin is generally destined to the furnace to power the manufacturing plant. Higher-value outlets would definitely help the economics of lignocellulosic biorefineries. Besides such yield loss, lignin may also cause operational problems. Its depolymerization is generally only partial and leads only to a minor fraction of monomeric and soluble components. Most of the rest remains at high molecular weight and fairly insoluble. When precipitated, it comes as a fine powder that melts at modest temperature (~ 150 °C) and, thereby, can cause quite some fouling challenges. The same largely applies to humins that are produced upon acidic or thermal degradation of the sugars and end up in the so-called "lignin" stream of lignocellulosic refineries.

23.4.3.3 ***Material of Construction*** As stated earlier, biorefineries operate in corrosive media. These are polar, often aqueous, acidic, and contaminated with chlorine and ash that come from the biomass. Such conditions may require equipments of resistant and expensive materials, based on premium alloys (e.g., Ni-rich alloy) or covered with ceramic, tantalum, or zirconium cladding. Biorefineries could therefore benefit from novel materials or novel treatments that provide corrosion resistance in a cost-effective way.

23.5 CONCLUSIONS

Shell was one of the first energy companies to invest in advanced biofuel research and has been active in this area now for more than a decade. Our experience so far is that developing cost-effective production processes for advanced biofuels is clearly very challenging, both in terms of chemistry and process technology.

In the early years of this technical journey, many options were investigated. Via an active and continuous portfolio approach, they were narrowed down to a few, which are best aligned with Shell's core competences. However, these choices still face significant challenges. Shell, though, remains committed to deliver economically attractive processes for the production of advanced biofuels. The biggest challenge in the route from biomass to advanced biofuels via platform molecules – such as sugars and sugar derivatives – is the efficient production of these platform molecules. In developing process routes via intermediate biocrudes, the main challenge is to produce biocrudes with improved characteristics to avoid as much as possible the need and cost of upgrading prior to coprocessing in existing refinery process units. Conversion of both platform molecules and biocrudes to fuels and fuel components is close to Shell core capabilities.

Sustainable biofuels start with the selection of sustainable feedstock that provides significant net CO_2 emission savings but also benefits local communities, preserves the quality of soil and water on/with which it is grown, and does not endanger local biodiversity. However, sustainable biofuels also imply the use of manufacturing processes that are clean, safe, efficient, and affordable. A lot of fundamental research will still be required to accelerate the development of the cost-effective production of advanced biofuels. Research in catalysis, process, and equipment engineering could focus on catalytic materials, catalytic deoxygenation, fermentation, separation, lignocelluloses fractionation, and handling of ash, solids and sticky lignin, and materials of construction, to mention a few.

REFERENCES

[1] http://www.shell.com/global/future-energy/scenarios/new-lens-scenarios.html, accessed Apr 2015.

[2] G.J. Kramer, M. Haigh, *Nature* **2009**, *462*, 568–569.

[3] J.-P. Lange, I. Lewandowski, P. Ayoub In "*Sustainable Development in the Process Industry—Cases and Impacts*" (Eds. J. Harmsen, J.B. Powell). Wiley, New York, **2010**, 171–208.

[4] J.-P. Lange, *ChemSusChem* **2009**, *2*, 587–592.

[5] U.R. Fritsche, H. Fehrenbach, S. Koppen, J. Adolf, D. Liebig, In "*Biofuels What Role in the Future Energy Mix? Facts, Trends and Perspectives*", Commissioned by Shell Deutschland Oil, Darmstadt, Heidelberg, Hamburg, October 2012; M.R.L.V. Leal, L. A. Horta Nogueira, L.A.B. Cortez, Applied Energy **2013**, *102*, 266–271.

[6] J.-P. Lange In "*Catalysis for Renewables: From Feedstock to Energy Production*" (Eds. G. Centi, R.A. van Santen). Wiley-VCH, Weinheim, **2007**, 21–51; J.-P. Lange, *Biofuels, Bioprod. Biorefin.* **2007**, *1*, 39–48.

[7] http://ethanolrfa.org/pages/World-Fuel-Ethanol-Production, accessed May **2015**.

[8] (a) H.G. Lawford, J.D. Rousseau, *Appl. Biochem. Biotech.* **2003**, *105–108*, 457–469; (b) J.S. Tolan, In "*Biorefineries – Industrial Processes and Product*", Vol. 1, (Eds. B. Kamm, P.R. Gruber, M. Kamm). Wiley-VCH, Weinheim, **2006**, 193–208.

[9] www.iogen.ca (pilot capacity).

[10] www.codexis.com (collaboration).

[11] M. Mackay, E. van Der Heide, US20130071900 A1, **2013**.

[12] (a) E. van Der Heide, T. Zhang, EP2336193 A1. **2011**; (b) E. van der Heide, M. MacKay, T. Zhang, WO2009080737 A2, **2009**.

[13] W.M. Bos, E. van der Heide, J.P. de Jong, M. Mackay, US2012156748 A1, **2012**.

[14] E. van der Heide, M. Mackay, T. Zhang, EP 2336196 A1. **2011**.

[15] D. Brown, A.J. Murphy, WO 2013083816 A1, **2013**.

[16] A continuation of the b-basic program can be found on: www.be-basic.org.

[17] A. Oudshoorn, "Recovery of bio-butanol', PhD thesis TU Delft 2012".

[18] (a) J. Tejero, F. Cunill, M. Iborra, J.F. Izquierdo, C. Fite, *J. Mol. Catal. A: Chem.* **2002**, *182–183*, 541–554; (b) A. Martinec, K. Setinek, L. Beranek, *J. Catal.* **1978**, *51*, 86–95.

[19] T. Tsuchida, S. Sakuma, T. Takeguchi, W. Ueda, *Ind. Eng. Chem. Res.* **2006**, *45*, 8634–8642.

[20] D.J. Hayes, S. Fitzpatrick, M.H.B. Hayes, J.R.H. Ross, In "*Biorefineries:* Industrial Processes and Products", Vol. 1 (Eds. B. Kamm, P.R. Gruber, M. Kamm). Wiley-VCH, Weinheim, **2006**, 139–164.

[21] (a) P.M. Ayoub, WO2005070867. **2005**;(b) J.L. Den Boestert, J.P. Haan, A. Nijmeijier, WO2006/089873, **2006**;(c) J.P. Haan, WO2011/161141, **2011**.

[22] I.T. Horvath, H. Mehdi, V. Fabos, L. Boda, L.T. Mika, *Green Chem.* **2008**, *10*, 238–242.

[23] (a) J.-P. Lange, R. Price, P.M. Ayoub, J. Louis, L. Petrus, L. Clarke, J. Gosselink, *Angew. Chem. Int. Ed.* **2010**, *49*, 4479–4483; (b) P.J. vd Brink, K.L. v Hebel, J.-P. Lange, L. Petrus, WO200667171, **2006**;(c) P.M. Ayoub, J.-P. Lange, WO 08142127, **2008**;(d) R.J. Haan, J.P. Lange, WO 11015645, **2011**.

[24] (a) J.-P. Lange, J.Z. Vestering, R.J. Haan, *Chem. Commun.* **2007**, 3488–3490; (b) R.J. Haan, J.-P. Lange, WO 05/58793, **2005**.

[25] J.-P. Lange, E. van der Heide, J. van Buijtenen, R.J. Price, *ChemSusChem* **2012**, *5*, 150–166.

[26] J.Q. Bond, D.M. Alonso, D. Wang, R.M. West, J.A. Dumesic, *Science* **2010**, *327*, 1110–1114.

[27] J. C. Serrano-Ruiz, D. Wang, J.A. Dumesic, *Green Chem.* **2010**, *12*, 574–577.

[28] J.-P. Lange, J. van Buijtenen, WO 09/156439, **2009**.

[29] (a) J.-P. Lange, W.D. van de Graaf, R.J. Haan, *ChemSusChem* **2009**, *2*, 437–441; (b) W.D. van de Graaf, J.-P. Lange, WO 07/23173, **2007**.

[30] R.J. Haan, J.-P. Lange, WO 2009/77606, **2009**.

[31] J. van Buijtenen, J.-P. Lange, R.J. Price, WO 2011/076736, **2011**.

[32] Y. Roman-Leshkov, C.J. Barrett, Z.Y. Liu, J.A. Dumesic, *Nature* **2007**, *447*, 982–986.

[33] C. H. Barrett, J. N. Chheda, G. W. Huber, J. A. Dumesic, *Appl. Catal. B: Environ.* **2006**, *66*, 111–118.

[34] A. Corma, O. de La Torre, M. Renz, N. Villandier, *Angew. Chem. Int. Ed.* **2011**, *50*, 2375–2378.

[35] K.J. Zeitsch, WO 2002/22593, **2002**.

[36] (a) J. van Buijtenen, J.-P. Lange, L. Espinosa Alonso, W. Spiering, R.F. Polmans, R.J. Haan, *ChemSusChem* **2013**, *6*, 2132–2136;(b) J. van Buijtenen, J.-P. Lange, R.J. Price, WO 2012/041990, **2012**.

[37] G.T. Klaghatgi, SAE 2005-01-0239. http://papers.sae.org/2005-01-0239/.
[38] R.H. Wijffels, M.J. Barbosa, *Science* **2010**, *329*, 796–799.
[39] P.J. le B. Williams, L.M.L. Laurens, *Energy Environ. Sci.* **2010**, *3*, 554–590.
[40] H. van Kaathoven, J.P. Haan, W. Bond, J.L.W.C. den Boestert, WO 2012/085210, **2012**.
[41] H. van Kaathoven, EP1426089, **2004**.
[42] K. Bus, J.P. Haan, J.L.W.C. den Boestert, J.L. Bravo, WO 2012/084864, **2012**.
[43] K. Bus, J.P. Haan, J.L.W.C. den Boestert, J.L. Bravo, WO 2012/084856, **2012**.
[44] C. Botella-Franco, H. de Lang, L. Meesala, A. Moppi, C.J. Schaverien, WO 2011/073427, **2011**.
[45] A.A. van Heuzen, H. de Lang, P. Moureaux, D. Reinalda, M.S. Rigutto, WO 2008/101945, **2008**.
[46] J. van Beijnum, E.J. Creyghton, A.H. Janssen, WO 2009/156452, **2009**.
[47] C.J. Schaverien, N.W.J. Way, WO 2012/083999, **2012**.
[48] J.W. Gosselink, C.J. Schaverien, N.W.J. Way, WO 2012/084000, **2012**.
[49] S.D. Doig, J.P. Haan, WO 2012/038332, **2012**.
[50] M.T. Croft, G.R. Lee, R.J. Price, R. McDaniel, US8518130, **2011**.
[51] O. Alvizo, R. McDaniel, B. Mijts, US20130011888, **2013**.
[52] R. McDaniel, G. Meshulam-Simon, O. Alvizo, X. Zhang, US20130143291 A1, **2013**.
[53] N. Ladygina, E.G. Dedyukhina, M.B. Vainshtein, *Process Biochem.* **2006**, *41*, 1001–1014.
[54] M.-O. Park, M. Tanabe, , K. Hirata, K. Miyamoto, *Appl. Microbiol. Biotechnol.* **2001**, *56*, 448–452.
[55] M. O. Park, *J. Bacteriol.* **2005**, *187*, 1426–1429.
[56] M.O. Park, K. Heguri, K. Hirata, K. Miyamoto, *J. Appl. Microbiol.* **2005**, *98*, 324–331.
[57] J. Han, W.-S. Chan, M. Calvin, *J. Am.Chem. Soc.* **1969**, *91*, 5156–5169.
[58] A. Felix-Moore, G.R. Lee, R.J. Price, US Patent 20110154728 A1, **2011**.
[59] M.K. Akhtar, N.J. Turner, P.R. Jones, *Proc. Natl. Acad. Sci. U. S. A.* **2013**, *110*, 87–92.
[60] H. Alper, G. Stephanolpoulos, *Nature Reviews* **2009**, *7*, 715–723.
[61] D.A. Hattendorf, A. Sero, US20120164686, **2012**.
[62] B. Behrouzian, R. McDaniel, X. Zhang, L. Clark, US2010135624, **2010**.
[63] D.A. Hattendorf, J.L. Shock, L. Clark, US20120165562, **2012**.
[64] R. McDaniel, B. Behrouzian, L. Clark, D. Hattendorf, F. Valle, US2011008535, **2011**.
[65] B. Behrouzian, L. Clark, Y. Zhu, M. Clay, K. Karlshoej, US20120009640, **2012**.
[66] L.P. Wackett, J.A. Frias, J.L. Seffernick, D.J. Sukovih, S.M. Cameron, *Appl. Environ. Microbiol.* **2007**, *73*, 7192–7198.
[67] T.M. Lux, R. Lee, J. Love, *J. Bacteriol.* **2011**, *193*, 1487–1488.
[68] A. Schirmer, M.A. Rude, X. Li, E. Popova, S. del Cardayre, *Science* **2010**, *329*, 559–562.
[69] T.P. Howard, S. Middelhauffe, C. Edner, D.M. Kolak, D.A. Parker, R. Lee, S.J. Aves, J. Love, *Proc. Natl. Acad. Sci. USA* **2013**, *110*, 7636–7641.
[70] C. Andre, S.W. Kim, X.-H. Yu, J. Shanklin, *Proc. Natl. Acad. Sci. U. S. A.* **2013**, *110*, 3191–3196.
[71] R.D. Cortright, R.R. Davda, J.A. Dumesic, *Nature* **2002**, *418* (6901), 964–967.

[72] J.W. Shabaker, R.R. Davda, G.W. Huber, R.D. Cortright, J.A. Dumesic, *J. Catal.* **2003,** *215*, 344–352.

[73] G.W. Huber, J.W. Shabaker, J.A. Dumesic, *Science* **2003,** *300* (5628), 2075–2077.

[74] G.W. Huber, R. D. Cortright, J. A. Dumesic, *Angew. Chem. Int. Ed.* **2004**, *43*, 1549–1551.

[75] J.W. Shabaker, G.W. Huber, J. A. Dumesic, *J. Catal.* **2004,** *222*, 180–191.

[76] R.R. Davda, J.W. Shabaker, G.W. Huber, R.D. Cortright, J. A. Dumesic, *Appl. Catal. B: Environ.* **2005**, *56,* 171–186.

[77] G.W. Huber, S. Iborra, A. Corma, *Chem. Rev.* **2006**, *106*, 4044–4098.

[78] G.W. Huber, J.A. Dumesic, *Catal. Today* **2006**, 111, 119–132.

[79] http://www.virent.com.

[80] E.L. Kunkes, D.A. Simonetti, R.M. West, J.C. Serrano-Ruiz, C.A. Gärtner, J.A. Dumesic, *Science* **2008**, *322*, 417–421.

[81] S.T. Sie, M.M.G. Senden, H.M.H. van Wechem, *Catal. Today* **1991**, *8*, 371–394.

[82] (a) M. Rudloff, *Proc. 12th Eur. Biomass Conf. Berlin*, **2007**, 2250–2251; (b) B. Hoffmann, *Aufbereitungs Technik* **2008**, *41*, 6–17; (c) C. Kiener, *Erdoel, erdgas, Kohle* **2008**, *124*, 68–71; (d) http://www.choren.com.

[83] J.C. Serrano-Ruiz, J.A. Dumesic, *Energy Environ. Sci.* **2011**, *4*, 83–99.

[84] F. de Miguel Mercader, M.J. Groeneveld, S.R.A. Kersten, N.W.J. Way, C.J. Schaverien, J.A. Hogendoorn, *Appl. Catal. B: Environ.* **2010**, *96,* 57–66.

[85] G. Fogassy, N. Thegarid, G. Toussaint, A.C. van Veen, Y. Schuurman, C. Mirodatos, *Appl. Catal. B: Environ.* **2010**, *96*, 476–485.

[86] Van Ngoc Bui, G. Toussaint, D. Laurenti, C. Mirodatos, C. Geantet, *Catal. Today* **2009**, *143*, 172–178.

[87] G. Fogassy, N. Thegarid, Y. Schuurman, C. Mirodatos, *Energy Environ. Sci.* **2011**, *4*, 5068–5076.

[88] C.J. Schaverien, N. Way, L. Meesala, A. Moppi, G. Fogassy, Y. Schuurman, C. Geantet, C. Mirodatos, *Proceedings of the Nordic Wood Biomass Conference* **2011**, Stockholm, March 22–24.

[89] R.H. Venderbosch, A.R. Ardiyanti, J. Wildschut, A. Oasmaa, H. J. Heeres, *J. Chem. Technol. Biotechnol.* **2010**, *85*, 674–686.

[90] R.H. Venderbosch, A.R. Ardiyanti, H.J. Heeres, V. Yakovlev, D. Ermakov, S. Khromova, V.N. Parmon, WO 2012030215 A1, **2012**.

[91] V.A. Yakovlev, S.A. Khromova, O.V. Sherstyuk, V.O. Dundich, D.Y. Ermakov, V.M. Novopashina, M.J. Lebedev, O. Bulavchenko, V.N. Parmon, *Catal. Today* **2009**, *144*, 362–366.

[92] K.L. von Hebel, J.-P. Lange, WO 2011/141545, **2011**.

[93] D. Martin Alonso, S.G. Wettstein, M.A. Mellmer, E.I. Gurbuzab, J.A. Dumesic, *Energy Environ. Sci.* **2013**, *6*, 76–80.

[94] K.L. von Hebel, J.-P. Lange, WO 2011/141546, **2011**.

[95] A.Q.M. Boon, L. Espinosa Alonso, J.W. Gosselink, J.W. Harris, A. Janssen, J.-P. Lange, C.J. Schaverien, N.W. Way, WO 2013/72391, **2013**.

[96] F. Goudriaan, B. van de Beld, F.R. Boerefijn, G.M. Bos, J.E. Naber, S. van der Wal, J.A. Zeevalkink, in A.V. Bridgwater (Ed.) *5th conf. on Progress in Thermochemical Biomass Conversion*, Tyrol, Austria, **2000**, Vol. 2, 1312–1325.

[97] (a) G. van Rossum, W. Zhao, M. Castellvi Barnes, J.-P. Lange, S.R.A. Kersten, *ChemSusChem* **2014**, *7*, 253–259; (b) M. Castellvi Barnes, S.R.A. Kersten, J.-P. Lange, W. Zhao, WO 2013/072383, **2013**;(c) S. Kumar, J.-P. Lange, G. van Rossum, S.R.A. Kersten, *Ind. Eng. Chem. Res.* **2014**, *53*, 11668–11676.

[98] A.Q.M. Boon, J.W. Gosselink, J.W. Harris, A.H. Janssen, M.J.P.C. Nieskens, E.S. Ranganathan, C.J. Schaverien, N.W.J. Way, WO 2013160253, **2013**.

[99] A.Q.M. Boon, J.W. Gosselink, J.W. Harris, A.H. Janssen, S. van Paasen, C.J. Schaverien, N.W.J. Way, WO 2012143551, **2012**.

[100] A.Q.M. Boon, J.W. Gosselink, J.W. Harris, A.H. Janssen, C.J. Schaverien, S. van Paasen, N.W.J. Way, WO 2012143573, **2012**.

[101] http://www.catchbio.com.

[102] J. Dai, H. Cui, J.R. Grace, *Prog. Energy Combus. Sci.* **2012**, *38*, 716–736.

INDEX

Note: Page numbers followed by *f* and *t* indicate figures and tables, respectively.

Industrial Biorenewables: A Practical Viewpoint, First Edition. Edited by Pablo Domínguez de María.